MASS SPECTROMETRY AND HYPHENATED TECHNIQUES IN NEUROPEPTIDE RESEARCH

MASS SPECTROMETRY AND HYPHENATED TECHNIQUES IN NEUROPEPTIDE RESEARCH

Edited by

Jerzy Silberring
Rolf Ekman

WILEY-INTERSCIENCE

A John Wiley & Sons, Inc., Publication

This book is printed on acid-free paper. ♾

Published simultaneously in Canada.

For ordering and customer service, call 1-800-CALL-WILEY.

Library of Congress Cataloging-in-Publication Data is available.

ISBN 0-471-35493-7

Printed in the United States of America.

10 9 8 7 6 5 4 3 2 1

CONTENTS

PREFACE

The purpose of this book today and in the future (application in the field of bioscience) is to present up-to-date use of mass spectrometry of peptides. There is a considerable need for reference works concerning peptide applications since mass spectrometry has been practiced by the experts. The volume is intended as an essential contribution not only for those researchers working with peptides but also for scientists from the mass spectrometry laboratories who would like to learn about challenges in neuropeptide identification and quantitation. The book includes three main areas: instrumentation, hyphenated techniques, and selected applications and presents the rapid progress in each area. This book is devoted to people working in research, both industry and academia and also to students.

This book may serve as a compendium and as a good starting point to enter the area of peptides, including their function and identification. For researchers the book will also contain a valuable knowledge on strategies for identification of peptides of endogenous origin and description of a necessary instrumentation (features of particular techniques, drawbacks, possible applications). We would like to emphasize that this book has been written by mass spectrometry *users* and not by mass spectrometrists. That means that we did our best to describe applications of this technique rather than the technique itself. Certainly, the readers should know how mass spectrometry works but it was more important to guide them on how to apply this technique in their research, which types of instrumentation should be used for their particular requirements, and how to plan experiments with the use of MS, which demands a certain level of knowledge that can't be provided by a mass spectrometrist.

In addition, this book is also devoted to people working in the pharmaceutical industry, where there is a rapidly growing interest in peptidergic drugs such as enzyme inhibitors, and in the chemical/biotechnology industry, involving the preparation of, for example, combinatorial peptide libraries and hormones (e.g., recombinant insulin, growth hormone), where quality control of the products is more than crucial. Other target groups may be people making decisions for future directions in their companies on how to select the optimal quality control methods and why should they apply mass spectrometry only.

Despite all our efforts, we know that no book is perfect, so we welcome suggestions and comments. Send your email to *silber@chemia.uj.edu.pl* or *rolf@neuro.gu.se*.

Finally, we would like to thank all our colleagues who contributed their expertise.

Kraków, Göteborg

JERZY SILBERRING
ROLF EKMAN

CONTRIBUTORS

Per E. Andrén. Department of Pharmaceutical Biosciences, Uppsala University, Box 591 Biomedicum, SE-75124 Uppsala, Sweden

Douglas F. Barofsky, Oregon State University, 153 Gilbert Hall, Corvallis, Oregon 97331-4003, USA

Ann Westman-Brinkmalm, Institute of Clinical Neuroscience, Experimental Laboratory Section, Göteborg University, Sahlgrenska University Hospital, S-431 80, Mölndal, Sweden

Rolf Ekman, Neurochemistry Section, Institute of Clinical Neuroscience, Göteborg University, Sahlgrenska University Hospital, SE-431 80 Mölndal, Sweden

Jonas Bergquist, Uppsala University, Institute of Chemistry, Department of Analytical Chemistry, Box 531, SE-751 21 Uppsala, Sweden

Johan Gobom, Max Planck Institute for Molecular Genetics, Ihnestrasse 73, 14195 Berlin, Germany

Kristina Hakansson, Florida State University, Ion Cyclotron Resonance Program, National High Magnetic Field Laboratory, 1800 East Paul Dirac Drive, Tallahassee, Florida 32310, USA

Ulf Hellman, Ludwig Institute for Cancer Research, Box 595, SE-751 24 Uppsala, Sweden

Simon J. Gaskell, Department of Chemistry, P.O. Box 88, Sackville Street, Manchester M60 1QD, United Kingdom

Peter Juhasz, Applied Biosystems, 500 Old Connecticut Path, Framingham, Massachusetts, 01701, USA

Juergen Kast, University of British Columbia, Department of Chemistry and Biomedical Research Centre, 2222 Health Sciences Mall, Vancouver, BC V6T 1Z3, Canada

Randall W. Nelson, Intrinsic Bioprobes Inc., 625 South Smith Road, Suite 22, Tempe, Arizona 85281, USA

Carol L. Nilsson, Göteborg University, Department of Medical Biochemistry, Medicinaregatan 9 A, Box 440, SE-405 30, Göteborg, Sweden

Fred Nyberg, Uppsala University, Department of Pharmaceutical Sciences, Box 591, SE-751 24 Uppsala, Sweden

Manfred Raida, CellZome GmbH, Mass Spectrometry, Meyerhofstrasse 1, Heidelberg D-69117, Germany

Krzysztof Rolka, University of Gdansk, Faculty of Chemstry, Sobieskiego Str. 18, 80-952 Gdansk, Poland

Jerzy Silberring, Jagiellonian University, Neurobiochemistry Unit, Faculty of Chemistry and Regional Laboratory, Ingardena Str. 3, 30-109 Krakow, Poland

Richard D. Smith, Environmental Molecular Sciences Laboratory, Mail Stop K8-98, Pacific Northwest National Laboratory, P.O. Box 999, Richland, Washington 99352, USA

Roman A. Zubarev, University of Southern Denmark, Department of Chemistry, Campusvej 55, DK-5230, Odense, Denmark

Peter Roepstorff, University of Southern Denmark, Department of Biochemistry and Molecular Biology, Campusvej 55, DK-5230, Odense, Denmark

Agnieszka Sciubisz, Jagiellonian University, Faculty of Chemistry and Regional Laboratory, Ingardena 3, 30-060 Krakow, Poland

MASS SPECTROMETRY AND HYPHENATED TECHNIQUES IN NEUROPEPTIDE RESEARCH

PART 1

INTRODUCTION AND GENERAL DESCRIPTION

INTRODUCTION

Mass Spectrometry of Peptides and Proteins: a Personal Historical View

PETER ROEPSTORFF

Department of Biochemistry and Molecular Biology, University of Southern Denmark, Odense University, Denmark

The 1990s explosion in the application of mass spectrometry in protein studies has caused many young scientists to believe that the history of mass spectrometry in protein chemistry just started around 1990. This is not true. The present advanced state of the field is the result of the efforts of a large number of scientists since 1960. Mass spectrometry has contributed to many important discoveries about proteins and especially their posttranslational modifications in a period when peptide analysis by mass spectrometry required extensive derivatization, because the only ionization technique available at that time, electron-impact ionization, required evaporation of the analyte into the ion source. Mass spectrometers in the 1960s were also highly intricate instruments requiring the operators to be vacuum experts, plumbers, electronics engineers, and chemists at the same time. Computers for data acquisition were nonexistent, and the spectra were recorded either on standard ink recorders or on—what was a big step forward because it allowed fast scanning—galvanometer recorders by photosensitive but, unfortunately, nondurable paper. Mass and peak intensity assignment was a time-consuming and boring manual job, which could easily take more than an hour per spectrum. I have been involved in peptide and protein mass spectrometry since the late 1960s, and in the following account I will try to give my personal view on the historical highlights, which have contributed to the present success of mass spectrometry in protein science.

In the first phase, 1965–1980, volatile peptide derivatives were needed. The first reports on peptide mass spectrometry appeared in the early 1960s. These

Mass Spectrometry and Hyphenated Techniques in Neuropeptide Research, Edited by Jerzy Silberring and Rolf Ekman.
ISBN 0-471-35493-7

were all based on electron-impact (EI) mass spectrometry, thus requiring preparation of volatile derivatives of the peptides. In spite of the fact that small peptides could be volatilized, the spectra were very complex as a result of numerous fragmentation and rearrangement reactions. The much more stable acylpeptide esters were investigated by Biemann's group and by McLafferty and co-workers in The United States, by Heins and Crützmacher in Germany, and by the group at the Shemyakin Institute in Moscow (reviewed in Ref. 1). The first programs for automated interpretation of the mass spectra of peptides appeared [2] at that time, and Klaus Biemann stated at the end of his paper that since a few micrograms per analysis are sufficient, and since the complete analytic procedure including computer treatment can be performed with great speed, the method seems promising for routine determination of the amino sequence of small peptides produced by partial hydrolysis of oligopeptides derived by enzymatic digestion of proteins. The most spectacular achievements in that period were the structure determination of a number of natural peptidolipids in a collaboration between M. Barber at AEI in Manchester and Lederer's group in Gif in France (reviewed in Ref. 3) starting with the structure determination of a heterogeneously acylated and heavily modified peptidolipid, Fortuitin, with a molecular mass of 1350 [4]. The reason for the success with these natural compounds was that they were all naturally *N*-methylated. This led to the development of the permethylation procedure [5]. In the following decade permethylation was the most common procedure for mass spectrometric analysis of pure peptides and simple peptide mixtures [6], whereas Biemann and his group advocated analysis by GCMS (gas chromatographic mass spectrometry) of complex mixtures of small peptides (2–5 residues) derived by partial acid analysis of larger peptides after reduction to aminoalcohols and trimethylsilylation [7]. During this period a number of new ionization methods were developed and the first attempts to perform LCMS (liquid chromatographic mass spectrometry) were made. These new methods included field desorption, which was the first method that allowed analysis of underivatized peptides, and chemical ionization. They were immediately tried for peptide analysis but did not move the limits for mass spectrometric peptide analysis beyond what could be achieved by EI. In spite of the limited peptide size (up to decapeptides) and sensitivities in the low nanomolar range, EIMS resulted in a number of important achievements. These included the structure determination of the neuropeptides TRH [8], enkephalin [9], and adipokinetic hormone [10] available in only minute amounts and often *N*-, and *C*-terminally blocked and thus not accessible by Edmann degradation; and also the determination of posttranslationally modified amino acid residues in proteins, such as γ-carboxyglutamic acid [11] and β-hydroxyaspartic acid [12] in the proteins involved in the blood clotting cascade.

The second phase, 1980–1990, PD and FAB MS opens the way to protein analysis by MS. At the end of the 1970s protein chemistry and peptide mass spectrometry were at a standstill. At the same time new genetic techniques started to appear, with the result that most protein chemists changed fields. It was, however, obvious to at least some of us that the new genetic techniques would soon lead to an increased need for protein chemistry and also that this protein chemistry would have to meet new demands. The importance of complete de novo protein sequencing would decrease, because the amino acid sequence of many proteins would come from cDNA sequencing. Instead, the focus would be on co- and posttranslational protein modification, that is, all the characteristics of a protein that could not be obtained from genomic information. Precise protein and peptide molecular weight information would meet many of the new demands, but the existing mass spectrometric techniques, unfortunately, did not fulfil these needs. The mass range was too limited and the sensitivity insufficient, mainly because extensive derivatization was required. Were there any openings? Yes, as a matter of fact a mass spectrometric technique, termed *plasma desorption mass spectrometry* (PDMS) [13], had existed for a number of years. However, its potential was realized by only a few scientists. In PDMS, desorption/ionization of the analyte deposited as a solid film on a thin aluminum foil was achieved by bombardment with high-energy fission fragments from a ^{252}Cf source placed behind the foil. Mass determination was obtained with a simple linear time-of-flight mass analyzer. Macfarlane and co-workers had demonstrated that it was possible to obtain mass spectra of peptides of up to a molecular weight of approximately 3500 [14] and protected nucleic acids of up to more than 6 kDa [15]. In 1981 in studies of the mechanism of fast heavy-ion-induced desorption using beams of fast ions from accelerators, Bo Sundqvist and collaborators at the Uppsala University in Sweden obtained the first spectrum of a protein, insulin, using bombardment with a beam of 90 MeV $^{127}I^{20+}$ ions from a tandem accelerator [16]. The studies in the Uppsala group were continued on a prototype ^{252}Cf PDMS instrument, and shortly afterward the possibility of obtaining mass spectra of much larger proteins, including 14-kDa phospholipases and the 23.5-kDa trypsin, was demonstrated [17,18]. The huge potential of this new mass spectrometric technique for protein analysis was clearly stated in these papers: "Our results demonstrate that PDMS is a viable method for MW determination up to 25,000. The precision obtained is sufficient to monitor even minor protein modifications. Therefore possible applications are control of products from genetic engineering and protein modification reactions in vivo and in vitro. We have also shown that the MW (molecular weight) of components in a mixture can be determined and this for mixtures which are difficult or impossible to analyze with conventional techniques. The precision of the MW determinations also makes PDMS useful for verifying sequence analysis."

PDMS was not the only alternative for protein analysis by MS. Soon after, Macfarlane's group started to use fission fragments for desorption/ionization in mass spectrometric studies, Benninghoven et al. [19] investigated the application of secondary-ion mass spectrometry (SIMS) using bombardment with low-energy ions (10–40 keV) for the analysis of solid organic molecules. They analyzed only small molecules such as amino acids. However, an extension of the SIMS technique using a liquid matrix (glycerol), was published some years later by Barber and co-workers [20] (the same Mickey Barber who 16 years earlier had determined the structure of Fortuitine by MS). They called the technique *fast-atom bombardment* (FAB) because they used fast-atom instead of ion bombardment. Since it was the use of a liquid matrix that was the essential discovery, and not FAB, the technique was later often referred to as *liquid SIMS*. The use of a liquid matrix also opened for LCMS with ionization by FAB. FABMS, like PDMS, made the analysis of small proteins such as insulin [21,22] and the 3-kDa larger proinsulin [23] possible. FAB quickly gained widespread use because it could readily be installed on existing magnetic sector and quadrupole mass spectrometers, but the full impact was not evident until instruments equipped with high-field magnets became commercially available a few years later. FABMS of proteins up to 24 kDa became a reality with these instruments [24]. The development of PDMS was somewhat slower than that of FABMS because commercial instrumentation was not immediately available. It was, however, clear that analysis of the very large molecules was favored by the high-energy bombardment in PDMS and the sensitivity as well as accessible mass range were further improved by the discovery of nitrocellulose as matrix for PDMS [25]. When a small company, BioIon, founded by the Uppsala group, in 1984 launched a commercial PDTOF (plasma desorption, time of flight) mass spectrometer this instrument quickly gained wide acceptance in biochemistry and biotechnology laboratories in academia and industry because it was the first commercial mass spectrometer, which was fully automated and could be used without skilled operators. In this respect it has served as inspiration for many of the contemporary mass spectrometers. Traditional mass spectrometry laboratories favoured the more sophisticated sector and quadrupole mass spectrometers with the result that the development of protein mass spectrometry in the following years followed two parallel paths: one, dominated by the PDMS community, concentrated on the exploitation of the molecular weight information as a supplement in protein chemistry and the other, promoted by the FAB community, directed toward the generation of structural information by tandem mass spectrometry.

In the 1980s, two parallel series of symposia on desorption of involatile molecules, the IFOS and the St. Malo meetings, were initiated by Benninghoven and Lebeyec, respectively. The two series were later fused into the "Desorption Meetings." The attendance at these meetings was (and still is) true interdisciplinary with the participation of a rather limited number of scientists ranging from theoretical physicists over sputtering and ion physicists, instrument developers, to chemists and biologists with a common belief in these new techniques. These meetings were very creative and inspiring, characterized by intense debates, strong arguments, and wild ideas and as a result, the fundamental understanding of the processes involved in the desorption of large biomolecules and the applications of the techniques moved rapidly forward. Thus, most of todays commonly used methods in protein mass spectrometry were already developed or conceived in that period. MS/MS of peptides with high-energy CID was pioneered by Biemanns group [26] and used in many groups simultaneously with development of the now obsolete four sector mass spectrometers. Low-energy collision using triple quadrupole instruments was being investigated in several groups [27]. The concept of peptide mass mapping was described using FAB [28] and PDMS [29]. Mass spectrometric MW information as a supplement to traditional protein sequencing [30] and for the characterization of recombinantly produced [31] and engineered proteins [32] were widely used. Numerous posttranslational modifications in proteins, including the hitherto inaccessible glycosylations and phosphorylations, were characterized by mass spectrometry. Studies leading to the now widespread use of peptide mass data for identification of proteins and their modifications in proteomics were initiated by matching peptide and protein MW information with DNA sequence data using FAB [33] and PDMS [34,35]. Of special interest for the scope of this book was a concept called *peptide charting*, which was used to identify neuropeptides relative to the precursor protein DNA sequences [36], a concept that in today's terminology would have been called *peptidomics*. Desiderio's pioneer work in the field of quantitative analysis of neuropeptides, first by field desorption [37] and later by FABMS [38] using ^{18}O-labeled or deuterated internal standards is another milestone interest.

1990 to the present day. MALDI and ESI conquer protein mass spectrometry. In spite of all the exciting applications during the 1980s, it became clear that the sensitivity and mass range obtainable by PDMS and FABMS could not meet the demands of the coming decade. A new miracle was needed—and it came in 1988 when two new ionization methods, electrospray ionization (ESI) [39] and matrix-assisted laser desorption/ionization (MALDI) [40], were described by John Fenn at the ASMS Conference in San Francisco in June and by Franz Hillenkamp at the 11th International Mass Spectrometry Conference in September in Bordeaux, respectively. MALDI was based on some 20 years' experience with laser desorption mass spectrometry (LDMS), which, in spite of a good potential, never succeeded in the analysis of large involatile molecules. Franz Hillenkamp had been an active participant in the meetings on the desorption and ionization of involatile molecules described earlier and had developed the laser microprobe mass analyzer (LAMMA). It was while searching for new applications for this instrument that he, together with Michael Karas, discovered the potential of using a matrix in LDMS. (For an historical review of the development that led to MALDI, see Ref. 41.) ESI MS is fundamentally different from the earlier used desorption methods. The analyte is introduced in a solution and the ions formed by desolvation of liquid droplets in a high electric field at atmospheric pressure, resulting in the formation of highly charged ions, which are then transported into the high vacuum of the mass spectrometer through a series of skimmers or a capillary. The principle was first described by Dole in 1968 [42] and developed into a practical ionization method by Fenn and collaborators [39]. (For an historical review of nearly 40 years of development that led to ESI, see Ref. 43.) These new ionization methods allowed the analysis of much larger proteins than hitherto possible with an unprecedented mass accuracy and sensitivity and ESI opened for the routine use of LCMS. ESI, as was the case with FAB, had the advantage that it could be implemented on existing commercially available mass spectrometers. It therefore "took off" a little faster than did MALDI, where the full exploitation required the development of new TOF mass spectrometers. However, the instrument manufacturers realized the huge market potential of mass spectrometry entering the protein and biotechnology fields. In the following years instrument development exploded, and especially TOF analyzers have reached a performance level, in terms of mass resolution and sensitivity, which far exceeds the dreams of almost all the scientists working in this field. ESI and MALDI, combined with the amazing instrument development, have opened for full exploitation of all the applications and concepts that had been developed since 1980, and have changed the scope of mass spectrometry for protein and peptide studies from being a promising technique pursued by a limited number of scientists to be an effective tool used by a very large and fast-growing scientific community. At present the performance of mass spectrometry is well ahead of the demands from biochemistry and molecular biology [44], and the only limit for further development is our imagination. The remaining chapters of this book, describe some of all the fascinating applications and new perspectives of mass spectrometry in protein studies with focus on the challenges in neuropeptide research. So I will end my historical overview at the beginning of this new era.

REFERENCES

1. K. Heyns and H. F. Grützmacher, *Fortschr. Chem. Forsch.* **6**, 536 (1966).
2. K. Biemann, *J. Am. Chem. Soc.* **88**, 2597 (1966); M. Senn, R. Venkataraghavan, and F. W. McLafferty, *J. Am. Chem. Soc.* **88**, 5593 (1966).
3. E. Lederer, *Pure Appl. Chem.* **17**, 489 (1968).
4. M. Barber, P. Jolles, E. Vilkas, and E. Lederer, *Biochem. Biophys. Res. Commun.* **18**, 469 (1965).
5. E. Vilkas and E. Lederer, *Tetrahedron Lett.* 3089 (1968).
6. P. Roepstorff and K. Brunfeldt, *Adv. Mass Spectrom.* **6**, 199 (1974).
7. K. Biemann, in *Biochemical Applications of Mass Spectrometry*, G. R. Waller and O. C. Dermer, eds., Wiley, New York, 1980, p. 469, and references cited herein.
8. R. Burgus, T. F. Dunn, D. Desiderio, D. N. Ward, W. Vale, and R. Gullemin, *Nature* **226**, 321 (1970).
9. J. Hughes, T. W. Smith, H. W. Kosterlitz, L. A. Fothergill, B. A. Morgan, and H. R. Morris, *Nature* **258**, 577 (1975).
10. J. V. Stone, W. Mordue, K. E. Batley, and H. R. Morris, *Nature* **263**, 207 (1976).
11. J. Stenflo, P. Fernlund, W. Egan, and P. Roepstorff, *Proc. Natl. Acad. Sci.* (USA) **71**, 2730 (1974).
12. T. Drakenberg, P. Fernlund, P. Roepstorff, and J. Stenflo, *Proc. Natl. Acad. Sci.* (USA) **80**, 1802 (1983).
13. D. F. Thorgerson, R. P. Skowronski, and R. D. Macfarlane, *Biochem. Biophys. Res. Commun.* **60**, 616 (1974).
14. C. J. McNeal and R. D. Macfarlane, *J. Am. Chem. Soc.* **103**, 1609 (1981).

15. C. J. McNeal, K. K. Ogilvie, N. Y. Theriault, and M. J. Nemer, *J. Am. Chem. Soc.* **104**, 981 (1982).
16. P. Håkansson, I. Kamensky, B. Sundqvist, J. Fohlmann, P. A. Peterson, C. J. McNeal, and R. D. Macfarlane, *J. Am. Chem. Soc.* **104**, 2948 (1982).
17. B. Sundqvist, I. Kamensky, P. Håkansson, J. Kjellberg, M. Salehpour, S. Widdiyasekera, J. Fohlmann, P. A. Peterson, and P. Roepstorff, *Biomed. Mass Spectrom.* **11**, 242 (1984).
18. B. Sundqvist, P. Roepstorff, J. Fohlmann, A. Hedin, P. Håkansson, I. Kamensky, M. Lindberg, M. Salehpour, and G. Säwe, *Science* **226**, 696 (1984).
19. A. Benninghoven, D. Jaspers, and W. Sichtermann, *Appl. Phys.* **11**, 35 (1976).
20. M. Barber, R. S. Bordoli, R. D. Sedgwick, and A. N. Tyler, *J. Chem. Soc. Chem. Commun.* 325 (1981).
21. M. Barber, R. S. Bordoli, G. J. Elliot, R. D. Sedgwick, A. N. Tyler, and B. N. Green, *J. Chem. Soc. Chem. Commun.* 936 (1982).
22. A. Dell and H. Morris, *Biochem. Biophys. Res. Commun.* **106**, 1456 (1982).
23. M. Barber, R. S. Bordoli, G. J. Elliot, N. J. Horoch, and B. N. Green, *Biochem. Biophys. Res. Commun.* **110**, 753 (1983).
24. M. Barber and B. N. Green, *Rapid Commun. Mass Spectrom.* 936 (1987).
25. G. P. Jonsson, A. B. Hedin, P. L. Håkansson, B. U. R. Sundqvist, B. G. S. Säve, P. F. Nielsen, P. Roepstorff, K.-E. Johansson, I. Kamensky, and M. S. L. Lindberg, *Anal. Chem.* **58**, 1084 (1986).
26. K. Biemann and S. A. Martin, *Mass Spectrom. Rev.* **6**, 1 (1987).
27. D. F. Hunt, J. Shabanowitz, P. R. Griffin, J. R. Yates, P. A. Martino, and A. L. McGormack, in *Biological Mass Spectrometry*, A. L. Burlingame and J. A. McCloskey, eds., Elsevier, Amsterdam, 1990, p. 337.
28. H. R. Morris, M. Panico, and G. W. Taylor, *Biochem. Biophys. Res. Commun.* **117**, 299 (1983).
29. A. Tsarbopulos, G. W. Becker, J. L. Occolowits, and I. Jardine, *Anal. Biochem.* **171**, 113 (1988).
30. P. Roepstorff, K. Klarskov, and P. Hoejrup, in *Methods in Protein Sequence Analysis*, 198••, p. 191.
31. P. Roepstorff, P. F. Nielsen, K. Klarskov, and P. Hoejrup, *Biochem. Soc. Trans.* **17**, 249 (1989).
32. P. F. Nielsen, P. Roepstorff, I. G. Clausen, E. B. Jensen, P. Jonassen, A. Svendsen, P. Balschmidt, and F. B. Hansen, *Protein Eng.* **2**, 449 (1989).
33. B. W. Gibson and K. Biemann, *Proc. Natl. Acad. Sci.* (USA) **81**, 1956 (1984).
34. P. Roepstorff, P. F. Nielsen, and P. Hoejrup, *Biochem. Soc. Trans.* **15**, 155 (1987).
35. H. V. Scheller, J. S. Okkels, P. B. Høj, I. Svendsen, P. Roepstorff, and B. L. Moeller, *J. Biol. Chem.* **264**, 18402 (1989).
36. G. J. Feistner, P. Hoejrup, C. J. Evans, C. F. Barofsky, and P. Roepstorff, *Proc. Natl. Acad. Sci.* (USA) **86**, 6013 (1989).
37. D. M. Desiderio and M. Kai, *Biomed. Mass Spectrom.* **10**, 471 (1983).
38. J. J. Kusmierz, R. Sumarada, and D. M. Desiderio, *Anal. Chem.* **62**, 2395 (1990).
39. J. B. Fenn, M. Mann, C. K. Meng, S. F. Wong, and C. M. Whitehouse, *Science* **246**, 64 (1989).

40. M. Karas and F. Hillenkamp, *Anal. Chem.* **60**, 2299 (1988).
41. F. Hillenkamp and M. Karas, *Int. J. Mass Spectrom.* **200**, 71 (2000).
42. M. Dole, L. L. Mack, R. L. Hines, R. C. Mobley, L. O. Ferguson, and M. B. Alice, *J. Chem. Phys.* **49**, 2240 (1968).
43. J. B. Fenn, *Int. J. Mass Spectrom.* **200**, 459 (2000).
44. P. Roepstorff, *Curr. Opin. Biotechnol.* **8**, 6 (1997).

CHAPTER 1

Neuropeptides in CNS Tissue and Body Fluids: Functional Aspects and Strategies for Their Identification, Purification, and Characterization

FRED NYBERG

Department of Pharmaceutical Biosciences, Uppsala University, Uppsala, Sweden

CONTENTS

Mass Spectrometry and Hyphenated Techniques in Neuropeptide Research, Edited by Jerzy Silberring and Rolf Ekman.
ISBN 0-471-35493-7

1.1. INTRODUCTION

The neuropeptides constitute a large family of bioactive compounds participating in the transmission or modulation of signals in the central nervous system (CNS) [59]. They are present at very low concentrations in the various CNS tissues and body fluids as well. A variety of neuropeptides have been identified and characterized. They are shown to be involved in several neurologic functions, including those related to food intake, memory, mood, pain, reward, stress, and neuroendocrinologic as well as neuroimmunologic regulation. The neuropeptides are released from nerve terminals of peptidergic neurons, and they induce their effects by binding to and activation of pre- or postsynaptically located receptors, which are embedded in the nerve cell membrane. A general pathway for the neuropeptide biosynthesis starts at the ribosomes in the nerve cell bodies, where the messenger RNA is translated into an amino acid sequence. Generally, the neuropeptides are synthesized as large, biologically inert protein precursors (prepropeptides). The neuropeptide precursors are subsequently cleaved by a series of proteolytic steps in a sequence-specific and tissue-specific fashion to generate smaller active peptides. During this process additional modifications such as amidation and truncation of the released sequence may also occur. Following secretion and interaction with its receptor, the active peptide is inactivated by enzymatic degradation. In some cases the released active peptide is converted to fragment with retained or different biological activity. The present knowledge about the functional relevance of most neuropeptides is still limited, and there is a real need to further explore this field. Extensive studies of these compounds in the area of biomedical science is thus a general desire, which at present engages many researchers, and their ambitions will certainly result in additional information in the near future. However, to accomplish this it will be necessary to have access to efficient, reliable, and highly reproducible techniques. This chapter discusses neuropeptide function and strategies for its identification, quantification, purification, and characterization. In particular, it focuses on peptides present in the CNS and various body fluids. It also discusses techniques that have been used in the past and on requirements for future investigations.

1.2. NEUROPEPTIDE BIOSYNTHESIS, PROCESSING, AND INACTIVATION

The endogenously produced neuroactive peptides are synthesized as large, biologically inert precursor proteins (prepropeptides) that undergo posttranslational modifications to generate the active peptide molecule. Biosynthesis of the prepropeptide occurs at the ribosomes located on the endoplasmic reticulum in the cell body of the peptidergic neuron [12,82]. The prepropeptide, usually 100–250 amino acid residues in length, consists of an *N*-terminal *signal sequence* adjacent to a variable stretch of unknown function, which is followed by a peptide-containing region, including several copies of active peptide units. The first cleavage consists of removal of the signal peptide, to yield the propeptide, before core glycosylation, glucose trimming, and a suitable folding of the propeptide occur. Additional modifications of the sugar content take place within the Golgi compartment, wherein the release of intermediate-sized peptides is initiated. The amino acid sequence of the propeptides indicates that the biologically active units are nearly always bracketed by basic amino acids. These residues, Lys and/or Arg, often arrange as pairs and sometimes even as quadruplets or singlets, and are recognized by specific enzymes that can hydrolyze the peptide bonds and release the active peptides, or rather the *N*- or *C*-terminal extension thereof.

A number of enzymes responsible for this endoproteolytic processing have been identified. They are all characterized by a catalytic domain with structural homology to subtilisin. Subtilisin is classified as a bacterial and calcium-dependent serine proteinase. Because of their similarities with the bacterial subtilase, all the propeptide cleaving endoproteases identified so far are considered to belong to the enzyme family of subtilases. Up to now at least seven mammalian subtilases capable of hydrolyzing peptide bonds at single or pairs of basic residues have been described [134]. The subtilases are generally named as precursor convertases (PC), and the individual members of this enzyme family are known as PC1, PC2, PACE (also furin), PACE4, PC4, PC5, and PC7. They cleave the propeptide sequence at the carboxy-terminal side of the paired basic residues (Lys–Arg or Arg–Arg) and in some cases of a single Arg residue [13,134]. However, precursor convertases cleaving the propeptide *N* terminal to or in between the dibasic stretch are also described [32,39]. Other proteases capable of releasing peptide fragments containing bioactive units are also known. Examples of these are an aspartic protease (PCE) acting as a primary proopiomelanocortin (POMC)-converting enzyme [76] and a protease belonging to a new class of subtilisin–kexin-like convertases (SKI-1) [93]. The propeptide convertases are believed to be copacked with their substrates in secretory granules, where they are optimally active at pH 5.5 [52,145]. According to the specificity of the subtilases [134] and similar propeptide convertases, their cleavage actions result in the production of peptide intermediates containing basic residue extensions at their *N*- and *C*-terminal ends, which, in turn, are removed by various amino- and carboxypeptidases before the active peptide is formed. In addition, other modification steps, such as *N*-terminal acetylation, *C*-terminal amidation, sulfation, and phosphorylation, and also formation of *N*-terminal pyroglutamyl residues, may occur.

N-Terminal acetylation occurs in several neuroactive peptides, including the endogenous opioid β-endorphin and in truncated forms of this peptides [76]. The significance of this modification is not fully understood, although it certainly affects the peptide stability toward peptidases acting from the amino terminus [56]. A number of neuropeptides are amidated at their *C* terminus, and for a major part of these peptides, this modification is essential for their activity. Among the enzymes responsible for this modification are peptidylglycine-α-hydroxylating monooxygenase, which hydroxylates the *C*-terminal Gly, the motif for amidation, and peptidyl-α-hydroxyglycine α-amidating lyase, which cleaves the hydroxyglycine to release glyoxylate and finally form the amidated peptide [60]. Both these enzymes are encoded by the single peptidylglycine α-amidating monooxygemase (PAM) gene. An amidating α-monooxygenase activity has also been identified in human CSF [100]. Another posttranslational modification of importance for some neuropeptides is accomplished by tyrosylpeptide sulfotransferase, an enzyme that catalyzes the transfer of a sulfate 3′-phosphoadenosine 5′-phosphosulfate to tyrosine [11]. A well-known peptide with a tyrosine *O*-sulfation is represented by cholecystokinin (CCK) [118]. For this peptide and also for elongated sequences of CCK, the sulfation is essential for the binding and its bioactivity [10].

An important phenomenon in neuropeptide processing is that in many cases an active matured peptide may be converted to a fragment with retained or modified biological activity. For instance, the tachykinin substance P (SP) may be converted to its bioactive *N*-terminal heptapeptide SP(1–7) by the action of more or less specific endopeptidase activities [115]. The *N*-terminal SP metabolite SP(1–7) shares some effects with SP but opposes others [64,171]. A number of endopeptidases capable of releasing the bioactive heptapeptide fragment of SP have been described [86,100,102,115]. Moreover, the dynorphins with affinity for kappa opioid receptors are converted by endoproteases releasing the enkephalin hexapeptide Leu–enkephalin–Arg, selective for the delta opioid receptor. Several so-called dynorphin-converting enzymes have been characterized [103,115,139,142]. An additional example of this kind of biotransformation is represented by the angiotensin system; angiotensin II, which, in its action on the cardiovascular system, acts on AT1 and AT2 receptors, is converted to a fragment (angiotensin IV) with selectivity for AT4 receptors associated with different biological activities [45].

The enzymes responsible for the inactivation of the peptide signal may be common for many neuropeptides. A common route of neuropeptide degradation is accomplished by exopeptidases. The enkephalins are inactivated by aminopeptidase M [19,42,48], and SP is degraded by a dipeptidyl prolylpeptidase, known as DP-IV [19,168]. The removal by aminopeptidases of the *N*-terminal tyrosine residue in opioid peptides will abolish the ability of the endogenous opioid peptides to activate opioid receptors. The enkephalins are also effectively inactivated by neutral endopeptidase (NEP) and angiotensin-converting enzyme (ACE) [100,123]. Both ACE and NEP are shown to hydrolyze and deplete the tachykinin effect of SP [67,158]. The neuropeptide-inactivating enzymes have represented importance targets for drug development strategies. Specific inhibitors directed toward individual enzymes and also against a combination of enzymes have thus been designed. For instance, the design of a mixed inhibitor (RB-101), preventing the degradation of enkephalins and thereby inducing pain-relieving effects, has receivied particular attention [122].

1.3. NEUROPEPTIDE RECEPTORS

Following release, the neuroactive peptides bind to and activate specific receptors located on their target cells. Most neuropeptide receptors belong to the superfamily of *G-protein-coupled* receptors [70]. Some peptides bind to only one known receptor, whereas others, such as galanin and melanocortin, bind to several receptor subtypes, all of them known to be linked to heterotrimeric guanine nucleotide-binding proteins (G proteins). The G protein consists of alpha, beta, and gamma subunits. Following agonist activation of a G-protein-coupled receptor, the alpha subunit dissociates from its beta–gamma subunits, leading to activation of ion channels or enzymes located in the cell membrane. Stimulation of G-protein-coupled receptors results in excitatory or inhibitory transmission but in some cases also in modulation of the actions on other transmitters.

In the regulation of ion channels the G-protein-coupled receptors, depending on cell type involved, activate specific inwardly rectifying K^+ channels or inhibit voltage-gated Ca^{2+} channels. Both these types of channels are believed to be regulated by the beta–gamma subunits, which directly limit the opening of Ca^{2+} channels and gate the K^+ channels. Examples of neuropeptide receptors regulating these channels are represented by those of the opioid peptides.

Other effector systems activated by the G proteins include several enzymes that regulate the formation of second messengers as cAMP, cGMP, and diacylglycerol. These second messengers stimulate or inhibit the action of various protein kinases and protein phosphatases, which, in turn, act on ion channels, leading to a subsequent generation of postsynaptic electrical signals. The second messengers, cAMP and cGMP, may also directly bind to and affect the activity of certain ion channels.

In conformity with most G-protein-coupled receptors, sustained agonist binding to the neuropeptides receptors result in internalization; that is, the peptide–receptor complex is transported into the cell. The receptor internalization is a multistep process that leads to either a recycling of the receptor back to the cell membrane or degradation [70,112]. The internalization process, which is initiated by a phosphorylation step, results in receptor desensitization [70]. The recycled receptor may subsequently repeat this cycle of ligand binding, phosphorylation, and internalization.

The interaction between neuropeptides and their receptors appears to involve a complex mechanism. Peptides are highly flexible entities, built up by a number of amino acids, with a dynamic structure in a water solution. Only a part of the peptide molecule may fit into the binding pocket of a G-protein receptor. A certain conformation may yield the peptide the highest affinity for the receptor, and certain amino acid residues are critical for binding. An interesting feature is that removal of a sequence segment may alter the receptor preference of a given peptide from one receptor subtype to another [99].

1.4. NEUROPEPTIDE FUNCTIONS

A substantial knowledge of neuropeptide function has emerged from studies using small-molecule agonists and antagonists. Using these compounds, it has been possible to explore the effects of the endogenous neuroactive peptides on behavior in whole animals as well as at the cellular level, by mimicking or blocking their actions. However, because of the limited availability of peptide agonists and antagonists that penetrate the blood–brain barrier and get access to the CNS, there is a lack of functional investigations of many neuropeptide systems. Another factor, which has limited the availability of functional research on neuropeptide research, is the difficulty to interpret data emerging from quantification of neuropeptide levels in CNS tissues. It has thus been difficult to discriminate between accumulation of peptides within neuronal cells due to inhibited release or increased biosynthesis versus accumulation of peptides in the extracellular compartment due to enhanced release. However, application of the microdialysis techniques has provided a basis for overcoming this difficulty. Other tools that have been used in functional studies of neuropeptides are microinjections of peptides in specific CNS areas and genetically modified mice. Transgenic mice that overexpress or lack the gene for a given peptide or receptor have significantly contributed to improvement in the understanding of neuropeptide function.

The number of known neuropeptides is still increasing. Up to now about 80 different mammalian neuropeptides (cytokinins excluded) are known, and this number may be doubled within five years. Regarding their functional significance, this chapter focuses on only a minor part. Among the best-characterised neuropeptides are, in addition to the abovementioned angiotensin II, opioid peptides, tachykinins, and cholecystokinins, including several peptides related to the HPA axis such as corticitrophin-releasing factor (CRF) but also calcitonin-gene-related peptide (CGRP), and neuropeptide Y(NPY). Some of the proposed functions of these peptides are reviewed below.

1.4.1. The Opioid Peptides

The opioid peptides (endorphins) have been recognized not only for their modulatory effect on pain transmission but also for their involvement in reward mechanisms, in stress, in food intake, and in learning and memory [160]. The classic endorphins are coded by three distinct genes, whose products are represented by preroopiomelanocortin (POMC), preproenkephalin, and preprodynorphin. The best-known peptides released from these prepropetides are β-endorphins, enkephalins, and dynorphin, respectively. At least three opioid receptor subtypes have been cloned and characterized: mu (μ), delta (δ), and kappa (κ). While the enkephalins have a high affinity for and stimulate δ-opioid receptors, the dynorphins preferentially bind to and activate κ-opioid receptors. β-Endorphin has long been considered as the endogenous μ-opioid receptor ligand. Enkephalins and dynorphins are present in the spinal cord neurons and shown to be involved in modulation of nociceptive signals. The enkephalins are putative transmitters in long descending pathways from the midbrain to the dorsal horn, whereas dynorphins occur mainly in spinal interneurons. The nociceptive information reaching the spinal cord dorsal horn is subjected to modulation by local circuit neurons and descending inhibitory systems, which includes monoamine and end enkephalinergic pathways that originate in periaqueductal gray (PAG), rostral ventral medulla (RVM), and other supraspinal areas [5,7]. Moreover, opioid receptors are located on peripheral primary afferents terminals and have been shown to be upregulated under inflammatory conditions [146,147].

It has been further shown that endogenous opioid peptides, present in resident immune cells within subcutaneous inflamed tissue, can be released to interact with opioid receptors on peripharal sensory nerves to inhibit nociception in inflamed tissue [146]. In the brain opioid peptides are found in areas related to various types of behavior, including reward. The activity of the mesolimbic dopamine pathway, originating in the ventral tegmental area (VTA) and projecting to the nucleus accumbens, is controlled by the opioid peptides [61,62]. Thus, opioid agonists (e.g., enkephalins and β-endorphin) stimulate dopamine release in the nucleus accumbens by acting on μ/δ receptors in the VTA, an effect that is opposed by dynorphin acting on κ receptors in the nucleus accumbens. Enkephalins and β-endorphins may also produce reinforcement through direct dopamine-independent action on μ/δ opioid receptors expressed in nucleus accumbens neurons. Regarding food intake, β-endorphin is also shown to stimulate the intake of preferred food by acting on this brain reward system [38,41]. The role of the endogenous opioids in learning and memory processes is complex. Depending on types of learning tasks, the observed effects of the opioid peptides on learning and memory differ. For instance, enkephalins were demonstrated to facilitate acquisition using a 12-choice mace [58], whereas these peptides were found to impair acquisition of of a one-way active avoidance task [54,121]. β-Endorphin has been shown to enhance acquisition of an inhibitory avoidance task but was reported to impair acquisition in an active avoidance task [85]. Dynorphins have generally been associated with an ability of inducing impairments in the learning process, although some exceptions have been described [159].

Among the atypical endogenous opioid peptides, the β-casomorphins and hemorphins are shown to mimic many effects opioid alcaloids, although their effects appear to be much weaker [111]. The highly potent μ-opioid receptor selective tetrapeptides, endomorphin 1 and endomorphin 2, discovered in the late 1990s, are known to produce strong analgesic effects [53,169]. However, endomorphin 1 was found to be more active than endomorphin 2 [169]. Endomorphin 1 was also found to be equipotent to morphine to inhibit the inflammatory responses in a vacuum-induced blister model for studies of inflammation processes [57]. In general it has been demonstrated that the endomorphins mimic most effects demonstrated for opioid alcaloids [53].

1.4.2. The Tachykinins

Substance P, neurokinin A (NKA), and neurokinin B (NKB) belong to the tachykinin family of neuropeptides. Members of this family share a common *C*-terminal sequence: Phe–X–Gly–Leu–Met–NH_2. The tachykinins, like other neuropeptides, are formed by cleavage of larger protein precursors, preprotachykinins [80,81]. SP and NKA are shown to be encoded by the preprotachykinin A gene and are produced by alternative splicing. Tissue-specific splicing patterns allow either SP alone, or both peptides, to be formed in different conditions. NKB can be generated from preprotachykinin B and is encoded by the gene of this prepropeptide. Three tachykinin receptors NK1, NK2, and NK3, belonging to the family of G-protein-coupled receptors are known [80]. SP exhibits highest affinity for the NK1 receptor, whereas NKA preferentially binds to the NK2 receptor and NKB seems to represent the endogenous ligand for the NK3 receptor.

Among the peptides belonging to the tachykinin family, substance P (SP) is the best characterized. As mentioned above, SP is widely distributed throughout the CNS, and its preferred receptor is highly expressed in areas that are critical for the regulation of pain influx, affective behavior, and stress [5,20,63,73]. Both SP and its receptor have been implicated in opioid dependence [14,24,83,91]. SP is colocalized with glutamate in the synaptic terminals of primary sensory nerve fibers (C fibers). The peptide is shown to facilitate transmission of the amino acid and thereby contribute to transmission of pain signals from the periphery into the CNS. SP is released not only from nerve terminals in the dorsal horn but also in a retrograde fashion from the free ending of nociceptive neurons. Release of SP from nociceptors is a phenomenon that occurs during neurogenic inflammation [57,80]. The localization of both SP and its receptor (NK1) in areas related to regulation of affective behavior and neurochemical response to stress has raised the question of a possible involvement of SP in mechanisms mediating these forms of behavior. In fact, central administration of SP or SP agonists in experimental animals has been found to induce conditioned place aversion and to produce an anxiogenic profile [25]. Moreover, NK1 receptor antagonists are currently being tested as antidrepessant and as anxiolytic medications. For example, the NK1 receptor antagonist (MK869) was reported to be effective in the treatment of depressive patients [63]. Accumulating data also suggest that SP may have an important role in opioid dependence. Thus, it has been demonstrated that genetic disruption of the NK1 receptor in mice is associated with loss of rewarding effects of morphine [91]. Furthermore, previous studies have shown that SP may enhance the reaction to opioid withdrawal in rats [137] and that NK1 receptor antagonists are capable of attenuating opioid withdrawal in rats [83] and guineapigs [14].

For activation of the NK1 receptor it seems that the *C*-terminal sequence is essential. However, it appears that several CNS effects of SP are not mediated through its *C*-terminal sequence. *N*-terminal SP metabolites, such as SP(1–7) and SP(1–8), have been shown to modulate or oppose several behaviors elicited by SP. For instance, SP(1–7) is reported to induce antinociception in mice [144], inhibit SP-induced rotational behavior in rats [47], and reduce naloxone-precipitated withdrawal in opioid-dependent mice [65].

1.4.3. The Angiotensin System

The renin–angiotensin system (RAS) has traditionally been viewed solely as a hormonal circulating system involved in the regulation of blood pressure and salt and fluid homeostasis [44]. However, it has become evident that peptides formed within this system also are present in the CNS and elicit actions on brain function [130]. Angiotensin II (AngII) is considered as the major biologically active peptide of the RAS. It is formed by the action of angiotensin-converting enzyme (ACE) on angiotensin I (AngI), an inactive peptide derived from circulating angiotensinogen. However, it is now recognized that there are a number of biologically active peptides formed from the metabolic processing of AngI and AngII, some of which have been found in the CNS. Among them are AngIII, AngIV, Ang(1–7), and Ang(3–7) [130]. The main actions of AngII are mediated via a specific membrane-bound G-protein-coupled receptor, the AT1 receptor. Most effects of AngII on the cardiovascular system are mediated through the AT1 receptor; however, the AT1 receptor is also expressed in certain brain nuclei, such as the hypothalamus and the pituitary. Some known effects of AngII on pituitary hormones, such as prolactin, result from stimulation of the AT1 receptor. AngII also recognizes an AT2 receptor, which is expressed in distinct brain regions and also in peripheral tissues. The AT2 receptor has been suggested to be involved in growth development and exploratory behavior. Several peripheral effects mediated through the AT2 receptor oppose those of the AT1 receptor.

AngII can be degraded further by enzymes that remove single amino acids sequentially from the *N*-terminal end, giving rise to AngIII and AngIV. Although these fragments have been thought to be of little importance, it has now been demonstrated that AngIII stimulates aldosterone secretion. Angiotensin IV has also been attributed to distinct action, such as on cerebral bloodflow and memory. A distinct receptor for AngIV has been identified in peripheral tissues and also in the brain. More recent evidence has suggested that both AngI and AngII may be converted to Ang(1–7). This heptapeptide was shown to elicit release of Arg–vasopressin (AVP) from neurohypophyseal–hypothalamic explants [132]. It also produces effects on the kidney and the cardiovascular system [130]. AngII and Ang(1–7) have been shown to have opposing actions suggesting the heptapeptide as a counterregulatory angiotensin compound. Ang(1–7) is shown to be generated in the brain, although its function in this compartment is still unclear [130]. Although specific binding sites for Ang(1–7) have been identified, no successful attempts to clone the receptor has yet been reported.

1.4.4. Calcitonin Gene-Related Peptide (CGRP)

Calcitonin-gene-related peptide (CGRP) is derived from the same primary gene transcript as calcitonin [124]. It is the first one of the first neuropeptides to be identified as a product of alternative splicing. Many cell types in the nervous system, including primary afferent sensory neurons, express CGRP. It is also expressed in spinal motorneurons [2,92], and CGRP expression is regulated during axonal growth [149]. In small-diameter sensory neurons CGRP has been found to be coexpressed with SP [33]. The biosynthesis of CGRP in dorsal root ganglia (DRG) seems to be regulated by nerve growth factor (NGF), and the high-affinity NGF receptor, trkA, is selectively expressed on nociceptive sensory neurons containing SP and CGRP [15,90].

Human and rat CGRP are found in two forms, α- and β-CGRP, which are encoded in two different genes [166]. These peptides are shown to activate at least two type of receptors, CGRP1 and CGRP2 [166]. The CGRP receptor subtypes are characterized as entities belonging to the family of G-protein-coupled receptors.

Although the precise function of CGRP is not well characterized, its putative role as a mediator of neurogenic inflammation is well established. The expression of CGRP is shown to be increased in DRG, on both the mRNA and protein levels, under conditions of peripheral inflammation [21,46,163].

The release of the first seven *N*-terminal amino acids from CGRP generates a product, CGRP(8–37), with antagonistic properties [55,166]. CGRP may also interact with the metabolism of SP [108]. It is shown to be cleaved into two fragments by the abovementioned substance P endopeptidase (SPE) [69].

1.4.5. Cholecystokinins (CCKs)

Cholecystokinin (CCK) is a gastrointestinal peptide known to be secreted by the duodenum in response to the presence of certain nutrients in the gut. It acts on G-protein-coupled receptors (CCK-A) in the gastrointestinal tract to decrease food intake. Circulating CCK does not cross the blood-brain barrier but is synthesized in the brain, where it acts mainly on CCK-B receptors to function as a satiety factor. Several sizes of CCK-like peptides have been found in peripheral tissues, and the main peptide produced by the intestine appears to consist of 58 amino acid residues [116]. In the brain a peptide of eight amino acids in length (CCK-8) is mainly produced.

In the mammalian midbrain CCK is shown to coexist with dopamine. The peptide is reported to modulate dopamine-mediated behavior [18]. Dopaminergic neurons in the midbrain are implicated in human neuropsychiatric disorders. For instance, studies of postmortem samples revealed high levels of CCK mRNA in midbrain dopamine neurons of schizophrenics [131]. Altered levels of the CCK peptide were earlier found in CSF samples collected from schizophrenic patients.

1.4.6. Corticotropin-Releasing Factor (CRF)

CRF is the principal regulator of the stress response. It represents a hypothalamic peptide of 41 amino acids [114] that stimulates the secretion and synthesis of pituitary ACTH, leading to production of glucocorticoid steroids from the adrenals. Originally, studies on CRF were focused on its role in the HPA axis. Interest in CRF has now extended to include its function in the CNS. The peptide has been found to be distributed broadly not only within CNS but also in extraneuronal tissues, such as placenta, testis, ovary, gut, immune system, adrenal gland, and heart. Within the CNS, CRF is believed to be involved in mechanisms mediating the behavior and immune response to stress.

In the hypothalamus CRF is synthesized by a subset of neurons in the paraventricular nucleus (PVN), from which it is released into the portal circulation. However, PVN neurons that express CRF also project to structures within the brain. The peptide is also known to be produced in regions outside the hypothalamus, such as the central nucleus of amygdala. In this area CRF seems to play a critical role in the brain response to stress and to also be involved in fear and anxiety [1].

Two types of CRF receptors are known: CRF-1R and CRF-2R. These receptors are characterized as G-protein-coupled receptors and share considerable sequence homology with one another [114]. Whereas CRF-1R is widely distributed in the brain, the CRF-2R appears to be concentrated in the lateral septal nuclei of the forebrain, another area associated with emotionality and fear.

1.4.7. Neuropeptide Y (NPY)

Neuropeptide Y (NPY) belongs to the pancreatic polypeptide family, which also include pancreatic polypeptide (PP) and polypeptide YY (PYY). NPY is present in the cerebral cortex, hypothalamus, and spinal cord [17,89]. NPY is colocalized with norepinephrine in the CNS and in the sympathetic nervous system [78].

NPY is shown to act on a series of receptors with varying selectivity for peptides belonging to the pancreatic polypeptide family. These receptors have been designated Y1–Y6 and are characterized by region-specific distributions in the CNS and are located on both presynaptic and postsynaptic sites [68,161].

A large number of studies have addressed possible functions of NPY and related structures in the CNS. NPY has been shown to be involved in the regulation of multiple functions; for instance, particular attention has been directed to feeding behavior and mood regulation. Activation of the Y1 receptor is proposed to cause a decrease in anxiety and may also produce a mild antinociceptive effect [17], whereas activation of the Y5 receptor has been demonstrated to stimulate feeding [164].

1.5. DISTRIBUTION OF NEUROPEPTIDES

According to the observations mentioned above, it seems that the majority of known neuroactive peptides are found in the neurvous system, including the central nervous system (CNS), the gastrointestinal tract (GI), and endocrine organs. However, some peptides are known to be produced in the plasma, vascular endothelium, heart, and cells of the immune system [51]. The following sections of this chapter focus mainly on neuropeptides found to be present in the CNS and various body fluids.

1.5.1. Neuropeptide Distribution in CNS Tissues

Apparently, neurons producing neuroactive peptides are widely distributed within the CNS [50]. According to the location of sites of the various steps involved in the biosynthetic pathway, high concentrations of the peptide preprohormones are generally found in the cell bodies, whereas the matured active peptides are found at high levels in areas near the nerve terminals. Often, dense concentrations of a given peptide are localized to areas related to its known function. For instance, opioid peptides are found in high concentrations in pain-processing pathways and in areas related to brain reward, whereas SP is found, for instance, in areas associated with aggression, depression, reward, and locomotion, as well as pain [50,51,120,162]. Calcitonin-gene-related peptide (CGRP), known to potentiate the effect of SP, is found to be colocalized with P in dorsal spinal neurons, but CGRP is also found in descending motor neurons of the spinal cord [92]. One of the most abundant peptides in the CNS is cholecystokin 8 (CCK-8) [162]. Moreover, within the CNS neuropeptides and their prestages and fragments as well, could also be found in the extracellular fluid, which is in constant exchange with the CSF. In addition to other nerve cells the peptidergic neurons are surrounded by glial cells in the various CNS tissues. The CNS tissue is normally easy to homogenize, and following disruption of the cell membranes, the various peptides are entering the buffer solution. However, the solubility of various peptides is related mainly to the degree of hydrophobicity, which may differ depending on peptide structure.

1.5.2. Neuropeptides in Body Fluids

Studies of neuropeptides in human subjects have been limited mainly to analysis of these compounds in body fluids, where they generally are present at extremely low concentrations. A variety of neuropeptides have been documented to be present in, for instance, plasma, breastmilk, urine, amnion fluid, synovial fluid, and cerebrospinal fluid (CSF). In studies of neuropeptides in patients with neurologic diseases, the most relevant fluid to be analyzed is the CSF [23,95,107]. This fluid is in direct contact with the extracellular fluid of the CNS, and the peptide levels in the CSF are much more likely to reflect events that occur in neuronal pathways than are those detected in plasma. Therefore, in studies of neuropeptides in CNS disorders many investigators have probed the CSF levels for the peptides of interest. However, it is also known that some neuropeptides may leave the CNS compartment and find their way into the circulatory system, and it has been possible to correlate plasma levels of certain neuropeptides with particular symptoms. The low levels of neuropeptides in the various body fluids make them difficult to detect, and highly sensitive techniques are required for their quantification. This fact also means that in attempts to purify a peptide from any of these body fluids, the availability to extremely large volumes is necessary. This difficulty could, however, be overcome by the access to highly sensitive techniques for peptide detection and methods with high recovery rates for peptide purification.

Another important fact to consider when dealing with quantification of peptides in body fluids is that in the CSF compartment it appears that the dominating peptides are represented more by prestages or fragments than by the active peptide [23,27,94,97,98,105,155,156]. In the case of opioid peptides, enkephalin-containing peptides with molecular size up to 25 kDa were identified [105] and the dominating form of dynorphin A immunoreactivity eluted as a polypeptide of 5 kDa [98]. The major CSF component recognized by antibodies directed toward substance P behaved as a polypeptide of approximately 6 kDa [155]. Therefore, in the choice of method for detection or purification of peptides in this fluid, this divergency should be taken into account.

1.6. METHODS FOR NEUROPEPTIDE DETECTION AND QUANTIFICATION

The procedures most frequently applied for neuropeptide measurements are based on immunologic methods, in particular radioimmunoassay (RIA), which has been used to assess the levels of a variety of neuropeptides in, for instance, CSF and plasma from patients with neurologic diseases and other disorders, as well [26,106]. Other procedures used are radioreceptor assays (RRAs), enzyme-linked immunosorbent assay (ELISA), and various chemical methods. Attempts to use mass spectrometric techniques for peptide quantification have been reported. In many cases the detection technique is combined with a preseparation procedure, such as high-performance liquid chromatography (HPLC), liquid chromatography (LC), capillary electrophoresis (CE), or other modes of electrophoresis.

1.6.1. Radioimmunoassay (RIA)

The RIA used in neuropeptide research is based on the specific interaction between a given peptide (the antigen) and its antibody. In general the antibodies are raised in rabbits against a peptide–thyroglobulin conjugate, and often the radioactive iodine-labeled peptide is used as tracer [106]. Separation of antibody-bound and free peptide is often based on the charcoal adsorption method employed, although use of the double-antibody procedure is also common for this purpose [16,98]. The RIA technique still represents the method of choice in most laboratories directing their research toward neuropeptide quantification. The advantages with this technique is that it is in most cases associated with sufficient sensitivity. It allows a large number of samples to be processed simultanously and is relatively cheap. RIA kits for a large number of neuroactive peptides are now commercially available. One drawback with the RIA technique is that it requires a freshly labeled radioactive tracer, which could affect the interassay coefficient of variance. Cross-reaction within the RIA with similar structures related or not related to the peptide to be measured contributes another difficulty [26,106]. Moreover, because of differences such as between antibodies and in the procedure for separating antibody-bound and free peptide, the recorded RIA values may differ between different laboratories. The fact that the RIA tends to detect prestages or fragments of a given peptide should also be considered [97,98,153,155]. Further difficulties with the RIA method have been discussed elsewhere [26,106]. To overcome some difficulties with the RIA, such as to remove cross-reacting substances, the technique has been combined with various preseparating procedures, including minicoloumns for ion-exchange chromatography [6,16], with silicagel cartridges (e.g., SepPak) for reversed-phase separation [135] or with both [37]. The use of both minicolumns and cartridges allows a simultaneous assay of many samples. However, the reliability and quality of the RIA are significantly improved in conjunction with HPLC or CE. In this case it is possible to detect and quantitate peptides with a high degree of accuracy and precision. On the other hand, these appoaches are often time-consuming, and the number of samples to be analyzed within a proper time is significantly decreased.

In order to detect and measure precursors and fragments of various neuropeptides, the RIA technique could be combined with a prior enzymatic reaction. For instance, a large precursor of Leu– and Met–enkephalin was identified following trypsin and carboxypeptidase degradation combined with RIAs for the enkephalin units [71,72]. Later a similar procedure was applied in studies of opioid peptide precursors in human CSF [105]. In the last study the carboxypeptidase step was omitted as the RIA used antibodies that recognized the enkephalin hexapeptides Met–enkephalin–Arg and Leu-enkephalin–Arg. This procedure was also applied for quantification of prodynorphin-derived opioid peptides in CSF from women at term pregnancy [79]. The enzyme–RIA approach has been used to detect substance P precursors in human brain tissue [104] and β-casomorphin-5 in human breastmilk [3]. In the former study [104] an enzyme cleaving the substance P structure to generate its *N*-terminal fragment substance P(1–7) [102] was used in combination with a RIA specific for the heptapeptide fragment. In the latter study [3] an enzyme cleaving the peptide at the *C*-terminal side of a glutamate residue was used.

1.6.2. Radioreceptor Assay (RRA)

An alternative to the RIA as a method for the detection of neuropeptide activity is provided by the RRA. This technique requires a suitable membrane preparation containing the desired receptor population. As radioligand the tritium-labeled or iodine-labeled peptide is often used. Sometimes these assays also use an organic nonpeptide compound as radiolabel. For separation of receptor-bound and free ligand in the RRA, centrifugation or filtration techniques are often used [34,133]. The RRA appeared early on to be an important tool for measuring opioid peptide activity in CSF. The technique was used to quantify the opioid activity in CSF from patients with chronic pain but also in subjects with other neurologic disorders, such as schizophrenia and postpartum psychosis [95,151,152,154]. However, it was found that the measured activity with RRA is far greater than that expected from RIA. The likely explanation for this is that in the CSF specimens there are a large number of different opioid peptides, which contribute to the activity detected by RRA; whereas in the case of using a RIA only a single structure or fragments thereof could be detected. Nevertheless, in many studies it was possible to find significant correlations between various symptoms of the specific disorders with the detected RRA activity [151,152,154]. The major drawback with the RRA technique is that it is unclear what particular opioid system the observed changes in the measured activity are related to. In addition, the outlined procedure for analysis of individual CSF samples consumes relatively large volumes of the fluid. In attempts to purify and characterize the opioid activity present in CSF, the RRA procedure has been of great value [101].

1.6.3. Bioassays

Currently the use of bioassay for activity screening of biological samples in peptide research is not common. Most peptides could be analyzed by more simple assays, such as the abovementioned RIA or RRA. However, in studies of unknown structures and during purification of peptides with unknown sequences, the bioassay is still essential. Also, in research attempting to synthesize peptide analogs to act as potent agonists or antagonists, the access to proper bioassays is of importance. For instance, in the opioid peptide field the guinea pig ileum (GPI) bioassay is still in frequent use in order to establish the biopotency of opioid peptides and their synthetic analogs [34,169]. In order to confirm the agonist activity of synthetic angiotensin II analogs, a bioassay measuring the contraction of rabbit aorta strips was used [133]. A variety of bioassays are also used to explore the possibility of further physiological actions of known neuropeptides. For instance, the ability of nociceptin to interact with memory and learning was studied in a rat model [128]. A whole-animal model was used to study the action of the substance P fragment SP(1–7) on the expression of NMDA receptor subunits in rat brain [171].

1.6.4. Chromatographic Techniques

A limited number of studies have used pure chromatographic techniques for neuropeptide quantification in biological samples. Several studies using chromatographic techniques to quantify synthetic peptides have been described, but none of them has been proved to be sufficient for detecting and quantifying peptides at their low levels in CSF. For instance, under certain pathologic conditions, when peptide levels reach abnormally high concentrations they are measurable by HPLC and UV detection. An example of this is a study on the opioid peptide hemorphin in CSF from patients with ischemia. Using the Pharmacia SMART System for reversed-phase HPLC with UV detection it was indeed possible to discern and quantify this peptide [126]. However, in most cases these techniques are used in conjunction with RIA or RRA. In this mode both HPLC and CE have proved to be of great use. On the other hand, for time-saving purposes, the application of the chromatographic procedures is often limited to structure verification [74,136,170]. However, studies have been reported in which HPLC combined with RIA was used for analysis of all individual samples. For instance, an *N*-terminal extended form of substance P and neuropeptide K was quantified in CSF collected from individual healthy subjects [156]. For identification and purification of neuropeptides, all the abovementioned chromatographic techniques are of great uses.

1.6.5. Electrophoresis

Electrophoreic techniques have been useful both for preparative and analytic purposes in neuropeptide research. Column electrophoresis in agarose suspension was earlier developed for both preparative and analytic use [96]. In preparative runs the procedure proved efficient in recovering peptide material for further purification or analysis [101]. In analytic electrophoresis the technique was used to confirm the structure of assayed immunoreactive material [104]. More recently, capillary electrophoresis (CE) systems were developed and became commercially available. These techniques have proved to be of great importance in several aspects. CE has been used not only to confirm the structure of isolated peptide [8] but also as an interface prior to MS analysis of peptides [43].

1.6.6. Mass Spectrometry (MS)

It has become evident that mass spectrometry (MS) is a key analytic tool in neuropeptide research [5,27,28,94,138]. With the introduction of soft ionization techniques in the early 1980s, MS became applicable for the analysis of many biologically active compounds. Today there are a variety of MS-based strategies available for peptide analysis; the most commonly used techniques are matrix-assisted laser desorption/ionization (MALDI) MS and electrospray (ES) MS. MALDI is naturally coupled to time-of-flight (TOF) mass analyzers because of the pulsed properties of these devices. The sensitivity of this combined instrument is currently in the attomolar to femtomolar range. In the ES-MS technique, gas-phase protonated molecules are generated by spraying the sample solution from the tip of a capillary held at a high potential, and the solvent evaporates from highly charged droplets. The ions are directed into the MS through an orifice leading to the mass analyzer. The development of micro-ES ion sources [28] has improved the ionization efficiency, and analysis of peptides with nano-ES sources has been carried out. MS-based strategies for peptide sequence analysis can now be used independently of the Edman degradation-based approach as a complement. The identification of a known peptide sequence can easily be performed using data generated with ES-MS/MS or by peptide ladder sequence analysis with MALDI [113].

An important objective of MS in neuropeptide research is the possibility to use this technique in direct analysis of peptides in biological samples. As mentioned above, the limitation by methods currently in use for the analysis of neuropeptides in body fluids stresses the need of new methods of high specificity, high reproducibility, and high sensitivity. The MS technique, with all its capabilities, provides facilities that should meet all these requirements. However, although it has been possible to use the technique for analysis of bioactive peptides present in tissues at comparatively high concentrations [77], the very low concentration of neuropeptides in body fluids, such as CSF, gives rise to difficulties. Using diluted samples with synthetic peptides, the MS technique is powerful enough to detect and quantify these compounds down to attomolar concentrations, but these levels are hardly detectable in biological samples, probably because of interfering compounds and background noise. An insufficient recovery before introducing the sample into the MS may also be a problem. However, as time goes by the techniques will certainly improve. In one study on opioid peptides in plasma, liquid chromatography was used online in combination with ES-MS/MS for quantitative analysis [84]. In this case the peptides were intravenously injected into rats and collected arterial blood samples were extracted and quantified by LC-ESI-MS/MS. The added peptides were recovered and quantified with comparatively high accuracy in levels down to concentrations of around 5 ng/mL, or approximately 5 pmol/mL. However, although this concentration range is suitable for pharmacokinetic analyses, it is far from that of endogenous opioid peptides. As mentioned above, in most body fluids such as plasma and CSF the neuropeptides are present in extermely low concentrations, varying within 5–100 pg/mL (i.e., ~5–100 fmol/mL) [106]. Furthermore, using in vivo microdialysis combined with micro-ESI-MS/MS it was possible to detect endogenous enkephalins down to 4–6 fmol per sample (10 μL) [28]. Another method based on immunoprecipitation followed by analysis using MALDI-TOFMS was successfully applied for studies of neurotensin in human brain tissues. By this approach it was possible to detect and quantify this peptide at concentrations down to the low femtomole range [40] not surpassing the detection of RIA. Taken together, although the current facilities of the MS technique are impressive and are of great value for identification and characterization of purified or partly purified neuropeptides, a further breakthrough still awaits before it can by used for quantification of these compounds in pure CSF or plasma samples.

1.7. METHODS FOR NEUROPEPTIDE IDENTIFICATION

As mentioned in the previous section, a great need for neuropeptide identification arises in connection with their quantification in various samples. Most analyses of a given peptide in tissue samples or in samples collected from various body fluids are completed with a chromatographic step, where pooled active material from the individual samples is characterized with regard to its elution profile. After completed chromatography, eluate fractions are subjected to analysis with RIA or RRA. In most cases the activity analyzed by RIA behaves identically to the synthetic peptide, and this observation leads the investigator to confirm an identity between the measured activity and the authentic peptide. However, sometimes the analyzed activity is resolved in several chromatographically separable active components, which indicates that the activity detected in addition to the native peptide could be due to prestages or fragments thereof [98,155]. Moreover, a heterogenous activity pattern following chromatography may indicate the presence of oxidized forms of the peptide [136]. However, in cases where RRA is employed to assess neuropeptide activity in individual samples a subsequent run on HPLC or electrophoresis for structure identification may give rise to several active peaks, which could be indicative of the presence of different peptides sharing affinity for the same receptor [e.g., 101]. It should be emphasized that, although a reliable HPLC or CE combined with RIA assay has confirmed that the detected immunoractivity behave identically to the synthetic analog of the analyzed peptide, this is not an ultimate confirmation of structure identity. Cross-reacting RIA active peptides could behave similarly on HPLC but differ in structure. To finally settle the question about the identity between a peptide detected by RIA and the authentic compound, an HPLC or CE run should be complemented by additional analysis using MS or sequence analysis. In most cases the peptide is not pure enough for Edman degradation, for instance, and in these cases the access to facilities such as ES-MS/MS would be very helpful.

Another area of neuropeptide research where the access of reliable techniques for peptide identification is of significant importance is represented by studies on neuropeptide processing and inactivation enzymes [138]. To identify the specific cleavage site of any particular peptidases, a common approach is to determine the nature or structure of the resulting products. For this purpose chromatographic techniques as well as immunologic procedures have proved to be of great value. For instance, in studies of opioid peptide processing enzymes or opioid peptide convertases, reversed-phase HPLC combined with RIA or amino acid analysis has been useful for product identification [103,128,139,142]. A similar combination of techniques was applied for the identification of processing and converting enzymes acting on substance P and prestages thereof [100,102]. The MS technique combined with HPLC has frequently been used for assessment of the structure of peptide fragments generated by proteolytic degradation [31,127,143].

1.8. METHODS FOR NEUROPEPTIDE PURIFICATION

1.8.1. Choice of Method for Guiding Purification

Before outlining the different steps in a purification scheme for a given peptide, it is necessary to find a suitable method to guide the purification. In this context there are two major lines to follow. The first is to probe a given bioactivity associated with the peptide to be purified, and the second method is outlined from already known chemical properties of the actual peptide. In the first case an assay based on the biological effect of interest should be established. For instance, in works focusing on the purification of opioid peptide activity a commonly used bioassay is the GPI assay [34,169]. A more simple procedure is given by RRA. In the past the RRA was frequently used for guiding purification of opioid active peptides [e.g., 101]. A most important tool for guiding purification is, however, provided by the RIA. The RIA has been used to follow the purification of a large number of neuroactive peptides from CNS tissue and body fluids. It was used, for instance, for the purification of neurotensin from human CSF [30] and β-casomorphin-8 from human breast-milk [119]. In addition, the RIA has also been shown to be useful for guiding neuropeptide purification when it is combined with an enzymatic procedure.

The enzyme–RIA combination was successfully used to identify and purify opioid peptide precursors as well as substance P precursors in CNS tissue and in human CSF [71,87,97,104,155].

Among the various chemical methods used for neuropeptide purification are those recognizing certain structural elements within a bioactive peptide. The elements could be represented by, for instance, acetylated *N*-terminals, certain di- or tri-peptide sequences, and amidated *C*-terminal residues. In this context the procedure originally developed by Mutt and co-workers [154] based on the identification of peptides containing an amidated *C*-terminal sequence should be acknowledged. This procedure has been an important tool for the identification and isolation of a large number of neuropeptides from various tissues. However, because of their low concentration in CSF, very few peptides present in this compartment have been purified by methods based on chemical assays.

1.8.2. Choice of Method for Purification

The techniques chosen for the purification of a given peptide may depend on several factors. In the case of body fluids the choice of techniques may be related to the purity of the fluid. For instance, the CSF contains relatively small amounts of proteins and other biomolecules and could be considered as a quite pure liquid, whereas plasma with huge amounts of proteins reflects the contrary. Tissue samples also contain large amounts of proteins and other contaminating material. Pure liquids, such as amnion fluid and CSF, may limit the purification procedure to a few separation steps, whereas more crude sources need to be subjected to several and much more sophisticated fractionation techniques.

Some peptide structures are hydrophilic (basic or acidic), whereas others are more hydrophobic. Depending on these properties peptides are more or less susceptible to sticking to the walls of vessels, tubings, and chromatogarphic columns. For peptides with basic properties, extraction at low pH is preferable, whereas neutral or acidic peptides are often extracted in solvents maintaining neutral pH. The order of introducing the various separation steps should also be chosen in a proper way. An ion-exchange chraomatography step yielding the peptide material in a fraction with high salt concentration should preferentially be followed by gel-filtration removing salts. For the gel filtration the buffer should be chosen in such a way that it can also be used as starting buffer in the following purification step. When an eluate sample needs to be evaporated, the preceding step should preferentially be carried out with volatile buffer.

1.8.3. Extraction Procedures

The access to efficient procedures for peptide extraction is of great importance in neuropeptide research. This need is essential not only in connection with quantitative analysis of peptides in biological fluids but also for their purification. In the following, we focus on both preextraction before RIA or RRA analyses of individual samples and suitable startegies for extraction as a first step in peptide purification.

Because of the presence of salts and large proteins, most samples need to pass a preextraction or preseparation procedure prior to quantitative analysis by RIA, RRA, or any other similar detection procedure used for peptide quantification. To accomplish this requirement, a number of different techniques have been developed and used in routine analysis of neuropeptides in clinics as well as in experimental animal laboratories. In analysis of peptides in CSF samples, preseparation on reversed-phase silicagel cartridges has frequently been used. Another preseparation procedure has used minicolumns with SP Sephadex prior to RIA [6]. Both these methods use volatile buffers for elution, which means that the eluates are easy to evaporate before RIA analysis. In the early procedure used in connection with measuring opioid activity by RRA [101], CSF samples were fractionated on columns with Sephadex G10, using diluted acetic acid as eluent, and lyophilized [154]. However, all these techniques are more or less connected with losses of peptide material. A more recently developed procedure, based simply on liquid–liquid extraction, was found to provide a basis for a very high recovery of CSF peptides, even in samples where they are present at low concentrations. This method has been applied successfully in a number of studies [75].

In studies aimed at purifying peptides from CNS tissues or body fluids, the extraction procedure is generally the first step. Prior to extraction all tissue material needs to be homogenized, which often can be carried out with an ultraturmix, by using a glass Teflon homogenizer, or by ultrasonification. The first-mentioned procedure is applied in cases of large amounts of tissue, whereas ultrasonification is in general used to homogenize small tissue amounts, in particular those representing discrete regions in the brain. Following homogenization the tissue homogenate is diluted in buffer (often at low pH) and heated before stirring for a couple of hours. Subsequent to extraction the homogenate is centrifuged and the extract (supernatant) is collected for further purification. In the case of body fluids the first extraction step is often omitted or substituted by repeated centrifugation. For clean fluids, such as CSF, the fluid speciemens can be directly diluted with buffer and applied to a chromatographic step.

1.8.4. Chromatographic Purification

Following extraction and centrifugation the collected extract is suitable for the next purification step. At this stage it is appropriate to find a procedure where the peptide fraction is concentrated. The supernatant extracts collected after centrifugation are often represented by large volumes and it is necessary to find a separation step where the resulting active fraction could be collected in a small volume. This objective can be accomplished by the introduction of a chromatographic matrix in which the peptide is adsorbed. A technique commonly used for this purpose is the ion exchanger. As most neuropeptides display a basic tendency to contain several basic amino acids, a cation exchanger, such as Sephadex gels carrying sulfopropyl groups (SP Sephadex), should be used. This separation should be performed under mild acidic conditions (e.g., pH 3–5), and the column can be eluted with stepwise increase in ionic strength or with a gradient with increasing buffer concentration. In cases where the volume of the original extract is comparatively high, it is appropriate to use stepwise elution, yielding the eluate fraction in a small volume, which could be suitable for the next purification step [110]. When starting with small volumes, this first concentrating step could be avoided and the extract could be directly applied on a gel filtration coulmn [104]. As an alternative to a cation exchanger the first concentrating step could also be achieved by using affinity columns. If the starting material appears clean, such as in the case of CSF, the first separation and concentration can be obtained by using large reversed-phase silica gel columns or an Aminex Q15 column [105]. In some cases it might be suitable to use a tandem-column system, combining ion-exchange chromatography with, for instance, reversed-phase separation. This technique could be used to collect peptides with a less basic and more acidic nature. These structures would not be attached to the cation exchanger but would be firmly adsorbed to the column containing the reversed phase matrix.

Following the first concentrating steps, preferentially performed using cation-exchange chromatography, the collected fraction would be further separated by moleclar sieve chromatography (gel filtration). This step would remove large proteins and salts present in the previous eluent buffer. The choice of gel matrix for this purpose depends on the size and nature of the peptide to be purified. Often Sepahedex G50 columns are used for this purpose [e.g., 30,34–36,103,105,110]. Such a separation column, with a length of about 100 cm, would allow proper separation of peptides in the molecular range of 500–20,000 Da. The buffer used in this separation step should be volatile to allow evaporation in a vacuum concentrator or lyophilization. The active material could then be dried and redissolved in a volume proper for additional purification.

Depending on the amount of material collected in the active peptide fraction from the gel filtration step a subsequent purification could be achieved by HPLC using a column suitable for preparative reversed-phase separation [e.g., 110]. Thereafter, additional reversed-phase HPLC could follow until the peptide of interest is obtained in a fraction associated with a homogenous UV peak. A final purification may also be achieved by various electrophoretic techniques. For instance, in the case of minute amounts of peptide material CE could be connected with less losses of peptide material than with RP-HPLC. It should also be stressed that the choice of chromatographic system for HPLC and CE is essential. The tubing material and the pathways that the eluate has to move prior to entering and after leaving the separation matrix should be connected with a minimum of loss.

In many cases, particularly in those when neuroactive peptides are recovered and purified from comparatively clean sources, such as CSF, it should be noted that the purification scheme can be limited to only one or two steps. For instance, this was shown for a hemoglobin-derived opioid peptide, LVV-hemorphin-7, which was present in CSF from a patient with cerebral bleeding. After a single separation on a column for HPLC gel filtration the decapeptide was recoverd and clearly detectable by MS [140].

Another example of this is the recovery of β-casomorphin-8 from human breastmilk by a two-step reversed-phase HPLC procedure [119].

1.8.5. Characterization of Purified Peptides

In characterization of the purified peptides, there are several approaches and techniques to consider. A large number of efficient strategies and methods for determination of physicochemical properties, including molecular size and primary structure, are now available. The choice of biological tools for the assessment of physiological function, metabolism, and receptor interaction can be determined for each individual peptide.

After being obtained as an apparently pure component, the isolated peptide should be properly collected and stored until further analysis. A first step of characterization concerns the elucidation of the peptide structure. For this purpose the choice of technique will depend on the amount of purified peptide recovered from the isolation procedure. Enough material will allow amino acid analysis and automatic sequencing that may readily give the full amino acid sequence of the peptide. If only small amounts of the peptide is available as a pure product sequence elucidation may be accomplished by combining Edman degradation or sequential exopeptidase cleaving using the MS/MS technique [9,113,125,157,167] or by direct FAB-MS analysis [36]. *C*-Terminal sequence analysis of peptides using carboxypeptidase digestion in combination with MALDI-MS is convenient for characterization of both neuropeptides and their prestages. Depending on the individual sequence, up to 20 residues may be identified by this method [9].

Access of enough material of an isolated peptide should also allow for studies on its metabolic stability, including attempts to identify more or less specific enzymes responsible for its inactivation [e.g., 66]. Subsequent studies should focus on attempts to identify the peptide preprohormone and the enzymatic pathways for its processing and maturation [10,114,117].

Elucidation of the amino acid sequence of a certain neuropeptide enables a subsequent determination of the DNA sequence of the gene encoding the corresponding prepropeptide. Knowledge about the primary structure of the prepropeptide and the subsequent nucleotide sequence of the encoding gene provide a basis for the production of knockout mice overexpressing or lacking this gene. These animals will in turn contribute an important model for studies of the functional relevance of the actual peptide. For instance, knockout mice overexpressing corticotropin-releasing factor (CRF) allowed studies of the stress-response circuit between the brain and immune system [148]. Using transgenic mice overexpressing preproenkephalin, the importance of enkephalins in the testicular function was studied [22].

Determining the structure of the isolated peptide should be followed by attempts to synthesize the whole-length peptide and fragments and analogs thereof, in order to confirm its biological potency [e.g., 34,88]. It is of interest to localize the part of the sequence that could be responsible for its biological action. In subsequent studies several other effects of the peptide could, of course, also be examined. For this purpose there is an obvious need of suitable bioassays. It is of particular importance to have access to experimental animal models for behavioral studies. In this context, the possibility of using transgenic mice lacking the gene expressing the particular neuropeptide or its receptor should be considered. For instance, access to knockout mice lacking the SP receptor NK1 was useful for exploring the importance for this peptide in opioid reward [91].

1.9. CONCLUDING REMARKS

Numerous neuroactive peptides are obviously known, although the discovery of still more compounds awaits. Thus far, relatively little is known about the function of most neuropeptides, a fact that motivates additional and extensive research in this area also in the future. It is obvious from the abovementioned observations that the choice of method for neuropeptide detection and purification depends highly on the source of the particular peptide to be studied. If the source of peptide is pure enough (e.g., as in the case of CSF), the purification scheme can be limited to a few simple chromatographic steps. In the case of plasma or CNS tissues, due to a large amount of contaminating peptides and proteins, additional separation steps are often required. Another important factor to consider is connected with the nature and properties of the particular neuropeptide. Furthermore, it is clear that development of new techniques is accelerating and will probably be connected with access to significantly improved procedures for detection, purification, and characterization of peptides in the near future. In this regard the MS technique will represent a key tool. A new concept in biomedical science, *functional genomics* or *proteomics*, has been introduced, and research within this field will certainly include identification, purification, and characterization of neuropeptides. Therefore, further advances in the development of techniques and procedures applicable for this type of research should be expected.

ACKNOWLEDGMENTS

This study was supported by the Swedish Medical Research Council (Grant 9459).

REFERENCES

1. L. Arborelius, M. J. Owens, P. M. Plotsky, and C. B. Nemeroff, The role of corticotropin-releasing factor in depression and anxiety disorders, *J. Endocrinol.* **160**, 1–12 (1999).
2. U. Arvidsson, F. Piehl, H. Johnson, B. Ulfhake, S. Cullheim, and T. Hökfelt, The peptidergic motorneuron, *Neuroreport* **4**, 849–856 (1993).
3. U. Assarhed, C. Larsson, U. Nöreby, L. Lindström, and F. Nyberg, Human β-casomorphin-5 containing peptides in human body fluids, in *β-Çasomorphins and Related Peptides*, V. Brantl and H. Teschemacher, eds., VCH, Weinhem, 1994, pp. 247–258.
4. S. Beranova-Giorgianni and D. M. Desiderio, Fast bombardment mass spectrometry of synthetic peptides, *Meth. Enzymol.* **289**, 478–499 (1997).
5. A. Basbaum and H. Fields, Endogenous pain control systems: Brainstem spinal pathways and endorphin circuitry, *Annu. Rev. Neurosci.* **7**, 309–338 (1984).
6. L. Bergström, I. Christensson, R. Folkesson, B. Stenström, and L. Terenius, An ion exchange chromatography and radioimmunoassay procedure for measuring opioid peptides and substance P, *Life Sci.* **33**, 1613–1619 (1983).
7. J.-M. Besson and A. Chaouch, Peripheral and spinal mechanisms of nociception, *Physiol. Rev.* **67**, 67–186 (1987).
8. V. Bonetto, H. Jörnvall, V. Mutt, and R. Sillard, Two alternative processing pathways for a preprohormone: A new bioactive form of secretin, *Proc. Natl. Acad. Sci.* (USA) **92**, 11985–11989 (1995).
9. V. Bonetto, A.-C. Bergman, H. Jörnvall, and R. Sillard, C-terminal sequence analysis of peptides and proteins using carboxypeptidases and mass spectrometry after derivatization of Lys and Cys residues, *Anal. Chem.* **69**, 1315–1319 (1997).
10. V. Bonetto, H. Jörnwall, M. Andersson, S. Renlund, V. Mutt, and R. Sillard, Isolation and characterissation of sulphated and nonsulphated forms of cholecystokinin-58 and their action on gallbladder contraction, *Eur. J. Biochem.* **264**, 336–340 (1999).
11. J. R. Bundgaard, J. Vuust, and J. F. Rehfeld, New consensus features for tyrosine O-sulfation determined by mutational analysis, *J. Biol. Chem.* **272**, 21700–21705 (1997).
12. E. Burger, Peptide hormones and neuropeptides: Proteolytic processing of the precursor regulatory peptides, *Arz. Forsch.* **38**, 754–761 (1988).
13. L. Canaff, H. P. J. Bennett, and G. N. Hendy, Peptide hormone precursor processing: Getted sorted? *Mol. Cell. Endocrinol.* **156**, 1–6 (1999).
14. L. A. Chahl and P. A. Johnston, Effect of nonpeptide NK-1 receptor antagonist CP-96,245 on the morphine withdrawal response of guinea pigs, *Regul. Pept.* **46**, 373–375 (1993).
15. M. D. Christensen and C. E. Hulsebosch, Spinal cord injury and anti-NGF treatment results in changes in CGRP density and distribution in the dorsal horn in the rat, *Exp. Neurol.* **147**, 463–475 (1997).
16. I. Christensson-Nylander, F. Nyberg, U. Ragnarsson, and L. Terenius, A general enzymatic radioimmunoassay procedure for quantitiative analysis of proenkephalin B derived peptides, *Regul. Pept.* **11**, 65–76 (1985).

17. W. F. Colmers and C. E. Wahlestedt, *The Biology of Neuropeptide Y and Related Peptides*, Humana Press, Ottawa, 1993.
18. J. N. Crawley, Cholecystokinin modulates dopamine-mediated behaviors, *Ann. NY Acad. Sci.* **713**, 138–142 (1994).
19. E. Csuhai, S. S. Little, and L. B. Hersh, Inactivation of neuropeptides, *Progr. Brain Res.* **104**, 131–142 (1995).
20. C. De Felipe, J. F. Herrero, J. A. O'Brien, J. A. Palmer, C. A. Doyle, A. J. H. Smith, J. M. A. Laird, C. Belmonte, F. Cervero, and S. P. Hunt, Altered nociception, analgesia and aggression in mice lacking the receptor for substance P, *Nature* **392**, 394–397 (1998).
21. L. F. Donaldson, A. J. Harmar, D. S. McQueen, and J. R. Seckle, Increased expression of preprotachykinin, calcitonin gene-related peptide, but not vasoactive intestinal peptide messenger RNA in dorsal root ganglia during development of adjuvant monoarthritis in the rat, *Mol. Brain Res.* **16**, 143–149 (1992).
22. D. M. Donovan, L. L. Miner, H. J. Sharpe, M. Takemura, E. Unterwald, F. Nyberg, M. J. Kreek, and G. R. Uhl, Transgenic mice: Preproenkephalin and the mu opiate receptor, *Ann. NY Acad. Sci.* **780**, 19–28 (1996).
23. L. Edvinsson, L. Minthon, R. Ekman, and L. Gustafson, Neuropeptides in cerebrospinal fluid of patients with Alzheimer's disease and dementia with frontotemporal lobe degeneration, *Dementia* **4**, 167–171 (1993).
24. P. J. Elliott and S. D. Iversen, Behavioural effects of tachykinins and related peptides, *Brain Res.* **381**, 68–76 (1986).
25. P. J. Elliot, G. S. Mason, E. A. Graham, M. P. Turpijn, and R. M. Hagan, Modulation of the rat mesolimbic dopamine pathway by neurokinins, *Brain Res.* **51**, 77–82 (1992).
26. R. Ekman, Radioimmunoassay of neuropeptides in the CSF: Problems and pitfalls, *Psykiatr. Tidskr.* **39** (Suppl. 11), 31–34 (1985).
27. R. Ekman, P. Juhasz, M. Heilig, H. Agren, and C. E. Costello, Novel neuropeptide Y processing in human cerebrospinal fluid from depressed patients, *Peptides* **17**, 1107–1111 (1996).
28. M. Emmett and R. M. Caprioli, Micro-electrospray mass spectrometry: Ultra-high-sensitivity analysis of peptides and proteins, *J. Am. Soc. Mass Spectrom.* **5**, 605–613 (1994).
29. M. Emmet, P. Andrén, and R. M. Caprioli, Specific molecular mass detection of endogenously released neuropeptides using in vivo microdialysis/mass spectrometry, *J. Neurosci. Meth.* **62**, 141–147 (1995).
30. U. Eriksson, P. E. Andrén, J. Silberring, F. Nyberg, and F.-A. Wiesel, Characterization of neurotensin-like immunoreactivity in human cerebrospinal fluid by high-performance liquid chromatography combined with mass spectrometry, *Biol. Mass Spectrom.* **23**, 225–229 (1994).
31. U. Eriksson, P. E. Andrén, R. M. Caprioli, and F. Nyberg, Reversed-phase high-performance liquid chromatography combined with tandem mass spectrometry in studies of a substance P-converting enzyme from human cerebrospinal fluid, *J. Chromatogr.* **743**, 213–220 (1996).
32. T. Foulon, S. Cadel, V. Chesneau, M. Draoui, A. Prat, and P. Cohen, Two novel metallopeptidases with a specificity for basic residues: Functional properties, structure, and cellular distribution, *Ann NY Acad. Sci.* **780**, 106–120 (1996).

33. I. L. Gibbins, J. B. Furness, and M. Costa, Pathway-specific patterns of the coexistence of substance P, calcitonin gene-related peptide, cholecystokinin and dynorphin in neurons of the dorsal root ganglia of the guinea pig, *Cell Tissue Res.* **248**, 417–437 (1987).

34. E.-L. Glämsta, A. Marklund, U. Hellman, C. Wernstedt, L. Terenius, and F. Nyberg, Isolation and characterization of a hemoglobin-derived opioid peptide from the human pituitary gland, *Regul. Pept.* **34**, 169–179 (1991).

35. E.-L. Glämsta, B. Meyerson, J. Silberring, L. Terenius, and F. Nyberg, Isolation of a hemoglobin-derived opioid peptide from cerebrospinal fluid of patients with cerebrovascular bleedings, *Biochem. Biophys. Res. Commun.* **184**, 1060–1066 (1992).

36. E.-L. Glämsta, F. Nyberg, and J. Silberring, Application of fast atom bombardment mass spectrometry for sequencing of a hemoglobin fragment naturally occurring in human cerebrospinal fluid, *Rapid Commun. Mass Spectrom.* **6**, 777–780 (1992).

37. E.-L. Glämsta, L. Mörkrid, I. Lantz, and F. Nyberg, Concomitant increase in blood plasma levels of immunoreactive hemorphin-7 and β-endorphin following long distance running, *Regul. Pept.* **49**, 9–18 (1993).

38. M. J. Glass, C. J. Billington, and A. S. Levine, Opioids and food intake: Distributed functional neural pathways? *Neuropeptides* **33**, 360–368 (1999).

39. P. Gluschankof, S. Gomez, A. Lepage, C. Créminon, F. Nyberg, L. Terenius, and P. Cohen, Role of substrate structure in selective processing of peptide hormones at basic amino acid pairs by endopeptidases, *FEBS Lett.* **234**, 149–152 (1988).

40. J. Gobom, K.-O. Kraeuter, R. Persson, H. Steen, P. Roepstorff, and R. Ekman, Detection and quantification of neurotensin in human brain tissue by matrix-assisted laser desorption/ionization time-of flight mass spectrometry, *Anal. Chem.* **72**, 3320–3326 (2000).

41. L. Grandison and A. Guidotti, Stimulation of food intake by muscimol and β-endorphin, *Neuropharmacology* **16**, 533–536 (1979).

42. C. Gros, B. Giros, and J.-C. Schwartz, Identification of aminopeptidase M as an enkephalin-inactivating enzyme in cerebral membrane, *Biochemistry* **24**, 2179–2185 (1985).

43. M. Gucek, M. Gaspari, K. Walhagen, R. J. Vreeken, E. R. Verheij, and J. van-der-Greef, Capillary electrochromatography/nanospray mass spectrometry for attomole characterization of peptides, *Rapid Commun. Mass Spectrom.* **14**, 1488–1454 (2000).

44. E. Haber and W. Carlson, Are inactive renins in plasma and kidney related to prorenin? *Clin. Exp. Hypertens. A* **4**, 1995–2006 (1982).

45. R. K. Handa, L. T. Krebs, J. W. Harding, and S. W. Handa, Ang IV AT4-receptor system in the rat, *Am. J. Physiol.* **274**, F290–299 (1998).

46. U. Hanesch, U. Pfrommer, B. D. Grubb, B. Heppelmann, and H. G. Schaible, Acute and chronic phases of unilateral inflammation in rat's ankle are associated with an increase in the proportion of calcitonin gene-related peptide-immunoreactive dorsal root ganglion cells, *Eur. J. Neurosci.* **5**, 154–161 (1983).

47. M. Herrera-Marschitz, L. Terenius, T. Sakurada, M. S. Reid, and U. Ungerstedt, The substance P(1-7) fragment is a potent modulator of substance P action in the brain, *Brain Res.* **521**, 316–320 (1990).

48. L. B. Hersh, Characterisation of membrane-bound aminopeptidases from rat brain: Identification of the enkephalin-degrading aminopeptidase, *J. Neurochem.* **44**, 1427–1435 (1985).

49. T. Hökfelt, Neuropeptides in perspective: The last ten years, *Neuron* **7**, 867–879 (1991).

50. T. Hökfelt, L. A. Ljungdahl, L. Terenius, R. P. Elde, and G. Nilsson, Immunohistochemical analysis of peptides pathways related to pain and analgesia: Enkephalin and substance P, *Proc. Natl. Acad. Sci.* (USA) **74**, 3081–3085 (1977).

51. T. Hökfelt, M.-N. Castel, P. Morino, X. Zhang, and A. Dagerlind, General overview of neuropeptides, in *Psychopharmacology: The Fourth Generation of progress*, F. E. Bloom and D. J. Kaplan eds., Raven Press, New York, 1994.

52. V. Y. H. Hook, M. R. Schiller, A. V. Azaryan, and N. Tezapisis, Proenkephalin-processing enzymes in chromaffin granules: A model for neuropeptide biosynthesis, *Ann. NY Acad. Sci.* **780**, 121–133 (1996).

53. G. Horvath, Endomorphin-1 and endomorphin-2: Pharmacology of the selective endogenous μ-opioid receptor agonists, *Pharmacol. Ther.* **88**, 437–483 (2000).

54. I. B. Introini, J. L. McGaugh, and C. M. Baratti, Pharmacological evidence of a central effect of naltrexone, morphine, and beta-endorphin and a peripheral effect of met- and leuenkephalin on retention of an inhibitory response in mice, *Behav. Neural Biol.* **44**, 434–446 (1985).

55. K. L. Jansen-Olesen and L. Edvinsson, Characterisation of CGRP-1 receptors in the guinea pig basilar artery, *Eur. J. Pharmacol.* **414**, 249–258 (2001).

56. H. Jörnwall, Acetylation of protein *N*-terminal amino groups: Structural observations on alpha-amino acetylated proteins, *J. Theor. Biol.* **55**, 1–12 (1975).

57. Z. Khalil, K. Sanderson, M. Modig, and F. Nyberg, Modulation of peripheral inflammation by locally administered endomorphin-1, *Inflamm. Res.* **48**, 550–556 (1999).

58. A. J. Kastin, E. L. Scollan, M. G. King, A. V. Schally, and D. H. Coy, Enkephalin and a potent analog facilitate maze performance after intraperitoneal administration in rats, *Pharmacol. Biochem. Behav*. 691–695 (1976).

59. A. J. Kastin, J. E. Zadina, R. D. Olson, and W. A. Banks, The history of neuropeptide research: Version 5.a, *Ann. NY Acad. Sci.* **780**, 1–18 (1996).

60. A. G. Katopodis, D. S. Ping, C. E. Smith, and S. W. May, Functional and structural characterisation of pepdidylamidoglycolate lyase, the enzyme catalyzing the second step in peptide amidation, *Biochemistry* **30**, 6189–6194 (1991).

61. G. Koob and E. Nestler, The neurobiology of drug addiction, in *The Neuropsychiatry of Limbic and Subcortical Disorders*, S. Salloway, P. Malloy, and J. L. Cummings, eds., American Psychiatric Press, Washington, DC, 1997, pp. 179–194.

62. G. Koob, P. P. Sanna, and F. E. Bloom, Neuroscience of addiction, *Neuron* **21**, 467–476 (1998).

63. M. S. Kramer, N. Cutler, F. Feighner, R. Shrivastava, J. Carman, J. J. Sramek, S. A. Reines, G. Liu, D. Snavely, E. Wyatt-Knoweles, J. J. Hale, S. G. Mills, M. MacCoss, C. J. Swain, T. Harrison, R. G. Hill, F. Hefti, E. M. Scolnick, M. A. Cascieri, G. G. Chicchi, S. Sadowski, A. R. Williams, L. Hewson, D. Smith, and N. M. Rupniak, Distinct mechanism for antidepressant activity by blockade of central substance P receptors, *Science* **281**, 1640–1645 (1998).

64. J. S. Kreeger and A. A. Larson, Substance P(1-7), a substance P metabolite, inhibits withdrawal jumping in morphine-dependent mice, *Eur. J. Pharmacol.* **238**, 111–115 (1993).

65. J. S. Kreeger and A. A. Larson, The substance P amino-terminal metabolite substance P(1-7), administred peripherally, prevents the development of acute morphine tolerance and attenuates the expression of withdrawal in mice, *J. Pharmacol. Exp. Ther.* **279**, 662–667 (1996).

66. I. Lantz, E.-L. Glämsta, L. Talbäck, and F. Nyberg, Hemorphins derived from hemoglobin have an inhibitory action on angiotensin converting enzyme activity, *FEBS Lett.* **285**, 39–41 (1991).

67. I. Lantz, M. Thörnwall, J.-E. Kihlström, and F. Nyberg, A comparison of human lung and brain angiotensin converting enzyme with regard to neuropeptide metabolism, *Biochem. Int.* **26**, 415–426 (1992).

68. D. Larhammar, Structural diversity of receptors for neuropeptide Y, peptide YY and pancreatic polypeptide, *Regul. Pept.* **65**, 165–174 (1996).

69. P. Le Grevés, F. Nyberg, F. Hökfelt, and L. Terenius, Calcitonin gene related peptide is metabolized by an endopeptidase hydrolyzing substance P, *Regul. Pept.* **25**, 277–286 (1989).

70. R. J. Lefkowikz, G-protein-coupled receptors III: New roles for receptor kinases and beta-arrestins in receptor signaling and desensitization, *J. Biol. Chem.* **273**, 18677–18680 (1998).

71. R. Lewis, S. Stein, L. Gerber, M. Rubenstein, and S. Udenfiend, High molecular weight opioid-containing proteins in striatum, *Proc. Natl. Acad. Sci.* (USA) **75**, 4021–4023 (1978).

72. R. V. Lewis, A. S. Stern, S. Kimura, J. Rossier, and S. Udenfriend, An about 50,000 dalton protein in adrenal medulla that may be a common precursor of Met- and Leu-enkephalin, *Science* **208**, 1459–1461 (1980).

73. H. Liu, P. W. Mantyh, and A. I. Basbaum, NMDA-recptor regulation of substance P release from primary afferent nociceptors, *Nature* **386**, 721–724 (1997).

74. Z. Liu, S. Lyrenäs, M. Thörnwall, P. Le Grevès, and F. Nyberg, Elevated cerebrospinal fluid level of subsance P and decreased undecapeptidase activity at term pregnancy, *Neuropeptides* **31**, 415–421 (1997).

75. Z. Liu, M. Welin, B. Bragee, and F. Nyberg, A high-recovery procedure for quantitative analysis of substance P and opioid peptides in human cerebrospinal fluid, *Peptides* **21**, 853–860 (2000).

76. Y. P. Loh, Peptide precursor processing enzymes within secretory vesicles, *Ann. NY Acad. Sci.* **493**, 292–307 (1987).

77. J. L. Lovelace, J. J. Kusmierz, and D. M. Desiderio, Analysis of methionine-enkephalin in human pituitary by multi-dimensional reversed-phase high performance liquid chromatography, radioreceptor assay, radioimmunoassay, fast atom bombardment mass spectrometry, and mass spectrometry-mass spectrometry, *J. Chromatogr.* **562**, 573–584 (1991).

78. J. Lundberg, L. Terenius, T. Hökfelt, and M. Goldstein, High levels of neuropeptide Y in peripheral noradrenergic neurons in various mammals including man, *Neurosci. Lett.* **42**, 167–172 (1983).

79. S. Lyrenäs, F. Nyberg, B. Lindberg, and L. Terenius, Cerebrospinal fluid activities of dynorphin converting enzyme at term pregnancy, *J. Obstet. Gynecol.* **72**, 54–58 (1988).

80. C. A. Maggi, R. Pattachini, P. Rovero, and A. Giachetti, Tachykinin receptors and tachykinin receptor antagonists, *J Autonom. Pharmacol.* **13**, 23–93 (1993).

81. J. E. Maggio, *Tachykinins. Ann. Rev. Neurosci.* **11**, 13–28 (1988).

82. R. E. Mains and B. A. Eipper, Peptides, in *Basic Neurochemistry: Molecular, Cellular, and Medical Aspects*, 6th ed., G. J. Siegel, B. Agranoff, R. W. Albers, S. K. Fischer, and M. D. Uhler, eds., Lippincott-Raven, Philadelphia, 1999, pp. 363–382.

83. R. Maldonado, D. Girdlestone, and B. P. Roques, RP 67580, a selective antagonist of neurokinin-1 receptors, modifies some of the naloxone-precipitated morphine withdrawal signs in rats, *Neurosci. Lett.* **156**, 135–140 (1993).

84. C. D. Márquez, S. T. Weintraub, and P. C. Smith, Quantitative analysis of two opioid peptides in plasma by liquid chromatography-electrospray ionization tandem mass spectrometry, *J. Chromatogr.* **694**, 21–30 (1997).

85. J. J. Martinez and H. Righter, Endorphins alter acquisition and consolidation of an inhibitory avoidance response in rats, *Neurosci. Lett.* **19**, 197–201 (1980).

86. R. Matsas, I. S. Fulcher, A. J. Kenny, and A. J. Turner, Substance P and [Leu]enkephalin are hydrolyzed by an enzyme in pig caudate synaptic membranes that is identical with the endopeptidase of kidney microvilli, *Proc. Natl. Acad. Sci.* (USA) **80**, 3111–3115 (1983).

87. L. McLoughlin, P. J. Lowry, S. J. Ratter, J. Hope, G. M. Besser, and L. H. Rees, Characterization of the proopiomelanocortin family of peptides in human cerebrospinal fluid, *Neuroendocrinology* **32**, 209–212 (1981).

88. J. C. Meunier, C. Mollereau, C. Suaudeau, C. Moisand, P. Alvinerie, J. L. Butour, J. C. Guillemot, P. Ferrara, B. Monsarrat, H. Mazarguil, P. Vassart, M. Parmentier, and J. Costentin, Isolation and structure of the endogenous agonist of opioid receptor like ORL1 receptor, *Nature* **337**, 532–535 (1995).

89. B. J. Morris, Neuronal localisation of neuropeptide Y gene expression in rat, *J. Comp. Neurol.* **290**, 358–368 (1989).

90. P. K. Mulderry, Neuropeptide expression by newborn and adult rat sensory neurons in culture: Effects of nerve growth factor and other neurotrophic factors, *Neuroscience* **59**, 673–688 (1994).

91. P. Murtra, A. M. Sheasby, S. P. Hunt, and C. De Felipe, Rewarding effects of opiates are absent in mice lacking the receptor for substance P, *Nature* **405**, 180–183 (2000).

92. D. Nohr, M. K-H. Schäfer, S. Persson, H. Romeo, F. Nyberg, C. Post, G. Ekström, and E. Weihe, Calcitonin gene-related peptide gene expression in collagen-induced arthritis is differentially regulated in primary afferents and motorneurons: Influence of glucocorticoids, *Neuroscience* **93**, 759–773 (1999).

93. T. Nagase, N. Miyajima, A. Tanaka, T. Sazuka, N. Seki, S. Sato, S. Tabata, K.-I. Ishikawa, Y. Kawarabayasi, H. Kotani, and N. Nomuea, III. The coding sequences of 40 new genes (KIA0081-KIA0120) deduced by analysis of cDNA clones from a human cell line KG-1, *DNA Res.* **2**, 37–43 (1995).

94. C. L. Nilsson, G. Karlsson, J. Bergquist, A. Westman, and R. Ekman, Mass spectrometry of peptides in neuroscience, *Peptides* **19**, 781–789 (1998).

95. F. Nyberg, CSF opioids in pathophysiology, in *Handbook of Experimental Pharmacology*, Vol. 104, 1993, pp. 653–672.

96. F. Nyberg and L. Terenius, Electrophoresis of opioid peptides on columns with agarose suspension, *Anal. Biochem.* **134**, 245–253 (1983).

97. F. Nyberg and L. Terenius, Identification of high molecular weight enkephalin precursor forms in human cerebrospinal fluid, *Neuropeptides* **5**, 537–540 (1985).

98. F. Nyberg and I. Nylander, Characterization of dynorphin A immunoreactivity in human cerebrospinal fluid, *Regul. Pept.* **17**, 159–166 (1987).

99. F. Nyberg and J. Silberring, Conversion of the dynorphins to Leu-enkephalin in human spinal cord, *Progr. Clin. Biol. Res.* **328**, 261–265 (1990).

100. F. Nyberg and L. Terenius, Enzymatic inactivation of neuropeptides, in *Degradation of Bioactive Substances, Physiology and Pathophysiology*, J. H. Henriksen, ed., CRC Press, New York, 1991, pp. 189–200.

101. F. Nyberg, A. Wahlström, B. Sjölund, and L. Terenius, Characterization of electrophoretically separable endorphins in human CSF, *Brain Res.* **259**, 267–274 (1983).

102. F. Nyberg, P. Le Grevés, C. Sundquist, and L. Terenius, Characterisation of substance P(1-7) and (1-8) generating enzyme in human cerebrospinal fluid, *Biochem. Biophys. Res. Commun.* **125**, 244–250 (1984).

103. F. Nyberg, K. Nordström, and L. Terenius, Endopeptidase in human cerebrospinal fluid which cleaves proenkephalin B opioid peptides at consecutive basic amino acids, *Biochem. Biophys. Res. Commun.* **131**, 1069–1074 (1985).

104. F. Nyberg, P. Le Grevès, and L. Terenius, Identification of substance P precursor forms in human brain tissue, *Proc. Natl. Acad. Sci.* (USA) **82**, 3921–3924 (1985).

105. F. Nyberg, L. Nylander, and L. Terenius, Enkephalin-containing polypeptides in human cerebrospinal fluid, *Brain Res.* **371**, 278–286 (1986).

106. F. Nyberg, I. Nylander, and L. Terenius, Measurements of opioid peptides in human body fluids by radioimmunoassay, in *Handbook of Experimental Pharmacology* Vol. 82, 1987, pp. 227–253.

107. F. Nyberg, H. Vaeroy, and L. Terenius, Opioid peptides and substance P in cerebrospinal fluid: Regulation and significance to pain, in *Basic Mechanism of Headache*, J. Olesen and L. Edvinson eds., Elsevier, Amsterdam, 1988, pp. 241–257.

108. F. Nyberg, P. Le Grevés, and L. Terenius, Modulation of endopeptidase activity by CGRP—a mechanism affecting substance P action, *Biochimie* **41**, 81–84 (1988).

109. F. Nyberg, M. Törnwall, and J. Hetta, Aminopeptidase in human CSF which degrades delta-sleep inducing peptide (DSIP), *Biochem. Biophys. Res. Commun.* **167**, 1256–1262 (1990).

110. G. Nyberg, K. Sanderson, P. Andrén, M. Thörnwall, M. Einarsson, B. Danielson, and F. Nyberg, Isolation of haemorphin-related peptides from filter membranes collected in connection with haemofiltration of human subjetcs, *J. Chromatogr.* **723**, 43–49 (1996).

111. F. Nyberg, K. Sanderson, and E.-L. Glämsta, The hemorphins: A new class of opioid peptides derived from the blood protein hemoglobin, *Biopolymers* **43**, 147–156 (1997).

112. R. H. Oakley, S. A. Laporte, and J. A. Holt, Association of beta-arrestin with G protein-coupled receptors during clathrin-mediated endocytosis dictates the profile of receptor resensitization, *J. Biol. Chem.* **274**, 32248–32257 (1999).

113. D. H. Patterson, G. E. Tarr, F. E. Regnier, and S. A. Martin, C-Terminal ladder sequencing via matrix-assisted laser desorption mass spectrometry coupled with carboxypeptidase Y time-dependent and concentration-dependent digestions, *Anal. Chem.* **67**, 3971–3978 (1995).

114. M. H. Perrin and W. W. Vale, Corticotropin releasing factor receptors and their ligand family, *Ann. NY Acad. Sci.* **885**, 312–328 (1999).

115. S. Persson, P. Le Grevés, M. Thörnwall, U. Eriksson, J. Silberring, and F. Nyberg, Neuropeptide converting and processing enzymes in the spinal cord and cerebrospinal fluid, *Progr. Brain Res.* **104**, 111–130 (1995).

116. J. R. Reeve, V. E. Eysselein, F. J. Ho, P. Chew, S. Vigna, R. Liddle, and C. Evans, Natural and synthetic CCK-58—novel reagents for studying cholecystokinin physiology, *Ann. NY Acad. Sci.* **713**, 11–21 (1994).

117. J. F. Rehfeld and H. F. Hansen, Characterisation of preprocholecystokinin products in the porcine cortex: Evidence of different processing pathways, *J. Biol. Chem.* **261**, 5832–5840 (1986).

118. J. F. Rehfeld, L. Bardram, S. Blanke, P. Cantor, L. Fries-Hansen, L. Hilstedt, A. H. Johnsen, H. J. Monstein, W. Wouter van Solinge, L. Odum, and C. Orskov, The expression of peptide hormones in normal cells and tumour cells, *Acta Oncol.* **30**, 429–433 (1991).

119. S. Renlund, I. Erlandsson, U. Hellman, J. Silberring, L. Lindström, and F. Nyberg, Micropurification of immunoreactive beta-casomorphins present in milk from women with post-partum psychosis, *Peptides* **14**, 1125–1132 (1993).

120. A. Ribero-da-Silva and T. Hökfelt, Neuroanatomical localisation of substance P in the CNS and sensory neurons, *Neuropeptides* **34**, 256–271 (2000).

121. H. Righter, R. Jensen, J. J. Martinez, R. Messing, B. Vasquez, K. Liang, and J. McGaugh, Enkephalin and fear-motivated behavior, *Proc. Natl. Acad. Sci.* (USA) **77**, 3729–3732.

122. B. P. Roques, Active site structure and design of selective and mixed inhibitors: new approaches in the search for analgesics and anti-hypertensives, *Biochem. Soc. Trans.* **21**, 678–685 (1993).

123. B. P. Roques, F. Noble, V. Daugé, M.-C. Fournié-Zaluski, and A. Beaumont, Neutral endopeptidase 24.11: Structure, inhibition and experimental and clinical pharmacology, *Pharmacol. Rev.* **45**, 87–146 (1993).

124. D. Rossum, U.-K. Hanisch, and R. Quirion, Neuroanatomical localization, pharmacological characterisation and functions of CGRP, related peptides and their receptors, *Neurosci. Behav. Rev.* **21,** 649–678 (1997).

125. M. Salzet, M. Verger-Bocquet, P. Bulet, J.-C. Beauvillain, and J. Malecha, Purification, sequence analysis, and cellular localisation of a prodynorphin-derived peptide related to the alpha-neo-endorphin in the rhynchobdellid leech theromyzon tessulatum, *Am. Soc. Biochem. Mol. Biol.* **271**, 13191–13196 (1996).

126. K. Sanderson, M. Thörnwall, G. Nyberg, E.-L. Glämsta, and F. Nyberg, Reversed-phase high performance chromatography for the determination of haemorphin-like immunoreactivity in human cerebrospinal fluid, *J. Chromatogr.* **676**, 155–160 (1994).

127. K. Sanderson, P. E. Andrén, R. M. Caprioli, and F. Nyberg, In vitro metabolism of LVV-hemorphin-7 in human plasma studied by reversed-phase high-performance chromatography, *J. Chromatogr.* **743**, 207–212 (1996).

128. J. Sandin, J. Georgieva, P. A. Schött, S. Ögren, and L. Terenius, Nociceptin/ Orphanin FQ microinjected into hippocampus impairs spatial learning in rats, *Eur. J. Neurosci.* **9**, 194–197 (1997).

129. J. Sandin, I. Nylander, and J. Silberring, Metabolism of beta-endorphin in plasma studied by liquid chromatography-electrospray ionization mass spectrometry, *Regul. Pept.* **73**, 67–72 (1998).

130. R. A. S. Santos, M. J. Compagnole-Santos, and S. P. Andrade, Angiotensin-(1-7): An update, *Regul. Pept.* **91**, 45–62 (2000).

131. M. Schalling, K. Friberg, K. Seroogy, P. Riederer, E. Bird, S. N. Schiffmann, P. Mailleux, J.-J. Vanderhaeghen, S. Kuga, M. Goldstein, K. Kitahama, P. H. Kuppi, M. Jouvet, and T. Hökfelt, Analysis of expression of cholecystokinin in dopamine cells in mesencephalon of several species and in human with schizophrenia, *Proc. Natl. Acad. Sci.* (USA) **97**, 8427–8431 (1990).

132. M. T. Schiavone, R. A. Santos, K. B. Brosnihan, M. C. Khosla, and C. M. Ferrario, Release of vasopressin from the rat hypothalamo-neurohypophysial system by angiotensin-(1-7) heptapeptide, *Proc. Natl. Acad. Sci.* (USA) **85**, 4095–4098 (1998).

133. B. Schmidt, S. Lindman, W. Tong, G. Lindeberg, A. Gogoll, Z. Lai, M. Thörnwall, B. Synnergren, A. Nilsson, C. J. Welch, M. Sohtell, C. Westerlund, F. Nyberg, A. Karlén, and A. Hallberg, Design, synthesis, and biological activities of four angiotensin II receptor ligands with β-turn mimetic replacing amino acid residues 3–5, *J. Med. Chem.* **40**, 903–919 (1997).

134. N. G. Seidah and M. Chretien, Proprotein and prohormone convertases: A family of subtilases generating diverse bioactive polypeptides, *Brain Res.* **848**, 45–62 (1999).

135. H. S. Sharma, F. Nyberg, Y. Olsson, and P. K. Dey, Alteration of substance P after trauma to the spinal cord, *Neuroscience* **38**, 205–212 (1990).

136. H. S. Sharma, F. Nyberg, M. Thörnwall, and Y. Olsson, Met-enkephalin-Arg6-Phe7 in spinal cord and brain following traumatic injury to the spinal cord: Influence of the 5-HT synthesis inhibitor p-chorophenylalanine, *Neuropharmacology* **32**, 711–717 (1993).

137. L. G. Sharpe and J. H. Jaffe, Captopril and capsaicin modify opioid withdrawal in the morphine dependent rat, *Pharmacol. Biochem. Behav.* **33**, 899–902 (1989).

138. J. Silberring, Characterisation of Neuropeptide processing by fast atom bombardment mass spectrometry, in *Methods in Molecular Biology, Neuropeptide Protocols*, G. B. Irvine and C. H. Williams, eds., Humana Press, Totowata, NJ, 1996, pp. 113–128.

139. J. Silberring and F. Nyberg, A novel endoprotease in bovine spinal cord with high specificity for dynorphin B, *J. Biol. Chem.* **264**, 11082–11086 (1989).

140. J. Silberring and F. Nyberg, Rapid analysis of endogenous' LVV-hemorphin-7 in cerebrospinal fluid by size-exclusion chromatography and electrospray ionization mass spectrometry, *J. Chromatogr*. **777**, 41–45.

141. J. Silberring and F. Nyberg, Application of photodiode array detection and fast atom bombardment mass spectrometry for the identification of the arginine residue in neuropeptides, *Biomed. Chromatogr.* **5**, 240–247 (1991).

142. J. Silberring, M. Castello, and F. Nyberg, Characterization of a dynorphin A converting enzyme in human spinal cord: An endoprotease related to a distinct conversion pathway for the opioid heptadecapeptide? *J. Biol. Chem.* **267**, 21324–21328 (1992).

143. J. Silberring, Y. M. Li, and S. Hjerten, Strategies in studies on neuropeptide processing using mass spectrometry, *Biochem. Soc. Trans.* **22**, 136–140 (1994).

144. S. R. Skilling, D. H. Smullin, and A. A. Larson, Differential effects of C- and N-terminal substance P metabolites on the release of amino acid neurotransmitters from the spinal cord: A potential role in nociception, *J. Neurosci.* **10**, 1309–1318 (1990).

145. S. P. Smeekens, Processing of protein precursors by a novel family of subtilisin-related mammalian endoproteases, *Biotechnology* **11**, 182–186 (1993).

146. C. Stein, The control of pain in peripheral tissue by opioids, *New Engl. J. Med.* **332**, 1685–1690 (1995).

147. C. Stein, A. H. S. Hassan, R. Przewlocki, C. Gramsch, K. Peter, and A. Herz, Opioids from immunocytes interact with receptors on sensory nerves to inhibit nociception in inflammation, *Proc. Natl. Acad. Sci.* (USA) **87**, 5935–5939 (1990).

148. M. P. Stenzel-Poore, J. E. Duncan, M. B. Rittenberg, A. C. Bakke, and S. C. Heinrichs, CRH overproduction in transgenic mice: Behavioral and immune system modulation, *Ann. NY Acad. Sci.* **780**, 36–48 (1996).

149. O. Tarabal, J. Caldero, J. Ribera, A. Sorribas, R. Lopez, J. Molgo, and J. E. Esquerda, Regulation of motorneuronal calcitonin gene-related peptide (CGRP) during axonal growth and neuromuscular synaptic plasticity induced by botulinum toxin in rats, *Eur. J. Neurosci.* **8**, 829–836 (1996).

150. K. Tatemoto and V. Mutt, Chemical determination of polypeptide hormones, *Proc. Natl. Acad. Sci.* (USA) **75**, 4115–4119 (1978).

151. L. Terenius, Endorphins and substance P in chronic pain, *Adv. Pain Res. Ther.* **10**, 9–15 (1987).

152. L. Terenius and F. Nyberg, Opioid peptides in CSF of psychiatric patients, *Progr. Brain Res.* **65**, 207–219 (1985).

153. L. Terenius and F. Nyberg, Opioid peptides in man—analytical aspects, *Life Sci.* **41**, 805–808 (1987).

154. L. Terenius, F. Nyberg, and A. Wahlström, Opioid peptides in human cerebrospinal fluid, in *Opioids, Past, Present and Future*, J. Hughes, H. O. J. Collier, M. J. Rance, and M. B. Tyers, eds., 1983, pp. 179–191.

155. G. Toresson, E. Brodin, A. Wahlström, and L. Bertilsson, Detection of N-terminally extended substance P but not of substance P in human cerebrospinal fluid—quantitation with HPLC-RIA, *J. Neurochem.* **50**, 1701–1707 (1988).

156. G. Toresson, E. Brodin, C. de-las-Carreras, C. Nordin, A. C. Zachau, and L. Bertilsson, Quantitation of N-terminal extended tachykinins in cerebrospinal fluid from healthy subjects, *Regul. Pept.* (1994).

157. A. Tsugita, K. Takamoto, M. Kamo, and H. Iwadate, *C*-terminal sequencing of protein: A novel partial acid hydrolysis and analysis by mass spectrometry, *Eur. J. Biochem.* **206**, 691–696 (1992).

158. A. J. Turner, Endopeptidase-24.11 and neuropeptide metabolism, in *Neuropeptides and Their Peptidases*, A. J. Turner, ed., Horwood, Chichester, UK, 1987, pp. 183–201.

159. M. Ukai, T. Kobayashi, N. Shinkai, X. Shan-Wu, and T. Kameyama, Attenuation of memory with Tyr-D-Arg-Phe-beta-Ala-NH2, a novel dermorphin analog with high affinity for mu-opioid receptors, *Eur. J. Pharmacol.* **274**, 89–93 (1995).

160. A. I. Vaccarino, G. A. Olson, R. D. Olson, and A. J. Kastin, Endogenous opioids: 1998, *Peptides* **20**, 1527–1574 (1999).

161. C. E. Wahlestedt and D. J. Reis, Neuropeptide Y-related peptides and their receptors—are the receptors potential therapeutic drug targets? *Annu. Rev. Pharmacol. Toxicol.* **33**, 309–352 (1993).

162. K. J. Watling and J. E. Krause, The rising sun shines on substance P and related peptides, *Trends Pharmacol. Sci.* **14**, 81–84 (1993).

163. E. Weihe, ••. Nohr, M. J. Millan, C. Stein, S. Muller, C. Gramsch, and A. Herz, Peptide neuroanatomy of adjuvant-induced arthritic inflammation in rat, *Agents Actions* **25**, 255–259 (1988).

164. J. G. Wettstein, B. Earley, and J. L. Junien, Central nervous system pharmacology of neuropeptide Y, *Pharmacol. Ther.* **65**, 397–414 (1995).

165. S. J. Wimalawansa and I. MacIntyre, The presence of calcitonin gene-related peptide in human cerebrospinal fluid, *Brain* **110**, 1647–1655 (1987).

166. S. J. Wimalawansa, Calcitonin gene-related peptide and its receptors: Molecular genetics, physiology, pathophysiology, and therapeutical potentials, *Endocr. Rev.* **17**, 533–585 (1996).

167. A. S. Woods, A. Y. Huang, R. J. Cotter, G. R. Pasternak, D. W. Pardoll, and E. M. Jaffe, Simplified high-sensitivity sequencing of a major histocompatibility complex class I–associated immunoreactive peptide using MALDI MS, *Anal. Biochem.* **226**, 15–25 (1995).

168. A. Yaron and F. Naider, Proline-dependent structural and biological properties of peptides and proteins, *Crit. Rev. Biochem. Mol. Biol.* **28**, 31–81 (1993).

169. J. E. Zadina, L. Hackler, L.-J. Ge, and A. J. Kastin, A potent and selective endogenous agonist for the μ-opiate receptor, *Nature* **386**, 499–502 (1997).

170. Q. Zhou, Z. Liu, A. Ray, W. Huang, K. Karlsson, and F. Nyberg, Alteration in the brain content of substance P(1-7) during withdrawal in morphine-dependent rats, *Neuropharmacology* **37**, 1545–1552 (1998).

171. Q. Zhou, P. Le Grevés, F. Ragnar, and F. Nyberg, Intracerebroventricular injection of the N-terminal substance P fragment SP1-7 regulates the expression of the N-methyl-D-aspartate receptor NR1, NR2A and NR2B subunit mRNAs in the rat brain, *Neurosci. Lett.* **291**, 109–112 (2000).

CHAPTER 2

Mass Spectrometry Instrumentation

ANN WESTMAN-BRINKMALM

Institute of Clinical Neuroscience, Experimental Laboratory Section, Göteborg University, Sahlgrenska University Hospital/Mölndal, Mölndal, Sweden

GUNNAR BRINKMALM

AstraZeneca R&D Mölndal, Mölndal, Sweden

CONTENTS

2.1. MASS AND MASS SPECTROMETRY

From the analytical point of view, mass determination is an attractive way to characterize molecules. Unlike other physicochemical characteristics, molecu-

Mass Spectrometry and Hyphenated Techniques in Neuropeptide Research, Edited by Jerzy Silberring and Rolf Ekman.
ISBN 0-471-35493-7

lar mass does not depend on external conditions. Moreover, molecular mass is an additive quantity; it is equal to the sum of masses of all molecular subunits, and the total mass is preserved in chemical reactions. The most precise way to measure atomic and molecular masses is by mass spectrometry (MS), which is an analytic technique where sample molecules after volatilization and ionization are separated in space or time by their mass-to-charge ratios (m/z), and where both m/z and ion abundances are subjects of measurement. A mass spectrometer consists of three main parts: the ion source where gas-phase ions are generated, the mass analyzer where the ions are separated, and the detector.

Mass spectrometric analysis can identify small modifications, such as point mutations and posttranslational modifications (e.g., phosphorylation and glycosylation) of a peptide/protein. Moreover, mass spectrometers can be utilized to determine the primary structure (amino acid sequence) of a peptide/protein. Amino acid sequencing of peptides is often accomplished through so-called tandem mass spectrometry [mass selection, mass separation (MS/MS)] experiments. When operating in MS/MS mode, the mass spectrometer is set to transmit only a selected m/z (or rather a narrow m/z range). Transmitted ions will then decay either as a result of acquired internal energy, postsource decay (PSD) or through collisions with gas molecules, collision-induced dissociation (CID). The sequence-specific fragments are then separated according to their m/z and detected. Often the background level is significantly reduced in MS/MS mode compared to "normal" mode of operation, and consequently the sensitivity can be rather high.

2.2. ION SOURCES

An ion source transfers the analyte molecules into gas phase and ionizes them. One common method of desorption/ionization is to deposit energy into a solid or liquid sample by bombarding it with a beam of primary particles. Secondary particles, such as isolated atoms or large polyatomic molecules and their clusters in different charge states, are subsequently transferred from condensed phase into gas phase. The primary particles can be kiloelectronvolt (keV) energy atoms [fast-atom bombardment (FAB)], keV energy ions [secondary-ion mass spectrometry (SIMS), liquid secondary-ion mass spectrometry (LSIMS)], MeV energy ions [plasma desorption mass spectrometry (PDMS)], or photons [laser desorption/ionization (LDI), matrix-assisted laser desorption/ionization (MALDI)]. Another common way of creating molecular ions in gas phase is to induce the formation of small charged droplets [atmospheric pressure chemical ionization (APCI) and electrospray ionization (ESI)]. Other examples of ion sources are electron impact (EI), chemical ionization (CI), thermospray, and field desorption (FD). In this section three of the most commonly used types of ion sources for peptide analysis are described.

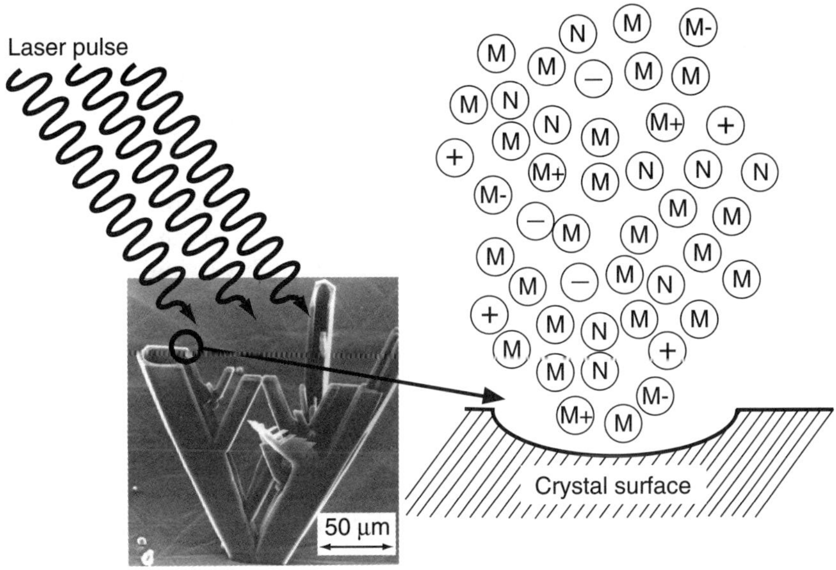

Figure 2.1. Schematic of a MALDI event. The SEM micrograph depicts a sinapinic acid–equine myoglobin crystal from a sample prepared according to the dried drop sample preparation method. In the desorption event neutral matrix molecules (M), positive matrix ions (M+), negative matrix ions (M–), neutral analyte molecules (N), positive analyte ions (+), and negative analyte ions (–) are created and/or transferred to gas phase.

2.2.1. MATRIX-Assisted Laser Desorption/Ionization

2.2.1.1. Principle. In MALDI-MS the sample consists of analyte molecules dilutely embedded in a solid or liquid matrix of highly light-absorbing, low-mass molecules [1,2]. The matrix molecules are resonantly excited by a laser pulse of typically nanosecond duration, and the absorbed energy causes translational motion and ionization of the analyte molecules. Consequently, a volume of the matrix and the trapped analyte molecules is ejected into the gas phase. The ejected material contains both neutral and charged species that interact with each other during the expansion of the plume in the ion source (Fig. 2.1). The MALDI ion source usually operates under high-vacuum conditions, typically a pressure of less than 10^{-6} torr, but atmospheric pressure MALDI has been demonstrated [3]. MALDI is frequently combined with time-of-flight (TOF) mass analyzers, but has also been used in conjunction with several other mass analyzers.

2.2.1.2. Sample Preparation. The matrix plays the key part of the MALDI method. Most of the matrices found so far are low-mass aromatic compounds such as the cinnamic acid derivatives, but several other compounds including

H_2O ice, glycerol, and urea have also been successfully applied as matrices [4–9]. Different matrix compounds are suitable for different classes of analytes. So far, the discovery of new matrix materials with desirable properties has to a large extent been a matter of trial-and-error experimentation, and the properties that separate most matrix candidates from the few that serve as effective matrices remain somewhat obscure. Briefly, a matrix has to absorb strongly at the appropriate wavelength [ultraviolet (UV), or infrared (IR)]; it must be possible to dissolve the analyte and the matrix in the same or in mixable solutions; and the matrix has to have the ability to ionize the analyte. It is also important that the matrix is relatively stable, that is, does not readily evaporate, under high-vacuum conditions.

Other key factors for obtaining useful MALDI spectra are the morphology of the matrix/analyte film and the nature of the molecular cocrystallization [10–12]. A straightforward sample preparation technique results in the polycrystalline samples that are most often utilized in MALDI-MS of peptides. The MALDI matrix is dissolved in a mixture of either ethanol or acetonitrile, and water at ~100 mM concentration. The matrix and analyte solutions are mixed in a test tube, resulting in a molar excess of matrix of typically 10,000:1. A droplet (0.1–20 μL) of the matrix/analyte mixture is deposited on a metal backing (target). If the mixture is allowed to dry completely in air, the result is a carpet of microcrystals of different size and orientation. The scanning electron microscopy (SEM) micrograph in Fig. 2.1 depicts a sinapinic acid–equine myoglobin crystal from a sample prepared according to the description above, the so-called dried drop sample preparation method. The relative analyte concentration and the local morphology at different points on the sample can differ considerably. Many MALDI users have reported that it is necessary to search for "sweet" spots on the sample, and the MALDI sample preparation is still widely considered as something of an art. However, the understanding of the crystallization and phase separation phenomena involved is improving and several methods to prepare more homogenous and reproducible samples have also been described [13–15]. Many MALDI mass spectrometers are also equipped with a videocamera to inspect the sample during data acquisition.

2.2.1.3. Sample Requirements. One feature of MALDI-MS, making it especially promising for mass spectrometric analysis of biological samples, is the ability to detect biomolecules in complex mixtures in the presence of large molar excesses of salts, buffers, and other species. Because of this ability, MALDI-MS has been utilized to study proteins and peptides in serum, cerebrospinal fluid, blood, tissue extracts, and whole cells [14,16–19]. However, sample contamination often disturbs the crystallization of the matrix/analyte, and causes peak broadening by fragmentation and adduct formation, thereby reducing sensitivity and mass accuracy. For example, phosphate buffers should, if possible, be replaced by Tris-HCl buffers, and zwitterionic detergents such as sodium dodecylsulfate (SDS) must be removed before MALDI mass analysis or exchanged by nonionic detergents such as octylglucoside [11,20,21].

MALDI-MS is extremely sensitive allowing detection levels of ≤10–100 fmol of peptide. However, in practice the sensitivity of MALDI is a function of concentration rather than of the amount of sample applied to the MALDI target. Typically, a 0.5–1-μL droplet containing 0.1–10 pmol/μL of sample is applied to the MALDI target, but the amount of sample actually consumed when recording a MALDI mass spectrum is considerably less [22,23]. By applying ~100-pL volumes of matrix/analyte mixture to the MALDI target, ~100 amol of peptide has been detected [24]. The mass range of MALDI-MS is strongly dependent on the mass analyzer, but proteins with masses of up to 1 MDa have been successfully analyzed with MALDI-TOFMS [25]. Because of the high background generated by the matrix, MALDI-MS analysis of small peptides (<600 Da) is often difficult.

2.2.1.4. Properties of the Ions. In positive ionization mode MALDI of peptides the most abundant species are usually singly protonated molecular ions $[M + H]^+$, where M corresponds to the molecular mass of the molecule. Doubly charged molecular ions $[M + 2H]^{2+}$ and dimeric species $[2M + H]^+$ also occur frequently. Salt adducts and other adducts, such as $[M + Na]^+$, $[M + K]^+$, and $[M + \text{matrix}]^+$, are also often present in MALDI mass spectra, especially when the sample is contaminated. In negative ionization mode the deprotonated molecular ions $[M - H]^-$ are usually the most abundant species. Because of the soft nature of the MALDI method, it is possible to desorb, ionize, and detect large, intact proteins. In many cases it is also possible to detect proteins or peptides intact with posttranslational modifications, including phosphorylation and glycosylation [26–28]. Only a few examples of the detection of intact noncovalent complex, such as protein–ligand complexes or protein–metal-ion complexes, have been reported [29].

It has also been demonstrated that molecular ions can decay in the ion source or after acceleration as a result of internal energy acquired in the desorption event or through collisions in the expanding matrix/analyte plume. Both in-source decay and metastable fragmentation (postsource decay) of the analyte ions can be of sufficient intensity to give partial peptide sequence information [30,31]. The likelihood of in-source and metastable fragmentation depends strongly on the laser fluence (laser energy deposited in the sample per unit area) and on the matrix used. When sequence information is required and the metastable fragmentation is insufficient, CID can be performed by employing a collision cell at a suitable position in the spectrometer. Depending on the type of analyzer, either high-energy CID (keV collision energy) or low-energy CID (<100 eV) can be performed. See the respective mass analyzer sections for descriptions of MS/MS procedures and Ref. 32 for a survey of CID processes. The molecular ions produced in the MALDI process have relatively high initial velocities, which can cause reduction in mass resolving power and transmission. Hence, the MALDI-MS mass resolving power depends strongly on laser fluence and is highest when the laser fluence is close to the threshold level.

2.2.1.5. Quantification. An illuminating example of the problem with quantification in MALDI-MS is the analysis of a sample containing an enzymatically digested protein. In the case of complete digestion, the sample is a mixture of equal amounts of the released peptides. The difference in mass between the peptides is small enough that the difference in transmission and detection efficiency of the mass spectrometer can be neglected. Nevertheless, the MALDI-MS signal intensities of the different peptides vary exceedingly, and frequently several released peptides present in the sample are not observed in the mass spectrum. The reasons behind the selectivity of the MALDI process has not yet been fully explained, but are probably a combination of varying ionization efficiencies and matrix/analyte cocrystallization efficiencies. Replacing the matrix or modifying the solvent system has been shown to alter the relative intensities between different components in mixtures [11]. However, within a limited concentration range, quantification can be achieved by using known quantities of the analyte to determine the concentration dependence of the MALDI-MS signal intensity [33]. Several research groups have also shown that quantitative data can be obtained with MALDI-MS by using a carefully chosen internal calibrant of known quantity and close chemical resemblance to the analyte of interest [18,34].

2.2.1.6. Coupling with Chromatography. Online coupling of MALDI to chromatography is difficult, but offline coupling, such as depositing directly on a plate covered with matrix, or fraction collection with subsequent mix with matrix, is frequently used [35,36].

2.2.2. Electrospray Ionization

2.2.2.1. Principle. In ESIMS the sample is dissolved in a polar, volatile solvent and transported through a needle placed at high positive or negative potential (relative to a nozzle surface) [37–39]. The high electric potential (1–4 kV) between the needle and the nozzle causes the fluid to form a Taylor cone, which is enriched with positive or negative ions at the tip. A spray of charged droplets is ejected from the Taylor cone by the electric field. The droplets shrink through evaporation, assisted by a warm flow of nitrogen gas, which passes across the front of the ionization source (Fig. 2.2). The exact mechanism of ion formation from charged droplets has still not been fully elucidated [40]. According to one proposed mechanism, the charge residue model, ion formation, is believed to result from an ion evaporation process, where the droplets undergo declustering, losing solvent molecules in the process and eventually producing individual ions [41]. In another proposed mechanism, the ion evaporation model, the droplet shrinks through evaporation until the repulsive Coulomb force overcomes the surface tension and the droplet splits into smaller offspring droplets [42]. Whatever the detailed mechanism, ions are formed at atmospheric pressure and pass through a cone-shaped orifice, into an intermediate-vacuum region, and from there through a small aperture

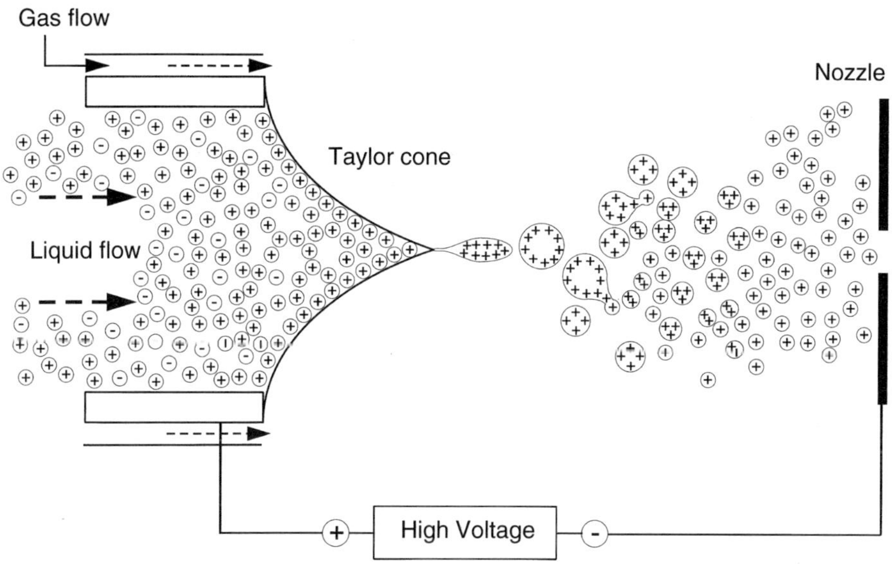

Figure 2.2. Schematic of an ESI source.

into the high vacuum of the mass analyzer. Electrospray has been used in conjunction with all common mass analyzers.

2.2.2.2. Sample Preparation and Requirements. Sample preparation requires only dissolution of the sample to a concentration of 0.1–10 pmol/μL in a mixture of water and organic solvent, commonly methanol, isopropanol, or acetonitrile. A trace of formic acid or acetic acid is often added to aid protonation of the analyte molecules in the positive ionization mode. In negative ionization mode ammonia solution or a volatile amine is added to aid deprotonation of the analyte molecules. Detergents, involatile substances, and salts should be avoided as far as possible. Salts disturb the spraying process and cause overlapping charge-state distributions of adduct ions, resulting in spectra of increased complexity and reduced sensitivity [43]. However, there are compound classes, including non-nitrogen-containing polar compounds, such as carbohydrates, that cannot easily bind a proton to form a charged adduct ion. In such cases it can be beneficial to add salt to induce other types of cations [44].

The sensitivity of ESIMS is good, with 100-fmol detection levels for many peptides. However, the sensitivity of ESI is a function of the concentration of the injected sample. High flow rates result in high sample consumption. It is therefore advantageous to use the lowest possible flow rate. Nanospray ionization is a low-flow-rate version of electrospray ionization [45,46], and is described in detail in a separate chapter. The flow rate when using this procedure is very low, 20–200 nL/min, compared to 1–1000 μL/min in conventional ESIMS, thereby decreasing the sample consumption and increasing the sensitivity considerably. Nanospray has also been shown to be more tolerant to salts than conventional ESIMS [45]. Subattomole levels of analyte have been detected by coupling capillary electrophoresis (CE) to ESI (see Section 2.3.5). ESIMS, the most versatile of the three techniques described in this section, can be used for analysis of polar molecules ranging from less than 100 Da to more than 1 MDa in molecular mass.

2.2.2.3. Properties of the Ions. A important feature of ESI is the capability to generate a distribution of multiply charged ions of the type $[M + n\text{H}]^{n+}$, $[M - n\text{H}]^{n-}$, $[M + n\text{Na}]^{n+}$, and $[M + n\text{K}]^{n+}$, where M corresponds to the molecular mass of the molecule and n represents the charge state. The maximum of the charge state distribution generally falls in the m/z range 500–2500 Da and allows the analysis of very large proteins using mass analyzers with limited m/z range. ESI produced peptides usually have 1–4 charges depending on size and the frequency and relative positions of amino acid residues with basic groups. Relatively small changes in analysis conditions, including pH, solvent composition, salt concentration, and partial denaturation of the analyte molecules, can alter the charge state distribution of a large molecule [47,48]. The complex pattern of multiple-charged ions makes interpretation of ESI spectra from protein mixtures difficult and in practice computers are used to transform the charge-state envelopes to single peaks at the respective molecular mass (or zero charge state).

ESI is an extremely gentle ionization method, practically not accompanied by fragmentation of the formed molecular ions. Consequently, weak bonds are often preserved and analysis of intact posttranslationally modified peptides/proteins and noncovalently bound complexes, including protein–ligand complexes, can be successfully performed with ESIMS [49–53]. Even though fragments are seldom produced in ESI, the ions generated are especially favorable for CID because the high charge state of the molecular ions increases the energy available for the collision event [54]. These CID fragments can be used to obtain partial peptide sequence information (sequence tags, described in detail in a separate chapter) and to sequence a peptide of unknown structure (de novo sequencing) [15,55,56]. Another method to obtain sequence specific fragments in ESIMS is to induce fragmentation in the ion source, so called nozzle-skimmer CID [57]. In practice, MS data from this method are more difficult to interpret since no ion selection is performed prior to the CID event.

2.2.2.4. Quantification. Analyte signal suppression caused by charge competition between electrolytes, including other analytes, is a major problem in ESI and may in practice prevent thorough analysis of complex mixtures if chromatographic prefractionation is not applied. These charge competition phenomena as well as the analyte signal's strong dependence on experimental conditions, such as pH, solvent composition, and salt concentration, make it risky to draw quantitative conclusions from ESIMS data. However, as in MALDI-MS, quantification can be achieved, within a limited concentration range, by using a carefully chosen internal calibrant of known quantity and close chemical resemblance to the peptide/protein of interest [58].

2.2.2.5. Coupling with Chromatography. The combination of ionization at atmospheric pressure and the continuous flow of solvent used in ESI allow direct coupling with separation techniques, such as liquid chromatography (LC) and CE, see more detailed descriptions in separate chapters in this book.

2.2.3. Fast-Atom Bombardment

2.2.3.1. Principle. In FABMS a liquid sample is bombarded with energetic atoms (typically Ar or Xe atoms of ~10 keV kinetic energy) [59,60]. Another technique that greatly resembles FAB is liquid secondary-ion mass spectrometry (LSIMS), where the liquid sample is bombarded with energetic ions [61,62]. Each high-energy particle impact is believed to start a cascade of collisions between the molecules in the sample, resulting in ejection of neutral molecules and ions through so-called sputtering (Fig. 2.3) [63]. The FABMS ion source can be combined with many different mass analyzers, but is most widely used together with quadrupole and sector instruments. A disadvantage with FAB is the rapid contamination of the ion source region, so frequent cleaning is required. This ionization technique is often available on older instruments and is gradually being replaced, mainly by ESI.

2.2.3.2. Sample Preparation. In order to keep the sample in the liquid state when it enters the high-vacuum ion source, the sample is usually dissolved in a viscous solvent with low freezing point, such as glycerol, 1-thioglycerol, a

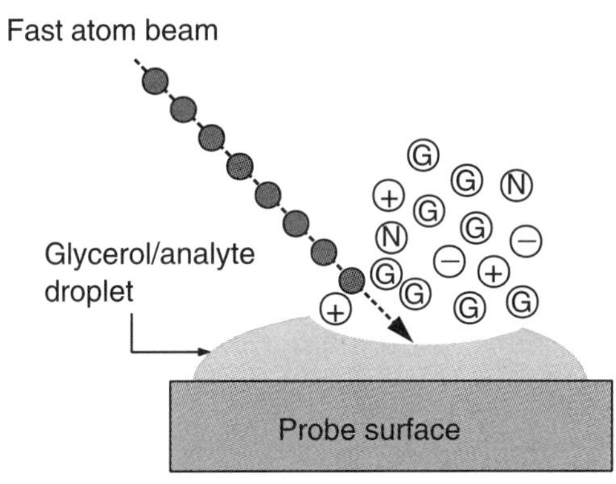

Figure 2.3. Schematic of a FAB source. In the sputtering process neutral analyte molecules (N), positive analyte ions (+), negative analyte ions (–), and glycerol ions and neutral molecules (G) are transferred to gas phase.

mixture of dithiothreitol and dithioerythritol, 3-nitrobenzyl alcohol, or triethanolamine [64–68]. The matrix also shields the sample molecules from damage caused by the impinging high-energy particles [69].

2.2.3.3. Sample Requirements. The detection limit of FABMS for peptides typically varies between 10 and 1000 pmol, depending on the physicochemical properties of the peptide [70]. Generally, the method works best for polar and thermally labile compounds. High salt concentration in the sample reduces sensitivity by ionization suppression and complicates the FAB spectra by producing cationized species.

2.2.3.4. Properties of the Ions. FABMS spectra are dominated by protonated molecular ions $[M + H]^+$ or other cationized species such as $[M + Na]^+$ in the positive mode and deprotonated molecular ions $[M - H]^-$ in the negative mode. Doubly charged molecular ions $[M + 2H]^{2+}$ and dimers $[2M + H]^+$ are also occasionally observed. Prompt fragmentation of the analyte ions often gives partial sequence information, but is seldom of sufficient intensity to fully sequence a peptide of unknown structure. However, fragmentation readily occurs with CID and provides further structural information. One major disadvantage with the FABMS technique is the intense chemical background due to matrix cluster ions and matrix fragment ions. The matrix can also react directly with the sample molecules, forming radical anions or causing reduction of the analyte. The FAB ion source is not as "soft" as ESI or MALDI, and the upper mass limit, which is strongly sample-dependent, seldom exceeds 10 kDa.

2.2.3.5. Quantification. Quantification is difficult in FABMS since the detected ions originate from the sample surface layer, and difference in analyte surface activity and hydrophobisity determines whether the surface layer may or may not be representative of the whole sample. Poor sensitivity is obtained with certain peptides as a consequence of their inability to be adequately distributed in the surface layer of the bombarded matrix droplet. In addition, like MALDI-MS and ESIMS, ion suppression effects exist in FABMS and further complicate quantification. However, as for MALDI-MS and ESIMS quantitative data can, within a limited concentration range, be obtained with FABMS by using a carefully selected internal standard [71,72].

2.2.3.6. Coupling with Chromatography. In continuous-flow FAB (CFFAB) sample solution is continuously delivered to the target, thereby making it possible to provide FABMS with online coupling to LC and CE [73,74]. Less organic matrix (~5% glycerol instead of >90% glycerol) is necessary to keep the sample liquid in CFFAB, resulting in greatly reduced chemical background. The reduced background and the constant refreshment of the surface layers makes CFFAB more sensitive than conventional FAB [75].

2.3. MASS ANALYZERS

A mass analyzer is a device that can separate species—atoms, molecules, or clusters—according to their mass. The separation should also be independent of the chemical conformation of the species. All present mass analyzers are based on electromagnetism, so ions are required to obtain separation. Therefore an ion source has to be coupled to the analyzer. The analyzer will then separate ions coming from the source according to their m/z. There are several types of mass analyzers used in mass spectrometric research, and they can be divided into different categories, such as magnetic or pure electric, scanning or nonscanning (pulse-based), and trapping or nontrapping analyzers. Furthermore, two (or more) analyzers of the same or of different type can be combined, which is very useful for fragmentation studies [mass selection/mass separation (MS/MS)]. In this section the following five mass analyzers are described: time-of-flight, magnetic/electric sector, quadrupole mass filter, quadrupole ion trap, and Fourier transform ion cyclotron resonance. After mass analysis the ions are registered by a detector, which in most cases is located outside the analyzer region.

2.3.1. Time of Flight

2.3.1.1. Principle. The TOF mass analyzer is very well suited for pulsed ion sources and is the method of choice for most MALDI instruments. The principle of the TOF analyzer was first published in 1946 [76], and a schematic is shown in Fig. 2.4. Ions can be transferred into gas phase by several methods, most commonly by particle impact (atoms or molecules, atomic or molecular ions, or photons, i.e., laser pulse irradiation). When this event occurs, a start signal is also generated. Normally the sample plate is floated on high positive or negative potential, typically 5–30 kV, and gas-phase ions will be accelerated toward ground potential. All ions with the same charge that are still intact after acceleration will then ideally have the same kinetic energy. After exiting the

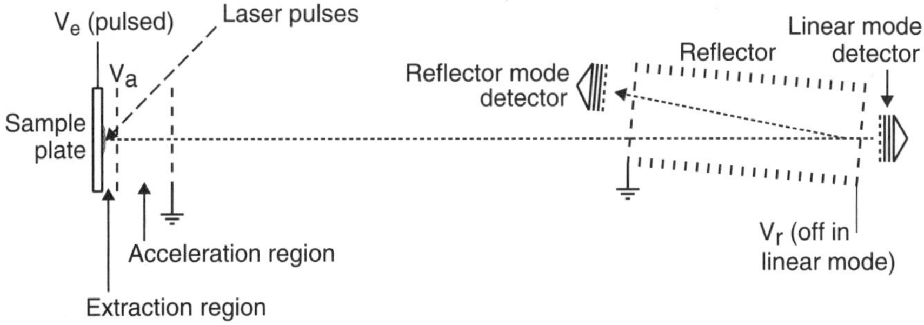

Figure 2.4. Schematic of a MALDI-TOF mass spectrometer with reflector and time-lag focusing.

acceleration region, the ions drift in a field-free region until they enter the stop detector region, where they may be postaccelerated to increase detection efficiency before they hit the stop detector (see also Section 2.4). The measured time of flight for an ion is the time difference between the start pulse and the stop pulse generated when the ion hits the detector. It is relatively easy to transfer the acquired time spectrum to a mass spectrum. The potential energy of the ion is transferred to kinetic energy

$$qU_a = \frac{mv^2}{2} \tag{2.1}$$

where v is the ion velocity after acceleration, m is the mass of the ion, q the charge of the ion, and U_a the potential difference over the acceleration region. For the field-free region the time of flight, t_{TOF}, for an ion is

$$t_{\mathrm{TOF}} = \frac{L}{v} = L\sqrt{\frac{m}{2qU_a}} \tag{2.2}$$

where L is the length of the field-free region in the linear mode (the expression becomes more complicated when accounting for acceleration, detector, and reflector regions but the principle is still the same). Since L and U_a are constants, the modified relationship between t_{TOF} and m/q, or more commonly m/z since the mass and charge are usually expressed in atomic units, can be written

$$t_{\mathrm{TOF}} = a + b\sqrt{\frac{m}{z}} \tag{2.3}$$

where a and b are constants. The obtained time spectrum can thus be calibrated with ions of known masses (i.e., there is no need to know the exact potentials and distances of the spectrometer, and the parameter a handles timing offsets for the start and stop events) and transformed into a mass spectrum. For reviews of TOF analyzers, see Refs. 77–79.

2.3.1.2. Time-Focusing Devices. The ejected ions have a spread in initial axial velocity, v_0, that limits the resolution of the TOF analyzer. However, there are two powerful devices that can compensate for this velocity spread, the electrostatic ion reflector and time-lag focusing (delayed extraction). The reflector was introduced by Mamyrin et al. in 1973 [80]. Mamyrin describes both a one-stage reflector that compensates for the initial energy spread to the first order and a two-stage reflector (with two regions with different electric field strength) that compensates for the spread to the second order. There are also examples of reflectors with a curved electric field [81]. This type of reflector has an advantage when performing MS/MS analysis (see discussion below).

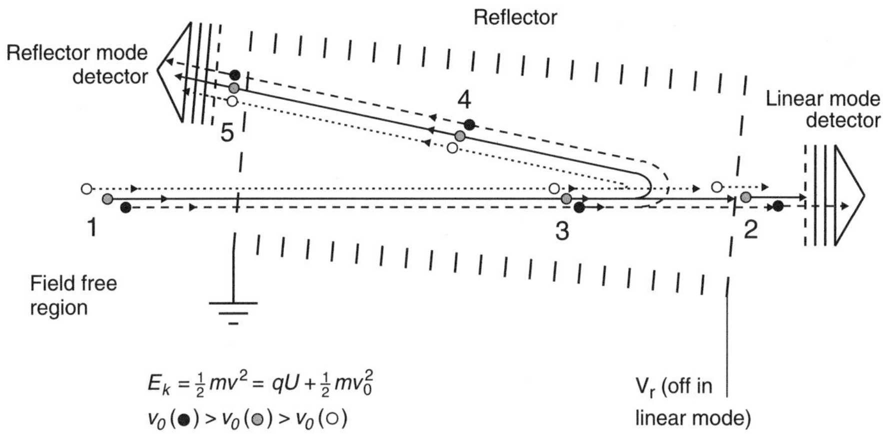

Figure 2.5. The working principle of a single-stage reflector. Desorbed ions will have different initial axial velocities, v_0. Consequently ions of the same *m/z* will at a certain time have a spread in space (1). With no reflector this spread will increase over time and result in peak broadening when registered by the linear mode detector (2). The electric field of the reflector will slow down the ions (3) and finally make them turn around (in noncoaxial designs the mirror is slightly tilted, typically ~1°, compared to the spectrometer axis, and the turnaround angle will not be exactly 180° but rather ~178°). Faster ions (higher v_0) will penetrate deeper into the reflector than slower ions (lower v_0), thus traveling a longer distance in the spectrometer. After reversing direction the ions will again gain speed (4) and if the mirror is properly set, ions with the same *m/z* will arrive at the reflector mode detector (5) with much less timespread than in the linear-mode case.

When applying the reflector the ions are not allowed to go all the way to the linear stop detector, but are instead reflected by a somewhat higher potential than the acceleration potential, after which they hit the reflector stop detector. The reflector not only increases the flight path, but compensates for the initial velocity spread by letting ions with different kinetic energy penetrate to different depths in the electric field, thus forcing the somewhat slower ions to take a somewhat shorter path in the spectrometer than the faster ions (see Fig. 2.5). When set properly, the reflector thus minimizes the spread in time of flight for ions of the same *m/z*, and better resolution is achieved. The reflector has the advantage of being mass-independent, so the optimum setting applies to the full mass range. The mirror can either be tilted versus the spectrometer axis or on axis. The latter arrangement requires a coaxial reflector mode detector, to allow the ions to enter the reflector. Two or more reflectors can be arranged in series [82–84], thus increasing the flight path and increasing the mass resolving power. The drawback is loss in sensitivity.

Time-lag focusing, or delayed extraction (DE), was introduced by Wiley and McLaren in 1955 [85] and has relatively recently become a standard feature of MALDI-TOF mass spectrometers [86–88]. The ion source is modified from

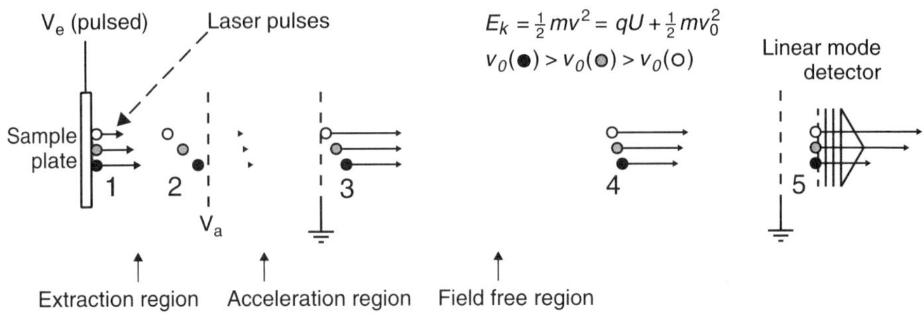

Figure 2.6. The principle of time-lag focusing. When the laser pulse desorbes the ions (1), the sample plate has the same potential as the first grid, so the ions do not experience any electric field but travel with their respective initial velocity, v_0. After a short delay time, τ, typically a few hundred nanoseconds, the sample plate potential is raised and the ions suddenly experience a force from the electric field (2). The slower ions (lower v_0) that are closer to the sample plate will be subjected to a higher potential than the faster ions (higher v_0) farther away and will therefore have gained more kinetic energy and speed when entering the field-free region (3). While traveling through the field-free region, ions with the same *m/z* will gradually get closer to each other (4), and if the source voltages and delay time are properly set, the ions will be detected (almost) simultaneously (5).

a single-stage to a two-stage acceleration region. At the time of desorption the first region (extraction region) is kept field-free. After a short time, typically a few hundred nanoseconds, the sample plate potential is raised instantaneously. Ions with lower initial axial velocity will have traveled a shorter distance toward the detector than those with higher initial axial velocity. Hence, when the electric field is switched on, the slower ions will suddenly be at higher potential than the faster ones, which means that the originally slower ions will instead be the faster ones when exiting the acceleration region. With properly set delay time and source voltages, ions of same *m/z* will now be focused in the sense that they will arrive at the detector at (almost) the same time. Time-lag focusing can compensate both for initial velocity spread and for initial spatial dispersion, although optimal compensation conditions for the two types of dispersions cannot be achieved simultaneously. The former feature is utilized for axial acceleration instruments like standard MALDI-TOF mass spectrometers (see Fig. 2.6), and the latter feature is utilized for orthogonal acceleration (OA) mass spectrometers (see discussion below). Time-lag focusing is unfortunately mass-dependent, so the optimum settings will vary depending on mass region of interest.

There are alternative ways of compensating for initial velocity or energy spread, such as employing a series of electric or magnetic sectors [e.g., 79]. However, reflectors and time-lag focusing are, by far, the most widespread techniques at present.

2.3.1.3. Space-Focusing Devices. To reduce transmission loss, additional space focusing devices, such as a guidewire or an einzel lens, can be employed. Both these devices handle the radial drift of the ions and can improve the transmission greatly, especially for larger molecular ions. The drawback is some loss in resolution.

2.3.1.4. Performance Parameters. There is no theoretical upper or lower limit in mass range for TOF analyzers. It becomes, however, more difficult to detect ions with increasing m/z. Since heavier ions hit the detector with lower speed, the detection efficiency decreases with increasing mass (see Section 2.4). Other factors also limit the practical upper mass limit, such as broader peaks due to wider isotopic distributions, more adduct formation, nonspecificity of the molecular species (e.g., glycoproteins or protein variants), and fragmentation (ion-source or postsource decay). As a result of postsource fragmentation of larger molecular ions, it is often difficult or impossible to detect a signal in reflector mode. This is not a problem for acquisition in linear mode since the fragments will arrive at virtually the same time as the intact species, thus still contributing to the signal intensity.

The resolution, or correctly the mass resolving power (both terms are frequently used), is usually defined as $m/\Delta m$ at full width at half maximum (FWHM); that is, the m/z of the peak divided by its width at half-height. The maximum mass resolving power of a TOF analyzer is mass-dependent and is also a function of the length of the flight path in the instrument. Other influencing factors are the detector surface, space focusing devices, voltage drifts in power supplies, jitter in the timing electronics, ringings, and mechanical misalignments (most of these factors are limiting for all analyzer types). The best commercial MALDI-TOF instruments are at present capable of (employing time-lag focusing and reflector) a mass resolving power of more than 25,000 for bovine insulin (5734 Da). When the mass increases above the point when the isotopes no longer can be separated, the isotopic envelope will determine the width of the peak and hence there will be a sudden drop in mass resolving power.

The mass accuracy is also highly dependent on the mode in which the instrument is operating. In reflector mode, with time-lag focusing, the best MALDI-TOF instruments are capable of achieving <10 ppm with internal standards (average of 5–10 calibrated spectra), provided the isotopes are resolved. In many cases it is not possible to mix in internal calibrants, and then the error in mass accuracy is often increased to 50–100 ppm.

The sensitivity depends mainly on the ionization method and the species (see the respective sections on ion sources above for examples). In general, a TOF analyzer is much more sensitive than a sector or quadrupole analyzer operating in scanning mode (see Sections 2.3.2 and 2.3.3). The dynamic range may be limited when utilizing fast digitizers that are common in MALDI instruments, but is excellent for data collection based on ion counting.

The TOF analyzer is, in general, the fastest mass spectrometric analyzer, even after considering the limitations described below. The repetition rate, that is, the number of start pulses per second, is determined fundamentally by the data collection time window; the larger the window, the lower the maximum repetition rate. In addition, other devices may limit the maximum repetition rate; for instance, the frequently employed nitrogen lasers often have a maximum pulse rate of 20 Hz, which is far less than the limit set by the collection window (a rather large window of 250 μs gives an upper limit of the repetition rate of 4000 Hz, system deadtime not included). Time-lag focusing depends on high voltage switches that may also limit the repetition rate, and for MALDI instruments transient recorders are used to collect data and will also limit the collection speed. For axial acceleration TOF mass spectrometers there is no intrinsic loss of ions due to scanning or source accumulation inefficiency.

TOF-based spectrometers are commonly equipped with microchannel plate (MCP) detectors or hybrids where one component is an MCP. These detectors are very sensitive. For very high ion currents, as in MALDI, detector saturation occurs. This can be dealt with in several, more or less efficient, ways. In MALDI, the main problem arises from very intense matrix signals. The low-mass ions can be deflected by employing timed deflection plates, which reduces the saturation greatly. Nevertheless, in the linear mode, many neutralized species will still hit the detector and affect it negatively. A better approach is then to activate the detector after the unwanted low-mass ions have already arrived, thus avoiding the saturation problem (see Section 2.4). For ion counting detection systems employing a time-to-digital converter (TDC), ions that have an intensity higher than one per collection cycle will be discriminated against. The data acquisition software can, to a great extent, adjust for this.

2.3.1.5. Orthogonal Acceleration Time-of-Flight. Orthogonal acceleration has made TOF analyzers suitable not only for desorption/ionization techniques (pulsed sources) but also for continuous ion sources, such as ESI [89–93]. Lately the MALDI technique has also been successfully coupled to OATOF [94,95], allowing construction of instruments equipped with both ESI and MALDI sources. For a comprehensive review of OATOF analyzers, see Ref. 96.

The ions generated by the source are typically accelerated over a potential of 5–20 V and propagate perpendicular to the TOF analyzer's main axis and enter the extraction/acceleration region at the end of the flight tube. To increase the mass resolving power in the TOF spectra it is essential to minimize the ion beam's divergence and lateral spread in the direction of the TOF analyzer axis. Tube lenses, quadrupoles, hexapoles, and octapoles are suitable focusing devices. In general, it is desirable to time the filling of the acceleration region with new ions with the TOF analysis, so that loss of analyte ions is minimized.

To improve the resolution, the analyzer can be equipped with time-lag focusing, compensating for lateral spread in the direction of the TOF analyzer axis, in combination with a reflector to compensate for the initial velocity spread. The standard method of detection is ion counting employing a TDC, and scanning rates of tens of kilohertz are achievable. With MALDI-produced ions a mass resolving power of ~10,000 (FWHM) for bovine insulin has been obtained [95], and even higher mass resolving power, ~15,000, can be reached for commercial instruments with an ESI source. Mass accuracy better than 10 ppm is attainable, and the sensitivity for ESI-produced ions is in the low femtomole range. By introducing a second, smaller reflector opposite to the main reflector, the ions can be reflected 3 times (first in the main reflector, then in the small one, and then in the main one again, so-called W configuration) before reaching the detector [97]. This will increase the flight path and consequently the mass resolving power, about 30,000 (FWHM) can be attained for proteins in the 5–20-kDa mass range. The drawback is, of course, the reduced transmission (and thus sensitivity), but the arrangement allows for operation also in the "normal" reflector mode if higher sensitivity is required. The acceleration potential (usually below 10 kV) for the TOF analysis is normally substantially lower than for an axial acceleration configuration, so there is a fairly steep sensitivity drop for higher *m/z*. For ESI-generated ions this seldom causes a problem, but in the MALDI case, detection of intermediate and large proteins may be difficult. OATOF analyzers are also very suitable for combination with a quadrupole or with magnetic/electric sectors as mass selector for MS/MS measurements (see Sections 2.3.1.6 and 2.3.6).

2.3.1.6. MS/MS—Mass Selection, Mass Separation. MS/MS can be performed in a TOF analyzer equipped with a reflector by utilizing PSD. For an extensive review of MALDI-PSD, see Ref. 31. To obtain fragmentation with MALDI the target is irradiated with much higher laser intensity than usual. This increases the internal energy of the desorbed analyte ions. Many ions will, of course, undergo fragmentation already in the acceleration region, but a sufficient number will survive until they enter the field-free region. Alternatively, fragmentation can be obtained by employing a collision gas somewhere in the field-free region, so-called high-energy CID.

The ion of interest can be selected by a timed deflection device before the reflector. The device allows unaffected passage for only a brief period of time. It is not necessary for the fragmentation to occur after the ion gate. The fragments will continue with approximately the same velocity as the precursor, and will therefore arrive at the selector at the same time. To get the best (narrowest) mass selection, the selector should be positioned as close to the reflector as possible (then the temporal dispersion of different *m/z* will be greatest). The reflector then acts as a second mass separator. All selected fragment ions will have the same velocity when entering the reflector, but their kinetic energies (which are mass-dependent) will be different. Fragment ions with different *m/z* ratios will now penetrate to different depths of the reflector, thus

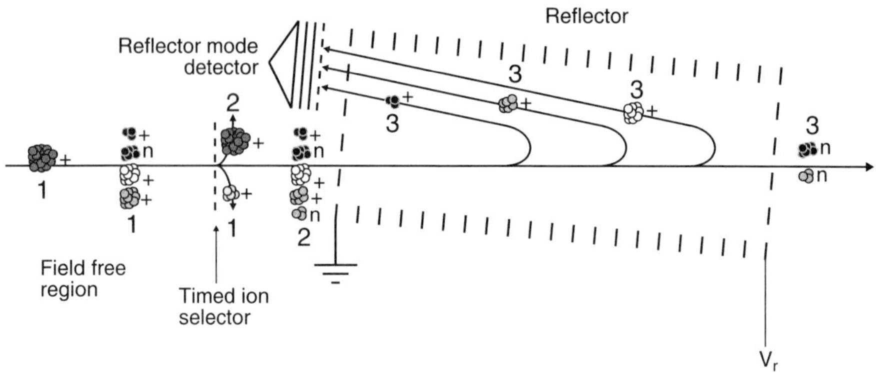

Figure 2.7. The PSD principle. Ions of different mass (or *m/z*) have been accelerated and entered the field-free region without fragmenting. The timed ion selector is active, so lighter ions arriving early are deflected away while ions of the chosen mass and heavier ions still have not reached the selector (1). The selector is then shut off for a brief period of time, allowing ions of the desired mass to pass and is then again activated so that heavier ions are deflected (2). Note that it is not necessary for the ions to remain intact all the way to the selector. Fragmentation can occur anywhere in the field-free region since the fragments will travel with essentially the same velocity as the precursor (black ion undergoes fragmentation before and gray are after passing the selector). The neutral fragments will not be affected by the reflector field and pass right through (and may be registered by the linear mode detector if the experimental setup so permits), while intact and fragment ions will be affected by the electric field of the reflector (3). Since the fragment ions have lost mass after acceleration, they also have lost kinetic energy. Therefore they will not penetrate as deep into the electric field and arrive earlier at the reflector mode detector than they would have had they still been intact. The lower the fragment mass, the shorter the flight time. The acquired PSD time spectrum can then be converted to a mass spectrum after some calculation.

acquiring different flight times (see Fig. 2.7). The conversion to a mass scale for the PSD spectrum is more complicated than for a normal spectrum. Moreover, the resolution for the PSD spectrum is strongly mass-dependent. This is remedied by collecting several spectra, typically about 10, at different reflector potentials. After collection, the parts of the spectra that have best resolution are stitched together.

The PSD method is somewhat cumbersome, and the resolution and mass accuracy in the resulting spectrum are not that good. By employing a reflector with a curved field [81], the fragment ions will be correctly focused onto the detector regardless of their kinetic energy. With this device there is no need to record several spectra to improve the resolution. However, curved field reflectors have generally been considered impractical; for example, axially curved fields inevitably also become radially inhomogeneous, thus creating conditions for spatial defocusing and thus transmission reduction. By applying correctional ion optics and large-area detection, performance can be substantially improved [98,99].

A tandem mass spectrometer with a pulsed linear TOF analyzer for mass selection and a pulsed reflector TOF analyzer for mass separation of the fragment ions has been constructed [100]. A collision cell is located between the two sections of the instrument. Typical mass selection resolving power is ~5000 (FWHM), and the fragment ion resolving power is generally ~3000, although higher values may be attained when optimizing on particular species. The use of hybrid instruments, like the quadrupole mass filter–time-of-flight (Q-TOF) mass spectrometer (see Section 2.3.6), is an alternative method of MS/MS analysis.

2.3.2. Magnetic/Electric Sector

2.3.2.1. Principle. In the 1910s Thomson used magnetic and electric fields to separate ions of different mass and energy [101]. A few years later Dempster employed a variable magnetic field to scan an *m/z* range [102]. High-resolution, double-focusing instruments were developed in the 1930s–1950s [103,104]. Sector analyzers are easily adaptable to continuous ion sources such as CFFAB and ESI. Although more difficult to implement, there are several examples of MALDI sector mass spectrometers [105–109]. As in the TOF case, ions leaving the source in a magnetic sector mass spectrometer are accelerated to a high velocity. The ions then pass through a magnetic sector in which the magnetic field is applied perpendicularly to the direction of the ion beam. The magnetic field will not change the speed of the ions, but force them into a circular motion. Therefore the vacuum tube in the magnetic sector follows an arc.

The magnetic sector will separate ions according to their mass-to-charge ratio, on the basis of the Lorentz force law

$$qvB = \frac{mv^2}{r} \tag{2.4}$$

where q is the ion charge, v is the ion velocity, B is the magnetic field strength (or more correctly the magnetic flux density), m is the ion mass, and r is the radius of the ion trajectory in the magnetic field. The kinetic energy of the ions is given by

$$qU_a = \frac{mv^2}{2} \tag{2.5}$$

where U_a is the potential difference over the acceleration region. Rearrangement gives

$$v = \sqrt{\frac{2qU_a}{m}} \tag{2.6}$$

Combining Eqs. (2.4) and (2.6) results in

$$\frac{m}{q} = \frac{B^2 r^2}{2U_a} \tag{2.7}$$

Conversion to atomic units gives

$$m/z = \frac{e}{m_u} \frac{m}{q} \tag{2.8}$$

where m_u is the atomic mass unit in kilograms and e is the elementary charge in coulombs.

A magnetic sector will separate ions according to their m/z, but it also acts as an ion optical lens, that is, it focuses a divergent beam. However, ions of the same m/z but with different energies will not be focused at the same point. Therefore the resolution will be limited by the initial velocity distribution of the ions. To obtain better resolution, it is necessary to add an electric sector that focuses ions according to their kinetic energy. The electric sector also applies a force perpendicular to the ion beam direction and is, therefore, also shaped like an arc. The electric sector provides an overall directional focusing, but ions of different energy will still have different foci. However, if the electric sector is designed so that the dispersion of ions due to their velocity spread is exactly equal and opposite to that of the magnetic sector, the result of combining the two types of sectors is a zero net velocity dispersion, namely, double focusing (Fig. 2.8). The electric sector radius is given by

$$qU_e = \frac{mv^2}{R} \tag{2.9}$$

where U_e is the potential between the electric sector electrodes and R is the electric sector radius. Combining Equations (2.5) and (2.9) gives

$$R = \frac{2U_a}{U_e} \tag{2.10}$$

No mass or charge appears in Eq. (2.10), so the ion flight path bends in an arc that depends only on the acceleration voltage and the electric sector voltage (and to a small but important extent on the ion's initial velocity).

The normal way of operating a magnetic sector mass spectrometer is to keep the acceleration potential and the electric sector potential constant and

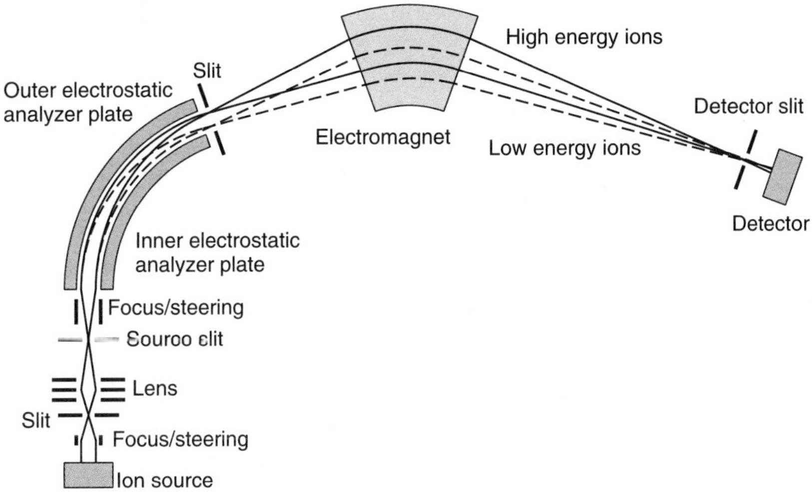

Figure 2.8. Schematic of a magnetic and electric analyzer with so-called forward geometry. Ions coming from the source are focused and steered into the electric analyzer. The electric analyzer does not disperse ions according to *m/z* but focuses ions of same kinetic energy to the same position at the exit slit. Ions with different kinetic energy will be focused to different positions. The magnetic sector then disperses the ions according to *m/z* but also (like the electric sector) focuses ions of different kinetic energy to different positions. In principle, both sectors create dispersions of the ions according to kinetic energy but in opposite directions. With a proper arrangement, these dispersions are balanced so that ions of only the desired *m/z* but with a small spread in kinetic energy are focused on the same position at the detector slit. The desired resolution (and transmission) is controlled by the opening of the detector slit.

vary the magnetic field strength. Ions with different *m/z* are focused at the collector slit in front of the detector at different magnetic field strengths.

Alternatively, the magnetic field strength can be kept constant and the electric field can be varied. The electric sector potential is coupled to the accelerating potential. The advantage with this mode of operation is that the electric field is not subject to hysteresis effects, so the relationship between *m/z* and the accelerating potential is linear. The disadvantage is that the sensitivity will now be roughly proportional to *m/z*. This mode of operation is commonly applied for accurate mass measurement at high resolution.

2.3.2.2. Space-Focusing Devices. To increase performance (i.e., increase transmission and correct for field imperfections caused by, e.g., fringe fields of the sectors), additional ion optical devices are utilized. These can consist of quadrupoles, hexapoles, octapoles, or ring electrodes for focusing; deflection plates for translational movements; and curvature and rotation lenses to correct for the ion beam's cross-sectional shape aberrations and rotation, respectively.

2.3.2.3. Performance Parameters. The upper m/z limit of a magnetic sector analyzer depends on the magnet. It is evident from Eq. (2.7) that both its maximum magnetic field strength [the larger commercial instruments have magnets with a maximum field strength of 1–2 T (teslas)] and its arc radius determine the maximum m/z at a certain acceleration potential. The higher the acceleration potential (and kinetic energy of the ions), the lower the upper m/z limit. The maximum transmission and sensitivity occur at the maximum working acceleration potential, so there is a tradeoff between high m/z and high sensitivity.

The mass resolving power and mass accuracy depend on the way the analyzer is operated. To obtain maximum performance, the magnetic field strength is kept constant and the electric sector and the acceleration voltages are scanned over a relatively small m/z range. This mode of operation is suitable when two peaks of the same nominal mass need to be separated or when the elemental composition of a molecule is to be determined. Under optimum conditions the more powerful instruments can achieve a mass resolving power well over 100,000 ($m/\Delta m$ at 10% valley definition, corresponding to approximately the double by the FWHM definition for a Gaussian peak shape) and a mass accuracy better than 1 ppm using internal calibrants. Sensitivity, resolution, and also scan speed are mutually limiting parameters. The resolution is determined by the slit widths. Higher mass resolving power is obtained by decreasing the slit widths, and thereby also reducing the number of ions reaching the detector.

When operating in magnetic scan mode, the performance is decreased and also dependent on the scan rate, due to the reluctance of the magnet. If the mass spectrometer is coupled to a chromatographic system, a relatively high scan rate is desired and the mass resolving power as well as the mass accuracy have to be compromised even further. The mass resolving power and mass accuracy obtained under such conditions are, however, usually good enough to provide information about the elemental composition of a molecule with a mass of several hundred daltons.

When operating in scanning mode, only one ion species is focused onto the detector at any given time, thus reducing the sensitivity. The wider the scan range, the lower the sensitivity. Furthermore, a short setting time after each scan is usually required to ensure return to the same starting magnetic field strength. Selected ion monitoring is one way to increase the sensitivity by increasing the time spent on recording ions of interest. The acceleration and electric sector voltages can be stepped with a fixed magnetic field strength to allow a few specific ions to be detected. However, these ions must be known in advance. The sensitivity and acquisition speed can be increased greatly, but all additional spectral information will be lost. By employing an array detector [110], a segment of the available m/z range can be detected simultaneously.

The magnetic field strength can then be stepped rather than scanned for recording of a wider *m/z* range, thus providing shorter acquisition time and, consequently, higher sensitivity. By utilizing such a detector, bovine ubiquitin (8565 Da) has been detected at a concentration of 0.3 fmol/μL [111]. Magnetic sector instruments generally have very high reproducibility, very good quantitative performance, and high dynamic range (depending on detector).

2.3.2.4. MS/MS—Mass Selection, Mass Separation. With a standard double-focusing instrument MS/MS, can be performed by so-called linked scan [112,113]. With this method both product (fragment) ion scans and precursor ion scans can be performed. The ions are fragmented in a collision cell (high-energy CID) located between the acceleration region and the first sector. For product ion scans the obtained resolution can be fairly good but the precursor selectivity is low, and in the precursor ion scan case the obtained peaks have rather poor resolution. Alternatively, for a reversed geometry spectrometer the precursor ion can be selected by the magnetic sector and then fragmented by CID, after which the electric sector is scanned to obtain the fragment ion spectrum. This method is referred to as mass-analyzed ion kinetic energy spectrometry (MIKES) and allows for rather good precursor selection but poor product ion resolution. If high performance is needed for the entire measurement, a four-sector instrument can be used [110,114]. In practice, such an instrument consists of two consecutive double-focusing sector instruments with a collision cell in between. With this arrangement not only fragment ion scans but also precursor ion scans, can be performed (see Fig. 2.11 for a general description). Four-sector instruments are, however, rather large and costly. Moreover, high mass resolving power also puts demands on the signal intensity. Alternatively, hybrid instruments that combines a double-focusing sector analyzer with a TOF or quadrupole configuration are common (see Section 2.3.6).

2.3.3. Quadrupole Mass Filter

2.3.3.1. Principle. The basic principles of the quadrupole mass filter were published in 1953 [115]. Like sector analyzers, quadrupole analyzers are well suited for continuous ion sources such as ESI or FAB. The analyzer employs a combination of direct-current (DC) and radiofrequency (RF) potentials. Four parallel rods are arranged symmetrically as is shown in Fig. 2.9. Ideally, these rods should have hyperbolic cross section so that true hyperbolic fields can be produced according to theory. In practice, cylindrical rods are often used to approximate the hyperbolic field. More recently, however, commercial instruments employing rods with hyperbolic cross sections have emerged, thus pushing the limits of performance.

Opposite rods are electrically connected in pairs and set at superimposed DC and RF potentials in a way so that the two pairs will, at any given time,

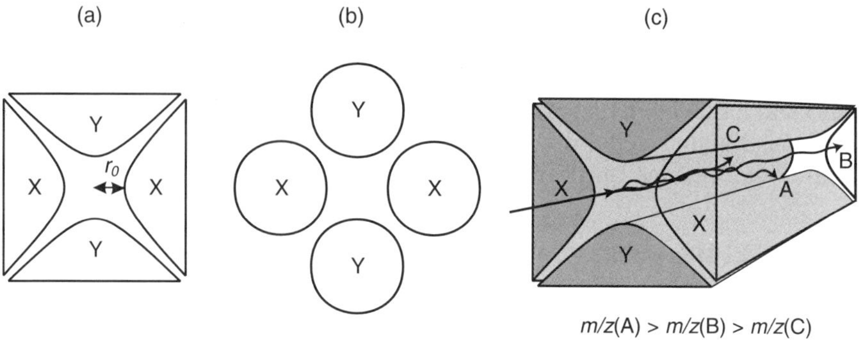

Figure 2.9. Schematic of a quadrupole analyzer: (a) a hyperbolic cross section; (b) the cross-section for cylindrical rods; (c) the operating principle of a quadrupole mass filter. The x-direction pair of rods acts like a highpass filter so ion C (with low m/z) is not allowed through, and the y-direction pair of rods acts like a lowpass filter and takes care of ion A (with high m/z). Only ion B having an m/z in the stable range (cf. Fig. 2.10) is allowed through the quadrupole mass filter for subsequent detection.

have potentials of the same magnitude, but of opposite sign. The rod potentials, ϕ_{0x} and ϕ_{0y}, are given by

$$\phi_{0x} = U - V\cos\omega t \tag{2.11a}$$

and

$$\phi_{0y} = -U + V\cos\omega t \tag{2.11b}$$

where U is the DC potential and $V\cos\omega t$ the time-dependent RF potential, in which V is the amplitude, $f = \omega/2\pi$ the radiofrequency, and t the time. Ions emerging from the source, typically accelerated over a potential of 5–20 V (the resolution is negatively affected by increased ion velocity), enter the analyzer region between the rods and travel parallel to the rods. At given values of U, V, and ω only certain ions will have stable trajectories through the quadrupole. The m/z range for ions allowed to pass through depends on the ratio of U to V. Ions that do not have a stable trajectory through the quadrupole will collide with the rods, never reaching the detector.

Qualitatively, for positive ions the DC potential U is positive and then the x-direction pair of rods will act as a highpass filter. High-mass ions that do not respond quickly to the alternating component of the electric field will respond mainly to the repulsive DC component, which forces them to the plane in the middle between the x electrodes. Low mass ions, on the other hand, respond faster to the RF component of the electric field. Once every cycle the sum of the DC and RF components will be negative for a short time, and the passing ions will experience an attractive force. If the mass of the ion is low enough the ion will be accelerated toward one of the x electrodes and hit it before the electrode potential becomes positive again. The y-direction pair of rods will instead act as a lowpass filter. The DC potential is negative and high-mass ions again tend to respond mainly to the DC component (because the radio-frequency is so high that the potential is more or less averaged out). Consequently, high-mass ions experience an attractive force and will finally hit one of the y electrodes. The low-mass ions also experience an attractive force most of the time, but they will also respond to the repulsive force when the electrodes are at positive potential and be focused toward the plane in the middle between the y electrodes. One can consider the analogy of a ball placed on the top of a curved surface (e.g., the side of a cylinder). If the ball is situated exactly on the top and undisturbed, it will remain there. However, since the ball is affected by gravity, the situation is unstable and any perturbation will cause the ball to roll to one of the sides. If the surface is instead wiggled back and forth in an appropriate way, the ball will not fall off and the situation can be mastered. The alternating field between the rods acts in a similar way.

The motion of an ion travelling in the quadrupole field is described by the Mathieu equation [116,117]

$$\frac{d^2u}{d\xi^2} + (a_u - 2q_u \cos 2\xi)u = 0 \tag{2.12}$$

where $\zeta = \omega t/2$ and u represents x or y. The Mathieu parameters a_u and q_u are defined as

$$a_u = \frac{8q_{\text{ch}}U}{\omega^2 r_0^2 m} \tag{2.13a}$$

and

$$q_u = \frac{4q_{\text{ch}}V}{\omega^2 r_0^2 m} \tag{2.13b}$$

where r_0 is half the distance between two opposite rods, q_{ch} the charge (not to be confused with the Mathieu parameter), and m the mass [use Eq. (2.8) for conversion to atomic units]. Substituting u for x and y gives $a_u = a_x = -a_y$ and $q_u = q_x = -q_y$. For a more extensive theoretical background, see, for example, Refs. 118–120. Only certain combinations of a and q give stable solutions to the Mathieu equation, that is, correspond to ions passing through the quadrupole. Moreover, only the a/q combinations that give stable solutions for *both* the x and the y directions will be useful.

Figure 2.10 shows the stable region closest to origin in a/q space. Scanning the m/z range corresponds to changing the voltages U and V at a constant

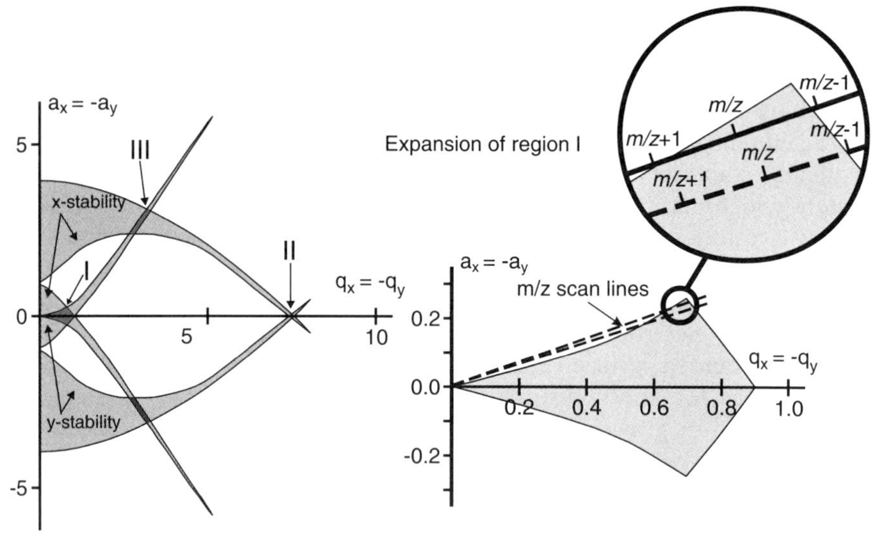

Figure 2.10. The *a/q* stability diagram. The lightly shaded parts are the regions of stability for the *x* and *y* directions, respectively, for the quadrupole field. The heavily shaded parts (regions I, II, and III) are the regions of simultaneous stability in *both* the *x* and *y* directions. Normally a quadrupole mass filter is operated in region I (enlarged). The upper *m/z* scanline represents unit resolution and lower transmission, while for the lower *m/z* scanline the mass resolving power is lower and the transmission is higher.

ratio, $a/q = 2U/V$, while keeping ω fixed. Since *a* and *q* are proportional to $U/(m/z)$ and $V/(m/z)$, respectively, it is obvious that for a certain setting of *U* and *V* only ions within a certain *m/z* range give values of *a* and *q* that are within the stable region, that is, are allowed to pass through the quadrupole. It can also be seen that changing the *U/V* ratio alters the *m/z* range allowed for passage; that is, the resolution and transmission depends on the *U/V* ratio.

2.3.3.2. Space-Focusing Devices. To achieve better sensitivity, several electrodes are utilized to adjust the beam on its way from the source to the quadrupole, and also between quadrupoles in instruments with more than one analyzer. These devices are more or less the same as those for sector instruments (see Section 2.3.2). Note that a quadrupole also can operate as a focusing device. When operated in RF-only mode (i.e., no DC voltage applied), a wide *m/z* range is allowed to pass through the filter.

2.3.3.3. Performance Parameters. Quadrupole mass spectrometers are generally substantially cheaper and less space-demanding than are sector instruments, but the mass resolving power is significantly lower. In practice the frequency, *f*, is fixed, typical values are of the order of 1 MHz. The DC voltage,

U, can be about 1 kV and the maximum RF voltage, V, a few kilovolts. Typical lengths of the quadrupole rods are 5–25 cm, depending on application, and the diameters usually vary from 6 to 15 mm. Mechanical accuracy in both the rods and their alignment is critical for high performance. For example, 0.1% tolerance in the alignment of the rods will limit the maximum mass resolving power to about 500 (FWHM). Scanning speeds of up to ~6 kDa/s (kilodaltons per second) are routine, and unit mass resolution (FWHM) of ≤2000 Da is attainable with cylindrical quadrupole rods. With hyperbolic rods, however, the transmission is significantly higher, and a mass resolving power of ~15,000 (at 1000 Da) can be achieved. Note that for a certain resolution setting, the peak width does not vary with m/z, so the relative resolution will improve with increasing mass. The mass range depends on the settings of the DC and RF voltages and the radiofrequency. Since the mass selectivity (i.e., resolution) of the quadrupole is dependent on the number of cycles an ion undergoes while in the analyzer, selecting a lower frequency in order to extend the m/z range means sacrificing resolution. Typical m/z values are 25–2000 Da with unit mass resolution or better and up to 4000 Da with less than unit resolution. Mass accuracy of ~20 ppm can be obtained with internal calibrants.

As is the case with sector instruments, a quadrupole analyzer operated in scanning mode only detects one ionic species at any given time, so quite a few of the ions produced are actually detected, thus decreasing the sensitivity. The sensitivity is vastly improved when scanning only a narrow m/z range or when monitoring a single ion species. Quadrupole-based instruments generally have good reproducibility, but the sensitivity is strongly mass-dependent.

2.3.3.4. MS/MS—Mass Selection, Mass Separation. A common type of instrument is the triple quadrupole mass spectrometer. In this case the first quadrupole can be used as a mass selector, that is, the quadrupole is set to transmit a particular m/z. Fragmentation occurs in the second quadrupole, which contains the collision cell (allowing for low-energy CID). Since the second quadrupole is used not as a filter but as a focusing device (RF-only mode), a hexapole or an octapole is often used instead. By scanning the third quadrupole, a fragment ion mass spectrum is generated (Fig. 2.11a). As with a four-sector spectrometer, precursor ion scans can also be performed. This is done by setting the *third* quadrupole to transmit a particular m/z while scanning the *first*, thus revealing if there is more than one precursor yielding fragments of a certain m/z (Fig. 2.11b).

2.3.4. Quadrupole Ion Trap

2.3.4.1. Principle. The quadrupole ion trap is based on the same principle as the quadrupole mass filter, but the geometry is different. The trap can be thought of as a linear quadrupole filter that has been bent into a torus, and in the process the inner electrode vanished, the top and bottom electrodes became caps, and the outer electrode became circular. The inner surfaces of

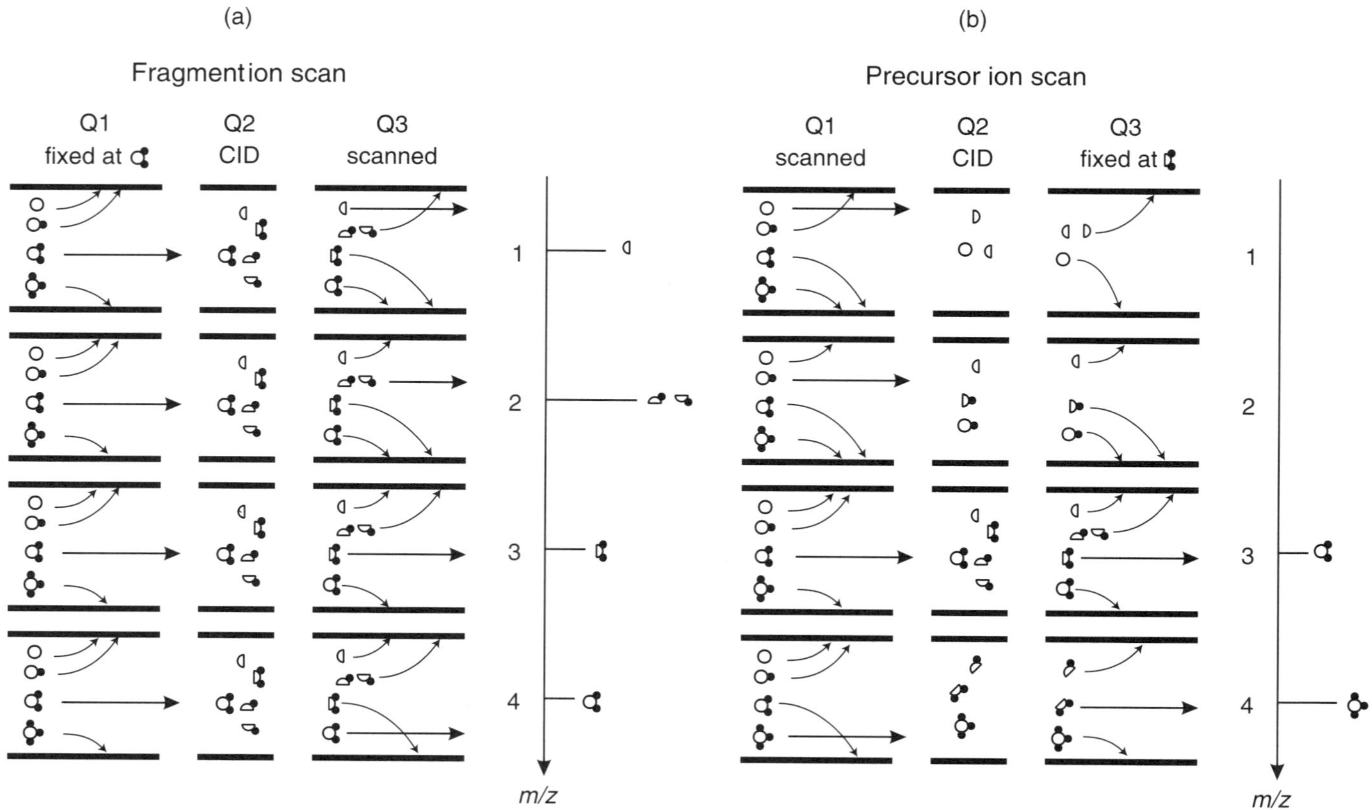
(a)
Fragmention scan
Q1
fixed at
Q2
CID
Q3
scanned
1
2
3
4
m/z
(b)
Precursor ion scan
Q1
scanned
Q2
CID
Q3
fixed at
1
2
3
4
m/z

the electrodes will still have hyperbolic cross sections (see Fig. 2.12). Under stable conditions ions moving around inside the trap will ideally continue to do that forever.

Following the same operating principles as for the quadrupole mass filter, the potentials applied to the electrodes would be

$$\phi_{0r} = U - V\cos\omega t \tag{2.14a}$$

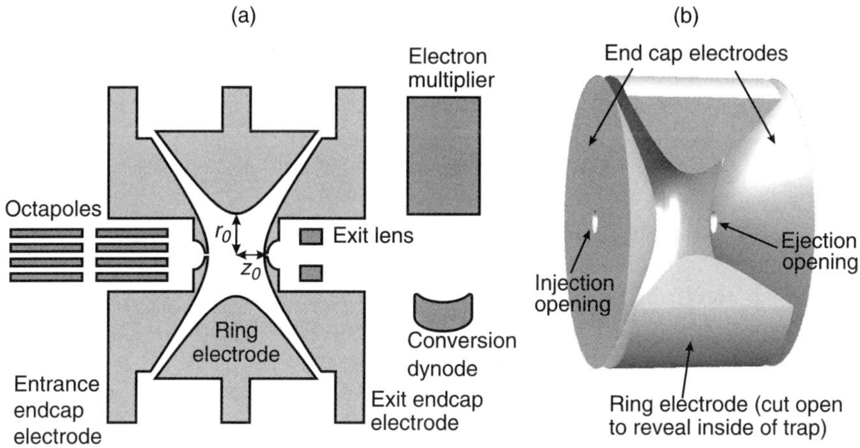

Figure 2.12. A cross-sectional schematic of a quadrupole ion trap mass spectrometer (a) with a three-dimensional perspective view of the quadrupole ion trap (b).

Figure 2.11. The principles of (a) fragment ion scanning and (b) precursor ion scanning. A triple quadrupole analyzer consists of three main regions, two analyzer quadrupoles (Q1 and Q3), and an intermediate quadrupole (Q2) that contains the collision cell. The voltages of Q1 and Q3 can be scanned independently to generate mass spectra. When operating in MS/MS mode, CID is achieved by letting gas into the collision cell. Q2 is operated in RF-only mode and acts as a focusing device to keep fragments (and unfragmented precursors) on axis. When performing a fragment ion scan (a), the first quadrupole (Q1) is set to transmit only a selected *m/z* (or more correctly a narrow *m/z* range). Transmitted ions will then fragment in the cell (Q2) and a fragment ion mass spectrum will be obtained by scanning the third quadrupole (Q3). The figure shows the transmission situation at four different points (1–4) of the voltage scan. A precursor ion scan (b), on the other hand, is performed by doing the opposite. Here the issue is to determine which original ions (precursors) that have fragmented into a species of a certain *m/z*. In this case Q3 is set to transmit only the *m/z* of interest, while scanning Q1 (transmission situations 1–4). During the scan different precursor ions enter Q2 and are fragmented therein. A signal is now observed only when there is an ion passing through Q1 that yields a fragment with the selected *m/z*, thus giving a precursor ion spectrum.

and

$$\phi_{0z} = -U + V\cos\omega t \tag{2.14b}$$

where U is the DC potential and $V\cos\omega t$ the time-dependent RF potential, in which V is the amplitude, $f = \omega/2\pi$ is the RF frequency, and t is the time. The Mathieu parameters for the cylindrical geometry of the trap are

$$a_r = -\frac{1}{2}a_z = \frac{8q_{\text{ch}}U}{\omega^2(r_0^2 + 2z_0^2)m} \tag{2.15a}$$

and

$$q_r = -\frac{1}{2}q_z = \frac{4q_{\text{ch}}V}{\omega^2(r_0^2 + 2z_0^2)m} \tag{2.15b}$$

where r_0 is the distance from the center of the trap to the inner surface of the ring electrode, z_0 is the distance from the center to the surface of the one of the endcap electrodes, q_{ch} the charge, and m the mass [again, use Eq. (2.8) for conversion to atomic units]. In an ideal quadrupole ion trap $r_0^2 = 2z_0^2$, but since a real trap has truncated and perforated electrodes, it does not have a pure quadrupolar field. To compensate for the deviations, the ion trap geometry is stretched so that $r_0^2 \neq 2z_0^2$. Moreover, there is no corresponding modification of the electrode shapes, which would be required to maintain true quadrupolar geometry. For more extensive descriptions of the quadrupole ion trap, see, for example, Refs. 121–124.

2.3.4.2. Ion Injection. Both continuous (e.g., ESI) and pulsed (e.g., MALDI) ion sources can be successfully coupled to an ion trap analyzer. Ions generated in the source have to be guided into the ion trap. This is done much in the same way as for a quadrupole mass filter. The ions arriving from the source are typically accelerated over a potential of 5–20 V. A set of beam steering lenses focuses the ions onto the axis of the inlet ion optics. Typical focusing devices are octapoles and tube lenses. In addition, a tube lens can be used as an ion gate in order to control the number of ions entering the trap, a feature that is useful for limiting the effects of space charging in the trap.

The ions may enter either through an opening in an endcap electrode (axial injection) or the ring electrode (radial injection) or along one of the hyperbolic asymptotes (asymptotic injection). During injection the amplitude of the main RF voltage may either be held constant or changed. With constant-voltage amplitude, where ions enter an active trapping field, ion kinetic energy is dissipated in repeated collisions with damping gas atoms. A normal operational background pressure in the trap is 1 mtorr of helium. With changing voltage amplitude, the increase in amplitude is synchronized with the ion injection. This method is particularly well suited for pulsed ion injection, such as MALDI, where the initial velocity spread is rather large. In this case the packet of ions to be injected will have *m/z* dispersion as in a TOF analyzer, and the RF voltage amplitude can be set to match ion energy and *m/z*, thus increasing the trapping efficiency [125,126].

The helium gas in the trap not only helps trap the ions but also cools them (i.e., the kinetic energy of a trapped ion is dissipated through repeated collisions with the He gas), thus forcing the ions to the center of the trap, where the quadrupole field is best defined. Both sensitivity and mass resolution are significantly enhanced by the presence of the He gas. Moreover, the He can also be used to induce fragmentation when working in MS^n mode (see text below).

2.3.4.3. Mass Analysis and Ion Detection. In most commercial ion trap instruments the endcap electrodes are held at ground potential (except for supplementary voltages; see text below). The net effect of this mode of operation is a 50% decrease in *m/z* range. Usually only an RF potential is applied to the ring electrode, which means that the Mathieu parameter a equals zero, that is, the trap is working along the q axis in the a/q stability diagram (see Fig. 2.13). The ions will line up along the q axis with the lowest *m/z* at the highest q. When the RF amplitude is set to a low—so-called storage—voltage, all ions above a certain *m/z* are trapped. This voltage is usually chosen so the

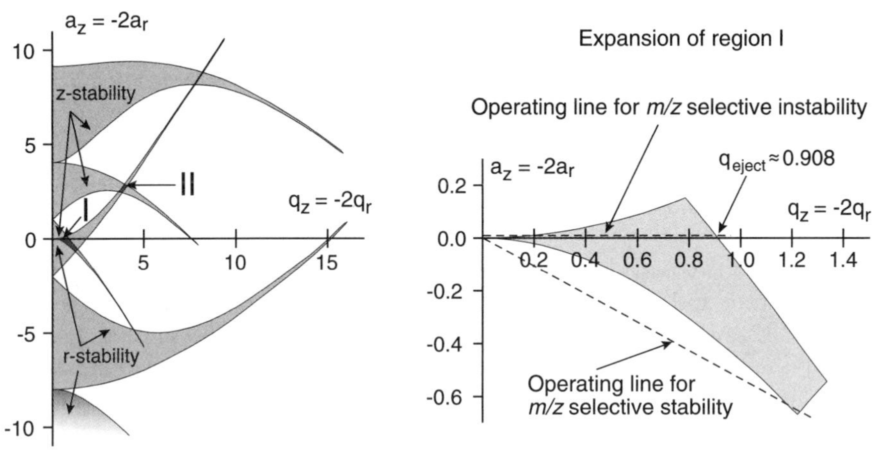

Figure 2.13. The a/q stability diagram. The lightly shaded parts are the regions of stability for the z and r directions, respectively, for the quadrupole field. The heavily shaded parts (regions I and II) are the regions of simultaneous stability in *both* the z and r directions. Normally a quadrupole ion trap is operated in region I (enlarged). The operating line for *m/z* selective stability corresponds to the same type of scan as with a quadrupole mass filter (cf. Fig. 2.10), while the operating line for *m/z* selective instability is used to for ion ejection (and subsequent detection if desired).

lowest trapped *m/z* is greater than those of water, air, and solvent ions (i.e., >100–150 Da).

One method of acquiring a mass spectrum is the mass-selective instability scan. As the RF voltage increases, the trapped ions move to higher *q* values. Eventually, the lowest *m/z* ion reaches and then crosses the stability boundary. This ion becomes unstable in the axial direction and is ejected through small holes in the endcap to a detector. As the RF voltage is further increased, ions with higher *m/z* become unstable and are ejected, thus yielding a mass spectrum.

An important additional technique is resonant ejection. In this technique, a supplementary RF voltage is applied to the endcaps, with a 180° phase shift between the two caps. The trapped ions oscillate with different frequencies related to their *m/z*. To eject ions of a certain *m/z*, a supplementary RF voltage with corresponding frequency (typically between a few and hundreds of kilohertz) is applied. The ions fall into resonance with the oscillating potential, their oscillation amplitudes in the axial direction increase, and finally they are ejected. When resonant ejection is combined with an RF scan, successively larger ions are moved to the position in the stability diagram, where they fall into resonance with the applied supplementary RF voltage and are ejected. This technique can be thought of as opening a hole in the stability diagram through which ions are scanned.

By applying the supplementary RF voltage during injection (without scanning the RF voltage on the ring electrode), it is possible to prevent certain ions from being trapped. This can be useful when having several species with large differences in abundance. Choosing not to trap the dominant species allows for higher concentration of the others, thus increasing sensitivity without decreasing performance due to space charging.

Finally, the supplementary RF voltage may be set to contain all frequencies except a narrow band, corresponding to a narrow *m/z* range, forcing all ions outside this range to leave the trap. This mode of operation, usually referred to as *ion isolation*, is the first step when performing MS/MS (see text below).

There are two ways to detect the ions. The most common is to eject them from the trap and have them hit a detector situated outside the trap, as seen in Fig. 2.12. A standard detector is the conversion dynode together with an electron multiplier. Ions ejected from the trap through the endcap are accelerated toward the detector and then amplified (see Section 2.4). Alternatively, the image current can be detected [127]. This is a nondestructive detection method utilized in *Fourier transform ion cyclotron resonance* (FTICR) mass spectrometers (see Section 2.3.5).

2.3.4.4. Performance Parameters. In practice, the frequency, *f*, is fixed, typically set to ~1 MHz, and the maximum main RF voltage, *V*, may be several kilovolts while the supplementary voltages may be up to tens of volts. The ion trap is normally a rather small device with a diameter and length of ~5 cm.

Mechanical accuracy in both the electrode surfaces and their alignment is critical for high performance. The mass range depends on the settings of the main RF voltage and frequency. Unfortunately, selecting a lower frequency in order to extend the *m/z* range means sacrificing resolution. Typical scan speeds are ~5000 Da/s (several commercial systems offer the possibility of different scan speeds), and typical *m/z* ranges are 15–3000 Da. The *m/z* range can be extended (up to *m/z* 20,000 for commercial instruments) with reduction in other performance parameters. These *m/z* ranges usually cover most analytes when employing an ESI source, but for MALDI an upper *m/z* limit of even 20,000 is still somewhat constraining. The quadrupole ion trap still suffers from mass accuracy problems (in comparison with the high resolution attainable), but ~100 ppm can be obtained. The maximum number of ions in the trap is limited to about 10^5 before space-charge effects seriously affect the performance, so the dynamic range is rather poor. By reducing the scan speed to ~0.02 Da/s, a mass resolving power of 3×10^7 (FWHM) has been achieved in a single scan for *m/z* 414 [124]. More typical resolving powers for commercial instruments are up to about 30,000 (FWHM) for *m/z* 1500. As the sensitivity of the trap is rather high, peptide levels below 5 fmol can be detected. With an ESI source, 50-fmol/μL peptide concentrations have been detected [128], and peptide amounts less than 10 fmol deposited on a target have been detected by utilizing an atmospheric pressure MALDI ion source [129]. The sensitivity would be increased even further with higher trapping efficiency; usually most of the ions entering the trap are never trapped, but go straight through or hit the endcap at the far end.

2.3.4.5. MS^n—Mass Selection, Mass Separation. The quadrupole ion trap is an excellent analyzer for performing MS/MS, or rather MS^n, experiments. Since the ions can reside in the trap for a long time, they can be manipulated several times, thus allowing consecutive fragmentation, namely, MS^n.

First ions of the desired *m/z* are isolated; that is, all other ions are ejected from the trap. Then a supplementary RF voltage is applied (see text above), but with amplitude low enough not to eject the ions but only excite them. Since the ions now have higher kinetic energies, repeated collisions with He atoms will cause fragmentation (the optimum excitation level depends strongly on the ions selected), so-called low-energy CID. All fragments within the *m/z* acceptance range of the trap will remain in the trap, and by then scanning the ring RF voltage, an MS/MS spectrum will be obtained. In practice, the low-mass cutoff for an MS/MS (or MS^n) scan is significantly higher than for the nontandem scan. The reason is that the trapping potential becomes shallower with wider *m/z* range. During the fragmentation phase the ions are excited and, thus, more prone to overcoming the trapping potential. By reducing the *m/z* range, loss in sensitivity can be significantly reduced.

The procedure can now be repeated, but instead of scanning the RF voltage to get a mass spectrum, one of the observed fragment ions can now be isolated and excited to obtain an MS^3 spectrum, and so on. Commercial ion traps can handle up to MS^{10}, but in practice the need for higher orders than MS^5 is rare. The ability to record mass spectra from multiple fragmentation is a powerful characterization tool.

2.3.5. Fourier Transform Ion Cyclotron Resonance

2.3.5.1. Principle. The principle of cyclotron resonance was developed in the early 1930s by Lawrence and others [130,131]. The utilization of the ion cyclotron resonance (ICR) technique for mass spectrometry was introduced around 1950 by Sommer et al. [132,133] and the combination with the Fourier transform (FT) technique was developed by Comisarow and Marshall in 1974 [134]. Coupling of external sources to an FTICR analyzer was first done in 1985 [135,136]. Like the quadrupole ion trap, ICR mass analyzers are capable of storing ions in a cell. The geometry of the cell varies between different instruments—a cylindrical cell is shown in Fig. 2.14. It consists of three pairs of opposing plates used for trapping, excitation, and detection.

The principle of ion cyclotron motion is shown in Fig. 2.15. The analyzer takes advantage of the same physical laws as the magnetic sector (one can think of the ICR analyzer as a sector with an arc of 360°). Ions with charge q and moving with velocity $\vec{\mathbf{v}}$ in a spatially uniform magnetic field of strength $\vec{\mathbf{B}}$, will experience a force

$$\vec{\mathbf{F}} = q\vec{\mathbf{v}} \times \vec{\mathbf{B}} \tag{2.16}$$

The force, $\vec{\mathbf{F}}$, is perpendicular to the plane determined by $\vec{\mathbf{v}}$ and $\vec{\mathbf{B}}$. If the ion motion is perpendicular to the magnetic field, the scalars can be used. Together with the relationship for centripetal force in circular motion

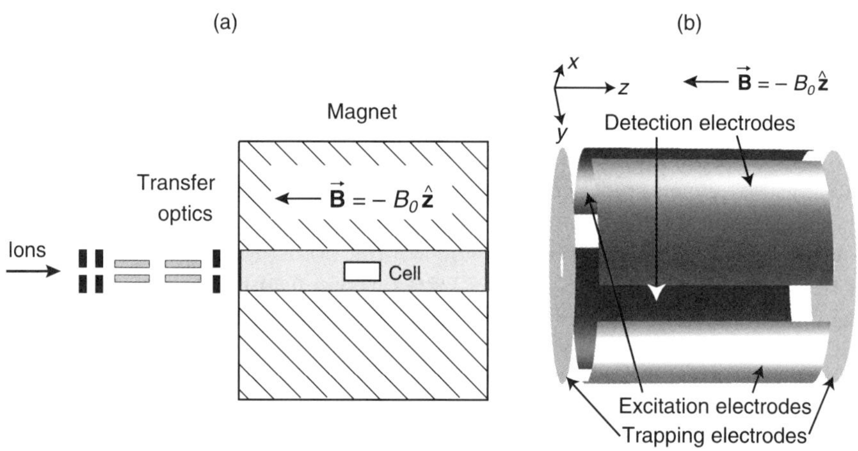

Figure 2.14. Schematic of an FTICR mass spectrometer (a) and a cylindrical cell (b).

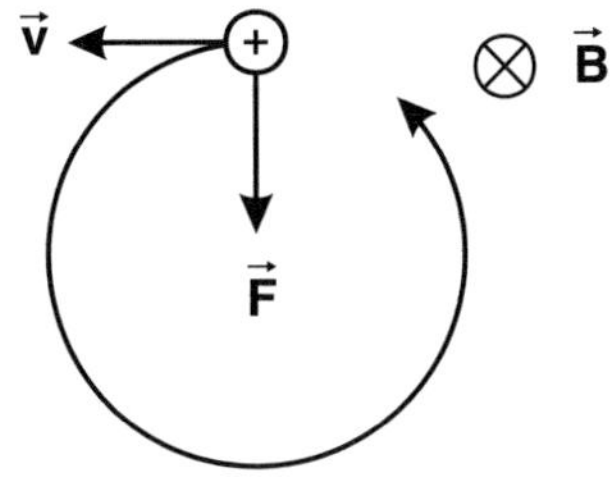

Figure 2.15. The trajectory of an ion moving in a uniform magnetic field. An ion with charge q and velocity $\vec{\mathbf{v}}$ moving non-parallel to the field lines of a spatially homogeneous magnetic field of strength $\vec{\mathbf{B}}$ will experience a force $\vec{\mathbf{F}} = q\vec{\mathbf{v}} \times \vec{\mathbf{B}}$ (the Lorentz force) directed perpendicular to the $\vec{\mathbf{v}}$-$\vec{\mathbf{B}}$-plane. Hence, the ion will begin a circular motion.

$$F = \frac{mv^2}{r} \tag{2.17}$$

where m is the ion mass, and r is radius of the ion trajectory, a relationship recognized from Section 2.3.2 emerges

$$qvB = \frac{mv^2}{r} \tag{2.18}$$

which is the same expression as in Eq. (2.4). Ions circulating in the cell will have an angular velocity

$$\omega = \frac{v}{r} \tag{2.19}$$

and by combining Eqs. (2.18) and (2.19):

$$q\omega rB = m\omega^2 r \tag{2.20}$$

After rearrangement the expression for the cyclotron frequency f_c is obtained:

$$\omega_c = 2\pi f_c = \omega = \frac{qB}{m} \tag{2.21}$$

For conversion to atomic units, see Eq. (2.8). An interesting feature of Eq. (2.21) is that all ions of a certain m/z have the same ICR frequency, independent of their velocity. Hence, energy focusing is not essential for precise determination of m/z. For comprehensive discussions of the properties of an FT-ICR analyzer, see for example Refs. 137–140.

2.3.5.2. Ion Injection. As with the quadrupole ion trap, both continuous (e.g., ESI) and pulsed (e.g., MALDI) ion sources can be successfully coupled to an ICR analyzer. Ions generated in the source have to be guided into the cell, much in the same way as with the ion trap. Ions coming from an ESI source typically have kinetic energies corresponding to an acceleration potential of ~1 V. A set of beam steering lenses focuses the ions onto the axis of the inlet ion optics. Typical focusing devices are octapoles and tube lenses. An additional challenge with the ICR analyzer is the very high magnetic field gradient that has to be overcome by the ions. If the radial (sideways) velocity component is large enough compared to the axial velocity, the ions will, when experiencing an increasing magnetic field strength, turn around and never enter the cell. The two principal ways to deal with this magnetic mirror effect are either to accelerate the ions to sufficient high velocity to pass the barrier or to hold the ions close to the magnet central axis since the magnetic mirror force is zero on-axis. The problem with high ion velocity is that it is then necessary to slow them down for capturing in the cell.

When the ions have entered the cell a small voltage (~1 V or slightly higher than the effective acceleration voltage) is applied to the two trapping electrodes. Ions that are inside the cell will now perform an oscillatory motion in the direction of the magnetic field (i.e., along the z axis). Before entering the cell, the ions can be accumulated [141,142] or an ion gate can be activated to limit the number of ions entering the cell, so as, for example, to avoid space-charge effects.

Simultaneously with the application of the trapping voltage a short burst of gas is allowed into the cell. When the ions collide with the gas molecules, they will change direction and will thus be affected by the magnetic field and begin their cyclotron motion (an ion with no radial movement will ideally be unaffected by the magnetic field). This increases the trapping efficiency and also cools the ions the same way as in the quadrupole ion trap, thus forcing them close to the center of the cell.

Since the cell volume is finite, the trapping voltage influences the radial motion (i.e., the cyclotron motion in the x–y plane of the ions). The shape of the electrode potentials causes an outward directed force, opposite to the inward-directed Lorentz magnetic force. As a result, the center of the cyclotron motion will also move around the z axis of the cell and the total ion motion in the x–y plane becomes a precession. The additional motion caused by the trapping potential is called *magnetron motion*. The frequencies of both the motion in the z direction and the magnetron motion are typically much lower than the cyclotron frequency. Also, during an experiment the magnetron radius is much smaller than the cyclotron radius. Nevertheless the magnetron motion perturbs the mass analysis, causing radial diffusion and sidebands, and also lowers the cyclotron frequency [138,139]. Hence, to obtain maximum resolution the trapping potentials should be as low as possible. The magnetron motion also causes problems during cooling by collisional damping (e.g., when trapping the ions). While the cyclotron radius quickly decreases, the magnetron radius, on the other hand, increases. To avoid losing all ions to the cell electrodes, quadrupolar excitation (RF voltages of the same amplitude, frequency, and sign are applied to the two excitation electrodes; see Fig. 2.14b) is applied to convert magnetron motion into cyclotron motion, causing both the cyclotron and magnetron radii to become near zero; thus the ions are axialized.

2.3.5.3. Mass Analysis and Ion Detection. The method for ion detection in an FTICR instrument is different from other mass spectrometers, where the ions hit a detector and are lost in the process. To detect the ions trapped in the ICR cell, so-called dipolar excitation is employed by applying RF voltages of the same amplitude and frequency but of opposite sign (cf. quadrupolar excitation above) to the two excitation electrodes (Fig. 2.16). Ions with the same ICR frequency as the RF voltage frequency (i.e., in resonance with the RF voltage) are excited to a larger orbital radius. The new radius depends on the temporal length and amplitude of the RF voltage. Moreover, this post-excitation ion cyclotron orbital radius is independent of m/z. This means that all ions in a given m/z range can be excited to the same ICR orbital radius. In an ideal, spatially homogeneous, electric field no mass discrimination will

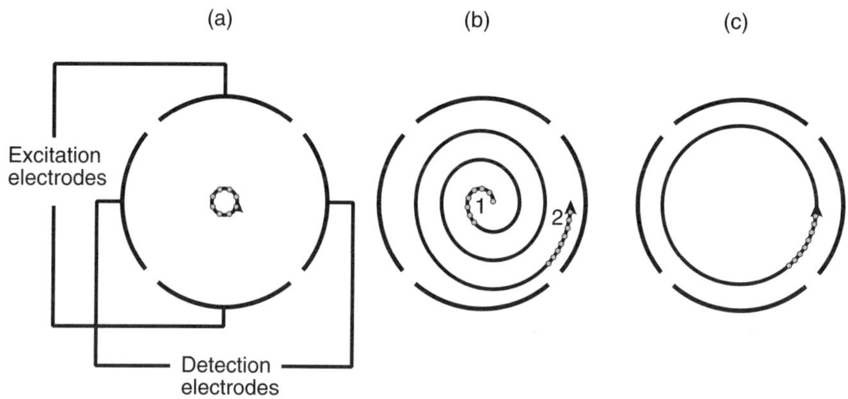

Figure 2.16. After trapping and cooling, the ions are orbiting close to the center of the cell and are spread out around the orbit, making detection impossible (a). When the excitation electrodes are activated, ions in resonance with the RF voltage are excited and obtain a larger orbital radius (b). During the process the ions will keep essentially the same distance to each other and thus the more or less uniform distribution around the orbit (1) will be transformed to a distinct packet of ions at the larger radius (2). After deactivation of excitation electrodes the ions will continue to orbit at the larger radius, and detection will now be possible (c).

result from the excitation. Typical excitation voltages are ~1–2 V with excitation times of ~1 ms.

By employing the excitation so that the ions will pass close enough to the detection plates (the strongest signal is obtained when their orbital diameter is slightly less than the distance between the plates, but the field is best defined at the center of the cell, which calls for a compromise), a small current, or image charge, will be induced in the plate each time an ion passes by. Since the ions have different ICR frequencies depending on their *m/z*, each generated current frequency will correspond to a certain *m/z*. However, a signal is observable only if there is a net difference between the two opposing detector electrodes. When the ions are injected, they are spread out more or less randomly around the orbit, and, therefore, the image charge induced in either of the two opposing plates will, on average, be balanced by an equal charge at the other plate, yielding a zero net signal. The preexcitation ICR orbital radius is rather small, and the ions are, although spread out along the orbit, fairly close together. When excited to a much larger orbital radius, they will all follow the same outward spiral during the time of excitation. During the process they will keep essentially the same distance to each other as before and travel together as a small ion packet in the larger orbit. Hence, resonant excitation also creates spatial coherence, and, therefore, a detectable signal is generated.

Ions of a certain *m/z* range can be excited to a larger orbital radius by applying an RF voltage that includes all the corresponding frequencies, to the excitation plates. This can be done in any of several ways. One way is to sweep the RF voltage frequency while keeping the amplitude constant, so-called chirp excitation. Then all ions with resonance frequency in this interval will be excited and subsequently detected. The chirp excitation, however, delivers nonuniform excitation amplitudes (resulting in a variation in apparent relative ion abundance) and also limited *m/z* selectivity at the low and high ends of the frequency sweep. Another way is to first define the *m/z* range to excite, convert to the corresponding frequency band, and then taking the inverse Fourier transform of the frequency spectrum, to produce a time domain excitation signal, so-called stored waveform inverse Fourier transform (SWIFT). The time-domain signal is more complicated than in the chirp case, but the excitation will be better executed.

The detector electrodes will now pick up signals generated by all the ion packages that circle in the excited orbit, so-called broadband detection. Each bundle will induce a signal every time it passes one of the electrodes and a complicated time domain signal is acquired. By Fourier transforming the time spectrum, a frequency spectrum is obtained, which in turn can be converted to a mass spectrum. Even if recorded for a very short time, all frequencies corresponding to excited ions will be present in the transformed data. However, the longer the acquisition time, the better the resolution.

Because long acquisition times are required for maximum resolution, it is essential that the ions can survive in the trap for extended periods of time. A main reason for ion loss is collisions with residual gas in the cell. Even if the ions are not lost from the cell, the spatial coherence is eventually lost. Therefore, it is essential to keep the pressure as low as possible, preferably in the 10^{-9}-torr region or below. Collisions will also perturb the cyclotron motion of the ions, causing peak broadening in the resulting mass spectrum. It is also important not to allow too many ions into the cell. When more than 10^6–10^7 ions are present in the cell with its typically spatially nonuniform trapping field, the effects of ion–ion repulsion have to be considered. Coulomb repulsion can, by pushing ions having like charge apart into regions of different applied fields, shift or broaden peaks in the mass spectra. By using an ion gate, the number of ions entering the cell can be limited in a controlled way. Moreover, peak coalescence (observation of a single peak having a narrower frequency width than the separation between the cyclotron frequencies of two ions with different *m/z*) occurs when two ion bundles have sufficiently large populations and their respective *m/z* ratios are very close. The two ion populations will overlap and then behave like one larger bundle yielding a single cyclotron frequency [143].

An alternative to broadband detection for recording spectra is heterodyne detection. Long sampling times are required to obtain maximum resolution. Detection in broadband mode can result in several hundred thousand data points per second, which puts some demands on recording equipment as well as on storage space. To record long transients, heterodyne detection can be employed to allow sampling at a lower rate; instead, the width of the recorded *m/z* range is sacrificed. With heterodyne (narrowband) detection, the image signal is multiplied with a reference frequency close to that of the signal of interest. The product signal then contains frequencies that equal the sum and the difference of the two original signals. The high-frequency part (the sum) can be filtered out, thus leaving only the difference frequency, which is much lower than the original frequencies. Hence, the sampling rate can be lowered, allowing for a much longer transient to be recorded with the same number of data points. The Fourier transform of the heterodyne signal gives peaks at the reduced frequencies, which are added to the reference frequency and then converted to a mass spectrum.

To enhance the signal-to-noise ratio, several scans can be acquired and averaged. The nondestructive detection of the FTICR analyzer allows for remeasurement of already detected ions (rather than repeating the whole injection and trapping procedure). As mentioned above, collisions between ions and residual gas molecules eventually ruins the detection. By employing quadrupolar excitation in combination with a short gas burst, thereby forcing the ions that still orbit (noncoherently) in the cell toward the center, and then once again perform dipolar excitation, the ions can be remeasured. This procedure can be repeated several times before the signal finally vanishes.

2.3.5.4. Performance Parameters. The dominant feature of an FTICR instruments is the magnet. Commercial instruments typically have superconducting magnets, and are available with different magnetic field strengths up to 9.4T, but there are also examples of instruments with up to 25-T field strength. The newer designs facilitate active shielding of the magnetic field, decreasing the size and weight considerably compared to previous passive shielding designs. The heart of the instrument, the cell, has typical dimensions of ~10cm, depending somewhat on the geometry. See Ref. 144 for recent description of a commercial 9.4-T instrument with examples of performance and applications.

The *m/z* range is limited by the magnetic field strength of the magnet and the kinetic energy of the ions being trapped. For commercial 9.4-T instruments, a typical *m/z* range is between ~30 and ~10,000. This range covers almost all ions produced by an ESI source, but puts rather severe constraints on MALDI-produced ions. For example, the heaviest molecule detected at present is an ESI-produced single DNA molecule of ~100MDa with ~30,000 charges [145]. The *m/z* of such a species is only ~3300, thus well below the upper limit for normal operation.

For most mass spectrometric applications, the mass resolving power of the FTICR analyzer is unprecedented. For single-*m/z* species, however, where Fourier transform methods are not needed, both higher mass resolving power and better mass accuracy have been achieved [146]. Different world records are listed in Ref. 140, including a mass resolving power of 8×10^6 (FWHM) for bovine ubiquitin (monoisotopic mass 8569.6Da) and 5×10^6 (FWHM) for a depleted p16 tumor suppressor protein (monoisotopic mass 15,792.1Da) [147]. In both cases the isotopic fine structure is revealed. Heterodyne detection and special care in the trapping process allowed for the measurement. In broadband mode unit resolution (i.e., isotopes are resolved) have been attained for Chondroitinase II, a 111.7-kDa protein [139,148]. The very high mass resolving power also allows for analysis of complex mixtures even in combination with ESI, which generates a multitude of charge states for each species.

A mass calibration for FTICR analyzers with superconducting magnets is very stable and is valid for many days for normal applications. Mass accuracy of ~1ppm can be obtained over a fairly wide mass range. Unique elemental compositions can be determined for masses over 800Da [149]. A particular problem that arises when analyzing larger proteins is that even if isotopic resolution is obtained, the correct mass assignment of the peaks is problematic. For bovine serum albumin (66.4kDa) the monoisotopic peak will have an intensity (area) of $<10^{-16}$ relative to the most intense isotopic peak, making it practically nonexistent. Moreover, the isotopic distribution becomes rather wide, making it virtually impossible to assign the first visible isotopic peak, and as a result a mass error of 1Da is easily obtained. Even for a smaller protein such as porcine trypsin (23.5kDa) the monoisotopic peak will have an intensity of $<10^{-5}$ relative to the most intense isotopic peak.

Since a minimum of about 100 ions are needed to generate a detectable signal under normal circumstances (ion counting is inherently more sensitive than image current detection) and space-charge effects become influential with more than 10^6–10^7 ions, the dynamic range is relatively poor. The minimum amount of analyte required for analysis depends strongly (as always) on the substance and on the properties of the ion source. Subfemtomole levels of human hemoglobin from a single blood cell have been detected [150] by coupling sheathless CE interface to an ESI source, and protein amounts just below 1 amol have been detected [151], also by sheathless CE-ESI coupling.

2.3.5.5. MS^n—Mass Selection, Mass Separation. Like the quadrupole ion trap, the FTICR analyzer is excellent for MS/MS or MS^n measurements. Perhaps even more so, since the ions remain in the cell after detection. In principle, one injection of ions is enough for a whole MS^n sequence, including acquisition of a mass spectrum of each step.

There are several ways of producing fragments; low-energy CID is one of the more common. The procedure is very similar to that of the quadrupole ion trap: first, all ions except those with the desired m/z are ejected from the cell by exciting them to a radius larger than the cell radius; second, or simultaneously with ejection, the selected ion species are excited to a larger cyclotron radius, giving them higher translational energy. Now the collision gas is introduced into the cell and the excited ions will undergo a number of collisions, gradually increasing their internal energy until fragmentation occurs. The particular energy required for fragmentation depends strongly on the ion species. As with the quadrupole trap, all fragments within the m/z acceptance range will remain trapped, and can, after further dipolar excitation, be detected, thus yielding an MS/MS spectrum.

The fragments can now be axialized and the procedure repeated, giving an MS^3 spectrum, and so on. As with the ion trap, there is no theoretical upper limit for the number of fragmentations that can be induced, but the need for orders higher than MS^5 is rare. As was stated in Section 2.3.4, the ability to record mass spectra from multiple fragmentation is a powerful tool for structural characterization.

To increase the MS/MS performance, several methods have been developed over the years; some common CID methods are sustained off-resonance irradiation (SORI), very low-energy (VLE), and multiple excitation collisional activation (MECA). Apart from CID, ions may be fragmented by photodissociation, surface-induced dissociation (SID), blackbody infrared radiative dissociation (BIRD), or electron-capture dissociation (ECD); see Ref. 140 for brief descriptions and references.

2.3.6. Hybrid Instruments

Hybrid mass spectrometers are instruments equipped with two or more different types of analyzers coupled together. The advantage of hybrid instruments is the possibility of MS/MS analysis with one high-performance analyzer and one less costly and less space-demanding analyzer for the part of analysis that can be compromised. Some combinations are more frequent than others, and some commercially available types are mentioned below. The double-focusing sector–TOF hybrid allows for very high precursor ion selectivity without the additional cost of a four-sector system. The mass resolving power of the TOF analyzer is high enough for most applications. For descriptions of such an instrument, see Refs. 108 and 152, and for some examples of applications, see Refs. 153–158. A double-focusing sector–quadrupole ion trap is described in Ref. 159. Such an instrument allows high selectivity of the precursor ion and also the possibility of MS^n; thus this is a powerful combination. A particularly fruitful combination is the quadrupole mass filter–TOF mass spectrometer with orthogonal acceleration into the TOF section (see Fig. 2.17). Although precursor ion selectivity is lower than for a sector instrument, the Q-TOF is less costly, and if the TOF analyzer is equipped with a reflector, fragment ions can be analyzed with relatively high mass resolving power. When operating in normal MS-mode, the quadrupole functions only as a focusing device, while the TOF analyzer is used for ion dispersion, thereby yielding spectra with high resolution. All the combinations described above are commercially available. Other examples of configurations are the double-focusing sector–quadrupole mass filter [159,160], the quadrupole ion trap–TOF [161,162] and the TOF–TOF [100]. A number of different instrument configurations are given in Table 2.1, with rough indications of the precursor ion selectivity and product ion resolution.

2.4. DETECTORS

The role of the detector is to convert the energy of the incoming ions into a current signal that is registered by the electronic devices and the computer of the acquisition system of the mass spectrometer. When the incoming ion strikes the detector, the energy from that impact causes emission of secondary particles, such as electrons or photons. The number of secondary particles created by an impact depends on the energy and the velocity of the incoming ion. Hence, if all ions are accelerated to the same kinetic energy, as in the TOF mass analyzer, then the detection sensitivity is lower for high-mass (slow) ions then for low-mass (fast) ions. To increase the detection sensitivity the ions are often postaccelerated before they strike the detector. A detector should have high efficiency for converting the exciting energy of the incoming ion to electrons or photons, a linear response, low noise, short recovery time, and minimal variations in transit time (narrow peak width). Some of the detectors currently used in mass spectrometers are described in this section. Hybrids of the different detectors are also frequently used. Ion detection in the FTICR instrument is, unlike that in most other mass analyzers where the ions hit a detector and are lost in the process, nondestructive and is briefly described in Section 2.3.5.

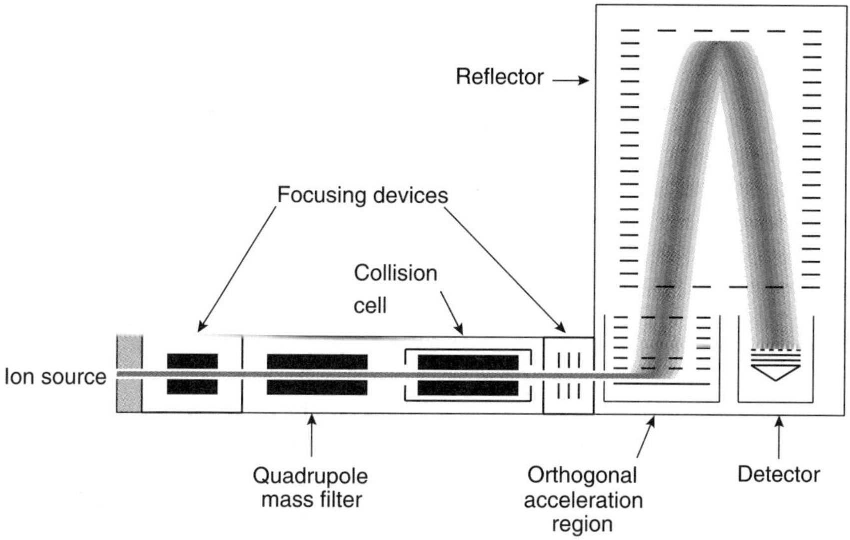

Figure 2.17. Schematic of a Q-TOF mass spectrometer.

TABLE 2.1. Some of the More Common Configurations for MS/MS Analysis[a]

Precursor Ion Selection—Product Ion Separation	Precursor Ion Selectivity	Product Ion Resolution
Fourier transform ion cyclotron resonance[b]	Very high	Unprecedented
Quadrupole ion trap[b]	High	High
Quadrupole mass filter—quadrupole mass filter	Medium–high	Medium–high
Quadrupole mass filter—time-of-flight	*Medium–high*	*High*
Sector—linked scan[c]	Poor	Medium
Sector/reversed geometry/electric sector scan[c]	Medium	Poor
Double-focusing sector—double focusing sector	Very high	Very high
Double-focusing sector—quadrupole mass filter	*Very high*	*Medium*
Double-focusing sector—quadrupole ion trap	*Very high*	*High*
Double-focusing sector—time-of-flight	*Very high*	*Medium–high*[d]
Time-of-flight/PSD/reflector "step scan"[c,e]	High	Poor–medium
Time-of-flight—time-of-flight	High	Medium–high[d]

[a] Hybrid configurations are shown in italics. The capabilities for mass selection and mass separation are indicated.

[b] Trapping analyzers can perform MS^n without additions or compromising performance.

[c] Performance compromised since part of the focusing capability is used for other task.

[d] Depending on analyzer configuration (linear, reflector, time-lag focusing).

[e] With a curved field reflector there is no need to acquire several spectra to improve resolution.

2.4.1. Electron Multipliers

An electron multiplier amplifies a weak current of incoming particles by using a series of secondary emission electrodes or dynodes to produce a considerably higher current at the anode [163]. When a particle impinges on the dynode, energy is transferred directly to the electrons in the dynode material and a number of secondary electrons are emitted. The dynode often consists of an alloy of an alkali or alkali-earth metal with a more noble metal. Thus, a thin insulating film of oxidized alkaline metal is formed on a conducting support. A good dynode material should emit many secondary electrons per primary incoming particle, have linear gain for high currents, and have low thermionic emission, namely, low noise. In a discrete dynode electron multiplier the string of dynodes are connected via a chain of resistors. A high voltage, typically 1–3 kV, is applied to the first dynode, and when a particle strikes the detector, secondary electrons are accelerated from one dynode to the next toward the anode at ground potential. The velocity of the secondary electrons and hence the gain of the electron multiplier depend on the voltage applied to the dynode string.

Dynode strings can be constructed in many ways, and the response time and range of linearity of the detector depend on the configuration. In the venetian-blind configuration (Fig. 2.18a) the dynodes are wide strips of material placed at an angle of 45° with respect to the electron cascade axis. This system offers a large input area to the incident primary particles. The advantage is that the dynodes are easily placed in line and the dimensions are not critical. The disadvantage is that it is impossible to prevent a fraction of the primary electrons from passing straight through. This results in low gain and large variations in transit time. In the box–grid configuration (Fig. 2.18b), the linear focused configuration, and the circular focused configuration the electrons are reflected from one dynode to the next, which isolates the cathode and anode well from each other, minimizing the risk of feedback. Another advantage of these latter detector configurations is that the efficient spacing allows many dynodes, giving high amplification. Very compact electron-multipliers that consist of one continuous dynode (Fig. 2.18c) are also often used [164]. The upper limit of the gain here is set by the onset of ion feedback. Ion feedback is caused by positive ions produced in the high-charge-density region at the output of the channel that drift back to the input. In the case of large single channels, ion feedback can be suppressed simply by bending or twisting the channels.

2.4.2. Microchannel Plates

Another example of a compact electron-multiplier configuration with short transit times is one of the most frequently used detectors in mass spectrometers, the microchannel plate [163,164]. An MCP consists of a parallel array of channel electron multipliers (Fig. 2.19). The inner surfaces of the channels are

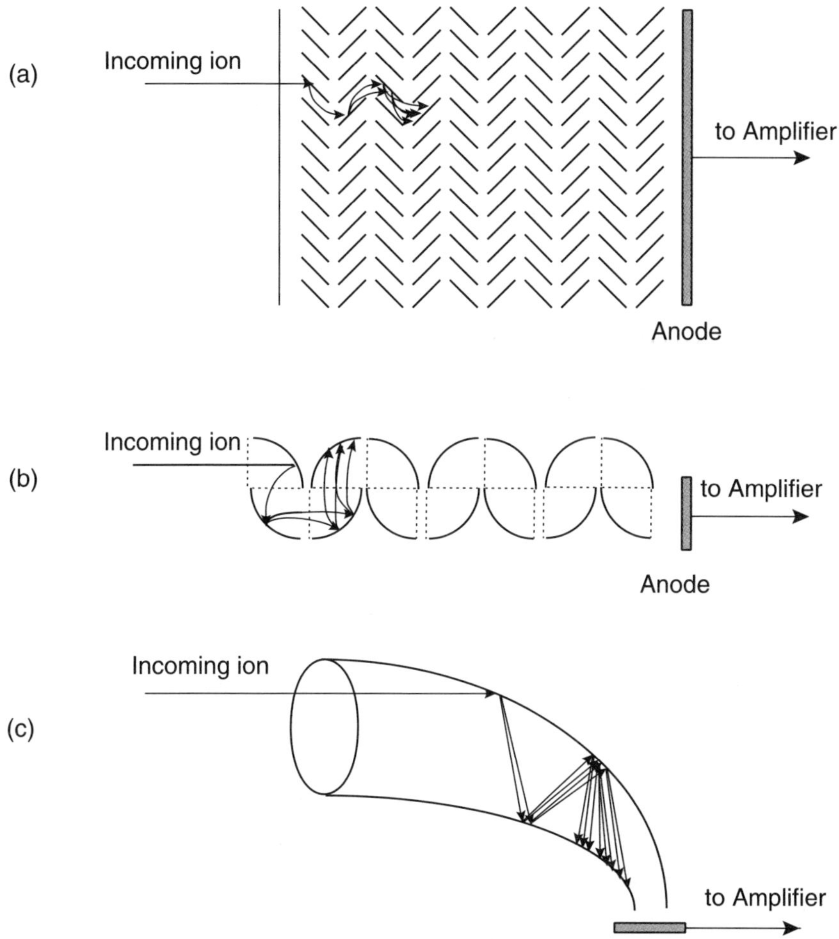

Figure 2.18. Various dynode configurations for electron multipliers: (a) venetian blind, (b) box–grid, and (c) continuous.

treated with a semiconductor material acting as a secondary electron emitter. The flat-end surfaces of the channels are coated with a metallic alloy to allow a potential difference to be applied over the MCP. When an ion strikes the channel wall, secondary electrons are generated and accelerated down the channel toward the output end. The secondary electrons also hit the channel wall, and an avalanche of electrons is generated. Each channel thus acts as a continuous dynode that generates up to 10^4 electrons per impact. Two or three MCPs are often stacked together to achieve higher gain.

A drawback of MCPs is that an electron cascade in one channel drains the neighboring channels for several microseconds leading to nonlinearities for

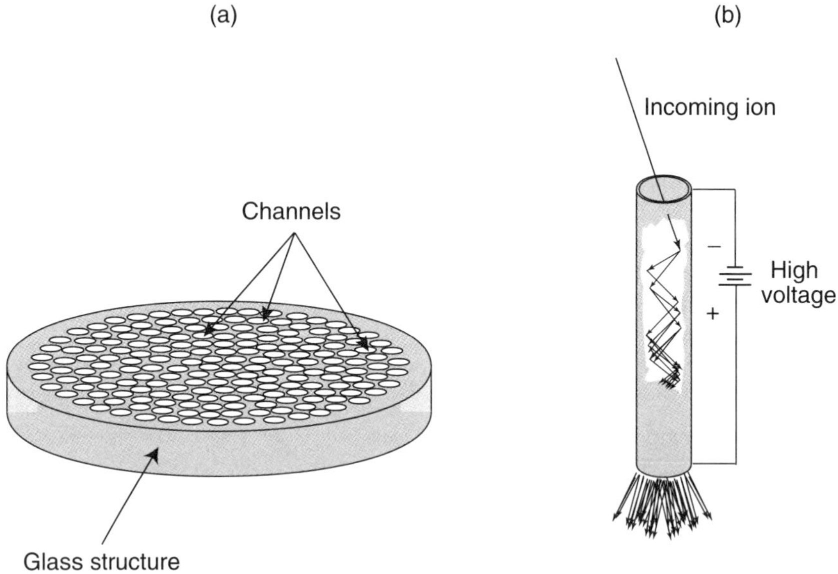

Figure 2.19. Schematic of an MCP (a) with a closeup of one channel (b). The size of the channels in the MCP schematic is greatly exaggerated compared to the size of the plate.

count rates above a few thousand incident particles per second (Fig. 2.20). This saturation effect is a major disadvantage when utilizing MCPs in a MALDI-TOF mass spectrometer. The yield of matrix ions is normally very high, and since the molecular mass of the matrix is low (~200 Da), their flight time is short, and they arrive at the detector before the analyte ions. Consequently, the gain of the detector will be low when the heavier (slower) analyte molecular ions arrive, thereby considerably reducing the sensitivity of the MCP detector. The saturation is more pronounced in the second MCP in a tandem MCP arrangement since each ion hitting the first MCP produces up to 10^4 electrons that strike the second MCP.

However, the gain of the MCP depends strongly on the potential applied, and a modification involving pulsing of the voltage over the first MCP in a tandem MCP detector has been successfully employed to reduce detector saturation [165,166]. When the matrix ions arrive at the first MCP the voltage, and consequently the amplification, are low. Thus, fewer electrons strike the second MCP, and fewer channels are saturated. The potential applied to the first MCP is then increased, and when the sample ions arrive at the MCP detector, both MCPs have maximum amplification. The time between the firing of the laser and the rise of the potential, that is, increase of amplification, can easily be varied and different portions of the spectrum are thus attenuated or amplified. Another way of increasing the dynamic range of the detector is to exchange the second MCP with a conventional electron multiplier.

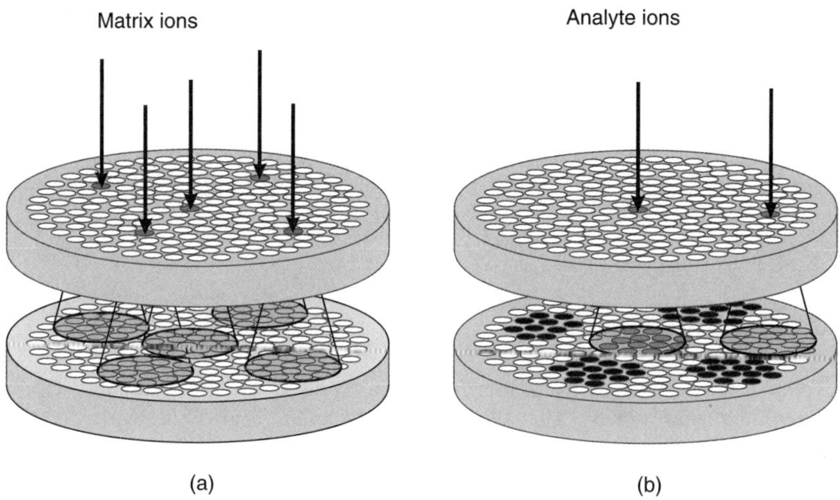

Figure 2.20. The principle of MCP detector saturation. Matrix ions strike the first (top) MCP (a), and up to 10^4 times more secondary electrons are produced and strike the second (bottom) MCP. Each channel that is hit by a secondary electron, and all its neighboring channels are drained and become inactive for several milliseconds. When the heavier (slower) analyte molecular ions arrive (b), large regions of the second MCP are depleted and the sensitivity of the MCP detector is reduced.

2.4.3. Scintillation Detector

The scintillation detector makes use of the fact that certain materials when struck by a particle or radiation emit a small flash of light, so-called scintillation [163]. Generally the detector consists of a scintillating material that is optically coupled to an amplifying device such as a photomultiplier. When an ion passes through the scintillator, it excites the atoms and molecules, causing light to be emitted. The light is transmitted to the photomultiplier, where it is converted into a weak current of photoelectrons, which is further amplified by an electron–multiplier system. The light output of a scintillator is directly proportional to the excitation energy of the incoming particle. Since the photomultiplier also is a linear device, the amplitude of the final electrical signal will also be proportional to the energy of the incoming particle. Scintillation detectors have fast recovery time and therefore accept high count rates.

2.4.4. Cryogenic Detectors

Energy-sensitive calorimetric detectors ("cryogenic detectors") have been coupled to TOF mass spectrometers [167]. Cryogenic detectors measure low-energy solid-state excitations (<5 meV) and must therefore be operated at temperatures typically below 2 K to avoid excessive thermal excitations. Compared to ionization-based detectors, which rely on electronvolt energies needed to produce secondary electrons or electronic excitations, cryogenic detectors are more sensitive to slow-moving (large) ions. Relatively large numbers of excitations are created for a given energy deposition, which allows the energy to be measured with smaller statistical error and, thus, greater precision. The disadvantages of cryogenic detectors are the extensive cooling necessary for operation and the practical problems when connecting the cooled detector with a spectrometer at room temperature.

ACKNOWLEDGMENTS

This work was supported by grants from The Swedish Medical Research Council (grants 13070 and 12575). The authors wish to thank Karl-Erik Karlsson, Mark Nicholas, and Jerzy Silberring for fruitful discussions and suggestions.

ABBREVIATIONS

a	atto, 10^{-18}
APCI	atmospheric pressure chemical ionization
BIRD	blackbody infrared radiative dissociation
CE	capillary electrophoresis
CF	continuous flow
CI	chemical ionization
CID	collision-induced dissociation
Da	dalton (or u, the atomic mass unit, defined by the mass of ^{12}C at exactly 12 u)
DC	direct current
DE	delayed extraction
ECD	electron-capture dissociation
EI	electron impact
ESI	electrospray ionization
eV	electronvolt
f	femto, 10^{-15}
FAB	fast-atom bombardment
FD	field desorption
FT	Fourier transform
FWHM	full width at half maximum
ICR	ion cyclotron resonance
IR	infrared
LC	liquid chromatography
LDI	laser desorption/ionization
LSIMS	liquid secondary ion mass spectrometry
m/z	mass-to-charge ratio (atomic units)

MALDI	matrix-assisted laser desorption/ionization
MCP	microchannel plate
MECA	multiple-excitation collisional activation
MIKES	mass-analyzed ion kinetic energy spectrometry
MS	mass spectrometry
MS/MS	mass selection, mass separation
MS^n	MS/MS of higher generations
n	nano, 10^{-9}
OA	orthogonal acceleration
p	pico, 10^{-12}
PDMS	plasma desorption mass spectrometry
ppm	parts per million
PSD	postsource decay
Q-TOF	quadrupole–time-of-flight
RF	radiofrequency, used here instead of AC (alternating current) to imply a high frequency
SDS	sodium dodecylsulfate
SEM	scanning electron microscopy
SID	surface-induced dissociation
SIMS	secondary ion mass spectrometry
SORI	sustained off-resonance irradiation
SWIFT	stored waveform inverse Fourier transform
TDC	time-to-digital converter
TOF	time-of-flight
UV	ultraviolet
VLE	very low-energy

REFERENCES

1. M. Karas and F. Hillenkamp, Laser desorption ionization of proteins with molecular masses exceeding 10,000 daltons, *Anal. Chem.* **60**, 2299 (1988).
2. K. Tanaka, H. Waki, Y. Ido, S. Akita, Y. Yoshida, and T. Yoshida, Protein and polymer analyses up to m/z 100,000 by laser ionization time-of-flight mass spectrometry, *Rapid Commun. Mass Spectrom.* **2**, 151 (1988).
3. V. V. Laiko, M. A. Baldwin, and A. L. Burlingame, Atmospheric pressure matrix-assisted laser desorption/ionization mass spectrometry, *Anal. Chem.* **72**, 652 (2000).
4. R. C. Beavis and B. T. Chait, Cinnamic acid derivatives as matrices for ultraviolet laser desorption mass spectrometry of proteins, *Rapid Commun. Mass Spectrom.* **3**, 432 (1989).
5. S. Berkenkamp, M. Karas, and F. Hillenkamp, Ice as a matrix for IR-matrix-assisted laser desorption/ionization: Mass spectra from a protein single crystal, *Proc. Natl. Acad. Sci.* (USA) **93**, 7003 (1996).
6. S. Ring and Y. Rudich, A comparative study of a liquid and a solid matrix in matrix-assisted laser desorption/ionization time-of-flight mass spectrometry and

collision cross section measurements, *Rapid Commun. Mass Spectrom.* **14**, 515 (2000).

7. J. Bai, X. Liang, Y. H. Liu, Y. Zhu, and D. M. Lubman, Characterization of two new matrices for matrix-assisted laser desorption/ionization mass spectrometry, *Rapid Commun. Mass Spectrom.* **10**, 839 (1996).
8. D. Schleuder, F. Hillenkamp, and K. Strupat, IR-MALDI-mass analysis of electroblotted proteins directly from the membrane: Comparison of different membranes, application to on-membrane digestion, and protein identification by database searching, *Anal. Chem.* **71**, 3238 (1999).
9. B. A. Budnik, K. B. Jensen, T. J. Jorgensen, A. Haase, and R. A. Zubarev, Benefits of 2.94 micron infrared matrix-assisted laser desorption/ionization for analysis of labile molecules by Fourier transform mass spectrometry, *Rapid Commun. Mass Spectrom.* **14**, 578 (2000).
10. I. D. Figueroa, O. Torres, and D. H. Russell, Effects of the water content in the sample preparation for MALDI on the mass spectra, *Anal. Chem.* **70**, 4527 (1998).
11. S. L. Cohen and B. T. Chait, Influence of matrix solution conditions on the MALDI-MS analysis of peptides and proteins, *Anal. Chem.* **68**, 31 (1996).
12. Y. Dai, R. M. Whittal, and L. Li, Confocal fluorescence microscopic imaging for investigating the analyte distribution in MALDI matrices, *Anal. Chem.* **68**, 2494 (1996).
13. J. Axelsson, A.-M. Hoberg, C. Waterson, P. Myatt, G. L. Shield, J. Varney, D. M. Haddleton, and P. J. Derrick, Improved reproducibility and increased signal intensity in matrix-assisted laser desorption/ionization as a result of electrospray sample preparation, *Rapid Commun. Mass Spectrom.* **11**, 209 (1997).
14. A. Westman, C. L. Nilsson, and R. Ekman, Matrix-assisted laser desorption/ionization time-of-flight mass spectrometry analysis of proteins in human cerebrospinal fluid, *Rapid Commun. Mass Spectrom.* **12**, 1092 (1998).
15. M. Mann and M. Wilm, Error-tolerant identification of peptides in sequence databases by peptide sequence tags, *Anal. Chem.* **66**, 4390 (1994).
16. A. M. Haag, J. Chaiban, K. H. Johnston, and R. B. Cole, Monitoring of immune response by blood serum profiling using matrix-assisted laser desorption/ionization time-of-flight mass spectrometry, *J. Mass Spectrom.* **36**, 15 (2001).
17. J. Yu, E. R. Butelman, J. H. Woods, B. T. Chait, and M. J. Kreek, In vitro biotransformation of dynorphin A (1–17) is similar in human and Rhesus monkey blood as studied by matrix-assisted laser desorption/ionization mass spectrometry, *J. Pharmacol. Exp. Ther.* **279**, 507 (1996).
18. J. Gobom, K. O. Kraeuter, R. Persson, H. Steen, P. Roepstorff, and R. Ekman, Detection and quantification of neurotensin in human brain tissue by matrix-assisted laser desorption/ionization time-of-flight mass spectrometry, *Anal. Chem.* **72**, 3320 (2000).
19. T. Krishnamurthy and P. L. Ross, Rapid identification of bacteria by direct matrix-assisted laser desorption/ionization mass spectrometric analysis of whole cells, *Rapid Commun. Mass Spectrom.* **10**, 1992 (1996).
20. U. Kallweit, K. O. Bornsen, G. M. Kreshbach, and H. M. Widmer, Matrix compatible buffers for analysis of proteins with matrix-assisted laser desorption/ionization mass spectrometry, *Rapid Commun. Mass Spectrom.* **10**, 845 (1996).

21. M. Puchades, A. Westman, K. Blennow, and P. Davidsson, Removal of sodium dodecyl sulfate from protein samples prior to matrix-assisted laser desorption/ionization mass spectrometry, *Rapid Commun. Mass Spectrom.* **13**, 344 (1999).

22. K. Strupat, M. Karas, and F. Hillenkamp, 2,5-Dihydroxybenzoic acid: A new matrix for laser desorption-ionization mass spectrometry, *Int. J. Mass Spectrom. Ion Proc.* **111**, 89 (1991).

23. A. Westman, T. Huth-Fehre, P. Demirev, and B. U. R. Sundqvist, Sample morphology effects in matrix-assisted laser desorption/ionization mass spectrometry of proteins, *J. Mass Spectrom.* **30**, 206 (1995).

24. R. M. Whittal, B. O. Keller, and L. Li, Nanoliter chemistry combined with mass spectrometry for peptide mapping of proteins from single mammalian cell lysates, *Anal. Chem.* **70**, 5344 (1998).

25. R. W. Nelson, D. Dogruel, and P. Williams, Mass determination of human immunoglobulin IgM using matrix-assisted laser desorption/ionization time-of-flight mass spectrometry, *Rapid Commun. Mass Spectrom.* **8**, 627 (1994).

26. R. S. Annan and S. A. Carr, Phosphopeptide analysis by matrix-assisted laser desorption time-of-flight mass spectrometry, *Anal. Chem.* **68**, 3413 (1996).

27. O. A. Mirgorodskaya, Y. P. Kozmin, M. I. Titov, R. Körner, C. P. Sönksen, and P. Roepstorff, Quantitation of peptides and proteins by matrix-assisted laser desorption/ionization mass spectrometry using ^{18}O-labeled internal standards, *Rapid Commun. Mass Spectrom.* **14**, 1226 (2000).

28. J. J. Lennon and K. A. Walsh, Locating and identifying posttranslational modifications by in-source decay during MALDI-TOF mass spectrometry, *Protein Sci.* **8**, 2487 (1999).

29. T. B. Farmer and R. M. Caprioli, Determination of protein-protein interactions by matrix-assisted laser desorption/ionization mass spectrometry, *J. Mass Spectrom.* **33**, 697 (1998).

30. R. S. Brown and J. J. Lennon, Sequence specific fragmentation of matrix-assisted laser-desorbed protein peptide ions, *Anal. Chem.* **67**, 3990 (1995).

31. B. Spengler, Post-source decay analysis in matrix-assisted laser desorption/ionization mass spectrometry of biomolecules, *J. Mass Spectrom.* **32**, 1019 (1997).

32. A. K. Shukla and J. H. Futrell, Tandem mass spectrometry: Dissociation of ions by collisional activation, *J. Mass Spectrom.* **35**, 1069 (2000).

33. R. P. Pauly, F. Rosche, M. Wermann, C. H. McIntosh, R. A. Pederson, and H. U. Demuth, Investigation of glucose-dependent insulinotropic polypeptide-(1-42) and glucagon-like peptide-1-(7-36) degradation *in vitro* by dipeptidyl peptidase IV using matrix-assisted laser desorption/ionization-time of flight mass spectrometry: A movel kinetic approach, *J. Biol. Chem.* **271**, 23222 (1996).

34. S. P. Gygi, B. Rist, S. A. Gerber, F. Turecek, M. H. Gelb, and R. Aebersold, Quantitative analysis of complex protein mixtures using isotope-coded affinity tags, *Nat. Biotechnol.* **17**, 994 (1999).

35. T. Miliotis, S. Kjellström, J. Nilsson, T. Laurell, L.-E. Edholm, and G. Marko-Varga, Capillary liquid chromatography interfaced to matrix-assisted laser desorption/ionization time-of-flight mass spectrometry using an on-line coupled piezoelectric flow-through microdispenser, *J. Mass Spectrom.* **35**, 369 (2000).

36. B. Finke, M. Mank, H. Daniel, and B. Stahl, Offline coupling of low-pressure anion-exchange chromatography with MALDI-MS to determine the elution order of human milk oligosaccharides, *Anal. Biochem.* **284**, 256 (2000).
37. M. Yamashita and J. B. Fenn, *J. Phys. Chem.* **88**, 4451 (1984).
38. M. L. Aleksandrov, L. N. Gall, V. N. Krasnov, V. I. Nikolaev, V. A. Pavlenko, and V. A. Shkurov, *Dokl. Akad. Nauk SSSR* **277**, 379 (1984).
39. J. B. Fenn, M. Mann, C. K. Meng, S. F. Wong, and C. M. Whitehouse, Electrospray ionization for mass spectrometry of large biomolecules, *Science* **246**, 64 (1989).
40. J. F. Mora, G. J. Van Berkel, C. G. Enke, R. B. Cole, M. Martinez-Sanchez, and J. B. Fenn, Electrochemical processes in electrospray ionization mass spectrometry, *J. Mass Spectrom.* **35**, 939 (2000).
41. M. Dole, L. L. Mack, R. L. Hines, R. C. Mobley, L. D. Ferguson, and M. D. Alice, *J. Chem. Phys.* **49**, 2240 (1968).
42. J. V. Iribane and B. A. Thomson, *J. Chem. Phys.* **64**, 2287 (1976).
43. T. L. Constantopoulos, G. S. Jackson, and C. G. Enke, Effects of salt concentration on analyte response using electrospray ionization mass spectrometry, *J. Am. Soc. Mass Spectrom.* **10**, 625 (1999).
44. K.-E. Karlsson, Cationization in electrospray microcolumn liquid chromatography-mass spectrometry, *J. Chromatogr.* **A794**, 359 (1998).
45. M. Wilm and M. Mann, analytical properties of the nanoelectrospray ion source, *Anal. Chem.* **68**, 1 (1996).
46. M. Wilm, A. Shevchenko, T. Houthaeve, S. Breit, L. Schweigerer, T. Fotsis, and M. Mann, Femtomole sequencing of proteins from polyacrylamide gels by nano-electrospray mass spectrometry, *Nature* **379**, 466 (1996).
47. L. Konermann and D. J. Douglas, Unfolding of proteins monitored by electrospray ionization mass spectrometry: A comparison of positive and negative ion modes, *J. Am. Soc. Mass Spectrom.* **9**, 1248 (1998).
48. A. T. Iavarone, J. C. Jurchen, and E. R. Williams, Effects of solvent on the maximum charge state and charge state distribution of protein ions produced by electrospray ionization, *J. Am. Soc. Mass Spectrom.* **11**, 976 (2000).
49. Y.-T. Li, Y.-L. Hsieh, J. Henion, M. W. Senko, F. McLafferty, and B. Ganem, Mass spectrometric studies on noncovalent dimers of leucine zipper peptides, *J. Am. Chem. Soc.* **115**, 8409 (1993).
50. S. A. Carr, M. E. Hemling, M. F. Bean, and G. D. Roberts, Integration of mass spectrometry in analytical biotechnology, *Anal. Chem.* **63**, 2802 (1991).
51. M. Karas, U. Bahr, and T. Dulcks, Nano-electrospray ionization mass spectrometry: Addressing analytical problems beyond routine, *Fresenius. J. Anal. Chem.* **366**, 669 (2000).
52. Y. Wang, M. Schubert, A. Ingendoh, and J. Franzen, Analysis of non-covalent protein complexes up to 290 kDa using electrospray ionization and ion trap mass spectrometry, *Rapid Commun. Mass Spectrom.* **14**, 12 (2000).
53. B. N. Pramanik, P. L. Bartner, U. A. Mirza, Y. H. Liu, and A. K. Ganguly, Electrospray ionization mass spectrometry for the study of non-covalent complexes: An emerging technology, *J. Mass Spectrom.* **33**, 911 (1998).

54. R. D. Smith, J. A. Loo, C. G. Edmonds, C. J. Barinaga, and H. R. Udseth, New developments in biochemical mass spectrometry: Electrospray ionization, *Anal. Chem.* **62**, 882 (1990).

55. A. Shevchenko, M. Wilm, and M. Mann, Peptide sequencing by mass spectrometry for homology searches and cloning of genes, *J. Prot. Chem.* **16**, 481 (1997).

56. A. Shevchenko, I. Chernushevich, M. Wilm, and M. Mann, De novo peptide sequencing by nanoelectrospray tandem mass spectrometry using triple quadrupole and quadrupole/time-of-flight instruments, in *Mass Spectrometry of Proteins and Peptides*, J. R. Chapman, ed., *Methods in Molecular Biology*, Vol. 146, Humana Press, Totowa, NJ, 2000, p. 1.

57. K. Håkansson, R. Zubarev, and P. Håkansson, Combination of nozzle-skimmer fragmentation and partial acid hydrolysis in electrospray ionization time-of-flight mass spectrometry of synthetic peptides, *Rapid Commun. Mass Spectrom.* **12**, 705 (1998).

58. M. Carrascal, K. Schneider, R. E. Calaf, S. van Leeuwen, D. Canosa, E. Gelpi, and J. Abian, Quantitative electrospray LC-MS and LC-MS/MS in biomedicine, *J. Pharm. Biomed. Anal.* **17**, 1129 (1998).

59. M. Barber, R. S. Bordoli, R. D. Sedgewick, and A. N. Tyler, Fast atom bombardment of solids (F. A. B.): A new ion source for mass spectrometry, *J. Chem. Soc. Chem. Commun.* 325 (1981).

60. S. Beranova-Giorgianni and D. M. Desiderio, Fast atom bombardment mass spectrometry of synthetic peptides, in *Methods in Enzymology: Solid-Phase Peptide Synthe*, G. B. Fields, ed., *Methods in Enzymology*, Vol. 289, San Diego, Academic Press, 1997, p. 478.

61. A. Benninghoven and W. K. Sichtermann, *Anal. Chem.* **50**, 1180 (1978).

62. A. Benninghoven, D. Jaspers, and W. Sichtermann, Secondary-ion emission of amino acids, *Appl. Phys.* **11**, 35 (1976).

63. P. Sigmund and C. Claussen, *J. Appl. Phys.* **52**, 990 (1981).

64. C. E. Heine, J. F. Holland, and J. T. Watson, Influence of the ratio of matrix to analyte on the fast atom bombardment mass spectrometric response of peptides sampled from aqueous glycerol, *Anal. Chem.* **61**, 2674 (1989).

65. H. Oka, Y. Ikai, T. Ohno, N. Kawamura, J. Hayakawa, K. Harada, and M. Suzuki, Identification of unlawful food dyes by thin-layer chromatography-fast atom bombardment mass spectrometry, *J. Chromatogr.* **A674**, 301 (1994).

66. H. Y. Lin, G. Gonyea, S. Killeen, and S. K. Chowdhury, Negative fast atom bombardment ionization of aromatic sulfonic acids: Unusual sample-matrix interaction, *Rapid Commun. Mass Spectrom.* **14**, 520 (2000).

67. B. J. Sweetman and I. A. Blair, 3-Nitrobenzyl alcohol has wide applicability as a matrix for FAB-MS, *Biomed. Environ. Mass Spectrom.* **17**, 337 (1988).

68. L. M. Mallis, H. M. Wang, D. Loganathan, and R. J. Linhardt, Sequence analysis of highly sulfated, heparin-derived oligosaccharides using fast atom bombardment mass spectrometry, *Anal. Chem.* **61**, 1453 (1989).

69. C. Dass, The role of a liquid matrix in controlling FAB-induced fragmentation, *J. Mass Spectrom.* **31**, 77 (1996).

70. P. C. Andrews and J. E. Dixon, Application of fast atom bombardment mass spectrometry to posttranslational modifications of neuropeptides, in *Methods in*

Enzymology: Hormone Action, Part K: *Neuroendocrine peptides*, P. Conn, ed., *Methods in Enzymology*, Vol. 168, Academic Press, San Diego, 1989, p. 72.

71. D. M. Desiderio, Quantitative analytical mass spectrometry of endogenous neuropeptides in human pituitaries, *Life Sci.* **51**, 169 (1992).
72. D. M. Desiderio and X. Zhu, Quantitative analysis of methionine enkephalin and beta-endorphin in the pituitary by liquid secondary ion mass spectrometry and tandem mass spectrometry, *J. Chromatogr.* **A794**, 85 (1998).
73. M. A. Moseley, L. J. Deterding, K. B. Tomer, and J. W. Jorgenson, Nanoscale packed-capillary liquid chromatography coupled with mass spectrometry using a coaxial continuous-flow fast atom bombardment interface, *Anal. Chem.* **63**, 1467 (1991).
74. R. M. Caprioli, T. Fan, and J. S. Cottrell, Continuous-flow sample probe for fast atom bombardment mass spectrometry, *Anal. Chem.* **58**, 2949 (1986).
75. R. M. Caprioli and W. T. Moore, Continuous-flow fast atom bombardment mass spectrometry, in *Methods in Enzymology: Mass Spectrometry*, J. A. McCloskey, ed., *Methods in Enzymology*, Vol. 193, Academic Press, San Diego, 1990, p. 214.
76. W. E. Stephens, A pulsed mass spectrometer with time dispersion, *Phys. Rev.* **69**, 691 (1946).
77. M. Guilhaus, V. Mlynski, and D. Selby, Perfect timing: Time-of-flight mass spectrometry, *Rapid Commun. Mass Spectrom.* **11**, 781 (1997).
78. C. Weickhardt, F. Moritz, and J. Grotemeyer, Time-of-flight mass spectrometry: State-of-the-art in chemical analysis and molecular science, *Mass Spectrom. Rev.* **15**, 139 (1996).
79. H. Wollnik, Time-of-flight mass analyzers, *Mass Spectrom. Rev.* **12**, 89 (1993).
80. B. A. Mamyrin, V. I. Karataev, D. V. Shmikk, and V. A. Zagulin, The mass-reflectron, a new non-magnetic time-of-flight mass spectrometer with high resolution, *Sov. Phys. JETP* **37**, 45 (1973).
81. T. J. Cornish and R. J. Cotter, A curved-field reflectron time-of-flight mass spectrometer for the simultaneous focusing of metastable product ions, *Rapid Commun. Mass Spectrom.* **8**, 781 (1994).
82. T. Cornish and R. J. Cotter, A compact time-of-flight mass spectrometer for the structural analysis of biological molecules using laser desorption, *Rapid Commun. Mass Spectrom.* **6**, 242 (1992).
83. T. J. Cornish and R. J. Cotter, A curved-field reflectron for improved energy focusing of product ions in time-of-flight mass spectrometry, *Rapid Commun. Mass Spectrom.* **7**, 1037 (1993).
84. C. K. G. Piyadasa, P. Håkansson, and T. R. Ariyaratne, A high resolving power multiple reflection matrix-assisted laser desorption/ionization time-of-flight mass spectrometer, *Rapid Commun. Mass Spectrom.* **13**, 620 (1999).
85. W. C. Wiley and I. H. McLaren, Time-of-flight mass spectrometer with improved resolution, *Rev. Sci. Instrum.* **26**, 1150 (1955).
86. R. S. Brown and J. J. Lennon, Mass resolution improvement by incorporation of pulsed ion extraction in a matrix-assisted laser desorption/ionization linear time-of-flight mass spectrometer, *Anal. Chem.* **67**, 1998 (1995).
87. M. L. Vestal, P. Juhasz, and S. A. Martin, Delayed extraction matrix-assisted laser desorption time-of-flight mass spectrometry, *Rapid Commun. Mass Spectrom.* **9**, 1044 (1995).

88. R. M. Whittal, L. M. Russon, S. R. Weinberger, and L. Li, Functional wave time-lag focusing matrix-assisted laser desorption/ionization in a linear time-of-flight mass spectrometer: Improved mass accuracy, *Anal. Chem.* **69**, 2147 (1997).

89. J. H. J. Dawson and M. Guilhaus, Orthogonal-acceleration time-of-flight mass spectrometer, *Rapid Commun. Mass Spectrom.* **3**, 155 (1989).

90. A. F. Dodonov, I. V. Chernushevich, and V. V. Laiko, Atmospheric pressure ionization time-of-flight mass spectrometer, *Book of Abstracts from 12th Int. Mass Spectrometry Conf.* Amsterdam, Aug. 26–30, 1991, p. 153.

91. C. H. Sin, E. D. Lee, and M. L. Lee, Atmospheric pressure ionization time-of-flight mass spectrometry with a supersonic ion beam, *Anal. Chem.* **63**, 2897 (1991).

92. J. G. Boyle and C. M. Whitehouse, Time-of-flight mass spectrometry with an electrospray ion beam, *Anal. Chem.* **64**, 2084 (1992).

93. A. N. Verentchikov, W. Ens, and K. G. Standing, Reflecting time-of-flight mass spectrometer with an electrospray ion source and orthogonal extraction, *Anal. Chem.* **66**, 126 (1994).

94. A. N. Krutchinsky, A. V. Loboda, V. L. Spicer, R. Dworschak, W. Ens, and K. G. Standing, Orthogonal injection of matrix-assisted laser desorption/ionization ions into a time-of-flight spectrometer through a collisional damping interface, *Rapid Commun. Mass Spectrom.* **12**, 508 (1998).

95. A. V. Loboda, A. N. Krutchinsky, M. Bromirski, W. Ens, and K. G. Standing, A tandem quadrupole/time-of-flight mass spectrometer with a matrix-assisted laser desorption/ionization source: Design and performance, *Rapid Commun. Mass Spectrom.* **14**, 1047 (2000).

96. M. Guilhaus, D. Selby, and V. Mlynski, Orthogonal acceleration time-of-flight mass spectrometry, *Mass Spectrom. Rev.* **19**, 65 (2000).

97. J. B. Hoyes, R. H. Bateman, and J. L. Wildgoose, High resolution analyser for an orthogonal TOF mass spectrometer, poster presented at 15th Int. Mass Spectrometry Conf. Barcelona, Spain, Aug. 26–Sept. 1, 2000.

98. U. N. Andersen, A. W. Colburn, A. A. Makarov, E. N. Raptakis, D. J. Reynolds, P. J. Derrick, S. C. Davis, A. D. Hoffman, and S. Thomson, In-series combination of a magnetic-sector mass spectrometer with a time-of-flight quadratic-field ion mirror, *Rev. Sci. Instrum.* **69**, 1650 (1998).

99. A. R. Botrill, A. E. Giannakopulos, A. Millichope, K. S. Lee, and P. J. Derrick, Combination of time-of-flight mass analysers with magnetic-sector instruments: In-line and perpendicular arrangements. Applications to poly(ethylene glycol) with long-chain end groups, *Eur. J. Mass Spectrom.* **6**, 225 (2000).

100. K. F. Medzihradszky, J. M. Campbell, M. A. Baldwin, A. M. Falick, P. Juhasz, M. L. Vestal, and A. L. Burlingame, The characteristics of peptide collision-induced dissociation using a high-performance MALDI-TOF/TOF tandem mass spectrometer, *Anal. Chem.* **72**, 552 (2000).

101. J. J. Thomson, *Rays of Positive Electricity and Their Application to Chemical Analysis*, Longmans, Green, London, 1913.

102. J. A. Dempster, *Phys. Rev.* **11**, 316 (1918).

103. J. Mattauch and R. Herzog, Double-focusing mass spectrograph and the masses of N^{15} and O^{18}, *Phys. Rev.* **50**, 617 (1936).

104. E. G. Johnson and A. O. Nier, Angular aberrations in sector shaped electromagnetic lenses for focusing beams of charged particles, *Phys. Rev.* **9**, 10 (1953).

105. R. S. Annan, H. J. Kochling, J. A. Hill, and K. Biemann, *Rapid Commun. Mass Spectrom.* **6**, 298 (1992).

106. R. S. Bordoli, K. Howes, R. G. Vickers, R. H. Bateman, and D. J. Harvey, *Rapid Commun. Mass Spectrom.* **8**, 585 (1994).

107. J. D. Lennon III, D. Shinn, R. W. Vachet, and G. L. Glish, Strategy for pulsed ionization methods on a sector mass spectrometer, *Anal. Chem.* **68**, 845 (1996).

108. K. F. Medzihradszky, G. W. Adams, A. L. Burlingame, R. H. Bateman, and M. R. Green, Peptide sequence determination by matrix-assisted laser desorption ionization employing a tandem double focusing magnetic-orthogonal acceleration time-of-flight mass spectrometer, *J. Am. Soc. Mass Spectrom.* **7**, 1 (1996).

109. V. S. K. Kolli and R. Orlando, A new strategy for MALDI on magnetic sector mass spectrometers with point detectors, *Anal. Chem.* **69**, 327 (1997).

110. M. L. Gross, Tandem mass spectrometry: Multisector magnetic instruments, in *Methods in Enzymology: Mass Spectrometry*, J. A. McCloskey, ed., *Methods in Enzymology*, Vol. 193, Academic Press, San Diego, 1990, p. 131.

111. J. A. Loo and R. Pesch, Sensitive and selective determination of proteins with electrospray ionization magnetic sector mass spectrometry and array detection, *Anal. Chem.* **66**, 3659 (1994).

112. K. R. Jennings and G. G. Dolnikowski, Mass analyzers, in *Methods in Enzymology: Mass Spectrometry*, J. A. McCloskey, ed., *Methods in Enzymology*, Vol. 193, Academic Press, San Diego, 1990, p. 37.

113. R. W. Vachlet, G. L. Glish, K. D. Ballard, and R. S. Orkiszewski, Prediction of artifact peak intensity in linked scans for dissociations occuring in the first field-free region of sector mass spectrometers, *J. Am. Soc. Mass Spectrom.* **8**, 554 (1997).

114. K. Rollins, J. H. Scrivens, R. C. K. Jennings, W. E. Morden, J. K. Welby, and R. H. Bateman, The use of tandem mass spectrometry as a problem-solving tool in the industrial environment, *Rapid Commun. Mass Spectrom.* **4**, 454 (1990).

115. W. Paul and H. Steinwedel, Ein neues Massenspektrometer ohne Magnetfeld, *Z. Naturforsch.* **8a**, 448 (1953).

116. E. Mathieu, *J. Math. Pures Appl.* **13**, 137 (1868).

117. D. W. McLachlan, *Theory and Applications of Mathieu Functions*, Clarendon, Oxford, 1947.

118. P. H. Dawson, *Quadrupole Mass Spectrometry and Its Applications*, Elsevier, New York, 1976.

119. P. H. Dawson, Quadrupole mass analyzers: Performance, design and some recent applications, *Mass Spectrom. Rev.* **5**, 1 (1986).

120. P. E. Miller and M. B. Denton, The quadrupole mass filter: Basic operating concepts, *J. Chem, Ed.* **63**, 617 (1986).

121. J. F. J. Todd, Ion trap mass spectrometer—past, present, and future (?), *Mass Spectrom. Rev.* **10**, 3 (1991).

122. K. R. Jonscher and J. R. Yates III, The quadrupole ion trap mass spectrometer—a small solution to a big challenge, *Anal. Biochem.* **244**, 1 (1997).

123. R. E. March, An introduction to quadrupole ion trap mass spectrometry, *J. Mass Spectrom.* **32**, 351 (1997).

124. R. E. March, Quadrupole ion trap mass spectrometry: Theory, simulation, recent developments and applications, *Rapid Commun. Mass Spectrom.* **12**, 1543 (1998).

125. V. M. Doroshenko and R. J. Cotter, A new method of trapping ions produced by matrix-assisted laser desorption ionization in a quadrupole ion trap, *Rapid Commun. Mass Spectrom.* **7**, 822 (1993).

126. G. C. Eiden, M. E. Cisper, M. L. Alexander, P. H. Hemberger, and N. S. Nogar, A method of increasing the sensitivity for laser desorption in an ion trap mass spectrometer, *J. Am. Soc. Mass Spectrom.* **4**, 706 (1993).

127. M. Soni, V. Frankevich, M. Nappi, R. E. Santini, J. W. Amy, and R. G. Cooks, Broadband Fourier transform quadrupole ion trap mass spectrometry, *Anal. Chem.* **68**, 3314 (1996).

128. A. Marina, M. A. García, J. P. Albar, J. Yagüe, J. A. L. de Castro, and J. Vázquez, High-sensitivity analysis and sequencing of peptides and proteins by quadrupole ion mass spectrometry, *J. Mass Spectrom.* **34**, 17 (1999).

129. V. V. Laiko, S. C. Moyer, and R. J. Cotter, Atmospheric pressure MALDI/ion trap mass spectrometry, *Anal. Chem.* **72**, 5239 (2000).

130. E. O. Lawrence and N. E. Edlefsen, *Science* **72**, 376 (1930).

131. E. O. Lawrence and M. S. Livingston, *Phys. Rev.* **40**, 19 (1932).

132. H. Sommer, H. A. Thomas, and J. A. Hipple, *Phys. Rev.* **76**, 1877 (1949).

133. H. Sommer, H. A. Thomas, and J. A. Hipple, *Phys. Rev.* **82**, 697 (1951).

134. M. B. Comisarow and A. G. Marshall, Fourier transform ion cyclotron resonance spectroscopy, *Chem. Phys. Lett.* **25**, 282 (1974).

135. R. T. McIver, Jr., R. L. Hunter, and M. T. Bowers, Coupling of a quadrupole mass spectrometer and a Fourier transform mass spectrometer, *Int. J. Mass Spectrom. Ion Proc.* **64**, 67 (1985).

136. P. Kofel, M. Alleman, H. P. Kellerhals, and K.-P. Wanczek, External generation of ions in ICR spectrometry, *Int. J. Mass Spectrom. Ion Proc.* **65**, 97 (1985).

137. A. G. Marshall and P. B. Grosshans, Fourier transform ion cyclotron resonance mass spectrometry: The teenage years, *Anal. Chem.* **63**, 215A (1991).

138. I. J. Amster, Fourier transform mass spectrometry, *J. Mass Spectrom.* **31**, 1325 (1996).

139. A. G. Marshall, C. L. Hendrickson, and G. S. Jackson, Fourier transform ion cyclotron resonance mass spectrometry: A primer, *Mass Spectrom. Rev.* **17**, 1 (1998).

140. A. G. Marshall, Milestones in Fourier transform ion cyclotron resonance mass spectrometry technique development, *Int. J. Mass Spectrom.* **200**, 331 (2000).

141. M. W. Senko, C. L. Hendrickson, M. R. Emmett, S. D.-H. Shi, and A. G. Marshall, External accumulation of ions for enhanced electrospray ionization Fourier transform ion cyclotron resonance mass spectrometry, *J. Am. Soc. Mass Spectrom.* **8**, 970 (1997).

142. M. E. Belov, E. N. Nikolaev, G. A. Anderson, H. R. Udseth, T. P. Conrads, T. D. Veenstra, C. D. Masselon, M. V. Gorshkov, and R. D. Smith, Design and performance of an ESI interface for selective external ion accumulation coupled to a Fourier transform ion cyclotron mass spectrometer, *Anal. Chem.* **73**, 253 (2001).

143. L. Paša-Tolic, Y. Huang, S. Guan, H. S. Kim, and A. G. Marshall, Ultrahigh-resolution matrix-assisted laser desorption/ionization Fourier transform ion cyclotron resonance mass spectra of peptides, *J. Mass Spectrom.* **30**, 825 (1995).

144. M. Palmblad, K. Håkansson, P. Håkansson, X. Feng, H. J. Cooper, A. E. Giannakopulos, P. S. Green, and P. J. Derrick, A 9.4T Fourier transform ion cyclotron resonance mass spectrometer: Description and performance, *Eur. J. Mass Spectrom.* **6**, 267 (2000).

145. R. D. Chen, X. H. Cheng, D. W. Mitchell, S. A. Hofstadler, Q. Y. Wu, A. L. Rockwood, M. G. Sherman, and R. D. Smith, Trapping, detection, and mass determination of coliphage T4 DNA Ions of 10^8 Da by electrospray ionization Fourier transform ion cyclotron resonance mass spectrometry, *Anal. Chem.* **67**, 1159 (1995).

146. F. Difilippo, V. Natarajan, M. Bradley, F. Palmer, and D. E. Pritchard, Accurate atomic mass measurements from Penning trap mass comparisons of individual ions, *Physica Scripta* **T59**, 144 (1995).

147. S. D.-H. Shi, C. L. Hendrickson, and A. G. Marshall, Counting individual sulfur atoms in a protein by ultrahighresolution Fourier transform ion cyclotron resonance mass spectrometry: Experimental resolution of isotopic fine structure in proteins, *Proc. Natl. Acad. Sci.* (USA) **95**, 11532 (1998).

148. N. L. Kelleher, M. W. Senko, M. M. Siegel, and F. W. McLafferty, Unit resolution mass spectra of 112 kDa molecules with 3 Da accuracy, *J. Am. Soc. Mass Spectrom.* **8**, 380 (1997).

149. R. P. Rodgers, E. N. Blumer, C. L. Hendrickson, and A. G. Marshall, Stable isotope incorporation triples the upper mass limit for determination of elemental composition by accurate mass measurement, *J. Am. Soc. Mass Spectrom.* **11**, 835 (2000).

150. S. A. Hofstadler, J. C. Severs, R. D. Smith, F. D. Swaneck, and A. G. Ewing, Analysis of single cells with capillary electrophoresis electrospray ionization Fourier transform ion cycloton resonance mass spectrometry, *Rapid Commun. Mass Spectrom.* **10**, 919 (1996).

151. G. A. Valaskovic, N. L. Kelleher, and F. W. McLafferty, Attomole protein characterization by capillary electrophoresis-mass spectrometry, *Science* **273**, 1199 (1996).

152. R. H. Bateman, A combined magnetic sector-time-of-flight mass spectrometer for structural determination studies by tandem mass spectrometry, *Rapid Commun. Mass Spectrom.* **9**, 1227 (1995).

153. T. Keough, M. P. Lacey, M. M. Ketcha, R. H. Bateman, and M. R. Green, Orthogonal acceleration single-pass time-of-flight mass spectrometry for determination of the exact masses of product ions formed in tandem mass spectrometry experiments, *Rapid Commun. Mass Spectrom.* **11**, 1702 (1997).

154. I. Lindh, W. J. Griffiths, T. Bergman, and J. Sjövall, Electrospray/collision-induced dissociation of derivatised peptides: Studies on a hybrid magnetic sector-orthogonal time-of-flight mass spectrometer, *Int. J. Mass Spectrom. Ion Proc.* **164**, 71 (1997).

155. A. Olling, M. E. Breimer, E. Peltomaa, B. E. Samuelsson, and S. Ghardashkhani, electrospray ionization and collision-induced dissociation time-of-flight mass spectrometry of neutral glycosphingolipids, *Rapid Commun. Mass Spectrom.* **12**, 637 (1998).

156. I. Lindh, L. Hjelmqvist, T. Bergman, J. Sjövall, and W. J. Griffiths, De novo sequencing of proteolytic peptides by a combination of C-terminal derivatization and nano-electrospray/collision-induced dissociation mass spectrometry, *J. Am. Soc. Mass Spectrom.* **11**, 673 (2000).

157. G. Tsaprailis, A. Somogyi, E. N. Nikolaev, and W. H. Wysocki, Refining the model for selective cleavage at acidic residues in arginine-containing protonated peptides, *Int. J. Mass Spectrom.* **195/196**, 467 (2000).

158. A. Weimann, P. Iannitti-Tito, and M. M. Sheil, Characterisation of product ions in high-energy tandem mass spectra of protonated oligonucleotides formed by electrospray ionisation, *Int. J. Mass Spectrom.* **194**, 269 (2000).

159. J. A. Loo and H. Muenster, Magnetic sector-ion trap mass spectrometry with electrospray ionization for high sensitivity peptide sequencing, *Rapid Commun. Mass Spectrom.* **13**, 54 (1999).

160. R. A. Yost and R. K. Boyd, Tandem mass spectrometry: Quadrupole and hybrid instruments, in *Methods in Enzymology: Mass Spectrometry*, J. A. McCloskey, ed., *Methods in Enzymology*, Vol. 193, Academic Press, San Diego, 1990, p. 154.

161. S. M. Michael, B. M. Chien, and D. M. Lubman, Detection of electrospray ionization using a quadrupole ion trap storage/reflectron time-of-flight mass spectrometer, *Anal. Chem.* **65**, 2614 (1993).

162. M. G. Qian and D. M. Lubman, Procedures for tandem mass spectrometry on an ion trap storage/reflectron time-of-flight mass spectrometer, *Rapid Commun. Mass Spectrom.* **10**, 1911 (1996).

163. W. R. Leo, *Techniques for Nuclear and Particle Physics Experiments*, Springer-Verlag, 1987.

164. J. Wiza, Microchannel plate detectors, *Nucl. Instrum. Meth.* **162**, 587 (1979).

165. O. Vorm and P. Roepstorff, Detector bias gating for improved detector response and calibration in matrix-assisted laser desorption/ionization time-of-flight mass spectrometry, *J. Mass Spectrom.* **31**, 351 (1996).

166. A. Westman, G. Brinkmalm, and D. F. Barofsky, MALDI induced saturation effects in Chevron microchannel plate detectors, *Int. J. Mass Spectrom. Ion Proc.* **169/170**, 79 (1997).

167. M. Frank, S. E. Labov, G. Westmacott, and W. H. Benner, Energy-sensitive cryogenic detectors for high-mass biomolecule mass spectrometry, *Mass Spectrom. Rev.* **18**, 155 (1999).

PART 2

HYPHENATED TECHNIQUES

CHAPTER 3

Capillary and Nano-LC Coupled with Mass Spectrometry

MANFRED RAIDA

Cellzome GmbH, Heidelberg, Germany

CONTENTS

3.1. REASONS FOR PERFORMING CAPILLARY OR NANO-LCMS

Any given compound within a complex mixture, as a peptide or a low-molecular-weight substance is determined by a limited number of partially or

Mass Spectrometry and Hyphenated Techniques in Neuropeptide Research, Edited by Jerzy Silberring and Rolf Ekman.
ISBN 0-471-35493-7

completely independent parameters. Some of them are easy to determine and to use, and therefore suitable for characterization. These are the molecular weight, or the mass, the hydrophobicity, and the overall charge of the molecule under given conditions. The exact mass can easily be determined by modern mass spectrometry; the hydrophobicity, by reversed-phase chromatography; and the charge, by ion-exchange chromatography. The characterization of the hydrophobicity and the exact mass alone allow one to define every compound even in a very complex mixture. In cases where the mixture still remains too complex, the third parameter, the charge, can be used to reduce the complexity for better characterization of the compounds. The orthogonal combination of the different chromatographic separation and the use of a mass spectrometer for exact determination of the mass of each eluting compound made this technique one of the most important analytic tools in pharmaceutical, biotechnological, and basic research.

But state-of-the-art mass spectrometry allows a much deeper insight into the structure of the analyzed compounds. By collision-induced dissociation (CID), also known as *collision-induced fragmentation*, compounds, which have been selected in a first mass filter, are partially "destroyed" under certain conditions in a gas-filled cell [1,2]. This is not a random destruction but follows exact rules. The masses of the resulting fragments are then determined in a second mass analyzer (MS/MS). Following the rules of fragmentation for the class of compounds being analyzed information about the structure of the compound are obtained. This technique can be applied for low molecular weight molecules as well as for peptides. Roepstorff proposed the nomenclature for the fragmentation of peptides [3]. MS/MS of peptides allows one to obtaining partial sequence information, sufficient for identification in protein or DNA databases [4] or in cases of yet unsequenced genomes evening the construction of primers to clone the new protein [5,6]. For detailed information about the analysis of peptides using CID see Kuster et al. 2001.

But for many years the sensitivity of the combined technique was low because of the performance of the available chromatographic separation and the mass spectrometers. The amounts of peptides needed were in the nanomole–picomole-ranges, far away from today's needs of modern bioanalysis.

The development of nanoelectrospray for the identification of proteins in the low femtomole range, obtained, for example, from silver-stained gels [7,8], has raised the question about the usefulness of LCMS for high-sensitivity analysis. More recent publications show that the state-of-the-art analysis moves from the both time- and manpower-consuming nanospray technique to capillary or nano-LCMS [9–11]. By LC the complexity of a mixture can be reduced dramatically, and fewer compounds have to be analyzed within a given time window. By modifying the slope of the gradient or by hyphenated techniques, such as multidimensional LC or peak parking, as described below, the number of coeluting compounds in the time window can be even more reduced. In addition, the concentration of compounds is increased in comparison to the nanospray technique. As mass spectrometry is a concentration-dependent technique, a high concentration of the analyte is required. In particular, capillary and nano-LC fulfils the requirements for high-sensitivity detection and, as shown below, to obtain further information. Beside the fractionation of complex mixtures, LC also increases the concentration of every compound in the eluent. In a nano-LC, all the peptides present in, for example, 10-μL injection volume are concentrated in about 100 nL: the average elution volume of a peptide from a nanocolumn. Using analytic HPLC, the same amount of peptide would elute in about 1 mL; in narrowbore HPLC, in about 50 μL. This results in an increase in sensitivity of about 500 compared to narrowbore HPLC and even 10,000 compared to analytic HPLC.

In addition, for the daily laboratory work, the scientist can automate LC/MS to a certain degree, such as by using an autosampler, while nanospray analysis requires a lot of direct attended manual work and cannot be automized.

Beside the developments in the research laboratories, a number of suppliers offer ready-made instrumentation to run capillary or nano-LC or components to modify a standard HPLC system for the needs of capillary or nano-LC. A list of suppliers is provided at the end of this chapter.

These facts make capillary or nano-LC/MS an indispensable and easily available tool in modern bioanalytical laboratories.

3.2. GENERAL ASPECTS

The availability of mass spectrometers for analytic laboratories created the necessity for direct coupling with a chromatographic separation system. For volatile substances the coupling with a gas chromatograph could be achieved easily (GC/MS). This technique is limited to volatile substances or to substances that can be modified to become a volatile compound, like amino acids. Nonvolatile compounds, like peptides, proteins, or nucleic acids, could not be analyzed in that way. But the importance of analyzing such substances has been recognized quite early, some peptides could be analyzed by mass spectrometry like TRH [12]. Klaus Bieman stated in 1962 in his book, *Mass Spectrometry, Organic Chemical Applications* that "It seems that a considerable amount of structural information can be obtained in this way using an extremely small amount of material." In 1962 this statement sounded rather futuristic, as for instance, sequence analysis of peptides or proteins using Edam degradation was even not established in many laboratories and protein sequencers were not commercially available. Mass spectrometry of peptides and proteins was far away from being a standard technique; only small peptides as tetrapeptides have been analyzed. In 1965 Barber et al. [13] described the analysis of small peptides by mass spectrometry. In the next ten years, dramatic developments in mass spectrometry in the field of protein and peptide analysis. It took quite some time until mass spectrometry and, moreover, the direct coupling with a solvent-based separation technique, such as HPLC, became an easy-to-use system in daily laboratory work. First FABMS allowed a coupling with liquid chromatography for the analysis of peptides, as described by Fridland and Desiderio [14] for neuropeptides. The analysis of peptides and proteins became routinely possible by the development of soft ionization techniques. For single proteins or peptides, analysis by mass spectrometry alone or in combination with liquid chromatography became a widely spread technique. But the view on biological processes has since changed dramatically. The completion of primary DNA sequences from a number of different species ranging from bacteria to mammals [15–21] has had a dramatic impact on the approaches to investigate biological organisms. To identify a protein involved in a cellular process less information and, more important, fewer amounts are necessary. In addition, the analysis of communication systems within the organism that regulate the reactions from the trigger to the answer has become feasible at a level that is compatible with studies in living organisms, as from the brain of freely moving animals by microdialysis. Finally, the state-of-the-art techniques led back to a >20-year-old idea, the mapping of all human proteins, the HUPO project [22], now renamed to "proteomics," the protein part of the genome [23–25].

The information drawn from the sequenced human genome resulted unexpectedly in only 35,000–40,000 different genes. On one hand, it could be expected that this limited number of proteins should simplify the analysis of biological processes, but because these proteins can be processed in different ways and a great number of single-nucleotide polymorphisms (SNPs) are present in the genome, the complexity of proteins is much higher. In addition, the question of protein–protein interactions, the formation of complexes as active units in the cell, showed that no simple view on the expressed proteins could be identified [26,27]. Detailed analysis of the biologically active structures is a prerequisite to understand and furthermore to influence the cellular functions.

The progress in analyzing the cellular proteins and their interactions is driven mainly by the development of high-sensitivity and high-resolution techniques in combination with bioinformatics. One of these techniques is mass spectrometry, a method that is over a century old. Mass spectrometry has been brought to the modern bioanalysis by the development of soft-ionization techniques such as plasma desorption ionization (PDMS) [28], atmospheric pressure ionization (API), electrospray ionization (ESI) [29], and matrix-assisted laser desorption and ionization (MALDI) [30]. In state-of-the-art bioanalysis only ESI and MALDI play a role, the other techniques are rarely used. In addition, the expanded mass range and the high resolution available in modern time-of-flight (TOF), quadrupole, ion traps, FTIC, and hybrid quadrupole time-of-flight mass analyzers (QqTOF) allows—in combination with the soft ionization techniques—the identification and characterization of proteins down to the low femtomole level and even below. The ability to select or to store one ion species selectively and to perform consecutive fragmentation and analysis of the fragment pattern (MS/MS) increases the power of mass spectrometry to the structural analysis [31]. This structural analysis plays an important role in the identification of proteins by partial sequence determination. But dramatic progress has been made not only in instrumentation but also in data interpretation and database analysis. With only little information proteins can be identified without any doubt. Programs to perform database searches now use raw MS data directly to identify the protein in DNA, EST, and protein databases [4,32–35].

Another technique with high impact for the analysis of biological systems is the development of high-resolution separation techniques as two-dimensional (2D) polyacrylamide gel electrophoresis (2D-PAGE) [36,37] and high-resolution liquid chromatography (HPLC). 2D-PAGE enables the separation of the proteins from a cell, a tissue, an organ, or a body fluid and the detection of up to 10,000 proteins by high-sensitivity staining. This has made 2D-PAGE the method of choice for the comprehensive analysis of complex protein mixtures [24,38–40]. By cutting protein spots from the gel, digestion of the protein in gel by specific proteases, extraction of peptides, and determination of the protein-specific peptide mass fingerprint, a technique for high-throughput analysis of state-dependent protein compositions of cellular systems became possible. The mass spectrometric method for protein identification at the level of silver-stained spots is either the MALDI-TOF or the nanospray ionization technique [8], which allows the determination of partial sequences.

This approach has led to the establishment of a new scientific direction, resulting in the founding of protein factories with the aim to identify thousands of proteins per day [40–44]. Nevertheless, 2D-PAGE is limited in some respects, no reliable method for separation of membrane proteins or large proteins has been developed yet, small proteins or peptides cannot be separated sufficiently, abundant proteins counteract with a high resolution of proteins with similar properties, and the dynamic range is limited. These facts limit the use of 2D-PAGE.

Peptides play an essential role in the regulation of the organism on the level of cells, tissues, and organs, and in the communication between compartments. Analysis of the peptides and understanding of their role and function is of indispensable importance for understanding the overall function of the organism and for the outcome of diseases.

These peptides, not accessible by 2D-PAGE, can be easily separated by liquid chromatography techniques. In contrast to many proteins, most of these peptides are easily soluble in aqueous solvents. They can be separated by different techniques, based on their overall charge using ion-exchange chromatography (IEX), by size using size-exclusion chromatography (SEC), or by hydrophobicity using reversed-phase chromatography (RP). In most applications RP chromatography uses solvent systems fully compatible with electrospray ionization. This makes the RP the method of choice to be coupled directly with the ESI source. Early work has been done for the analysis of neuropeptides by coupling of LC with field desorption mass spectrometry [45].

But not only for peptides, also for the analysis of proteins, LCMS plays an important role. Proteins in functional complexes as the ribosome have been successfully analyzed after digestion of the whole complex [46]. The highly complex peptide mixture was then separated by a combination of cation-exchange chromatography and reversed-phase chromatography directly coupled with mass spectrometric analysis. This approach shows that without prior separation on a gel, the composition of a complex functional system can be determined.

Beside the development of mass spectrometers, the HPLC has made great progress toward the high-resolution and high-sensitivity separation of peptides. RP materials have been developed that are compatible with the special physical and chemical features of peptides. The separation materials have been improved also by reducing the column diameter to as low as 50 μm i.d. (inner diameter), which has increased the sensitivity by a factor of several thousands, compared to the standard, still termed *analytic HPLC* using 4.6-mm-i.d. columns. The flow rates applied for the capillary or nano-HPLC have dropped from 1 mL/min used in analytic HPLC to 100–2000 nL/min. These flow rates are fully compatible with the ESI techniques, allowing a splitless coupling of the RP-HPLC with the mass spectrometer (RPLC-MS). The low flow rates in capillary or nano-LC results in a high concentration of the peptide in the effluent. In combination with modern mass spectrometers, allowing the fast data-dependent switching from MS to MS/MS mode, beside the exact molecular masses of the eluting peptides structural features as partial sequences or post-translational modifications can also be determined online. In the following section we describe state-of-the-art LCMS and related techniques with respect to peptides obtained from natural sources or enzymatic digests of proteins or protein complexes.

3.3. STARTING WITH CAPILLARY OR NANO-LCMS

Developments in modern bioanalytic laboratories are always driven by a combination of three requirements: more information in shorter time from less material. While traditional techniques in protein analysis, as Edman degradation or amino acid analysis, have reached their limits, modern mass spectrometry is still in development, driven by these requirements. In 1990 the specifications for a plasma-desorption mass spectrometer were set to 20 pmol of LHRH with an accuracy of 0.1%, which means at 1000 ± 1 amu (atomic mass units) or 1000 ppm. The average time for recording a spectrum was about 20 min, with a fresh ^{252}Cf source, going up to several hours with an old, less intensive source. Today's requirements are a mass accuracy at 50 ppm or even better, meaning 0.05 amu at 1000 m/z. The amount to be applied should be in the low-femtomole (fmol) range and the time to analyze, in case of MALDI-MS, seconds to minutes for one sample. In addition, resolution became an important factor, now ranging from about 10,000 for time-of-flight instruments to over 100,000 for FTICR instruments. These factors have a great impact on the identification of proteins present even in very low amounts [47].

The combination of high mass accuracy, high sensitivity, high resolution, and short analysis times also had an impact on the sample application. Whereas the use of 1- or 0.5-mm-i.d. columns with flow rates between 5 and 50 μL/min were state of the art in 1996 or 1997, now capillary, 0.15–0.3-mm-i.d., or nano, 0.05–0.075-μm-i.d., columns became necessary for the increasing demands. From analytic to microbore HPLC, sensitivity is increased by a factor of about 100, and from microbore to nano-HPLC by another factor of 100, finally resulting in a gain of sensitivity of several thousands. Using HPLC together with modern mass spectrometers, amounts of protein or peptide can be analyzed, which was not possible previously. This brings every modern laboratory to the decision to establish capillary or nanoflow HPLC coupled with mass spectrometry. But even an older mass spectrometer, such as an 8-year-old triple-stage quadrupole mass spectrometer, will gain a dramatic increase in sensitivity when connected online to a capillary or nano-HPLC.

As shown for the nanospray technique, identification of low abundant proteins, such as from silver-stained gels, have been carried out with these old instruments.

The step from microbore HPLC to capillary HPLC is not as big as one would imagine; nevertheless it must be carefully planned. The following will give advises how to set up a capillary or nano-HPLC and couple it to an electrospray mass spectrometer.

3.4. OVERVIEW ON INSTRUMENTATION

Capillary or nano-HPLC involves the use of low flow rates, often no longer easily visible on the column outlet and difficult to determine. This requires an optimized design of the system starting from the pump over the injection valve to the column and over the detector to the electrospray source. Points mentioned in the following sections are the pumps or pump modifications, including the low flow rate, the connectors, the injection valve, the column, the UV cell, and the electrospray source. Aspects such as multidimensional liquid chromatography, peak parking to increase the time to analyze the peptides, and aspects of mass spectrometry are also discussed. Suppliers for hardware components mentioned in the text are examples of new companies, and new developments may appear.

3.5. INSTRUMENT REQUIREMENTS

3.5.1. HPLC Pumps

It is obvious that capillary (0.3- or 0.15-mm-i.d. columns) or nano- (0.075- or 0.05-mm-i.d. columns) LC requires different instrumentation than analytic, microbore (2-mm-i.d. columns) or narrowbore (0.5–1-mm-i.d. columns) (see

TABLE 3.1. Column Diameter, Flow Rates, and Detection Range of Different LC Systems

Flow Range Category	Column ID (mm)	Flow Rate Range (μL/min)	Optimum Flow Rate (μL/min)	Analyte Capacity (gs)
Analytical	4.6 or 4.0	750–1500	1000	10^{-3}–10^{-5}
Narrowbore	2.0	100–1000	200	10^{-5}–10^{-9}
Microbore	1.0	20–200	50	10^{-6}–10^{-10}
Microbore	0.5	5.00–50.0	5.0	10^{-7}–10^{-11}
Capillary	0.3	1.00–20.0	2.0	10^{-8}–10^{-12}
Capillary	0.15–0.20	1.00–10.0	1.0–2.0	10^{-9}–10^{-13}
Nano	0.075 or 0.1	0.10–1.00	0.15–0.5	10^{-10}–10^{-14}
Nano	0.05	0.05–0.50	0.1	$<10^{-12}$

Table 3.1). For use of these columns, a number of compatible systems are available on the market. Flow rates in gradient mode down to 10μL/min are easily obtained by standard double pistons or double syringe pumps. Below 5–10μL/min, these systems fail to produce a reproducible and stable flow in a gradient manner. To create a gradient at low flow rates the precolumn split of the solvent flow has become the standard method. Almost any available analytic or microbore HPLC system can easily be converted into a capillary or nano-LC by using a flow splitter between the pump and the injection valve [48–50]. A flow splitter consists of a tee connector whereby the different backpressures of the separation column and a bypass line at the preset flow rate from the pump the required flow over the separation column is obtained. The bypass line can be constructed from a fused-silica capillary that is cut to the needed length or an outdated analytic or microbore HPLC column. Most of the solvent runs over the bypass line into the waste, for example, at a flow rate of about 150–200nL/min for a 15-cm × 75-μm-i.d. column the initial flow rate is between 15 and 150μL/min, depending on the split rate. The backpressure of the bypass line is generally significantly lower than that of the separation column. Beside in-house-built splitter systems, ready-made systems are also available. They are all based on the same principle. The flow can be determined using a microliter syringe attached to the end of the column via fused-silica tubing used as the injection needle and a Teflon sleeve. The fused-silica injection needle is available as a part for gas chromatography. The volume of liquid in the syringe is measured over a certain time and recalculated for the flow rate per minute. This determination should be made at different stages of the gradient to ensure a stable flow over the whole time.

3.5.2. Injection Valve and Connectors

The low flow rates in the separation path require the absence of any dead volumes that would result in peak broadening and loss of resolution. This requires a special low dead-volume injection valve with a small loop and zero dead-volume connectors.

The injection valve should be specified for low internal volumes, preferably in the nanoleter range. Injection valves are available from several suppliers, as listed in Section 3.11. These injection valves generally work as specified by the manufacturer. Because of the low flow rates, scratches in the rotor will create dead volumes. Therefore the rotor has to be checked from time to time to ensure that no scratches are present.

The connections between the valve and the fused-silica or peek capillaries and the column have to be free of any dead volumes. Using fused-silica capillaries in peek tubings, the cutting of both components must be absolutely planar, and the fused silica has to be neither longer nor shorter than the peek tubing. Any unequality will cause dead volumes, resulting in reduced separation quality and peak broadening.

Usually the column is directly attached to the injection valve with no additional tubing in between. As nanocolumns are generally without a frit on the inlet, a Teflon or glass fiber filter frit is placed in front of the column into the valve port. This prevents insoluble material to reach the columns and helps prolong their lifetimes. The connection is obtained by using standard ferrules that have to fit to the injection valve and either peek or Teflon sleeves to hold the capillary. It is important to mention that the different brands of injection valves require different ferrules, and consulting the manual or catalogs from the supplier helps in these cases. Low-dead-volume or even dead-volume-free connectors are available to link capillaries on the high-pressure side, upstream of the column. In all connections blunt-end-cut capillaries are a prerequisite for high resolution, sharp and symmetric peaks, and high sensitivity. Any dead volume will result in peak broadening and loss of resolution.

Leaks are hard to detect at flow rates below 1 μL/min as no visible liquid leaving the connections can be observed. Therefore it is difficult to give general advice to search for suspected leaks. One approach to identify leaks is the use of a pH indicator stick in case acidic solvent is used, such as with formic acid or TFA. Even very low amounts of liquid can be seen by the change in color. On the other side, if leaks occur upstream of the injection valve, no loss of the sample material will occur; therefore the flow at the outlet should be taken as a criterion for the quality of the solvent flow, and the leak itself as part of the separation pathway, and equalized by the pump flow, which may be slightly higher to compensate for the loss of solvent. Leaks at the injection valve or between the injection valve and the column should be taken more seriously as they may result in sample loss. As the pressure drops dramatically behind the nanocolumn, the connectors used here can be tight-fitting Teflon sleeves, available in all inner diameters compatible with fused-silica capillaries and even for connecting fused-silica capillaries of different outer diameters, such as from 280 to 375 μm. The fused-silica capillary behind the column to the ESI source should be of smaller diameter than the column itself; for a 75-μm-i.d. column, a 20-μm-i.d. capillary fulfills the requirements.

3.5.3. UV Detector

In some cases UV absorbance must be observable. In analysis of very low amounts of peptides, signals from eluting peptides are rarely observed in UV, while the mass spectrometer easily detects and fragments peptides. One reason for recording the UV trace is to control the injection rate and thereby the flow rate. The typical form of the injection signal and its retention time give information about the stability of the flow and the injection. To see even small amounts of peptides, UV detection is usually performed at 205 or 214 nm.

The UV cell has to be constructed in such a way that no additional volume is added to the flow path. Standard UV cells do not fulfill these requirements. This is obtained either by a measurement directly through the fused silica or by a U- or Z-shaped cell with the same inner diameter as in the fused-silica capillary. These cells are commercially available and can be incorporated into a large number of UV, diode array, and fluorescent detectors. Davis et al. have described the different UV cells used for capillary and nano-LC [49].

3.5.4. Electrospray Ionization Source

The standard ESI source requires flow rates between 1 and 20 μL/min to obtain a stable spray. Lower flow rates would not result in a stable spray. To be combined with capillary or nano-LC, a spray source has been developed that shares some similarities with the nanospray ionization source. The most reliable spray is obtained by using laser-pulled fused-silica needles with a tip diameter of 10 μm for 75-μm-i.d. columns and 5 μm for 50-μm i.d. columns. These needles are directly connected to the fused-silica capillary from the column or UV cell using either a low dead-volume tee connector to apply the voltage for ionization directly to the solvent or through a Teflon sleeve.

The tee connector is used to create a contact between a platinum wire and the liquid to transfer the voltage into the solvent flow. The application of the voltage via a tee connector can also be placed upstream before the column without influencing the quality of the separation. In this case the needle is used as the column itself, and the chromatographic material is held by a frit in the tip of the needle.

In the second approach the needle has to be covered with a noble metal to transfer the voltage to the liquid. Two different types of needles are available: one coated distal and one coated up to the tip. Depending on the mass spectrometer used, either one of these work well. In case of the 90° angle of the Micromass Z-spray, the distal coated needle works best while in the case of axially arranged ion sources, the needles with the coated tips show higher performance and better sensitivity. Further, more detailed information on electrospray sources compatible with capillary- or nano-LC is provided by several authors [8,51–58].

The needle has to be mounted into a device that allows a fine adjustment in all three directions. Commercial devices are available or can be built using a micromanipulator. The spray region and the inlet of the mass spectrometer have to be illuminated for better observation. This can be done by small lamps or by a glass-fiber light source.

The quality of the spray has to be controlled in two ways, observed as a stable spray with the Taylor cone and in the recorded mass spectra to avoid water clusters that interfere heavily with the other signals. Orientation of the needle and optimal voltage has to be set individually and may have to be changed after some time as the needle ages and the resistance increases. Some information about optimizing the spray is given by Geromanos et al. [59].

The following general hints may help to create a stable spray for high-sensitivity analysis. If the distance between the tip of the needle and the inlet to the mass spectrometer is below 1 cm, either an off-axis orientation or a 90° angle to the inlet leads to the best results. By observation using a stereomicroscope, a videocamera, or a simple magnifying device, the needle is set to a starting position as mentioned above. The flow is started and without applied voltage, drops should appear at the tip and stay there.

In case of a clotted needle or a clotted connection or tubing, no drops will appear. Clotted needles cannot be cleaned; they have to be discarded. Clotting can occur for a number of reasons. Inserting the fused-silica capillaries into the Teflon connectors may cause the release of small Teflon pieces that move by the solvent into the needle. Pieces of the polyimick covering of the fused silica can result in clotting. Touching the tip of the needle causes either destruction of the needle tip, resulting in an unstable spray, or clotting. As pieces are seldom observed, no time should be wasted to analyze the needle itself if the clotting has been localized here. To localize the clotting, the flow should be observed from the end of the column to the connection to the needle.

If the flow can be observed by formation of drops, the voltage should be increased carefully. Starting with a voltage of 1000–1500 V, the drops should become smaller, with single but still visible drops drawn to the inlet of the mass spectrometer. Further increasing the voltage in 200–300-V steps should result in optimal conditions in the stable spray with a good visible Taylor cone. To observe the spray, the power of light has to be increased often. An optimal spray is usually observed between 2500 and 3800 V for the Micromass Z-spray source, between 1400 and 2400 V for the PE-SCIEX QSTAR source, and about 2000 V for the ThermoFinnigan LCQ source, a significantly higher voltage than used for nanospray ionization. Increasing the voltage over the optimal setting does not result in a more reliable spray, as large drops or a solvent "beam" directly into the orifice may form.

Having obtained a stable spray with low background and lack of water clusters, are can employ a gradient without any sample to control it over the whole solvent composition range.

Depending on the composition of the solvent, the quality of the spray may change dramatically. At low organic percentage the formation of droplets can fail, resulting in formation of large drops on the needletip or in a straight solvent "beam" into the inlet of the mass spectrometer. In both cases very intensive water clusters can be seen in the mass spectra and signals from the compounds are suppressed.

To optimize the sensitivity of the mass spectrometric detection, the use of flow injection under equal flow rates is recommended. Syringe pumps suitable for flow rates of 100–200 nL/min are available. A selected peptide dissolved in the solvent comparable to the elution conditions from the column can be used to optimize all settings.

3.5.5 Solvents

Generally reversed-phase HPLC and electrospray are compatible with respect to the solvents used. Mixtures from water and volatile organic solvents (acetonitrile, methanol, or isopropanol) and a volatile acidic modifier as TFA or formic acid are best suited for the ionization of peptides and proteins by electrospray. Nevertheless, TFA may cause suppression of signals in the mass spectrometer; therefore, instead of TFA, formic acid must be used. Formic acid has two major drawbacks over TFA: a strong absorbance in the 205–214-nm range and a weak influence on the hydrophobicity of the peptides. At 0.1% formic acid in water as solvent A and 0.09% formic acid in acetonitrile as solvent B, the absorbance remains in a range in which signals can still be detected. But under these conditions most chromatographic materials loose their separation efficiency and even very different peptides elute within a short time. Increasing the formic acid concentration in both water and acetonitrile to 0.5% overcomes this problem and increases the quality of the separation. Care has to be taken to exchange the solvents frequently as formic acid degrades. This leads to a loss of separation and the increase of background ions in the mass spectrometric detection. Furthermore, while TFA as the ion-pair reagent increases the hydrophobicity of the peptide, formic acid has a weak influence, resulting in an early elution of the peptide from the column. This further results in a composition of the solvent during the elution with low concentration of organic solvent, creating problems in stability of the spray. Therefore, the setting of the voltage and the orientation of the needle have to be done with respect to the formic acid conditions.

In addition, the low concentration of the organic solvent can cause problems with the reversed-phase material. Some materials work fine with concentrations of organic solvent above 10% in the liquid phase may loose their separation quality completely at concentrations below 10%. This is caused by collapsing of the C18 chains in pure water or with low concentrated organic solvent. The binding of peptides to such collapsed C18 chains does not occur any more, and the column is often destroyed.

Several manufactures have developed pure-water-resistant reversed-phase materials by adding additional groups to the surface of the silicagel to prevent the collapsing. Another approach to overcome the problems is the use of polymer-based materials that have properties similar to those of silica-based C18 materials.

The selection of solvent and reversed-phase material has to be done with respect to the actual problem: this requires the testing of different combinations.

3.5.6. Columns

Separation of peptides is heavily influenced by the quality and selectivity of the column. Capillary and nanocolumns are available from several suppliers, packed with a wide range of different materials, and can even be obtained custom-packed. But commercially available columns may not fulfill the specific needs in all cases. The downscaling of an established separation method with a well-tested phase may result in the nonavailability of an equivalent nanocolumn. This requires the packing of columns in the laboratory by the user. In contrast to microbore and narrowbore columns, capillary and nanocolumns can be made in the lab with the required materials quite easily. Two approaches have been described for packing the fused-silica capillaries, usually with 75 or 50 μm i.d. and 380 μm o.d.:

1. A stainless-steel reservoir is connected to the fused-silica capillary and to a high-pressure syringe pump connected to the other side over a shutoff valve. The reservoir, which can be a stainless-steel capillary, is filled with slurry of the RP phase in an organic solvent, methanol, ethanol, acetonitrile, and isopropanol. The pump runs with the same organic solvent. An initial pressure to 400 psi is built up with the valve between pump and reservoir closed. After reaching the desired pressure, the valve is opened and the column packed.
2. A high-pressure cell and a thick-walled metal bomb are used. The fused-silica capillary enters the cell through a pressure-stable screw. The slurry is applied in a plastic tube and the inlet of the column dips into the slurry. A pressure of about 100 bar using helium is applied and the column is packed by this way. In both cases the progress and the homogeneity of the settling of the matrix can be observed microscopically. After packing the column is washed with organic solvent under the same pressure for about 1 h. Care is taken to avoid rapid release of the pressure, as backflow of the phase may occur, resulting in inhomogeneous packing and loss of separation quality.

The outlet of the column is closed by a frit. The construction of this frit using different approaches has been described in a number of publications. One way is the use of a glass-fiber disk (Whatman GCF) between the end of the column and the connected capillary. Another approach is the sintering of C18 phase into the end of the capillary. Detailed descriptions are given in the literature [49,60,61]. A temporarily frit can also be used. These frits, sold as inline filters with 0.5–2 μm pore size, are then removed after packing of the columns and then the glass-fiber frit is applied. The column does not need an inlet frit, but again care has to be taken to avoid sudden pressure drops during use.

The in-lab packing of columns allows the use of every available phase. The size of the particles must range between 3 and 5 μm, but larger particles can also be used as perfusion materials. As some C18 phase, mainly with 3- or 5-μm particles, are not available commercially in bulk, either a new column or outdated columns, after removal of the first 1–2-cm phase from the top, can be emptied. The phase has to be washed to remove fines that may block the outlet frit. This is done by suspending the phase in the organic solvent and removing the supernatant after settling of the particles for at least 2 times. Extensive treatment of ultrasonic has to be avoided because of partial destruction of the silicagel, resulting in bare silica surface with high peptide binding properties. General guidelines on preparation of capillary- or nanocolumns has been provided by a number of authors [49,62–69].

3.6. MULTIDIMENSIONAL CHROMATOGRAPHY COUPLED TO MASS SPECTROMETRY

LC/MS overcomes the limitations in the analysis of complex mixtures as present in the nanospray technique. But very complex peptide mixtures, as obtained from a natural source as blood [70], urine [71], lymphatic liquid or digests of large proteins [72], or complete protein complexes [73,74], cannot be separated sufficiently in one LC step. Coeluting peptides may avoid the detection of all components or even in the worst case may suppress other peptides that are no longer detected in mass spectrometry. In case of automatic MS/MS only a limited number of peptides can be "sequenced" simultaneously ranked by their intensity; therefore peptides present in low concentrations or low-ionizing peptides are missed or "overlooked" by the mass spectrometer. Several approaches can be taken to overcome these limitations, as reduction of the complexity by taking only peptides with certain properties. One approach is the capturing of cystein containing peptides in the first step and then the identification of these peptides alone by LCMS/MS [75]. One example of such complex mixtures are the peptides present in blood or urine. Within on single RP chromatography only a limited number of peptides can be detected. The method of choice to reduce the complexity that is analyzed by the mass spectrometer is the orthogonal multidimensional chromatography of the complex starting mixture. *Multidimensional chromatography* means the consecutive combination of chromatographic principles based on different specificities [74,76]. Complex samples applied to the first chromatographic separation system fractions are collected either by time or by elution behavior.

These samples are then applied to a second system, based on a different specificity. In most cases this is an RP chromatography directly coupled to a mass spectrometer. As the RP chromatography is based on the hydrophobicity of the peptides, the first separation should be based on an independent parameter, as size, isoelectric point or any kind of affinity. But even applying a different RP chromatography, for example, at a different pH or with an RP material with different selectivity, can be used in the first step.

The system most frequently described in the literature is the combination of ion exchange and RP chromatography. As peptides are easily converted at low pH into positively charged compounds, cation-exchange chromatography has turned out to be the method of choice for the first step in analysis of complex mixtures. Applying the crude, desalted sample onto a preparative cation exchange column can do this first step and either eluting the bound peptides in a salt gradient or with defined pH-steps, as described by Schulz-Knappe, and Richter [77,78]. This is similar to the well-established 2D electrophoresis using the isoelectric point of a protein in the first dimension in the isoelectric focusing and the molecular mass in the SDS polyacrylamide electrophoresis. The cation-exchange/RP chromatography method has been used to analyze the peptides from human blood. Human blood is composed concerning the peptides below 20 kDa of degradation products of larger proteins and biologically active peptides as insulin. To understand the composition of blood may help to understand or diagnose diseases in an early stage. As the concentration of the peptides range from the low femtomole to the nanomole range, one single RP chromatography cannot suffice to identify all components. The approach to analyze blood peptides is discussed in more detail in the chapter by M. Kellmann and M. Juergens in this book. In this procedure seven fractions were collected from a cation exchanger by applying buffers with increasing pH values. These fractions were then analyzed by RP chromatography coupled directly to an electrospray ionization mass spectrometer or off line with a MALDI-MS.

This approach applies the offline combination of the two chromatographic systems. This can be applied if enough starting material is available or if the fractions of the first separation should be used for further processing after identification, such as for purification of peptides from these fractions. The presence of only very small amounts of starting material, the collection of fractions may cause severe loss of peptides due to adsorption on plastic walls. Therefore the direct coupling of the two systems has been developed. As shown for the identification of proteins from functional complexes, based on MS/MS analysis of peptides after tryptic digest, the used of stacked cation exchange and RP materials in one nanoflow column overcomes this sample loss. Ribosomes of a large number of proteins are constructed to form the functional machine for transcription of the mRNA to the protein. Identification of these proteins was performed in the 1980s using a 2D gel system and sequencing of the proteins. In the 1960s ribosome was used to develop a technique of online orthogonal multidimensional chromatography to mass spectrometry to identify the proteins [46,79]. The chromatographic column consists of a tandem stack of a cation-exchange material behind an RP material in 75-μm columns. The outlet of the column is the spray needle itself, and a frit is placed in the tip of the needle. Desalted peptide mixtures after tryptic digestion of the isolated ribosomes are directly injected onto the cation exchanger, then eluted with short KCl gradients onto the RP material and from this are eluted by an acetonitrile gradient. During the gradient elution from the RP material the noneluted peptides from the first step are held back on the cation exchanger. Applying this approach, the very complex mixture of tryptic peptides from the large ribosomal complex can be separated into several subfractions that are further separated by RP chromatography, followed by identification of the peptides by MS/MS. Beside this stacking-column technique the two separation principles can be integrated into one chromatographic system using valves to separate them.

The first system is, as stated above, not limited to cation exchanger. Any kind of separation, independent on the buffer systems, can be used in the first dimension, as long as the final separation is compatible with electrospray ionization. This allows the use of affinity or pseudoaffinity columns to specifically trap peptides for detailed analysis. Examples are given for affinity purification of cystein-containing peptides, for removal of SDS, and for phosphopeptides [80–83].

3.6.1. Peak Parking

One of the limits of LC-MS/MS analysis of complex mixtures is the complexity of coeluting peptides. Modern mass spectrometers are able to fragment up to eight peptides on the basis of their intensity detected within a number of MS scans preset by the user. The problems are obvious, mainly that the time for consecutive fragmentation may not be sufficient to obtain all peptides with an intensity necessary to identify the peptides. In case of more than eight coeluting peptides the remaining ones may get lost, and important information will not be obtained. This limits the LC-MS/MS of complex mixtures. One approach to overcome these problems is the reduction of the complexity of the starting material by off- or online multidimensional chromatography as described above. A more recent technique is the so-called peak parking [84]. "Peak parking" means essentially the interruption of the HPLC run when a peptide to be analyzed is detected by the mass spectrometer. The solvent flow rate is reduced to about 10%, for example, 15 nL/min instead of 150 nL/min of the original flow rate, and the gradient itself is stopped. Using a microvalve, the flow from the column to the spray is interrupted by switching against a closed port. The content of the flow line between the valve and the spray needle is transferred to the spray using very low flow rates (e.g., 20 nL/min). Under conditions used in nanoflow LC the volume in the flow line between valve and spray needle is about the elution volume of a peptide. The length of this flow line has to be adapted for very long connections between the valve and the needle, which may create problems and loss of peptides. The time to analyze peptide mixture is therefore dramatically increased; beside the eight most dominant peptides, less prominent peptides can also be identified. After no more peptides are detected by the mass spectrometer or after a preset time to elute the capillary completely, the valve is switched back and the gradient continued with the normal flow rate.

This technique requires a mass spectrometer capable of controling the HPLC instrument and the microvalve under predefined conditions. The length of the separation for an unknown mixture itself cannot be predicted under these conditions, and UV detection does not give any useful signals.

This technique can be applied only to nanoflow HPLC separation, as under higher flow rates the blocking of the column outlet will cause backpressure that may damage the column or result in leaks.

3.7. DATA-DEPENDENT STATE SWITCHING, DATA-DEPENDENT ACQUISITION, AND INFORMATION-DEPENDENT ACQUISITION (SEVERAL PHRASES FOR ONE METHOD)

While in the beginning of LCMS of protein or peptide spectrometry, only the masses of the eluting compound could be recorded, modern mass spectrometers now offer the possibility of obtaining structural information in a data-dependent manner. Basically this is the automatic switching from one state to another when molecules of predefined masses are detected. In most cases it involves switching from MS mode to MS/MS mode [85–89]. First the mass of the eluting compound is detected, and then, depending on this mass, the instrument turns itself into fragmentation modus by increasing the collision energy and records the MS/MS spectra. The collision cell remains gas-filled at all times, and at low ion energy the molecules pass this area without significant fragmentation. The first quadrupole is used to select precursor ions for fragmentation, and analysis of the fragment ions is carried out in the second quadrupole or in the time-of-flight analyzer. In case of ion trap instruments the molecule to be analyzed is kept in the cell while all other molecules are emitted. Then the frequency is increased to create fragmentation. The fragments themselves are emitted from the cell and their masses determined. Because of the physical limitation of the ion trap, the lower 25% of the mass range are not detected, and immonium ions are lost. The advantage of ion traps over the other mass spectrometers is the short duty cycle, collecting the ions for analysis happens in the 100–200-ms range while one scan in a QqTOF takes about 1 s. For all instruments the user has to predefine the fragmentation energy, which is closely dependent on the mass of the molecule and the charge, by separate experiments. The number of coeluting peptides for one detection–fragmentation period is usually eight or fewer analyzed with respect to intensity.

After a predefined period or when the intensity of the selected precursor ion drops under a preset level, the instrument switches back to MS modus, waiting for the next ion that fulfills the conditions to be fragmented.

Data-dependent mode switching is one of the most powerful techniques for analysis of complex mixtures. In case of proteins that cannot be identified by their peptide mass fingerprint using MALDI-MS techniques, the additional information obtained through the fragmentation pattern helps, even if only one or two peptides are detected, to identify the protein without doubt.

In case of peptides not created by specific digestion of larger proteins, as present in the human blood, the identification by MS/MS has been shown for molecular weights up to 9000. This allows for the identification of circulating peptides without loss of information due to former fragmentation with proteolytic enzymes.

Modern capillary or nano-LC even allows sensitive analysis comparable to the sensitivity of MALDI-MS. The matrix/peptide spot from the MALDI plate itself can be reused for a second analysis. After dissolving the matrix and the peptides in organic/water solution and dilution to a composition that allows binding to the C18 phase, the mixture can be directly applied to the nanocolumn. The use of a desalting/preconcentration column to avoid the transfer of the matrix to the separation column is highly recommended.

3.8. SAMPLE PREPARATION AND APPLICATION

Sample preparation for capillary or nano-LC is equal to the techniques that are applied for standard HPLC. The sample has to be free of particles, and the organic solvent content has to be compatible with binding to the column. Some detergents such as SDS interfere heavily with the binding to the C18 surface of the column and therefore have to be avoided. The major limitation of the capillary or nano-LC is the low amount of sample that can be applied directly onto the column. At flow rates of 100–200 nL, samples in the microliter range would take hours to be loaded onto the column. The use of a capture/desalting precolumn helps reduce the application time and to remove the salt. This small (e.g., 0.3 × 1-mm) column can be used as a sample loop in an injection valve compatible with the low flow rates. The low backpressure of such a small column allows the sample to be injected by a normal microliter syringe. After washing with starting solvent or 0.1% TFA in water, the column is switched into the flow path to the column and the peptides are eluted in the gradient onto the separation column [90]. This procedure requires that the materials in the capture and the separation column be of equal or very similar selectivity. The use of a capture column also allows the high-volume application; samples taken from natural sources do not have to be preconcentrated or desalted before loading on the nanocolumn. This procedure minimizes sample loss.

Samples obtained by microdialysis from living organs can be directly concentrated on the precolumn prior to analysis without any additional step.

In addition to desalting and concentration, a precolumn can also be used to remove contaminants from the sample that interfere with reversed-phase separation, as SDS. Special materials have been developed for this purpose for use in precolumns.

3.9. GRADIENT FORMING

Normal single or double-piston HPLC-pumps having a specified flow rate from the low-microliter to the microliter range cannot be used directly to create a stable and reproducible gradient at flow rates of 100nL–2μL. Even syringe pumps do not fulfill the needs for the low flow rates. The reason for this is that the pump rate of the solvents at the beginning of the gradients is below the flow rates under those at which the pump is able to create a stable flow. The standard procedure for creating stable gradients with flow rates as low as 100nL/min is the use of postpump splitting. Splitters are connectors—connectors with different backpressures on the outlet side. One outlet side is the separation part, consisting of the injection valve, the column, the flow cell of the detector, and the electrospray needle. The other path is constructed from either an outdated column or a fused-silica capillary. The backpressure of this path is lower than that of the separation path. At a given flow rate, which has to be determined for the certain system, most of the solvent runs over the second path into the waste while only a very little flow runs over the separation path. The flow rate over the separation path is easily determined with a microliter syringe connected to the column outlet. For a 75-μm-i.d. column with 15–20cm length, the flow should be in the 150–250nL/min range. The flow rate of the pump has to be in the range in which the pump runs stable, for a standard analytic HPLC system 300–500μL/min, for a pump compatible with narrowbore HPLC 100–200μL/min.

Another approach is the preforming of gradients under standard flow conditions and application of this gradient at low flow rates to the separation column. This approach has been described in detail [50,91–93].

To optimize the gradient and the settings of the mass spectrometer, a tryptic digest of a known protein should be used. They are either available commercially or can easily be prepared in the laboratory. An optimized gradient and instrument should give close to 100% sequence coverage in case of a small protein as hemoglobin. Missing peptides are short di- or tripeptides that were excluded because of their mass range. In case of larger proteins such as albumin, less sequence coverage is observed. Very large proteins such as myosin can be used to optimize the system for very complex mixtures. Another test is the mixing of different proteins and the rate of identification.

A well-tuned system should be able to identify all proteins from a mixture of 10–20 different proteins.

To determine the sensitivity of the mass spectrometric detection, a quantified tryptic digest should be used. Sensitivity below 10–20 fmol in MS/MS mode should be reached with a nanoflow LC and a modern mass spectrometer.

3.10. SUMMARY

Since the late 1990s the direct coupling of capillary or nano-LC with electrospray mass spectrometry has become a widely used technique in many laboratories. Increasing sensitivity on both sides, on the LC and on the mass spectrometers have made this technique an indespensable tool in modern bioanalysis.

3.11. SUPPLIERS FOR CAPILLARY AND NANO-LC EQUIPMENT

The address of the supplier can normally be obtained through the Internet, using the given Web address. Local suppliers may be different from the original address but are usually given on the Web page. The following list may not be complete but will give an overview of the suppliers:

LCPackings/Dionex (*www.dionex.com* or *www.lcpackings.com*): complete capillary and nano-LC systems, including autosampler, column valves, pumps, detectors, columns, connectors, flow splitters, and nano-UV cells.

MiChrom (*www.michrom.com*): complete capillary- and nano-LC systems, including Autosampler, column valves, injectors, pumps, detectors, and columns. Michrom also sells quantified tryptic digests of a large number of proteins.

AGILENT (*www.agilent.com*): complete capillary LC systems, capillary columns.

MicroMass/Waters (*www.micromass.com*): complete capillary and nano-LC equipment.

MicroTech (*www.michrotech.*com): complete capillary LC system, columns.

NewObjective (*www.newobjective.com*): nanospray needles from fused silica, nanocolumns filled or empty or with frits and nanocolumns with direct spray outlet, nanospray source with observation unit, tools.

UpChurch Scientific (*www.upchurch.com*): a wide range of essential HPLC parts as connectors, tubings, and other components. Upchurch is developing a microvalve with pump for peak paking.

PolyMicro (*www.polymicro.com*): fused-silica capillaries.

REFERENCES

1. W. E. Wilson et al., Physalaemin-like immunoreactive peptides from rabbit stomach, *Int. J. Pept. Protein Res.* **28**(1), 58–66 (1986).
2. S. A. Carr et al., Integration of mass spectrometry in analytical biotechnology, *Anal. Chem.* **63**(24), 2802–2824 (1991).
3. P. Roepstorff and J. Fohlman, Proposal for a common nomenclature for sequence ions in mass spectra of peptides (letter), *Biomed. Mass Spectrom.* **11**(11), 601 (1984).
4. M. Mann and M. Wilm, Error-tolerant identification of peptides in sequence databases by peptide sequence tags, *Anal. Chem.* **66**(24), 4390–4399 (1994).
5. M. Muzio et al., FLICE, a novel FADD-homologous ICE/CED-3-like protease, is recruited to the CD95 (Fas/APO-1) death–inducing signaling complex, *Cell* **85**(6), 817–827 (1996).
6. A. Shevchenko, M. Wilm, and M. Mann, Peptide sequencing by mass spectrometry for homology searches and cloning of genes, *J. Protein Chem.* **16**(5), 481–490 (1997).
7. A. Shevchenko et al., Mass spectrometric sequencing of proteins silver-stained polyacrylamide gels, *Anal. Chem.* **68**(5), 850–858 (1996).
8. M. Wilm and M. Mann, Analytical properties of the nanoelectrospray ion source, *Anal. Chem.* **68**(1), 1–8 (1996).
9. P. Huang et al., On-line capillary liquid chromatography tandem mass spectrometry on an ion trap/reflectron time-of-flight mass spectrometer using the sequence tag database search approach for peptide sequencing and protein identification, *J. Am. Soc. Mass Spectrom.* **11**(2), 127–135 (2000).
10. J. Gobom et al., Sample purification and preparation technique based on nanoscale reversed-phase columns for the sensitive analysis of complex peptide mixtures by matrix-assisted laser desorption/ionization mass spectrometry, *J. Mass Spectrom.* **34**(2), 105–116 (1999).
11. P. Suder et al., Determination of nociceptin-orphanin FQ metabolites by capillary LC-MS, *Brain Res. Brain Res. Protoc.* **6**(1–2), 40–46 (2000).
12. R. Burgus et al., [Molecular structure of the hypothalamic hypophysiotropic TRF factor of ovine origin: mass spectrometry demonstration of the PCA-His-Pro-NH2 sequence], *C R Acad. Sci. Hebd. Seances Acad. Sci. D* **269**(19), 1870–1873 (1969).
13. M. Barber et al., Determination of amino acid sequences in oligopeptides by mass spectrometry. II. The structure of peptidolipin NA, *Tetrahedron Lett.* **18**, 1331–1336 (1965).
14. G. H. Fridland and D. M. Desiderio, Profiling of neuropeptides using gradient reversed-phase high-performance liquid chromatography with novel detection methodologies, *J. Chromatogr.* **379**, 251–268 (1986).
15. W. C. Nierman et al., Complete genome sequence of Caulobacter crescentus, *Proc. Natl. Acad. Sci.* (USA) **98**(7), 4136–4141 (2001).
16. J. C. Venter et al., The sequence of the human genome, *Science* **291**(5507), 1304–1351 (2001).
17. S. Broder and J. C. Venter, Whole genomes: The foundation of new biology and medicine, *Curr. Opin. Biotechnol.* **11**(6), 581–585 (2000).

18. S. Broder and J. C. Venter, Sequencing the entire genomes of free-living organisms: The foundation of pharmacology in the new millennium, *Annu. Rev. Pharmacol. Toxicol.* **40**(6), 97–132 (2000).
19. E. W. Myers et al., A whole-genome assembly of *Drosophila*, *Science* **287**(5461), 2196–2204 (2000).
20. H. Tettelin et al., Complete genome sequence of *Neisseria meningitidis* serogroup B strain MC58, *Science* **287**(5459), 1809–1815 (2000).
21. J. C. Venter, *E. coli* sequencing, *Science* **267**(5198), 601 (1995).
22. A. Bouchie, HUPO is underway, *Nat. Biotechnol.* **19**(4), 297 (2001).
23. J. E. Celis et al., Human and mouse proteomic databases: Novel resources in the protein universe, *FEBS Lett.* **430**(1–2), 64–72 (1998).
24. W. P. Blackstock and M. P. Weir, Proteomics: quantitative and physical mapping of cellular proteins, *Trends Biotechnol.* **17**(3), 121–127 (1999).
25. J. A. Loo et al., Application of mass spectrometry for target identification and characterization, *Med. Res. Rev.* **19**(4), 307–319 (1999).
26. A. Gottschalk et al., Identification by mass spectrometry and functional analysis of novel proteins of the yeast [U4/U6.U5] tri-snRNP, *Embo. J.* **18**(16), 4535–4548 (1999).
27. G. Neubauer et al., Identification of the proteins of the yeast U1 small nuclear ribonucleoprotein complex by mass spectrometry, *Proc. Natl. Acad. Sci.* (USA) **94**(2), 385–390 (1997).
28. R. D. Macfarlane and D. F. Torgerson, Californium-252 plasma desorption mass spectroscopy, *Science* **191**(4230), 920–925 (1976).
29. C. M. Whitehouse et al., Electrospray interface for liquid chromatographs and mass spectrometers, *Anal. Chem.* **57**(3), 675–679 (1985).
30. F. Hillenkamp and M. Karas, Mass spectrometry of peptides and proteins by matrix-assisted ultraviolet laser desorption/ionization, *Meth. Enzymol.* **193**, 280–295 (1990).
31. P. T. Kenny and R. Orlando, Tandem mass spectrometric analysis of peptides at the femtomole level, *Anal. Chem.* **64**(8), 957–960 (1992).
32. W. Zhang and B. T. Chait, ProFound: An expert system for protein identification using mass spectrometric peptide mapping information, *Anal. Chem.* **72**(11), 2482–2489 (2000).
33. J. A. Taylor and R. S. Johnson, Sequence database searches via de novo peptide sequencing by tandem mass spectrometry, *Rapid Commun. Mass Spectrom.* **11**(9), 1067–1075 (1997).
34. M. Mann, P. Hojrup, and P. Roepstorff, Use of mass spectrometric molecular weight information to identify proteins in sequence databases, *Biol. Mass Spectrom.* **22**(6), 338–345 (1993).
35. D. N. Perkins et al., Probability-based protein identification by searching sequence databases using mass spectrometry data, *Electrophoresis* **20**(18), 3551–3567 (1999).
36. P. H. O'Farrell, High resolution two-dimensional electrophoresis of proteins, *J. Biol. Chem.* **250**(10), 4007–4021 (1975).
37. J. Klose, Protein mapping by combined isoelectric focusing and electrophoresis of mouse tissues. A novel approach to testing for induced point mutations in mammals, *Humangenetik* **26**(3), 231–243 (1975).

38. P. Berndt, U. Hobohm, and H. Langen, Reliable automatic protein identification from matrix-assisted laser desorption/ionization mass spectrometric peptide fingerprints, *Electrophoresis* **20**(18), 3521–3526 (1999).
39. S. Beranova-Giorgianni and D. M. Desiderio, Mass spectrometry of the human pituitary proteome: Identification of selected proteins, *Rapid Commun. Mass Spectrom.* **14**(3), 161–167 (2000).
40. K. Gevaert and J. Vandekerckhove, Protein identification methods in proteomics, *Electrophoresis* **21**(6), 1145–1154 (2000).
41. S. D. Patterson et al., Mass spectrometric identification of proteins released from mitochondria undergoing permeability transition, *Cell Death Differ.* **7**(2), 137–144 (2000).
42. R. F. Service, Proteomics. Can Celera do it again? (news), *Science* **287**(5461), 2136–2138 (2000).
43. T. P. Conrads et al., Utility of accurate mass tags for proteome-wide protein identification, *Anal. Chem.* **72**(14), 3349–3354 (2000).
44. M. J. Chalmers and S. J. Gaskell, Advances in mass spectrometry for proteome analysis (in process citation), *Curr. Opin. Biotechnol.* **11**(4), 384–390 (2000).
45. D. M. Desiderio et al., High-performance liquid chromatographic and field desorption mass spectrometric measurement of picomole amounts of endogenous neuropeptides in biologic tissue, *J. Chromatogr.* **217**, 437–452 (1981).
46. A. J. Link et al., Direct analysis of protein complexes using mass spectrometry, *Nat. Biotechnol.* **17**(7), 676–682 (1999).
47. M. K. Green, M. V. Johnston, and B. S. Larsen, Mass accuracy and sequence requirements for protein database searching, *Anal. Biochem.* **275**(1), 39–46 (1999).
48. M. P. Balogh and C. C. Stacey, Chromatographic developments in low flow delivery for liquid chromatography-mass spectrometry, *J. Chromatogr.* **562**(1–2), 73–79 (1991).
49. M. T. Davis and T. D. Lee, Analysis of peptide mixtures by capillary high performance liquid chromatography: A practical guide to small-scale separations, *Protein Sci.* **1**(7), 935–944 (1992).
50. A. Ducret et al., A simplified gradient solvent delivery system for capillary liquid chromatography-electrospray ionization mass spectrometry, *Anal. Biochem.* **265**(1), 129–138 (1998).
51. K. W. Fong and T. W. Chan, A novel nonmetallized tip for electrospray mass spectrometry at nanoliter flow rate, *J. Am. Soc. Mass Spectrom.* **10**(1), 72–75 (1999).
52. M. T. Davis et al., A microscale electrospray interface for on-line, capillary liquid chromatography/tandem mass spectrometry of complex peptide mixtures, *Anal. Chem.* **67**(24), 4549–4556 (1995).
53. T. G. Heath and A. B. Giordani, Reversed-phase capillary high-performance liquid chromatography with on-line UV, fluorescence and electrospray ionization mass spectrometric detection in the analysis of peptides and proteins, *J. Chromatogr.* **638**(1), 9–19 (1993).
54. L. Jiang and M. Moini, Development of multi-ESI-sprayer, multi-atmospheric-pressure-inlet mass spectrometry and its application to accurate mass measurement using time-of-flight mass spectrometry (published erratum appears in *Anal. Chem.* **72**(4), 885 (2000)), *Anal. Chem.* **72**(1), 20–24 (2000).

55. A. Lundqvist, J. Pihl, and O. Orwar, Tantalum-filament pulling device for fabrication of submicrometer fused-silica tips (in process citation), *Anal. Chem.* **72**(22), 5740–5743 (2000).
56. R. E. Moore et al., A microscale electrospray interface incorporating a monolithic, poly(styrene-divinylbenzene) support for on-line liquid chromatography/tandem mass spectrometry analysis of peptides and proteins, *Anal. Chem.* **70**(23), 4879–4884 (1998).
57. C. L. Gatlin et al., Protein identification at the low femtomole level from silver-stained gels using a new fritless electrospray interface for liquid chromatography-microspray and nanospray mass spectrometry, *Anal. Biochem.* **263**(1), 93–101 (1998).
58. M. R. Emmett et al., Application of micro-electrospray liquid chromatography techniques to FT-ICR MS to enable high-sensitivity biological analysis, *J. Am. Soc. Mass Spectrom.* **9**(4), 333–340 (1998).
59. S. Geromanos, G. Freckleton, and P. Tempst, Tuning of an electrospray ionization source for maximum peptide-ion transmission into a mass spectrometer, *Anal. Chem.* **72**(4), 777–790 (2000).
60. B. G. Belenkii, Review. Use of laser detectors in capillary liquid chromatography, *J. Chromatogr.* **434**(2), 337–361 (1988).
61. R. M. Caprioli et al., Microbore high-performance liquid chromatography-mass spectrometry for the analysis of proteolytic digests by continuous-flow fast-atom bombardment mass spectrometry, *J. Chromatogr.* **443**, 355–362 (1988).
62. K. K. Unger et al., Synthesis of spherical porous silicas in the micron and submicron size range: Challenges and opportunities for miniaturized high-resolution chromatographic and electrokinetic separations (in process citation), *J. Chromatogr. A* **892**(1–2), 47–55 (2000).
63. G. S. Chirica and V. T. Remcho, A simple procedure for the preparation of fritless columns by entrapping conventional high performance liquid chromatography sorbents (in process citation), *Electrophoresis* **21**(15), 3093–3101 (2000).
64. G. S. Chirica and V. T. Remcho, Fritless capillary columns for HPLC and CEC prepared by immobilizing the stationary phase in an organic polymer matrix (in process citation), *Anal. Chem.* **72**(15), 3605–3610 (2000).
65. N. Ishizuka et al., Performance of a monolithic silica column in a capillary under pressure-driven and electrodriven conditions (in process citation), *Anal. Chem.* **72**(6), 1275–1280 (2000).
66. I. Gusev, X. Huang, and C. Horvath, Capillary columns with in situ formed porous monolithic packing for micro high-performance liquid chromatography and capillary electrochromatography, *J. Chromatogr. A* **855**(1), 273–290 (1999).
67. M. Petro et al., Use of porous pyrolytic carbon for analytical and microscale high-performance liquid chromatographic bioseparations, *J. Chromatogr. B Biomed. Sci. Appl.* **718**(1), 187–192 (1998).
68. J. E. MacNair, K. C. Lewis, and J. W. Jorgenson, Ultrahigh-pressure reversed-phase liquid chromatography in packed capillary columns, *Anal. Chem.* **69**(6), 983–989 (1997).
69. H. Nakayama et al., Capillary column high-performance liquid chromatographic-electrospray ionization triple-stage quadrupole mass spectrometric analysis of

proteins separated by two-dimensional polyacrylamide gel electrophoresis. Appli-cation to cerebellar protein mapping, *J. Chromatogr. A* **730**(1–2), 279–287 (1996).

70. M. Raida et al., Liquid chromatography and electrospray mass spectrometric mapping of peptides from human plasma filtrate, *J. Am. Soc. Mass Spectrom.* **10**(1), 45–54 (1999).
71. G. Heine, M. Raida, and W. G. Forssmann, Mapping of peptides and protein fragments in human urine using liquid chromatography-mass spectrometry, *J. Chromatogr. A* **776**(1), 117–124 (1997).
72. V. Nier et al., Variability in the ratio of mutant to wildtype myosin heavy chain present in the soleus muscle of patients with familial hypertrophic cardiomyopathy. A new approach for the quantification of mutant to wildtype protein, *FEBS Lett.* **461**(3), 246–252 (1999).
73. M. P. Washburn, D. Wolters, and J. R. Yates, Large-scale analysis of the yeast proteome by multidimensional protein identification technology, *Nat. Biotechnol.* **19**(3), 242–247 (2001).
74. K. Yamaguchi, K. von Knoblauch, and A. R. Subramanian, The plastid ribosomal proteins. Identification of all the proteins in the 30 S subunit of an organelle ribosome (chloroplast), *J. Biol. Chem.* **275**(37), 28455–28465 (2000).
75. C. S. Spahr et al., Simplification of complex peptide mixtures for proteomic analysis: Reversible biotinylation of cysteinyl peptides (in process citation), *Electrophoresis* **21**(9), 1635–1650 (2000).
76. M. T. Davis et al., Automated LC-LC-MS-MS platform using binary ion-exchange and gradient reversed-phase chromatography for improved proteomic analyses, *J. Chromatogr. B Biomed. Sci. Appl.* **752**(2), 281–291 (2001).
77. P. Schulz-Knappe et al., Peptide bank generated by large-scale preparation of circulating human peptides, *J. Chromatogr. A* **776**(1), 125–132 (1997).
78. R. Richter et al., Composition of the peptide fraction in human blood plasma: Database of circulating human peptides, *J. Chromatogr. B Biomed. Sci. Appl.* **726**(1–2), 25–35 (1999).
79. W. Tong et al., Identification of proteins in complexes by solid-phase microextraction/multistep elution/capillary electrophoresis/tandem mass spectrometry, *Anal. Chem.* **71**(13), 2270–2278 (1999).
80. J. P. C. Vissers, J. P. Chervet, and J. P. Salzmann, Sodium dodecyl sulphate removal from tryptic digest samples for on-line capillary liquid chromatography/electrospray mass spectrometry, *J. Mass Spectrom.* **31**(9), 1021–1027 (1996).
81. M. Quadroni and P. James, Phosphopeptide analysis, *Exs* **88**, 199–213 (2000).
82. P. Cao and J. T. Stults, Mapping the phosphorylation sites of proteins using on-line immobilized metal affinity chromatography/capillary electrophoresis/electrospray ionization multiple stage tandem mass spectrometry, *Rapid Commun. Mass Spectrom.* **14**(17), 1600–1606 (2000).
83. M. C. Posewitz and P. Tempst, Immobilized gallium(III) affinity chromatography of phosphopeptides, *Anal. Chem.* **71**(14), 2883–2892 (1999).
84. M. T. Davis and T. D. Lee, Rapid protein identification using a microscale electrospray LC/MS system on an ion trap mass spectrometer, *J. Am. Soc. Mass Spectrom.* **9**(3), 194–201 (1998).

85. M. G. Qian, Y. Zhang, and D. M. Lubman, Collision-induced dissociation of multiply charged peptides in an ion-trap storage/reflectron time-of-flight mass spectrometer, *Rapid Commun. Mass Spectrom.* **9**(13), 1275–1282 (1995).

86. T. Natsume et al., Combination of biomolecular interaction analysis and mass spectrometric amino acid sequencing, *Anal. Chem.* **72**(17), 4193–4198 (2000).

87. C. L. Gatlin et al., Automated identification of amino acid sequence variations in proteins by HPLC/microspray tandem mass spectrometry, *Anal. Chem.* **72**(4), 757–763 (2000).

88. M. Carrascal et al., Quantitative electrospray LC-MS and LC-MS/MS in biomedicine, *J. Pharm. Biomed. Anal.* **17**(6–7), 1129–1138 (1998).

89. M. D. McGinley et al., A simplified device for protein identification by microcapillary gradient liquid chromatography-tandem mass spectrometry (in process citation), *Electrophoresis* **21**(9), 1678–1684 (2000).

90. J. Roboz et al., On-line buffer removal and fraction selection in gradient capillary high-performance liquid chromatography prior to electrospray mass spectrometry of peptides and proteins, *Rapid Commun. Mass Spectrom.* **8**(8), 621–626 (1994).

91. M. H. Chen and C. Horvath, Temperature programming and gradient elution in reversed-phase chromatography with packed capillary columns, *J. Chromatogr. A* **788**(1–2), 51–61 (1997).

92. J. E. MacNair, K. D. Patel, and J. W. Jorgenson, Ultrahigh-pressure reversed-phase capillary liquid chromatography: isocratic and gradient elution using columns packed with 1.0-micron particles, *Anal. Chem.* **71**(3), 700–708 (1999).

93. R. S. Plumb et al., The application of fast gradient capillary liquid chromatography/mass spectrometry to the analysis of pharmaceuticals in biofluids, *Rapid Commun. Mass Spectrom.* **13**(10), 865–872 (1999).

CHAPTER 4

Capillary Electrophoresis Coupled to Mass Spectrometry for Peptide and Protein Analysis

MAGNUS WETTERHALL

Department of Analytical Chemistry, Institute of Chemistry, Uppsala University, Uppsala, Sweden

TORBJÖRN JOHNSON

Department of Psychiatry and Neurochemistry, Institute of Clinical Neuroscience, Göteborg University, Sahlgrenska University Hospital, Mölndal, Sweden

JONAS BERGQUIST

Department of Analytical Chemistry, Institute of Chemistry, Uppsala University and Department of Psychiatry and Neurochemistry, Institute of Clinical Neuroscience, Göteborg University, Sahlgrenska University Hospital, Mölndal, Sweden

CONTENTS

Neuropeptides represent the most diverse class of transmitter molecules. It is therefore a true challenge to characterize the role of neuropeptides in cellular communication and regulation. Recent published studies indicate that

Mass Spectrometry and Hyphenated Techniques in Neuropeptide Research, Edited by Jerzy Silberring and Rolf Ekman.
ISBN 0-471-35493-7

neuropeptides may act not only as autocrine, paracrine, or endocrine transmitters but also directly as transcription regulators via internalization and nuclear localization. Therefore we feel that direct analysis of the molecular environments in cellular systems as well as subcellular organelles is a new and interesting area for which sophisticated separation and detection methods are of great need. The characterization of unknown molecules—peptides—isolated by, for example, capillary electrophoresis (CE), requires a detector that can provide specific chemical information of the analytes. This is exactly what mass spectrometry can offer. Furthermore, the power of combining CE and mass spectrometry (MS) is that it provides a two-dimensional separation both in electrophoretic mobility and molecular weight. This chapter briefly explains the basic principles of capillary electrophoresis for peptide and protein separation, and exemplifies the recent use of mass spectrometry as detector for this group of substances.

4.1. CAPILLARY ELECTROPHORESIS

Capillary electrophoresis (CE) or the most common form of CE, capillary-zone electrophoresis (CZE), is a widely used analytical technique that effectively separates substances by their electrophoretic mobility. Several comprehensive textbooks have been published on this topic [1,2]. Areas where CE has been applied for protein and peptide separation and analysis are summarized in Table 4.1. Some benefits of CE are its high separation efficiency and low sample consumption. Briefly, the system consists of a buffer-filled narrow fused-silica capillary immersed into buffer vials, and a high voltage (5–60 kV) is applied across the capillary. The inner diameter of the capillary is usually between 5 and 100 μm, and the outer diameter is between 200 and 400 μm. Figure 4.1 shows the general setup for capillary electrophoresis.

The applied electric field will cause a migration of the ionic species in the capillary. This migration is called the *electrophoretic mobility* (μe). The cationic (positively charged) substances will move toward the cathode (– electrode), and the anionic (negatively charged) substances will move toward the anode (+ electrode). Neutral substances will be unaffected by the imposed electrical field. The rate of the electrophoretic mobility is dependent on the analyte's charge-to-size ratio. Small and highly charged analytes will migrate faster than will large analytes with a low charge. There is a second movement in the capillary called the *electroosmotic flow* (EOF) that is dependent on the buffer and the inner surface, of the capillary. The inner surface of the capillary will attract buffer ions with opposite charge to the surface, yielding an inner static layer (Stern layer) and an outer diffuse layer (outer Helmholtz plane). The buffer ions will move toward their counterelectrode, pulling the buffer solution and all analytes in that direction. In summary, the electrophoretic mobility will give rise to separation of ionic substances and the EOF will most often cause a net movement of all substances toward the detection end of the capillary, even if a reversed EOF may be used to increase the peak capacity of the separation.

TABLE 4.1. Areas Where CE has been Applied for Protein and Peptide Separation and Analysis

Area	Applications
Binding Studies	Antigen-antibody affinity analysis
	DNA binding peptides and proteins
	Drug-protein interactions
	Metal binding peptides and proteins
Clinical and Diagnostic Studies	Aminoacid analysis
	Bodyfluid assays
	Enzyme activity
	Neuropeptide analysis
	Protein variants, e.g. cytokines and serum proteins
Mobility Studies	Models for prediction of separation
Preparative Purification	Capillary gel electrophoresis
	Preparation prior to e.g. MS analysis
	Tryptic peptide digest purification
Process Monitoring	Derivatization control
	Modification monitoring, e.g. reduction
	Quantitation
	Tryptic peptide digest control
Purity Control	Product screening
	Quality control of process
	Quantitation and identification of contaminants
Stability Studies	Environmental factors
	Kinetics
	pH
	Temperature
Structural Studies	Higher order structures of large proteins
	Hyphenated techniques, e.g. HPLC-CE, CE-MS
	Peptide mapping
	Polymerization
	Post-translational modifications

The use of narrowbore capillaries allows injection of samples in the femtoliter (10^{-15} L) range. This makes CE an ideal method for analysis and separation of biological samples where the amount of sample often is limited. However, the analysis of small sample volumes requires efficient detection methods. The most general modes of detection are UV detection, electrochemical detection, and laser-induced fluorescence detection (LIF). The UV absorption method has poor sensitivity and requires sample concentrations in the micromolar (10^{-6} M) range. LIF detection is extremely sensitive, and molecules in the femtomolar (10^{-15} M) range can be detected. One drawback is the requirement of sample derivatization prior to detection. More recently successful attempts have been made to couple CE and mass spectrometry.

Mass spectrometry is a very powerful tool to measure molecular weight and to structurally characterize various substances. Since the mid-1990s there has

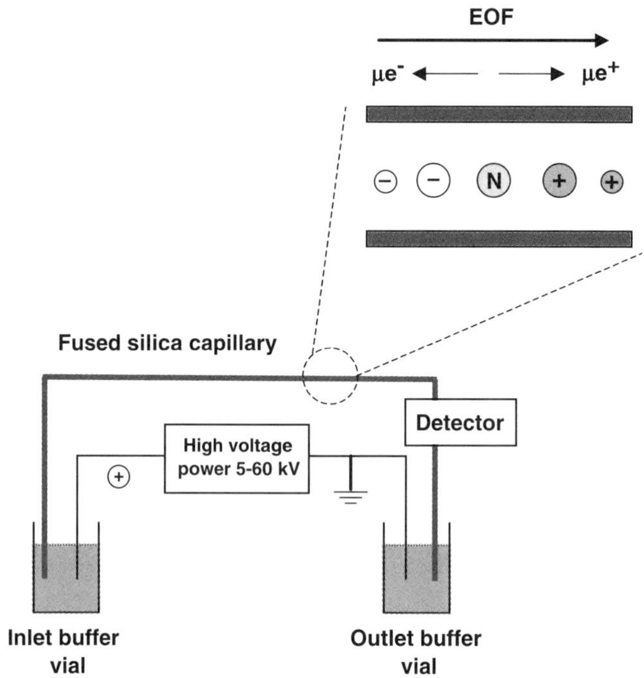

Figure 4.1. A schematic figure for the general setup of capillary electrophoresis. The enlarged section shows the direction of the electrophoretic mobilities (μe) and the electroosmotic flow (EOF) in a bare fused-silica capillary.

been an accelerating growth in application of mass spectrometry to biological analysis. MS can be used to determine substances in a broad mass range and implements the analysis of proteins and peptides in complex mixtures [3].

The coupling of CE to different MS ionization modes requires that the CE buffer be compatible with the MS requirements. Standard for a useful CE buffer is good buffering capacity, low conductivity, and detector compatibility (e.g., good UV transparency). The choice of buffer ions, the ionic strength, and the pH must be optimized for each specific separation. With untreated fused-silica capillaries, the use of high-pH buffers generally produces fast separations due to high EOF. At low pH, peptides migrate primarily on the basis of their m/z ratio, since the EOF is diminished. Buffers based on sodium phosphate, borate, citrate, acetate, or low-conductivity zwitterionic "Good's" buffers [4–6] are often used for separation of peptides and proteins. Typical buffer concentrations are 10–200 mM (pH range ~2–11). The optimum buffer pH depends strongly on the pI values of the peptides. For acidic peptides the neutral pH region is preferable, while basic and neutral peptides are best separated in the low pH region. Many of the separation strategies used for peptides also work for proteins. However, because they are more complex, proteins often require coated capillaries and modified conditions due to adsorption effects.

High salt concentration disturbs the MS analysis. The use of volatile buffers such as ammonium acetate, bicarbonate, formate, or small organic acids and bases can solve this problem. The addition of an organic modifier often increases the stability and performance of the electrospray ionization (ESI) process. Another issue to consider is the distribution of sample from the capillary outlet to the MS analyzer. Dilution of sample and noncocrystallization additives in the matrix-assisted laser desorption/ionization (MALDI) matrix will impair the MS analysis.

4.1.1. Capillary Electrophoresis Coupled to Electrospray Ionization Mass Spectrometry

Electrospray ionization (ESI) [7] is the most common ionization technique for MS analysis of liquid samples. Separation techniques such as capillary electrophoresis, capillary electrochromatography (CEC), and liquid chromatography (LC) of samples with clinical and biomedical origin are performed in liquids. ESI ideally suits the online coupling of these separation techniques to MS.

Some benefits of ESI are that it is a "soft" ionization technique in the sense that it yields no fragmentation (cleavage) products and that the analytes often are multiply charged. The multiple charge effect produces ions with relatively low m/z (often below 2000 amu). This is essential when analyzing large biomolecules, which often have a molecular weight of ≥10,000 Da. Equally important is the fact that the multiple charge effect yields a more accurate original molecular mass determination of the analytes. It should be added that ESI is an atmospheric pressure ionization technique. Other techniques such as fast-atom bombardment (FAB) [8] or MALDI can perform liquid ionization, but only under vacuum conditions. These techniques therefore have low compatibility with organic solvents that are used in chromatographic or electrophoretic separations. There is much debate and research regarding the mechanism of ESI [9–11]. The process for positive ESI is shown in Fig. 4.2.

A short description of the ESI mechanism is given here without going into too much detail. In ESI, a high voltage is commonly applied to the ESI capillary through which a sample solution is emerging. The counter electrode (inlet of the MS) is often grounded. Ionic species in the emerging sample solution undergo electrophoretic movement, moving toward the electrode with the opposite potential in comparison to their charge. This will give rise to a Taylor cone [12], where the counterbalancing force is the surface tension of the solution. If the applied electric field is strong enough, the Taylor cone will yield a Taylor jet, which in turn breaks into droplets that contain (in the case of positive ESI) positive ions. As the droplets move toward the counterelectrode, solvent will evaporate from the droplets, which causes an increase in surface charge density. Eventually the *Rayleigh stability limit* will be reached, which is the point when the force of electrostatic repulsion between like charges on the surface of the droplets becomes equal to the surface tension force holding the droplets together. At this point the droplets will undergo *Coulomb fission*

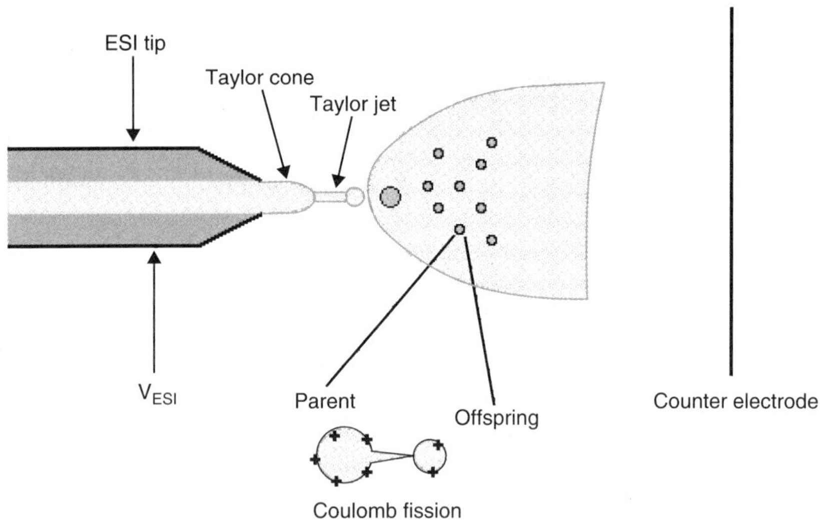

Figure 4.2. A schematic figure for the production of ions in electrospray ionization. The enlarged section shows the event of Coulomb fission.

and split into smaller "offspring" droplets. The droplet fission process is repeated and eventually the ions will be transferred to the gas phase.

The ESI process behaves as a constant current electrochemical cell [7], and thus it is affected by the concentration of the analytes and of the background electrolytes. In complex mixtures, the presence of a high amount of an easily ionized substance might suppress the signal of another analyte, and thus the analysis will be biased. To avoid this problem, it is often necessary to incorporate a separation step prior to ESI.

4.1.2. Interfacing Capillary Electrophoresis and Mass Spectrometry (CE-MS)

Several approaches have been designed regarding the online coupling of CE-MS via an ESI interface. The approaches include the use of (1) sheath flow interface [13] (2) liquid junction interface [14], and (3) sheathless interface [15,16]. In this section the features of these interfaces along with their advantages, disadvantages, and requirements, are briefly outlined.

4.1.2.1. Sheath Flow Interface. The sheath flow interface (Fig. 4.3) is the most widely used method for interfacing CE with ESI. In this setup the capillary outlet is introduced through a narrow metal tube, which at a constant rate delivers a sheath liquid to the end of the separation capillary. As the sheath liquid flows through the tube it mixes with the separation buffer that elutes from the CE capillary, thus providing the electrical contact between the CE buffer and the ESI needle and closing the CE circuit. In some setups a sheath gas may be added to assist the electrospray formation.

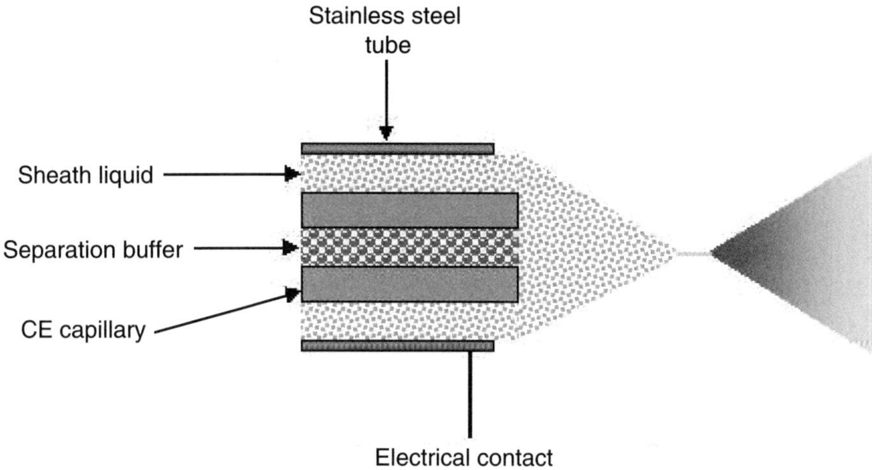

Figure 4.3. A schematic representation of the sheath flow interface. The eluting separation buffer mixes with the sheath liquid at the outlet end of the capillary thus providing the electrical contact for electrospray ionization.

The flow rate of the sheath liquid is typically in the μL/min (microliter per minute) range, whereas the EOF is in the nL/min (nanoliter per minute) range. The sheath liquid, due to the higher flow rate, dominates the ESI process. This results in relaxed limitations on the CE buffer and a higher salt content in samples, and less volatile buffer ions may be used. On the other hand, it implies requirements on the sheath liquid used. The sheath liquid should typically contain 60–80% organic solvent, modified with 1–3% acid in deionized water. Both the sheath liquid and the separation buffer should be filtered and degassed prior to analysis.

Although the addition of a sheath liquid may increase the stability of the ESI, it may also have negative effects on the analysis. The sheath liquid may increase the background noise [17]; hence the need for an off-axis spray setup in respect to the inlet of the mass spectrometer. The sheath liquid also acts as the outlet buffer reservoir for CE, and it will have a composition different from that in the inlet buffer reservoir. This can cause shifts in the migration order of the analytes and backmigration of sheath liquid counterions into the CE capillary [18]. The sheath liquid will also dilute the eluting analyte concentrations, thus causing lower sensitivity. Another cause of lower sensitivity is ion suppression of the analyte ions due to the fact that the sheath liquid ions will compete in the ESI process.

4.1.2.2. Liquid Junction Interface. In the liquid junction interface (Fig. 4.4) a liquid reservoir surrounding the junction of the separation capillary and a transfer capillary provides the ESI contact. The outlet end of the separation capillary and the inlet end of an ESI needle are positioned opposite each other with a gap of 10–20 μm, allowing makeup liquid from the reservoir to be drawn into the ESI needle.

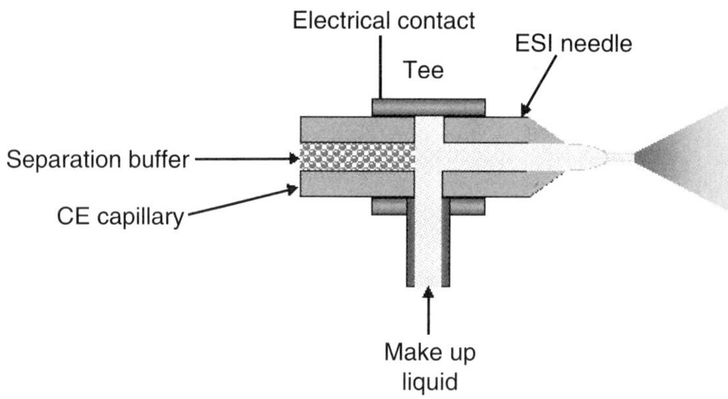

Figure 4.4. A schematic figure of the liquid junction interface. The eluting separation buffer mixes with a makeup liquid in a metal tee, where the electrical contact for electrospray ionization is established.

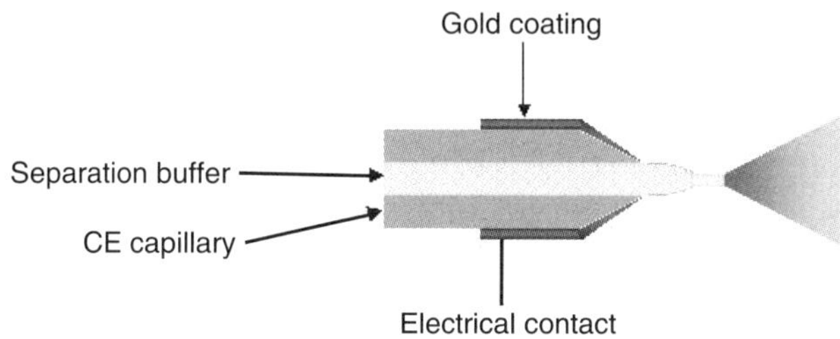

Figure 4.5. A schematic figure of the sheathless interface. Electrical contact is established at the tip of the separation capillary, where the separation buffer is eluting, thus resulting in electrospray ionization.

The liquid junction interface is seldom used because of difficulties in reproducible alignment of the capillaries and similar drawbacks suffered with the sheath flow interface. Problem may also arise with salt crystallization in the interface gap region.

4.1.2.3. Sheathless Interface. In the sheathless interface (Fig. 4.5) the outlet end of the capillary is either made of metal or tapered and metal-coated, thus providing the electrical contact between the separation buffer and the ESI potential. The tapering of the capillary can be accomplished by either hydrofluoric acid solution [17], a micropipette puller [19] (heat), or mechanical grinding [20]. The metal applied on the tapered tip is usually gold or in some cases platinum or silver. Different solutions for the metal contact have been described [19–23]. One alternative sheathless interface is the "upstream" interface. In this setup a small microhole is drilled near the outlet end of the capillary. The hole is then covered with gold epoxy [24] or a small platinum electrode is inserted [25] to provide the electrical contact for the ESI.

The sheathless interface implies stricter limitations regarding the separation buffer and salt contents in the sample. Volatile buffers with low surface tension and low conductivity are preferred [26–28]. A coating of the inner surface of the capillary wall is also often necessary [29–31] in order to reduce analyte–wall interactions and to produce a stable and, often less pH-dependent EOF. The inner coating should produce a stable negative charge or reverse the polarity of the inner capillary wall, thus reversing the EOF instead of eliminating it.

The advantage of a sheathless interface is higher sensitivity, due to no dilution of the eluting analytes. Charge-state distribution shifts are avoided and pure aqueous systems may be electrosprayed, thus facilitating the analysis of fragile and noncovalent bound analytes. The drawbacks of the system are the need of internally coated CE capillaries and a limited lifetime of the interface. A "plugging" of the capillary, rather than metal coating deterioration, often ends the lifetime of the interface. Filtering buffers and samples prolong the lifetime of the CE capillary.

4.1.2.4. Peptide and Protein Analysis by CE-ESI-MS. The combination of CE and ESI-MS is likely to become one of the most important improvements for proteomic research. CE-ESI-MS not only allows highly accurate determinations of molecular masses of peptides, proteins, and other biomolecules. It also provides important structural data on amino acid sequence, the sites of posttranslational modifications and peptide mapping and facilitates the analysis of noncovalent interactions of peptides and proteins.

4.1.2.5. Considerations When Analyzing Complex Samples. There are numerous parameters to take into account when analyzing complex biological samples. The substances of interest are often abundant at subpicomolar (10^{-12} M) concentrations in a complex sample matrix, which will disturb the analysis. Thus there is often a need for preconcentration of analytes and sample cleanup prior to CE-ESI-MS analysis. This is often conducted offline, which is time-consuming. The use of an online method will speed up the process [32–35]. The choice of mass analyzer has an impact on the CE-ESI-MS analysis. The peaks in CE have a very high efficiency; the typical width of a peak is a couple of seconds or less. The time-of-flight (TOF) mass analyzer has a rapid full mass range scanning speed, making it ideally suitable for CE separations. The use of a pressure-assisted CE-ESI-TOFMS analysis has shown significantly reduced analysis time without compromising the overall separation. It has, on the contrary, been shown to improve ESI and CE performance [36,37]. If the mass analyzer has a relative slow scanning speed (quadruple or Fourier transform ion cyclotron resonance mass analyzer) or if the sample is of a very complex nature, the reduced elution speed detection (RES) [38] might be an attractive alternative. In RES detection the applied CE voltage is decreased just prior to elution of the first analyte. The decreased

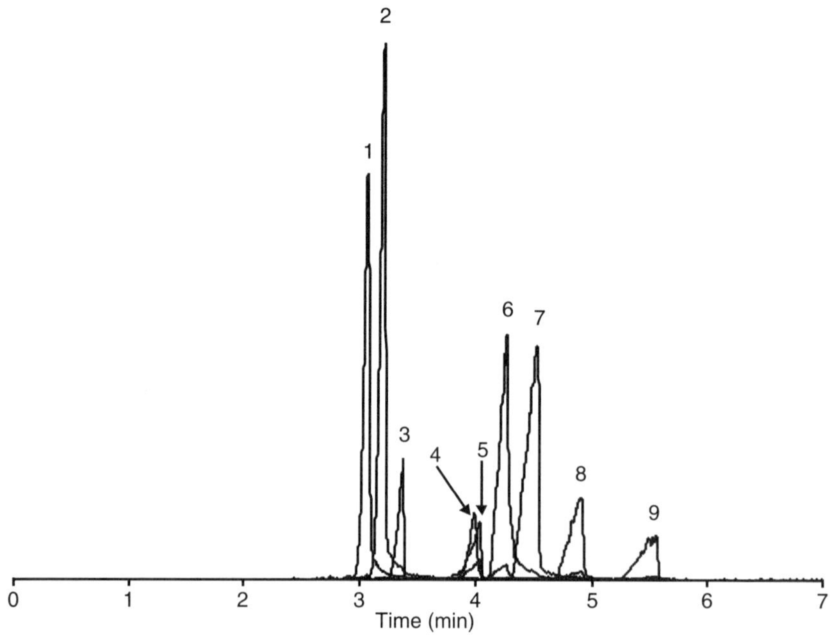

Figure 4.6. A CE-ESI-OTOFMS separation of nine neuropeptides (250 pg of each). The separation was performed in a 40-cm, 50-μm-i.d. MAPTAC [30]-coated sheathless ESI capillary [22]. The nine neuropeptides analyzed were 1 Met–enkephalin (573.23 u), 2 Leu–enkephalin (555.28 u), 3 oxytocin (1006.44 u), 4 bombesin (1619.82 u), 5 bradykinin fragment 1–5 (572.32 u), 6 luteinizing hormone releasing hormone (1181.58 u), 7 [Arg^8]–vasopressin (1083.44 u), 8 bradykinin (1059.57 u), and 9 substance P (1346.74 u). All masses are given as monoisotopic mass-to-charge ratio m/z.

CE voltage will reduce the EOF, thus allowing more scans per peak, and will reduce ion suppression effects.

4.1.2.6. Applications of CE-ESI-MS to Peptide and Protein Analysis.

The application of CE-ESI-MS in proteomic research has seen a breakthrough [39–41]. The rapid full mass scanning TOF analyzer has shown promising features when analyzing complex biological samples with CE-ESI-MS. In Fig. 4.6 a CE-ESI-OTOFMS separation of nine neuropeptides is shown. Li et al. [42] and Wu et al. [43] have shown the advantages of ion trap storage prior to TOF analysis. Determination of mutation sites in hemoglobin variants and identification of bovine cytochrome C and β-lactoglobulin was performed by analysis of tryptically digested samples. Cao and Moini have shown the use of CE-ESI-TOFMS for the identification of proteins in tryptic digests [44] and to detect α and β chains of hemoglobin in a single intact erythrocyte [45]. Ensing et al. [46] has demonstrated CE-ESI-MS separation and analysis of two fragments and their degradation products originating from neuropeptide Y.

Despite the fact that the Fourier transform ion cyclotron resonance mass analyzer (FTICR) has relatively slow scanning speed (~1–10 scans per

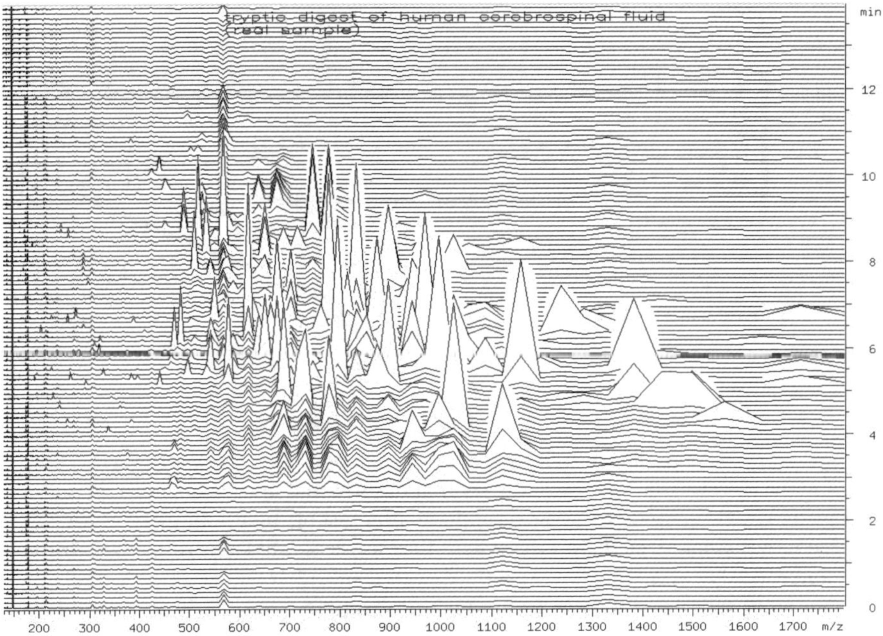

Figure 4.7. A CE-ESI-FTICRMS separation of a tryptically digested human cerebrospinal fluid sample. The separation was performed in a 97-cm, 25-μm-i.d. MAPTAC [30]-coated sheathless ESI capillary [22].

second), it is still very useful for CE-ESI-MS analysis of complex samples due to the high mass resolving power [47]. Valaskovic et al. have demonstrated sequence specific identification of carbonic anhydrase at attomole levels (10^{-18} mol) in erythrocytes using CE-ESI-FTICRMS analysis [48]. Hofstadler et al. have shown the use of CE-ESI-FTICRMS to characterize the α and β chains of hemoglobin in single human erythrocyte cells [49].

Although CE-MS analysis of peptide and protein samples is rapid and relatively straightforward, the analysis requires extensive data processing, which is often the most time-consuming and complicated step in the analysis. Figure 4.7 shows an electropherogram obtained from a CE-ESI-FTICRMS analysis of tryptically digested human cerebrospinal fluid (CSF). As can be seen, the electropherogram is very complex. The processed data must be matched to protein databases in order to obtain sequence coverage and identification of the abundant proteins in the sample. The characterization of proteins requires accurate mass measurements in order to obtain correct identifications.

4.2. MATRIX-ASSISTED LASER DESORPTION/IONIZATION MASS SPECTROMETRY (MALDI-MS)

The MALDI mass spectrometer consists of two main parts: an ion source and a mass analyzer. In the ion source the molecules are desorbed and ionized. In

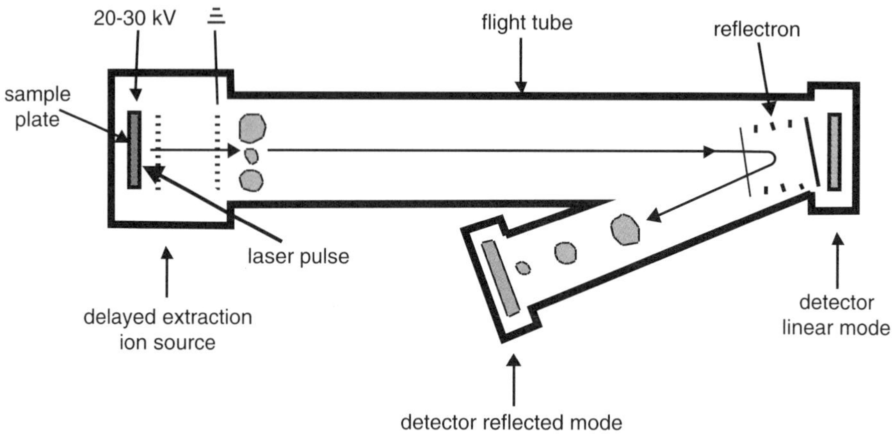

Figure 4.8. A schematic figure of a MALDI-TOF mass spectrometer. In the ion source the sample is irradiated with laser pulses, that desorbs and ionizes the molecules. In the delayed extraction mode a voltage is applied after a short time, and the ions are accelerated into the field-free flight tube, where they are separated according to their mass-to-charge (m/z) ratio. The ions can be detected in either linear or reflected mode. Detection in reflected mode gives the highest resolution, but the linear mode is normally more sensitive.

the mass analyzer ions are separated according to their mass-to-charge (m/z) ratio, detected and recorded (Fig. 4.8).

The analysis of a sample in a MALDI requires that the molecules be electrically charged (ionized) and in the gas phase. Heating a sample to get it into the gas phase is easily accomplished. However, many analytes, such as peptides and proteins, are decomposed on heating. Therefore a "soft" ionization/desorption technique is required to produce intact gaseous ions. MALDI, like electrospray ionization (ESI), is a "soft" ionization technique that is widely used for MS analysis of peptides and proteins [50,51]. In MALDI-MS, the analyte is mixed with a matrix solution and deposited on a sample plate. During evaporation a cocrystallization takes place where the analyte is incorporated in the solid matrix. The matrix consists of small UV-absorbing molecules, and often an organic acid, such as α-cyano-4-hydroxycinnamic acid (CHCA) or 2,5-dihydroxybenzoic acid (DHB). The sample is irradiated with laser pulses that are absorbed by the matrix molecules. Rapid heating causes an expansion of the matrix and transfers the analytes into gas phase, where the analyte molecules are ionized by the excited matrix.

MALDI-MS is a rapid method that facilitates molecular weight determination in a large mass range. Important features of MALDI-MS are its excellent capacity to analyze complex mixtures of proteins and peptides and its relative tolerance toward salts, buffers, lipids, and other species that are present in biological samples [52]. In contrast to ESI, MALDI creates primarily singly charged ions, making the interpretation easy with a one-to-one correspondence between ions in the mass spectrum and the peptides or proteins in the mixture.

The MALDI ion source requires a mass analyzer that is compatible with pulsed ionization techniques and is therefore often coupled to a TOF mass analyzer. Mass-to-charge ratios are determined by measuring the time it takes for ions to move from the ion source through a field-free flight tube to a detector. Given a constant accelerating voltage, the flight time for an ion is related to its m/z ratio. To correct for initial different kinetic energies and velocities of desorbed ions, the MALDI-TOF-MS is equipped with delayed extraction and a reflectron, resulting in significantly improved mass resolution.

State-of-the-art MALDI-TOF instruments provide sensitive detection (low femtomole), accurate mass determination (<50 ppm error), and the possibility to obtain structural information of analyte molecules by post-source-decay (PSD) fragment ion analysis [53].

Although MALDI-TOF-MS has a good capacity to analyze complex biological samples, great improvement in sensitivity and sample components detection can be achieved by implementing a separation step prior to MALDI-MS analysis. When ions are formed in the ion source, there is always a competition of charges among the analyte molecules. Molecules with low affinity for charges and molecules of low abundance will be differentiated in the ionization process (ion suppression). A separation step reduces the amount of components that are introduced into the mass spectrometer at each moment, and thereby reduces ion suppression. Capillary electrophoresis (CE) is a high-resolving separation technique, suitable for the analysis of complex mixtures and small sample amounts [54,55]. The combination of the two techniques provides a powerful multidimensional analytic tool for analysis of complex peptide mixtures [41,56]. CE coupled to MALDI-MS has been used for peptide [57–59], glycopeptide [60], and protein [57,61–64] analysis.

The sample amounts required for MALDI-TOFMS matches the amounts typically analyzed by CE (femtomole–picomole). After initial analysis, a sample can be reanalyzed, since only a small portion of the crystalline MALDI sample is ablated by the desorption laser during spectra acquisition. The sample can subsequently be analyzed in MS/MS mode, after on-probe chemical reactions [65,66], or even recovered for other experiments.

4.2.1. Offline CE-MALDI-TOF

The simplest method to combine the separation power of CE with MALDI-MS detection involves offline fraction collection. The offline approach allows the CE separation and MALDI-MS to function and operate as two independent systems. As a result, each system may be individually optimized. The CE fractions can be collected in vials or deposited directly onto MALDI target plates [56,67].

Problems associated with interfacing CE and MALDI include the difficulty of collecting the small volumes of CE effluent, without disturbing the separa-

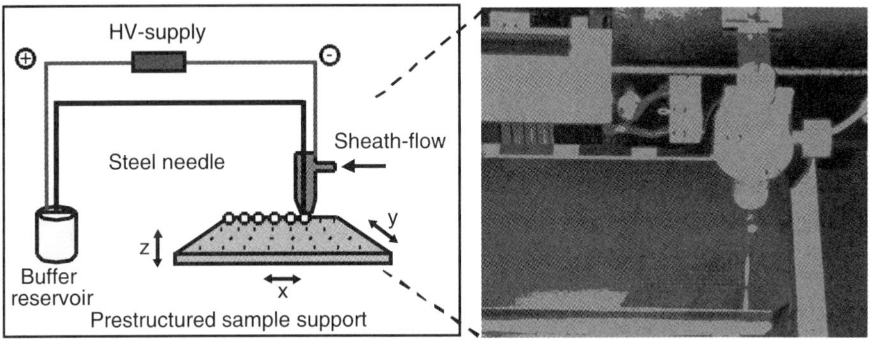

Figure 4.9. Experimental setup of the CE-MALDI-MS interface. The separation capillary enters the T connection from the top and exits at the outlet of the steel needle (bottom). CE effluent is transported onto the MALDI sample support by a liquid sheath flow supplied through the Teflon tubing to the right. At a set time interval, the sample support is lowered and moved to the next position. (Modified from Johnson et al. [69].)

tion. Furthermore, taking full advantage of the high resolving power of CE requires mass analysis of a large number of fractions, which is laborious if done manually. In another study [68], a new type of prestructured MALDI sample support was reported. The sample support is coated with a thin layer of hydrophobic Teflon that carries an array of small hydrophilic spots, that is, sample anchors. All the aqueous sample droplets (0.5–2 μL) make contact with one anchor. Because of the strong solvent repellent nature of the Teflon surface, the sample is, after solvent evaporation, directed exclusively onto the anchor, resulting in sample concentration. This gives significantly improved detection sensitivity. Furthermore, the hydrophilic anchors direct sample crystallization to a well-defined array, thus making efficient automated sample analysis possible.

Our group has developed a novel interface for CE-MALDI offline coupling, taking advantage of the properties of prestructured sample supports [69]. Separated sample components are fractionated onto a prestructured sample support by the use of a liquid sheath flow (Fig. 4.9). The sheath flow dilutes the analytes but ensures efficient removal of separated analytes from the capillary outlet. Fractionated analytes are then again concentrated on the sample support, allowing sensitive MALDI-TOFMS analysis. Prior to fraction collection, MALDI matrix solution is deposited onto the sample support, along with two standard peptides, which are later used for internal spectral calibration. The ability to automatically mass-analyze CE fractions in short time facilitates analysis of a large number of fractions, with the result that the number of sample components per fraction is highly reduced. Signal suppression, which hampers MALDI-TOFMS analysis of complex mixtures, is thereby reduced. The performance of the method was evaluated with a mixture of tryptic digests of proteins from a human fetal brain cDNA expression library (Fig. 4.10). The total number of peptides identified was increased from 47 to 211 when the

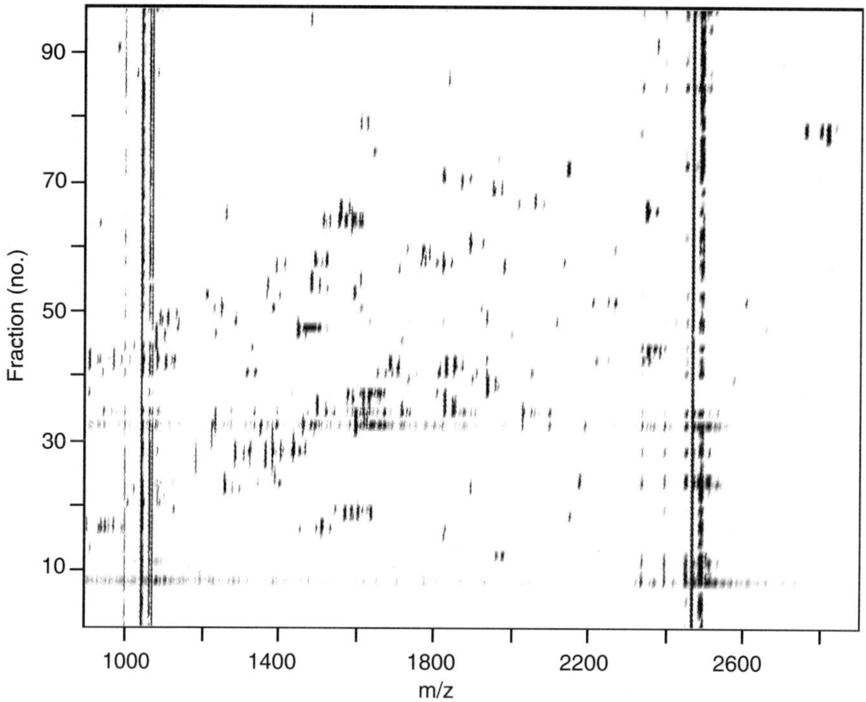

Figure 4.10. A 2D plot of CE-MALDI-MS data obtained from the analysis of 50 nL of a mixture of protein tryptic digests. The vertical lines in the figure result from the signals of the internal calibrants, angiotensin II and ACTH clip 18–39, present in all fractions. (Modified from Johnson et al. [69].)

CE-MALDI interface was used compared to direct MALDI-MS analysis. Sequence coverage with CE-MALDI was in the 25–60% range for the different proteins, corresponding to an increase of 1.3–4.9 times relative to MALDI-MS of the crude mixture. In the analyses of a standard peptide mixture, analytes of low femtomole amounts were detected and no noticeable carryover between fractions was discovered. Furthermore, isolation of precursor ions by a timed ion selector for post-source-decay (PSD) analysis was performed. Besides PSD analysis, this offline approach allows reinvestigation of the collected fractions by other methods following the mass spectrometric analysis.

4.2.2. Direct CE-MALDI-TOF

A direct offline coupling of CE with MALDI has been reported by the Caprioli group [57]. Sample was deposited continuously onto a moving target. The target consisted of a cellulose membrane coated with MALDI matrix. After CE eluent deposition and evaporation, the target was inserted into a mass spectrometer and laser scanning of the sample track monitored the entire CE separation. Mass spectra were taken every 250 μm. The reported sensitivity for neurotensin and cytochrome C was in the attomole range.

4.2.3. Online CE-MALDI-TOF

The on-line coupling of CE with MALDI detection involves continuous introduction of CE eluent and matrix into the mass spectrometer. Because of the difficulties in transferring liquids into the low pressure of the mass spectrometer, MALDI has been used primarily for the analysis of solid samples, and the technique cannot be readily combined with solution-based separation methods. Online coupling requires continuous solvent evaporation, regeneration of the interface, and absence of memory effects. Several attempts have been made to develop online MALDI analysis of liquid samples. One approach use continuous flow through a capillary that transports the analyte and matrix to the tip of the sample probe [70]. The mixture is desorbed directly from the probe tip by a laser. One drawback is that only a limited number of matrixes have been found suitable. Aerosol liquid introduction is another inlet approach for subsequent MALDI analysis [71,72]. The solution containing analyte and matrix is sprayed with a pneumatic nebulizer into the mass spectrometer where the solvent evaporates. The dried aerosol particles are ionized with a pulsed laser and analyzed by TOFMS. Drawbacks associated with the aerosol MALDI method are limited mass resolution and high sample consumption. Another category of online MALDI interfaces uses a moving surface inside the mass spectrometer for continuous vacuum deposition of matrix and analyte. The moving surface can be a rotating ball [73], a wheel [74], or a moving tape [75]. The liquid is deposited onto the moving surface through a capillary that makes contact with the surface. The solvent evaporates leaving a sample trace that is irradiated by a desorption laser. Cleaning the target surface from deposited matrix and analyte in between runs are of major importance, to avoid memory effects. Another drawback with online techniques is that they do not allow reinvestigation of the sample since it is consumed during mass analysis.

4.3. CONCLUSION

More recent developments in CE and ESI- or MALDI-MS allow direct analysis of a wide variety of compounds. The combination of CE-MS provides an effective tool to characterize complex samples. Detection limits in femto- to attomole levels have been achieved for complex biological samples. These advances in analytical techniques allows a fresh look at some of the time honored problems in the neurochemical laboratory, namely, structural characterization of low quantities of transmitter molecules in both neurologic and immunologic tissues. Both ESI and MALDI-MS will in the future support neuroscience research, each method having unique capabilities as well as some fundamental similarities. It is also noteworthy to briefly mention new techniques that are similar to CE, such as capillary electrochromatography (CEC) and microfabricated separation techniques. CEC combines the features of CE and liquid chromatography (LC). The analytes are separated on the basis of both electrophoretic mobilities as in CE and polar interactions with a stationary phase as in LC. This results in a separation of all analytes including neutrals, which cannot be resolved in plain CE. The idea of microfabricated based separation is that the whole analytic procedure (sample pretreatment, separation, and detection) can be performed automatically on chip, thus reducing sample loss and tedious sample handling, and furthermore facilitate high-throughput screening.

These techniques in combination with molecular biology can provide opportunities to study gene regulation, apoptosis, and their relation to disorders in the central nervous system. Hopefully many neuroscientists may soon see these separation techniques and mass spectrometers as new companions.

ACKNOWLEDGMENTS

We wish to thank all our collaborators for their extensive contribution to the work described in this review. We are also grateful to Professor Karin Markides for critical reading of the manuscript. This work was supported by the Swedish Foundation for Strategic Research, the Fredrik and Ingrid Thuring Foundation, the Wilhelm and Martina Lundgren Foundation, the Magnus Bergvall Foundation, the Swedish Alzheimer Foundation, the Syskonen Svensson Foundation, the Gamla Tjänarinnor Foundation, the Knut and Alice Wallenberg Foundation, the Swedish Lundbeck Foundation, the Swedish Society for Medical Research, the Swedish Natural Science Research Council (Grant K-AA/KU 12003-300), and the Swedish Medical Research Council (Grant 13123).

REFERENCES

1. K. D. Altria, *Capillary Electrophoresis Guidebook: Principles, Operation, and Applications*, Vol. 52, Humana Press, Totowa, NJ, 1996.
2. J. P. Landers, *Handbook of Capillary Electrophoresis*, 2nd ed., CRC Press, Boca Raton, FL, 1997.
3. C. L. Nilsson, G. Karlsson, J. Bergquist, A. Westman, and R. Ekman, *Peptides* **19**, 781–789 (1998).
4. W. J. Ferguson and N. E. Good, *Anal. Biochem.* **104**, 300 (1980).
5. N. E. Good, *Biochemistry* **5**, 467 (1966).
6. N. E Good and S. Izawa, *Meth. Enzymol.* **24**, 53 (1972).
7. R. B. Cole, *Electrospray Ionization Mass Spectrometry: Fundamentals, Instrumentation & Applications*, Wiley, New York, 1997.
8. M. A. Moseley, L. J. Deterding, K. B. Tomer, and J. W. Jorgenson, *Anal. Chem.* **63**, 109–114 (1991).

9. R. B. Cole, *J. Mass. Spectrom.* **35**, 763–772 (2000).
10. P. Kebarle, *J. Mass Spectrom.* **35**, 804–817 (2000).
11. J. F. Mora, G. J. Van Berkel, C. G. Enke, R. B. Cole, M. Martinez-Sanchez, and J. B. Fenn, *J. Mass Spectrom.* **35**, 939–952 (2000).
12. G. I. Taylor, *Proc. Roy. Soc. Lond.* **280**, 383 (1964).
13. R. D. Smith, C. J. Barinaga, and H. R. Udseth, *Anal. Chem.* **60**, 1948–1952 (1988).
14. E. D. Lee, W. Muck, J. D. Henion, and T. R. Covey, *Biomed. Environ. Mass.* **18**, 844–850 (1989).
15. J. A. Olivares, N. T. Nguyen, C. R. Yonker, and R. D. Smith, *Anal. Chem.* **59**, 1230–1232 (1987).
16. J. H. Wahl, D. C. Gale, and R. D. Smith, *J. Chromatogr. A* **659**, 217–222 (1994).
17. D. C. Gale and R. D. Smith, *Rapid Commun. Mass Spectrom.* **7**, 1017–1021 (1993).
18. F. Foret, T. J. Thompson, P. Vouros, B. L. Karger, P. Gebauer, and P. Bocek, *Anal. Chem.* **66**, 4450–4458 (1994).
19. R. S. Ramsey and S. A. McLuckey, *J. Microcolumn Sep.* **7**, 461–469 (1995).
20. D. R. Barnidge, S. Nilsson, K. E. Markides, H. Rapp, and K. Hjort, *Rapid Commun. Mass Spectrom.* **13**, 994–1002 (1999).
21. M. S. Kriger, K. D. Cook, and R. S. Ramsey, *Anal. Chem.* **67**, 385–389 (1995).
22. D. R. Barnidge, S. Nilsson, and K. E. Markides, *Anal. Chem.* **71**, 4115–4118 (1999).
23. L. L. Fang, R. Zhang, E. R. Williams, and R. N. Zare, *Anal. Chem.* **66**, 3696–3701 (1994).
24. J. H. Wahl and R. D. Smith, *J. Cap. Electrophor.* **1**, 62 (1994).
25. P. Cao and M. Moini, *J. Am. Soc. Mass Spectrom.* **8**, 561–564 (1997).
26. P. Kebarle and L. Tang, *Anal. Chem.* **65**, A972–A986 (1993).
27. P. Kebarle and M. Peschke, *Anal. Chim. Acta* **406**, 11–35 (2000).
28. Y. J. Lee, W. E. Price, and M. Sheil, *Analyst* **120**, 2689–2694 (1995).
29. M. A. Moseley, J. W. Jorgenson, J. Shabanowitz, D. F. Hunt, and K. B. Tomer, *J. Am. Soc. Mass. Spectrom.* **3**, 289–300 (1992).
30. J. F. Kelly, L. Ramaley, and P. Thibault, *Anal. Chem.* **69**, 51–60 (1997).
31. I. Rodriguez and S. F. Y. Li, *Anal. Chim. Acta* **383**, 1–26 (1999).
32. W. Tong, A. Link, J. K. Eng, and J. R. Yates, *Anal. Chem.* **71**, 2270–2278 (1999).
33. A. J. Tomlinson, L. M. Benson, S. Jameson, D. H. Johnson, and S. Naylor, *J. Am. Soc. Mass Spectrom.* **8**, 15–24 (1997).
34. S. Naylor and A. J. Thomlinson, *Talanta* **45**, 603–612 (1998).
35. M. B. Barroso and A. P. de Jong, *J. Cap. Electrophor.* **5**, 1–7 (1998).
36. P. Cao and M. Moini, *Electrophoresis* **19**, 2200–2206 (1998).
37. J. Hau and M. Roberts, *Anal. Chem.* **71**, 3977–3984 (1999).
38. D. R. Goodlett, J. H. Wahl, H. R. Udseth, and R. D. Smith, *J. Microcol. Sep.* **5**, 57–62 (1993).
39. R. D. Smith, J. H. Wahl, D. R. Goodlett, and S. A. Hofstadler, *Anal. Chem.* **65**, A574–A584 (1993).
40. R. D. Smith, *Int. J. Mass Spectrom.* **200**, 509–544 (2000).

41. D. Figeys, S. P. Gygi, Y. Zhang, J. Watts, M. Gu, and R. Aebersold, *Electrophoresis* **19**, 1811–1818 (1998).
42. M. X. Li, J. T. Wu, L. Liu, and D. M. Lubman, *Rapid. Commun. Mass Spectrom.* **11**, 99–108 (1997).
43. J. T. Wu, M. G. Qian, M. X. Li, L. Liu, and D. M. Lubman, *Anal. Chem.* **68**, 3388–3396 (1996).
44. P. Cao and M. Moini, *Rapid Commun. Mass Spectrom.* **12**, 864–870 (1998).
45. P. Cao and M. Moini, *J. Am. Soc. Mass Spectrom.* **10**, 184–186 (1999).
46. K. Ensing, T. de Boer, N. Schreuder, and R. A. de Zeeuw, *J. Chromatogr. B* **727**, 53–61 (1999).
47. J. C. Severs, S. A. Hofstadler, Z. Zhao, R. T. Senh, and R. D. Smith, *Electrophoresis* **17**, 1808–1817 (1996).
48. G. A. Valaskovic, N. L. Kelleher, and F. W. McLafferty, *Science* **273**, 1199–1202 (1996).
49. S. A. Hofstadler, J. C. Severs, R. D. Smith, F. D. Swanek, and A. G. Ewing, *Rapid Commun. Mass Spectrom.* **10**, 919–922 (1996).
50. M. Karas and F. Hillenkamp, *Anal. Chem.* **60**, 2299–2301 (1988).
51. F. Hillenkamp, M. Karas, R. C. Beavis, and B. T. Chait, *Anal. Chem.* **63**, 1193A–1203A (1991).
52. K. Strupat, M. Karas, and F. Hillenkamp, *Int. J. Mass Spectrom. Ion Process.* **111**, 89–102 (1991).
53. B. Spengler, D. Kirsch, R. Kaufmann, and E. Jaeger, *Rapid Commun. Mass Spectrom.* **6**, 105–108 (1992).
54. I. Messana, D. V. Rossetti, L. Cassiano, F. Misiti, B. Giardina, and M. Castagnola, *J. Chromatogr. B* **699**, 149–171 (1997).
55. H. G. Lee and D. M. Desiderio, *Anal. Chim. Acta* **383**, 79–99 (1999).
56. A. I. Gusev, *Fresenius J. Anal. Chem.* **366**, 691–700 (2000).
57. H. Zhang and R. M. Caprioli, *J. Mass Spectrom.* **31**, 1039–1046 (1996).
58. M. A. Winkler, S. Kundu, T. E. Robey, and W. G. Robey, *J. Chromatogr. A* **744**, 177–185 (1996).
59. L. Cruz, R. W. Garden, H. J Kaiser,. and J. V. Sweedler, *J. Chromatogr. A* **735**, 375–385 (1996).
60. H. J. Boss, M. F. Rohde, and R. S. Rush, *Anal. Biochem.* **230**, 123–129 (1995).
61. W. Weinmann, C. E. Parker, L. J. Deterding, D. I. Papac, J. Hoyes, M. Przybylski, and K. B. Tomer, *J. Chromatogr. A* **680**, 353–361 (1994).
62. K. L. Walker, R. W. Chiu, C. A. Monnig, and C. L. Wilkins, *Anal. Chem.* **67**, 4197–4204 (1995).
63. J. A. Chakel, E. Pungor, W. S. Hancock, and S. A. Swedberg, *J. Chromatogr. B Biomed. Sci. Appl.* **689**, 215–220 (1997).
64. N. Bihoreau, C. Ramon, M. Lazard, and J. M. Schmitter, *J. Chromatogr. B Biomed. Sci. Appl.* **697**, 123–133 (1997).
65. M. O. Glocker, S. H. Bauer, J. Kast, J. Volz, and M. Przybylski, *J. Mass Spectrom.* **31**, 1221–1227 (1996).

66. B. Thiede, A. Otto, U. Zimny-Arndt, E. C. Muller, and P. Jungblut, *Electrophoresis* **17**, 588–599 (1996).
67. J. S. Page, S. S. Rubakhin, and J. V. Sweedler, *Analyst* **125**, 555–561 (2000).
68. M. Schuerenberg, C. Luebbert, H. Eickhoff, M. Kalkum, H. Lehrach, and E. Nordhoff, *Anal. Chem.* **72**, 3436–3442 (2000).
69. T. Johnson, J. Bergquist, R. Ekman, E. Nordhoff, M. Schuerenberg, K. D. Kloeppel, M. Muller, H. Lehrach, and J. Gobom, *Anal. Chem.*, **73**, 1670–1675 (2001).
70. Q. Zhan, A. Gusev, and D. M. Hercules, *Rapid Commun. Mass Spectrom.* **13**, 2278–2283 (1999).
71. K. K. Murray and D. H. Russell, *Anal. Chem.* **65**, 2534–2537 (1993).
72. X. Fei, G. Wei, and K. K. Murray, *Anal. Chem.* **68**, 1143–1147 (1996).
73. H. Orsnes, T. Graf, H. Degn, and K. K. Murray, *Anal. Chem.* **72**, 251–254 (2000).
74. J. Preisler, F. Foret, and B. L. Karger, *Anal. Chem.* **70**, 5278–5287 (1998).
75. J. Preisler, P. Hu, T. Rejtar, and B. L. Karger, *Anal. Chem.* **72**, 4785–4795 (2000).

CHAPTER 5

Sequencing of Neuropeptides by Nanoelectrospray Mass Spectrometry

JÜRGEN KAST

Biomedical Research Centre, University of British Columbia, Vancouver, BC, Canada.

MATTHIAS WILM

European Molecular Biology Laboratory, Heidelberg, Germany

CONTENTS

5.1. HISTORICAL BACKGROUND

The idea to electrospray a liquid in order to generate ions of analyte molecules was formulated first in 1968 by Dole and his co-workers [1]. In the early 1970s, they were able to obtain ions of polymers in the gas phase, but lacked sufficient mass spectrometric equipment to demonstrate the feasibility of their theories [2]. They formulated a putative ionization model in which initially large droplets are generated that undergo repetitive Coulomb explosions caused by charge concentration due to solvent evaporation. Thus, nanometer-range droplets are formed that contain only one analyte molecule. After complete solvent evaporation, these are left as ions in the gas phase.

Mass Spectrometry and Hyphenated Techniques in Neuropeptide Research, Edited by Jerzy Silberring and Rolf Ekman.
ISBN 0-471-35493-7

In the early 1980s, small molecules were analyzed for the first time directly by mass spectrometry by two groups independently from each other [3,4]. The key discovery that electrospray can also be applied to macromolecules was made in the late 1980s by Fenn's group at Yale University [5]. They were able to detect single protein ions in a mass spectrometer by the use of countercurrent "drying" gas that removes the residual solvent shell around the protein without fragmenting it. Ionization was explained as a field evaporation mechanism from small, highly charged droplets generated by Coulomb explosions as proposed earlier for small molecules by Iribarne and Thomson [6]. Macromolecules partially extrude from the droplets' surface, become charged, and are emitted as single molecules due to Coulomb repulsion [7]. In contrast to Dole's theory of single molecules in single droplets, ions are generated from droplets still containing a large number of macromolecules that take an active role in the ionization process.

In 1994, a theoretical description of the electrospray process was published by Wilm and Mann lending support to Dole's theory. On the basis of these results, a more efficient ion source was constructed that initially was called "microelectrospray" [8] but later was renamed "nanoelectrospray" to avoid confusion with other microscale electrospray interfaces [9,10].

5.2. CHARACTERISTICS OF THE NANOELECTROSPRAY ION SOURCE

The theory underlying the nanoelectrospray ion source describes the size of the cone formed at the tip of the spraying liquid called *Taylor cone*, the onset voltage of the electrospray, and the emission of droplets. Their size is proportional to two-thirds of the power of the flow rate. Typically, electrospray ion sources are spraying from capillaries with about 100 μm inner diameter at flow rates of 1–10 μL/min. They produce droplets that have an initial diameter of more than 1 μm and contain more than 300 analyte molecules at concentrations of 1 pmol/μL [8]. In contrast, nanoelectrospray ion sources spray from gold-coated capillaries that are pulled to very narrow tips with an inner diameter of 1 μm. The flow rate of typically 10–20 nL/min is allowed to adjust itself to produce a stable continuous electrospray. The droplets thus produced are less than 200 nm in diameter and contain only one analyte molecule at 1 pmol/μL concentrations. They are quite monodisperse and evaporate more rapidly because of their small size, increasing the number of analyte molecules available for desorption. Additionally, the nanoelectrospray ion source can be mounted much closer to the vacuum orifice of the mass spectrometer, resulting in increased ionization, desolvation, and transfer efficiencies into the vacuum than with conventional electrospray ion sources. On a triple quadrupole instrument, the overall efficiency of the nanoelectrospray ion source was determined to be one ion detected from 1300 molecules sprayed at isotopic resolution [8]. Under optimized conditions with sacrificed resolution, even one

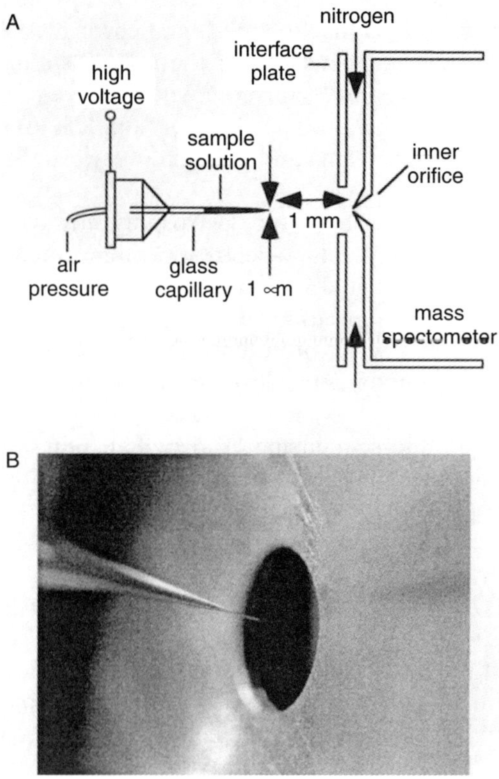

Figure 5.1. The nanoelectrospray ion source on a Sciex API-III electrospray mass spectrometer: (a) Schematic drawing of the setup; (b) snapshot under operating conditions.

ion from 390 sprayed molecules could be detected. This means a >500 times higher overall efficiency for the nanoelectrospray ion source than reported for a conventional electrospray source on the same instrument [11].

The smaller size of electrospray droplets produced with the nanoelectrospray ion source allows distinct changes in the source region. The emitter can be moved closer to the instrument's orifice with a typical distance of 1–2 mm to the vacuum orifice. The nebulizer gas is not used any more, and the flow rate of the countercurrent drying gas can be reduced to about 50% of its normal value. The potential difference between the emitter and the orifice is lowered to 500–800 V (Fig. 5.1). The resulting spray is very stable; only 6.1% standard deviation of the intensity over a period of 20 min using one needle, and 17% using 25 different needles has been reported [11]. The nanoelectrospray ion source is an offline system and therefore limited to the sample volume and concentration that can be loaded into the capillary. Although this can be overcome by an additional concentration and desalting step prior to analysis, its performance and separation power for complex mixtures is not comparable to other systems that rely on liquid chromatography separation. These disadvantages are compensated by the robustness of the system that results in almost no sample losses, the complete absence of cross-contamination effects due to the fact that one capillary per sample is used, and the applicability of a wide range of solvents and pH values that can be changed during one experiment [11].

The most important feature of nanoelectrospray, however, is the extended analysis time available with the nanoelectrospray ion source. At a flow rate of 20 nL/min, 1 µL of analyte solution lasts for 50 min with almost no change in signal intensity. This allows one to not only optimize the experimental parameters but also perform a series of different experiments and using different operating modes. For example, applications of the nanoelectrospray ion source include identification of subpicomole amounts of gel-separated proteins after digestion by peptide sequencing [12,13], detection and localization of posttranslational modifications [14], and de novo sequencing of peptides and proteins after isotopic labeling [15,16].

5.3. MANUFACTURING OF NANOELECTROSPRAY CAPILLARIES

The key feature of the nanoelectrospray ion source is the low and stable flow rate of 10–20 nL/min achieved by using dedicated glass capillaries. Therefore, the results that can be obtained by nanoelectrospray depend to a large degree on the shape of the needle. A set of two nanoelectrospray needles are produced by pulling one borosilicate glass capillary (GC120 F-10, Clark Electromedical Instruments, Pangbourne, UK) on a microcapillary puller (Model P-97, Sutter Instruments Co., Novato, CA, USA) using a two-step pulling cycle. In the first step (heat 450, pull strength 80, velocity 9, cooling time 200), the outer diameter is reduced to about 0.5 mm, and in the second step (heat 500, pull strength 100, velocity 10, cooling time 150) the needles are separated completely. The parameters of the second pulling cycle are adjusted to generate a needle with a tip of 1–2 µm in diameter and a hairline fine filament less than 0.8 mm in length (Fig. 5.2). The length of this filament determines the total flow resistance of the capillary. Longer filaments (200–500 µm; Fig. 5.2a) allow lower flow rates but increase the risk of occasional plugging of the tip. Shorter filaments (100–200 µm; Fig. 5.2b) are easier to handle but shorten the time for analysis because of a higher flow rate.

Pulled capillaries are arranged in trays for safe storage and metalized by gold deposition in a vacuum chamber (Polaron SC7610 sputter coater, Fisons Instruments, East Sussex, UK). The tray is positioned at about 5 cm distance to the gold target. The chamber is then evacuated to a pressure of 0.1 torr, and a current of 15–20 mA is applied for 2–3 min to obtain a complete, homogeneous, transparent thin layer of gold. Alternatively, ready-to-use nanoelectro-

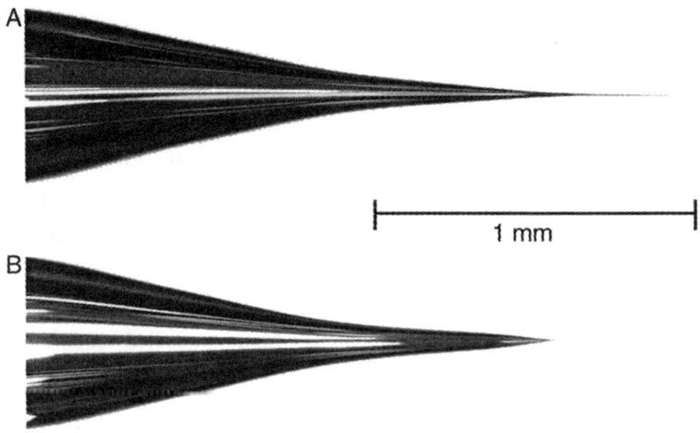

Figure 5.2. Pulled, gold-coated nanoelectrospray capillaries with (a) long and (b) short filaments. Longer filaments allow lower flow rates and extended analysis time, but are more difficult to handle and may plug more often than shorter filaments.

spray capillaries can be purchased from various commercial sources (e.g., MDS Proteomics, Odense, DK; New Objective, Boston, MA, USA).

5.4. OPERATION OF THE NANOELECTROSPRAY ION SOURCE

The main objectives in setting up a nanoelectrospray ion source to any instrument is to achieve perfect geometric alignment to the vacuum orifice of the instrument and to enable a continuous low flow rate. A typical nanoelectrospray setup consists of three devices: a pressurized capillary holder, a micromanipulator for *x*–*y*–*z* positioning of the holder, and an optical system to control the position of the capillary in front of the vacuum orifice of the mass spectrometer (Fig. 5.1). The individual design has to meet the geometric prerequisites of the instrument. Complete nanoelectrospray setups are available from different suppliers as an alternative.

For operation, the pulled, gold-coated capillary is filled with the sample and mounted in a gastight metal holder. The electrical contact can be ascertained by using graphite paste or a metal clamp. The holder is then moved forward to place the capillary in front of the orifice at a distance of 1–2 mm under visual control, and the electrospray is started. Pulled capillaries may have tip diameters as small as 100 nm that cause flow resistance too high to be overcome using only moderate pressure. Therefore, the opening of the needle is checked first by gently pressurizing the holder. If a droplet appears, the spray can be established right away. Otherwise, the capillary has to be moved gently against a flat surface to break the very tip until a droplet becomes visible. Etching the needles with hydrogen fluoride to open pulled capillaries has also been reported [17]. After opening the capillary, a potential difference of 500–800 V is applied to start the electrostatic spray. By observing the ion current on the mass spectrometer, the pressure applied to the capillary is reduced to minimize the flow rate while maintaining a stable spray. In case of needle plugging, the voltages can be dropped and the surface touched again to reopen the needle. Harsher conditions for reopening include reducing the voltage instead of dropping it; this may cause severe damage of the tip due to high local field strengths, but if all efforts to reopen the needle by mechanic means fail, the surface can be touched with the voltage set to 500 V. However, this procedure damages the gold layer on the tip of the needle. In any case, the tip of the needle should be kept sufficiently small to maintain a reasonable flow resistance and low flow rate.

5.5. SAMPLE PREPARATION FOR NANOELECTROSPRAY MASS SPECTROMETRY

Because of the low flow rate of the nanoelectrospray ion source, sample volumes of less than 1 μL are sufficient for most experiments. Highest sensitivity can be achieved by concentrating the sample to this small volume. Typically, samples originating from biological sources, such as body fluids, are present at low concentrations in large volumes under physiological conditions. In order to optimize nanoelectrospray sensitivity, such samples should be concentrated and desalted prior to analysis [11,18,19]. In contrast to all online electrospray ionization methods, nanoelectrospray has to rely on additional offline purification and concentration steps. This can be done by using commercially available microscale cartridges (ZipTip, Millipore Corporation, Bedford, MA, USA) or by employing self-packed microcolumns in GELoader tips or pulled capillaries (Fig. 5.3a). Self-packed columns have advantages over ready-to-use cartridges because they allow one to choose the ideal chromatographic material and to adjust the column size. If the bed volume is too high, the sample will not be eluted completely in volumes of 1 μL, resulting in sample losses. Typically, perfusion chromatography material (e.g., POROS, PerSeptive Biosystems, Framingham, MA, USA) is used that is designed for the use with low-pressure systems. It is available for a variety of purposes, including purification of proteins (POROS R1), peptides (POROS R2), oligonucleotides and hydrophilic peptides (Oligo R3), and for anion exchange (POROS HQ) or immobilized metal-affinity chromatography (POROS MC). A slurry of the appropriate beads is prepared in methanol and filled into the GELoader tip or the glass capillary and packed by eluting the methanol. If a glass capillary is used, a special holder to mount the capillary in a table centrifuge can be used for the entire purification process (Fig. 5.3b). For this purpose, the tip of the glass capillary is carefully opened to let the liquid pass freely through the capillary while the chromatographic material is retained. Flow is then ascertained by spinning the centrifuge at low speed. Columns no

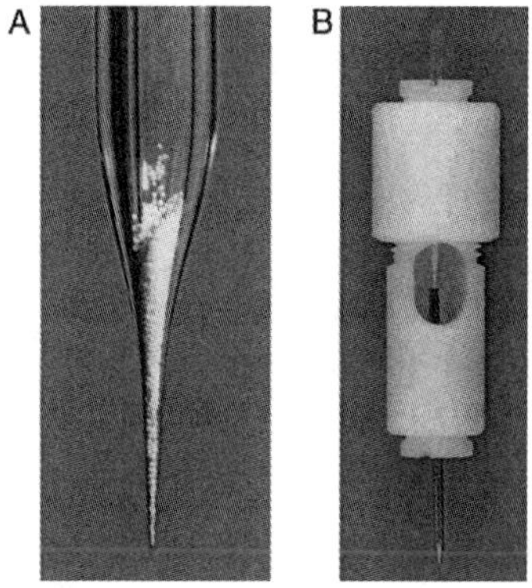

Figure 5.3. Purification and concentration for nanoelectrospray: (a) capillary column made from a pulled glass capillary, filled with POROS beads for perfusion chromatography; (b) holder for direct elution from capillary column into nanoelectrospray capillary by centrifugation.

larger than 100–200 nL in size can be handled safely this way. After preequilibration of the column, sample loading, and thorough washing, the sample is eluted with a small volume of 0.3–1 μL directly into a nanoelectrospray capillary. The buffers used depend on the analyte molecule; for peptides concentrated on a POROS R2 column, mixtures of formic acid, water, and methanol (5:90:5 for equilibration and washing, 5:30:65 for elution) are typically used. If complex sample mixtures are analyzed, it may be useful to fractionate the samples by eluting them in several steps while increasing the content of organic solvent.

5.6. PEPTIDE SEQUENCING BY NANOELECTROSPRAY TANDEM MASS SPECTROMETRY

The identification of gel-separated proteins after enzymatic digestion by peptide sequencing has become one of the main applications of nanoelectrospray mass spectrometry [20,21]. Because of the extended measurement time and the possibility to optimize the experimental parameters, the highest quality of results can be obtained. Sequencing of peptides consists of at least three different steps: (1) peptide candidates have to be identified as precursor ions for subsequent sequencing, (2) the precursor ions have to be selected and fragmented one by one, (3) and the resulting MS/MS data have to be interpreted to identify the peptide.

To screen for peptide candidates, a mass spectrum is acquired over a wide mass range. The spectrum is then inspected for peptides to identify their individual *m/z* values and their charge states by their isotopic spacing. Instruments with time-of-flight analyzers, such as Q-Tof instruments, possess enough reso-

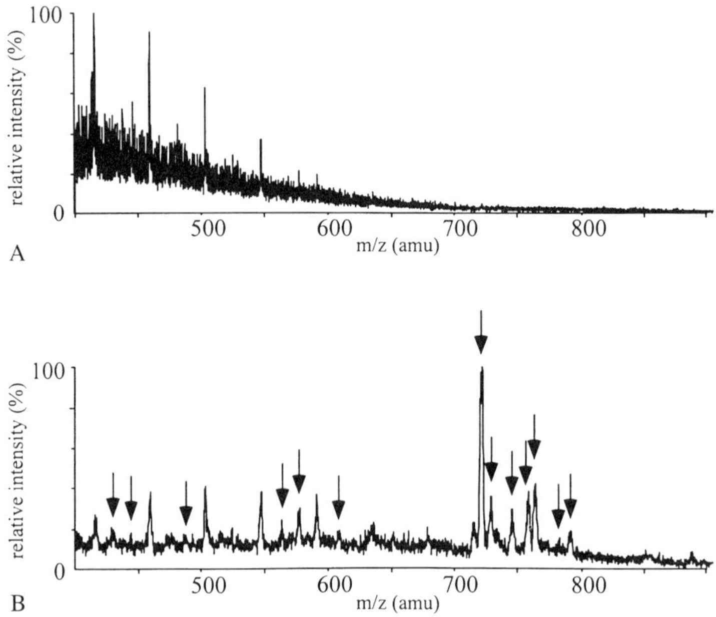

Figure 5.4. Detection of peptides below the level of chemical noise by precursor ion scanning on a triple quadrupole instrument: (a) nanoelectrospray mass spectrum of a peptide; (b) precursor ion scan of *m/z* 86 of the same sample. Leucine or isoleucine containing peptides are marked with arrows.

lution to determine both at the same time. For instruments with quadrupole analyzers, a number of peptide candidates and their approximate m/z values are determined first. To reveal their precise *m/z* values and charge states, multiple ion monitoring under high-resolution settings can then be performed. Once a number of peptide candidates has been identified, doubly and triply charged peptides are selected and fragmented first. These are the most promising candidates because doubly charged ions produce easy-to-interpret fragmentation spectra while triply charged ions allow readout of long sequence stretches.

If the amount of peptides is quite low, the sensitivity is limited by the signal-to-noise ratio. At least two different techniques have been developed to detect peptide ions that are below the level of chemical noise using the nanoelectrospray ion source. On triple quadrupole instruments, precursor ion scans for immonium ions of leucine and isoleucine residues after high-energy collisions can be performed [14]. This allows one to distinguish between leucine or isoleucine containing peptides and chemical noise because only the peptides will report the specific marker ion at *m/z* 86 (Fig. 5.4).

Excluding chemical noise by precursor ion scanning has proved to be an extremely sensitive method of identifying peptide candidates on triple quadru-

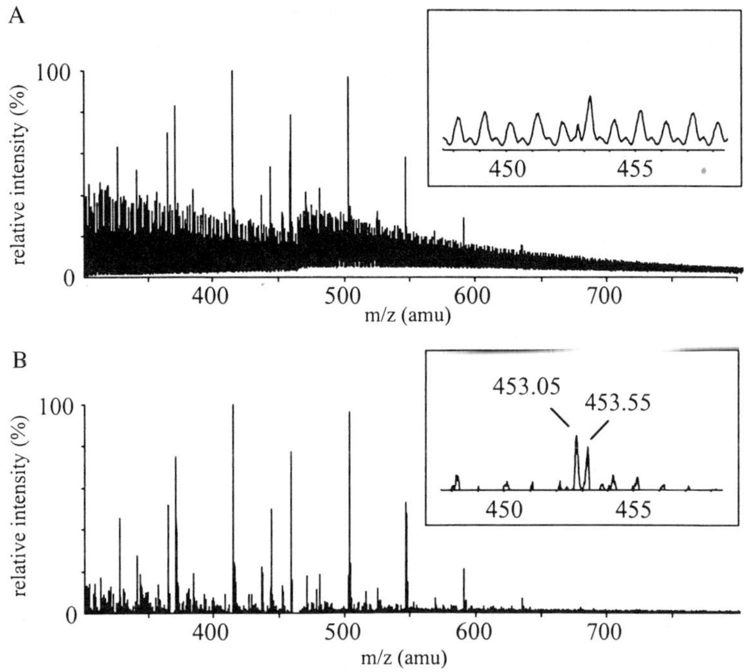

Figure 5.5. Detection of peptides below the level of chemical noise by filtering using Fourier transformation on a quadrupole time-of-flight instrument: (a) nanoelectrospray mass spectrum of a peptide; (b) the same spectrum after eliminating regular noise by Fourier transformation.

pole instruments because of its linear geometry avoiding ion losses. The different geometry of quadrupole time-of-flight instruments has made their precursor ion scanning mode less sensitive [22]. An alternative to improve the signal-to-noise ratio on quadrupole time-of-flight (TOF) type of instruments is to filter the chemical noise by an algorithm based on a Fourier transformation. This method relies on the observation that the noise in an electrospray spectrum is quite regular. These regularities are reflected by distinct peaks at frequencies of 0, 1, 2, etc. Hz in the Fourier spectrum. Removing these frequencies results in a spectrum where most of the chemical noise has been eliminated (Fig. 5.5).

Once a set of peptide candidates has been identified for subsequent MS/MS sequencing, each peptide is selected and fragmented individually. Care has to be taken to exclude as much as possible of the chemical noise surrounding the peptide signal. A higher amount of chemical noise will result in the presence of additional signals in the lower *m*/*z* range of the MS/MS spectrum and mask peptide fragment ions. Fragmentation is usually performed by adjusting the collision energy and varying it over the fragment spectrum. Large fragments in the upper part of the spectrum require a lower collision energy than do shorter fragments in the mid or lower part. Therefore, it can be useful to split the acquisition of the fragment spectrum into two or three different parts. On triple quadrupole instruments, the collision energy can be adjusted automatically within one scan calculating it from the mass of the precursor ion, its charge state, and the fragment mass acquired. In contrast, on quadrupole TOF instruments, the collision energy can be changed after only one scan has been completed; therefore, the entire fragment spectrum has to be acquired for each collision energy. The acquisition of precursor ion scans or the adjustment of the collision energy can be time-consuming, but the low flow rate of the nanoelectrospray ion source guarantees an extended analysis time. The nanoelectrospray ion source allows an interactive approach for the analysis and therefore often increases the quality of the acquired fragmentation spectra compared to completely automatic acquisition. Only the most recent developments in automatic acquisition software rival this quality by evaluating the acquired fragmentation spectrum in time and making further acquisition depending on this assessment.

Interpretation of MS/MS spectra is usually done offline after all experiments have been finished or the sample has been consumed completely. The amount of information that has to be read out of a fragmentation spectrum depends on the task that has to be performed. If proteins present in a protein sequence database have to be identified by peptide sequencing after tryptic digestion, obtaining sequences 2–4 amino acid residues in length will be sufficient to identify the protein unambiguously. For this purpose, a peptide sequence tag is created that consists of the sequence stretch obtained, the masses of the unsequenced *N*- and *C*-terminal parts of the peptide, the overall peptide mass, and the particular cleavage specificity of the enzyme used [23] (Fig. 5.6). If a protein cannot be identified, error-tolerant searches by omitting one or two parts of the given information can be performed to take modifications and sequence errors into account. Once an entry in the database is found and the complete peptide sequence is obtained, the identity of the peptide is reassured by matching the predicted fragment ions to the signals in the spectrum.

Assigning a sequence stretch can become a demanding task and depends to a large extent on the experience of the operator. However, there are some rules to facilitate the identification of continuous ion series. On triple quadrupole and quadrupole TOF instruments, mainly *b* and *y* ions are obtained upon fragmentation [24]. The *y* ions dominate the fragmentation spectrum of peptides derived from tryptic digestion due to charge localization on the *C*-terminal basic amino acid residue. In order to obtain a sequence stretch, an uninterrupted series of ions of the same type (*b* or *y*) has to be identified. Usually, a high signal in the upper *m*/*z* range is chosen to start interpretation, and additional fragment ions are sought with a mass difference that corresponds to an amino acid residue. Typically, *b* and *y* ions are flanked by a satellite ion with a 17 or 18 Da lower mass due to the loss of ammonia or water,

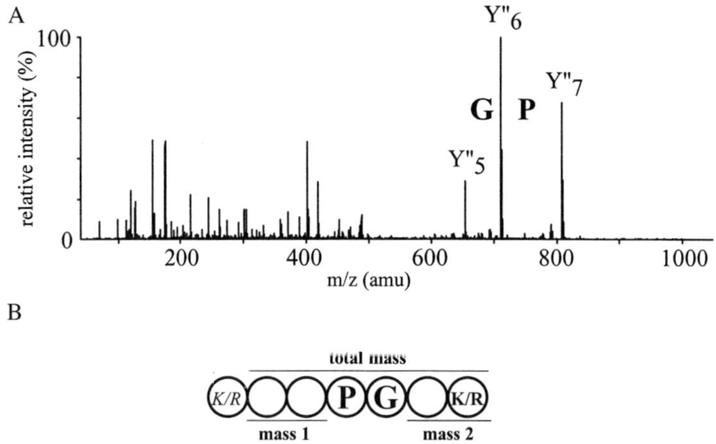

Figure 5.6. Creating a peptide sequence tag from a fragmentation spectrum: (a) assignment of three fragments of the *y* ion series; (b) the resulting peptide sequence tag.

which can be used for confirmation. Unfortunately, there are several pairs of amino acid residues with the same nominal mass: leucine and isoleucine, lysine and glutamine, and oxidized methionine and phenylalanine. While leucine and isoleucine are isobaric and cannot be distinguished, lysine/glutamine and oxidized methionine/phenylalanine residues have slightly different monoisotopic masses of 128.095 Da (K) versus 128.059 Da (Q) and 147.035 Da (oxidized M) versus 147.068 Da (F), respectively. The identification of the correct residue from the latter two pairs therefore depends on the resolution and mass accuracy of the instrument used. Additionally, fragment ions containing an oxidized methionine show characteristic satellite ions with 64 Da lower masses due to the loss of methylsulfoxide from the sidechain [25] that are not present with phenylalanine. Proline residues interrupt ion series due to a very characteristic fragmentation pattern. Their *N*-terminal amide bond is labile, giving rise to a very intense *y* ion, whereas their *C*-terminal bond is very stable, leading to none or a very small fragment ion. Additionally, the *y* ion generated by fragmentation of the *N*-terminal bond of the proline can fragment a second time, generating an additional *b* ion series called *internal fragment ions*.

The sequence-tag-based protein identification is a very specific and reliable way to identify a protein from its peptide's fragment spectra. It is a manual method, however. Commercially available programs that are accessible via the Internet can be used to identify proteins without any manual interpretation of the fragment spectra. They rely on probabilistic scoring routines when comparing the signal pattern of the given spectrum to predicted patterns based on the sequence of peptides derived from proteins in the database [26,27]. All methods described here are applicable for peptides derived from proteins that are present in a database. They fail, however, for sequencing of peptides coming from proteins whose sequence is not present in databases.

5.7. "DE NOVO" PEPTIDE SEQUENCING USING NANOELECTROSPRAY MASS SPECTROMETRY

Sequencing of peptides "de novo" for complete sequence readout has been used mostly for the identification of proteins that are not present in protein databases [15]. Most methods rely on the presence of one complete ion series, either *b* or *y* ions, and a means to confidently identify these ions. This can be done by modifying the chemical structure of the peptides. Peptide derivatization can be used to change the fragmentation pattern of the peptide favoring one specific ion type, or to identify one type of fragment ions by the mass shift on chemical modification. When trypsin is used as enzymatic cleavage reagent, the resulting tryptic peptides have a basic residue at their *C*-termini. Charge localization to the *C*-terminus of the peptide then occurs naturally favoring continuous *y* ion series. In contrast, the intensity of *b* ions can be increased by adding a positively charged chemical group to the *N* terminus of a peptide [28]. Identifying an ion series by shifting its masses has initially been developed for triple quadrupole instruments. The digestion solution is split into halves with one aliquot untreated and the second esterified by introducing a methyl group (+14 Da) into each acidic group: the *C*-terminus and each aspartic and glutamic acid residue [20,29]. For both subsets, peptide sequencing experiments of the unseparated peptide mixtures are performed individually and then compared to each other. The *y* ions in the fragmentation spectra are identified by distinct mass shifts of 14 Da for each acidic group present. This method often allows readout of long or even complete peptide sequences. A similar technique has been developed for high-resolution instruments, including quadrupole TOF mass spectrometers, relying on the incorporation of ^{18}O atoms into the *C*-terminal carboxyl group of the peptides [15,30]. This is achieved by performing tryptic digestions in the presence of 50% of ^{18}O water in the digestion buffer. Fragmentation of the isotopically labeled peptides then reveals the *y* ions by a characteristic isotopic pattern showing a 2-Da split that facilitates sequence readout (Fig. 5.7). In contrast to esterification, the ^{18}O method requires only one experiment, reducing the risk of sample losses; does not change the chemical nature of the labeled peptide; and results in a single, labeled product that co-localizes with the unlabeled peptide in the mass spectrum.

"De novo" sequencing from ^{18}O-labeled peptides relying on the 2-Da mass split on *y* ions is often time-consuming and limited in sensitivity. Low peptide amounts usually result in low *y* ion intensities in their fragmentation spectra and masked isotopic patterns due to increasing chemical background. The more recently developed "differential scanning" technique increases the reliability of the called sequences and allows for automation [16]. It is based on

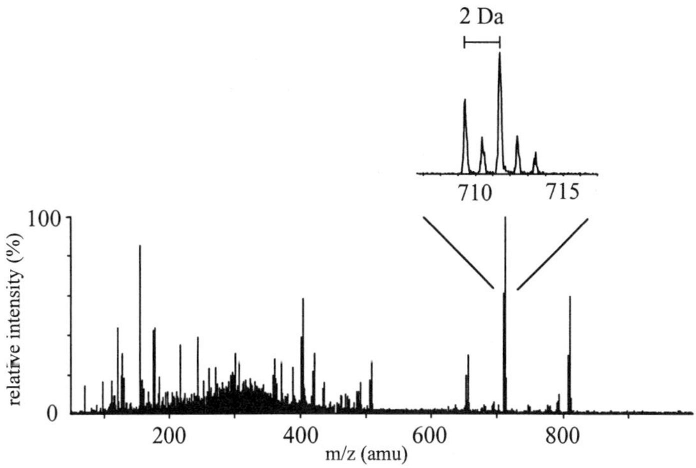

Figure 5.7. Identification of *y* ions by a 2-Da split upon incorporation of ^{18}O into the carboxy terminus of a peptide.

the creation of two fragmentation spectra for each isotopically labeled peptide. Transmitting all isotopes of a peptide and fragmenting them results in a MS/MS spectrum with *y* ions showing the characteristic 2-Da split. In contrast, excluding the ^{16}O part and exclusively transmitting the ^{18}O-labeled peptide isotopes results in suppression of the ^{16}O part of the *y* ions upon fragmentation, whereas *b* ions are not affected (Fig. 5.8a,b). A computer algorithm can then be used to score the *y* ions in the two fragmentation spectra on the basis of their isotopic distribution in the all-isotope spectrum and the suppression of the first isotope in the ^{18}O-isotopes-only spectrum. The resulting scored and deconvoluted spectrum shows intense signals for *y* ions and can be used for automated sequence readout (Fig. 5.8c).

So far, the differential scanning technique has been applied only to ^{18}O-labeled, tryptic peptides to identify complete *y* ion series. It is, however, also applicable to nontryptic peptides, such as neuropeptides. In such cases, an isotopically labeled chemical reagent could be introduced into the *N* terminus of the peptide that would result in isotopically labeled *b* ions with increased intensity in the fragmentation spectrum [28]. Producing two sets of fragmentation spectra by differential scanning would then allow specific identification of the *b* ion series for complete sequence readout.

5.8. PERSPECTIVES

Using the nanoelectrospray ion source for peptide sequencing will remain one of its main applications because of its high ionization efficiency and extended time for analysis. However, more recent developments in LC-MS/MS, partic-

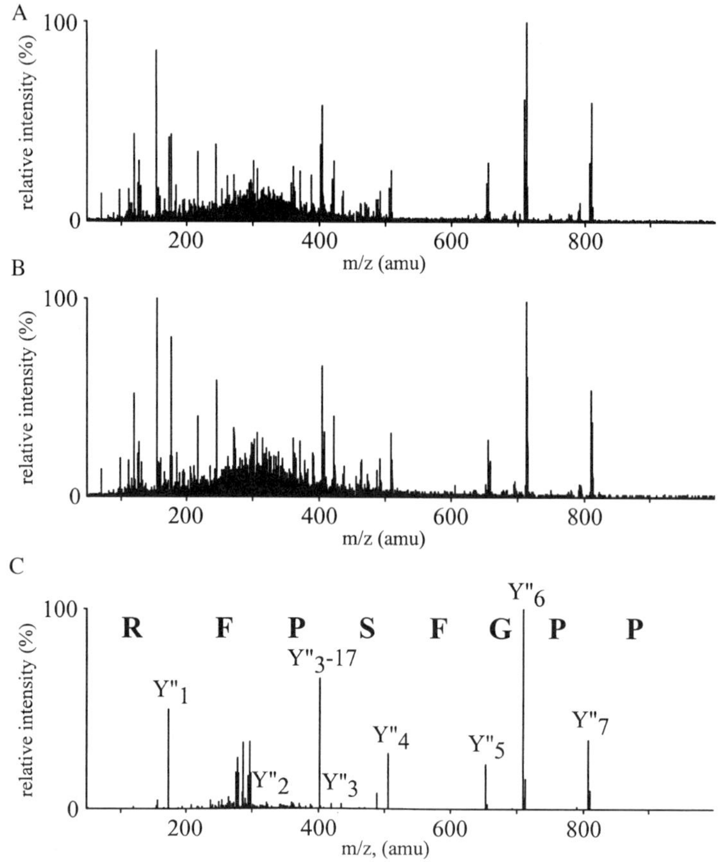

Figure 5.8. Differential scanning for enhanced sequence readout of ^{18}O-labeled peptides: (a) fragmentation spectrum when all isotopes are transmitted; (b) fragmentation spectrum when only ^{18}O-labeled isotopes are transmitted, resulting in isotopic suppression of the ^{16}O part of the *y* ions; (c) scored and deconvoluted spectrum identifies *y* ions.

ularly the introduction of low-flow systems that deliver gradients with stable flow rates of 100–250 nL/min, are now providing an alternative to nanoelectrospray for protein identification by peptide sequencing. The advantages of an LC-MS/MS system, namely, its ability to separate and concentrate individual analytes from complex mixtures and its potential for automation, are impaired by the loss of ionization efficiency due to the larger tip of the spraying needle and the forced flow rate. New emitters for LC-MS/MS have been reported [31,32] that hold the potential to bridge chromatographic separation and nanoelectrospray ionization efficiency. Eventually, peptide sequencing could be performed using a hybrid HPLC–nanoelectrospray ion source replacing conventional nanoelectrospray. Then, the answer to the question "LC-MS/MS or nanoelectrospray MS/MS" will be "both."

REFERENCES

1. M. Dole, L.L. Mack, R. L. Hines, R. C. Mobley, L. D. Ferguson, and M. B. Alice, Molecular beams of macroions, *J. Chem. Phys.* **49**, 2240–2248 (1968).
2. G. A. Clegg and M. Dole, Molecular beams of macroions. III. Zein and polyvinylpyrrolidone, *Biopolymers* **10**, 821–826 (1971).
3. M. Yamashita and J. B. Fenn, Electrospray ion source. Another variation on the free-jet theme, *J. Phys. Chem.* **88**, 4451–4459 (1984).
4. M. L. Alexandrov, L. N. Gall, N. V. Krasnov, V. I. Nikolaev, V. A. Panvlenko, V. A. Shkurov, G. I. Baram, M. A. Grachev, V. D. Knorre, and Y. S. Kusner, *Bioorg. Khim.* **10**, 710–712 (1984) (in Russian).
5. C. K. Meng, M. Mann, and J. B. Fenn, Of protons and proteins, *Z. Phys. D: Atoms, Mol. Clusters* **10**, 361–368 (1988).
6. B. A. Thomson and J. V. Field, Induced ion evaporation from liquid surfaces at atmospheric pressure, *J. Chem. Phys.* **71**, 4451–4463 (1979).
7. J. B. Fenn, Ion formation from charged droplets: Role of geometry, energy, and time, *J. Am. Soc. Mass Spectrom.* **4**, 524–535 (1993).
8. M. S. Wilm and M. Mann, Electrospray and Taylor-Cone theory, Dole's beam of macromolecules at last? *Int. J. Mass Spectrom. Ion Process.* **136**, 167–180 (1994).
9. M. R. Emmett and R. M. Caprioli, Micro-electrospray mass spectrometry: Ultra-high sensitivity analysis of peptides and proteins, *J. Am. Soc. Mass Spectrom.* **5**, 605–613 (1994).
10. J. H. Wahl, D. R. Goodlett, H. R. Udseth, and R. D. Smith, Attomole level capillary electrophoresis-mass spectrometric protein analysis using 5-µm-i.d. capillaries, *Anal. Chem.* **64**, 3194–3196 (1992).
11. M. Wilm and M. Mann, Analytical properties of the nanoelectrospray ion source, *Anal. Chem.* **68**, 1–8 (1996).
12. M. Wilm, A. Shevchenko, T. Houthaeve, S. Breit, L. Schweigerer, T. Fotsis, and M. Mann, Femtomole sequencing of proteins from polyacrylamide gels by nano electrospray mass spectrometry, *Nature* **379**, 466–469 (1996).
13. A. Shevchenko, M. Wilm, O. Vorm, and M. Mann, Mass spectrometric sequencing of proteins from silver stained polyacrylamide gels, *Anal. Chem.* **68**, 850–858 (1996).
14. M. Wilm, G. Neubauer, and M. Mann, Parent ion scans of unseparated peptide mixtures, *Anal. Chem.* **68**, 527–533 (1996).
15. A. Shevchenko, I. Chernushevich, W. Ens, K. B. Standing, B. Thomson, M. Wilm, and M. Mann, Rapid "de novo" peptide sequencing by a combination of nano-electrospray, isotopic labeling and a quadrupole/time-of-flight mass spectrometer, *Rapid Commun. Mass Spectrom.* **11**, 1015–1024 (1997).
16. S. Uttenweiler-Joseph, G. Neubauer, S. Christoforidis, M. Zerial, and M. Wilm, Automated de novo sequencing of proteins using the differential scanning technique, *Proteomics* (2001).

17. G. A. Valaskovic, N. L. Kelleher, D. P. Little, D. J. Aaserud, and F. W. McLafferty, Attomole-sensitivity electrospray source for large-molecule mass spectrometry, *Anal. Chem.* **67**, 3802–3805 (1995).
18. H. Erdjument-Bromage, M. Lui, L. Lacomis, A. Grewal, R. S. Annan, D. E. McNulty, S. A. Carr, and P. Tempst, Examination of micro-tip reversed-phase liquid chromatographic extraction of peptide pools for mass spectrometric analysis, *J. Chromatogr. A* **826**, 167–181 (1998).
19. J. Gobom, E. Nordhoff, E. Mirgorodskaya, R. Ekam, and P. Roepstorff, Sample purification and preparation technique based on nano-scale reversed-phase columns for the sensitive analysis of complex mixtures by matrix-assisted laser-desorption/ionization mass spectrometry, *J. Mass Spectrom.* **34**, 105–116 (1999).
20. D. F. Hunt, J. R. Yates III, J. Shabanowitz, S. Winston, and C. R. Hauer, Peptide sequencing by tandem mass spectrometry, *Proc. Natl. Acad. Sci.* (USA) **83**, 6233–6237 (1986).
21. K. Biemann and H. A. Scoble. Characterization by tandem mass spectrometry of structural modifications in proteins, *Science* **237**, 992–998 (1987).
22. H. Steen, B. Küster, M. Fernandez, A. Pandey, and M. Mann, Detection of tyrosine phosphorylated peptides by precursor ion scanning quadrupole TOF mass spectrometry in positive ion mode, *Anal. Chem.* (in press).
23. M. Mann and M. S. Wilm, Error tolerant identification of peptides in sequence databases by peptide sequence tags, *Anal. Chem.* **66**, 4390–4399 (1994).
24. P. Roepstorff and J. Fohlman, Proposed nomenclature for sequence ions, *Biomed. Mass Spectrom.* **11**, 601–602 (1984).
25. X. Jiang, J. B. Smith, and E. C. Abraham, Identification of a MSMS diagnostic for methionine sulfoxide, *J. Mass Spectrom.* **31**, 1309–1310 (1996).
26. D. N. Perkins, D. J. Pappin, D. M. Creasy, and J. S. Cottrell, Probability-based protein identification by searching sequence databases using mass spectrometry data, *Electrophoresis* **20**, 3551–3567 (1999).
27. J. R. Yates 3rd, A. L. McCormack, and J. Eng, Mining genomes with MS, *Anal. Chem.* **68**, 534A–540A (1996).
28. F. Karimi-Busheri, G. Daly, P. Robins, B. Canas, D. J. Pappin, J. Sgouros, G. G. Miller, H. Fakhrai, E. M. Davis, M. M. Le Beau, and M. Weinfeld, Molecular characterization of a human DNA kinase, *J. Biol. Chem.* **274**, 24187–24194 (1999).
29. A. Shevchenko, M. Wilm, and M. Mann, Peptide sequencing by mass spectrometry for homology searches and cloning of genes, *J. Protein Chem.* **16**, 481–490 (1997).
30. Y. Takao, H. Hori, K. Okamoto, A. Harada, M. Kamachi, and Y. Shimonishi, Facile assignment of sequence ions of a peptide labelled with ^{18}O at the carboxyl terminus, *Rapid Commun. Mass Spectrom.* **5**, 312–315 (1991).
31. M. S. Wilm and J. Kast, A true nanoelectrospray ion source compatible with on-line HPLC systems, poster presented at the 48th ASMS Conf. Mass Spectrometry and Allied Topics, Long Beach, CA, 2000.
32. G. Klesper, H. Klesper, and G. Fusshoeller, A new ES capillary design for improved detection sensitivity: Detailed characterization of the analytical qualities, poster presented at the 48th ASMS Conf. Mass Spectrometry and Allied Topics, Long Beach, CA, 2000.

CHAPTER 6

Laser-Machined Microdevices for Mass Spectrometry

YUEHE LIN, KEQI TANG, and RICHARD D. SMITH

Environmental Molecular Sciences Laboratory, Pacific Northwest National Laboratory, Richland, Washington, USA

CONTENTS

6.1. INTRODUCTION

Microfabricated analytical devices are of growing interest because of their potential to enhance sensitivity and sample speed, and to reduce instrument size, cost, and weight [1–4]. Initial attention has focused primarily on the development of miniaturized separation systems [5–9]. Most of the microanalytic devices reported have been fabricated on glass substrates using photolithog-

Mass Spectrometry and Hyphenated Techniques in Neuropeptide Research, Edited by Jerzy Silberring and Rolf Ekman.
ISBN 0-471-35493-7

raphy, wet chemistry, and chemical etching processes [1–9]. Issues associated with the use of glass chips for such devices include the need for time-consuming manual fabrication methods, high-temperature bonding requirements, and the highly fragile nature of the products. As a result, researchers have begun to evaluate polymer materials for microanalytic applications [10–12]. These materials are lightweight, chemically inert to common reagents used in analyses, easily fabricated and bonded, and robust. An additional advantage of using plastics as microchip substrates is their low surface charge, which often eliminates the need for time-consuming coating steps for certain types of separation.

Analyte detection remains an important issue for microdevices. Most analyses have been conducted using a laser-induced fluorescence (LIF) detector. However, LIF is not an ideal detection scheme, because it requires the analyte of interest to be fluorescent. Since most compounds are not natural fluorophores, a derivatization step is often required. Mass spectrometry (MS) has been investigated as an alternative method of detection for microfluidic devices. The mass spectrometer and the microchip separation device are well matched, because of the similarity between the flow rates generated by the microdevices and those required for electrospray ionization MS (ESIMS), despite the mismatch in physical dimensions [13]. One important advantage of using MS as microchip detector is that MS can provide molecular weight and structure information for positive identification of analytes.

The success of protein identification by ESIMS is critically dependent on the quality of the sample. Therefore, sample processing prior to mass spectrometric analysis is important to reduce the level of background contamination. The sensitivity of the analysis is also dependent on the type of ESI interface used. The low flow rates associated with microsprayers dramatically reduce sample consumption and enhance the sensitivity of the ionization process. Barry [14–15] and Aebersold et al. [16] and their respective co-workers were the first to demonstrate the advantages of microchips for protein sample preparation for ESIMS. Since then, several groups have developed microchip devices for MS sample processing [17–20]. The advantages of using microfabricated devices for sample delivery to ESIMS instruments include the following: (1) small quantities of analytes are consumed, (2) sample losses due to routine sample handling are reduced, (3) interference and contamination introduced by the experimenter are minimized, (4) the potential for sample multiplexing on a single device has been demonstrated, (5) integrated devices to perform several steps in an analysis can be generated, and (6) microfabricated devices can be operated automatically. Sample preparation, such as enzyme digestion of proteins, separation of a sample from other components such as salts and detergents during isolation and extraction, and separation of proteins or their digests by electrokinetic methods, have been demonstrated on chip [1–9,20]. The coupling of such sample processing with the additional separation and identification power of MS/MS may replace the slow and labor-intensive two-dimensional (2D) gel electrophoresis technology commonly used now for protein separation. In this chapter, we summarize the most recent work conducted in our lab on the development of laser-machined plastic microfluidic devices for rapid and automatic sample cleanup and separations of biological samples for mass spectrometric detection of proteins.

6.2. FABRICATION METHODS

Ultraviolet (UV) excimer laser direct-write micromachining systems (Potomac LMT-4000 and LMT-5000) were used to produce microchannels of various geometries in polymeric materials. The systems contained Potomac TGX-1000 KrF (248 nm) excimer lasers configured for focused (direct-write) operation at the machining surface. The minimum spot size of an unapertured laser beam at the work surface was approximately 10 µm. The maximum pulse repetition rate was 2 kHz with a maximum pulse energy of 45 µJ. Machining was performed on a set of 4-in. (LMT-4000) or 12-in. (LMT-5000) computer-controlled *X–Y* stages.

Microchannels in polymer devices were produced using two distinct methods. One approach was to use the focused excimer laser beam to directly mill channels into bulk polycarbonate or polyimide substrates. Because the unapertured laser beam width at the substrate was approximately 10 µm, multiple parallel passes were required to produce channels that were typically 50 or 100 µm wide. The depths of the channels produced by this method, commonly 50–150 µm, were controlled by adjusting the feed rate of the part being machined or by changing the repetition rate of the laser. Microchannels produced by this method were sealed by applying a thin polyethylene terephthalate (PET) sheet directly to the face of the part using a heat-activated adhesive, or by using a gasket compressed between the machined face and a second piece of holding block. Fluid connections to the microchannel were commonly made by mechanically machining the female part of a small chromatographic fitting (e.g., Upchurch Scientific) through the part from the back, and connecting to the microchannel through a small drilled hole.

A second approach to the formation of microchannels involved the use of a polymer (e.g., polyimide) gasket with the channel machined through the thickness of the material. This method was used to produce microchannels up to 500 µm wide and 225 µm deep. For this approach, only the perimeter of the channel needed to be machined, with the remaining center portion of the channel stripped out and discarded. When sandwiched between two solid pieces of polymer, the gasket formed a well-defined microchannel whose depth was determined by the thickness of the gasket material. These gaskets could be produced from sheet material having adhesive on one or both sides to improve sealing capabilities.

When feasible, devices were stacked together to form three-dimensional structure. In doing so, fluid flows were directed through a series of layers in the device. Minimum dead volume in the device was achieved by directing the fluid flow from one layer to the next using small-diameter drilled vials.

The micronozzle arrays (discussed in more detail in Sections 6.6 and 6.7) were fabricated from 1-mm-thick polycarbonate sheet material using a Resonetics Maestro UV excimer laser micromachining station (Resonetics, Nashua, NH). All machining was performed using a Lumonics 848 excimer laser operating at 248 nm. Through-holes were first machined at a high demagnification factor (~35×) using a small circular mask in the laser beam. The laser demagnification was then reduced to approximately 5× and a well was machined around the nozzle outlet. Because of the inherent taper of the laser at low demagnification factors, the nozzle tips produced by this method had a conical cross section recessed into the end of the chip. The openings of these micromachined nozzles were flush with the end of the polycarbonate chips, thereby reducing susceptibility to breakage or other damage. Nozzle diameters as small as 50 μm have been fabricated using this method.

6.3. MASS SPECTROMETRIC ANALYSIS

A Finnigan LCQ (San Jose, CA) quadrupole ion trap mass spectrometer with a modified micro-ESI source was used for microdialysis and isoelectric focusing/ESIMS experiments. The heated inlet capillary was maintained at 160°C. For online microdialysis/ESIMS experiments, the high voltage for ESI was applied to the metal syringe needle used for the buffer delivery and was carried to the sample side through the electrical conductivity of the buffer solution. Details of the membranes and flow rates used in specific experiments are discussed in the following section. A typical electrospray voltage was 2.2 kV.

6.4. MICROFABRICATED DIALYSIS DEVICES

6.4.1. Microfabricated Single-Stage Microdialysis Device for Fast Desalting of Biological Samples

ESIMS is amenable to analysis of nearly the entire range of biomolecules, and the use of mild ionization conditions allows noncovalent associations to be directly detected. However, sample matrix interference (generally due to low-molecular-weight salts) presents a major limitation to the use of ESIMS for analyzing large biopolymers. Biological samples extracted under physiological conditions typically contain significant amount of nonvolatile solvents and salts (e.g., sodium chloride), and other reagents are often added to stabilize samples or to maintain their enzymatic activity. The presence of small to moderate amounts of salts will not only affect electrospray stability but may also significantly reduce the analyte ion intensity due to the ion competition in the electrospray process and the formation of multiple sodium adduct species. Low signal-to-noise ratios in the mass spectra and poor reproducibility due to excessive adduction can result in inaccurate mass assignments and, in severe cases, even preclude spectrum interpretation. Although online liquid chromatography (LC)/MS can tolerate small to moderate amounts of salt and desalting columns are commercially available, they are not suitable for many ESIMS experiments where samples may precipitate on the column or interact with the solvent, or when very high salt concentrations are present. Direct ESIMS of biological samples often relies on centrifugation or offline microdialysis for desalting. Such a "batch cleanup" process can be time-consuming (taking several hours) and may suffer from adverse effects, such as sample losses during each step due to protein aggregation or precipitation on the filtering media. When only a small amount of sample is available, these effects can become even more pronounced. If noncovalent associations are of interest, slow desalting steps are more likely to result in denaturation and/or complex dissociation in solution.

We used a microdialysis approach in our lab to address complex biological samples [21–23]. This approach employs a small microdialysis tube through which the sample flows in conjunction with a countercurrent flow of a suitable buffer. The dialyzed samples can be directly interfaced with the ESI source at a flow rate of approximately 2–5 μL/min. While the microdialysis approach has demonstrated great promise for efficient sample cleanup capability, further improvement of this approach (i.e., to reduced further liquid flow rates and sample process time) is limited by the need to use dialysis fibers of even smaller diameter (the dialysis fibers of 200 μm were used in our earlier work). It is clear that miniaturization of the approach will not only further increase its speed, flexibility, and robustness but also allow the use of lower flow rates and result in enhanced sensitivity. More recently, we reported on a microfabricated microdialysis device constructed for this purpose, and demonstrated its desalting efficiency for protein samples [17].

Figure 6.1 shows the structure of the single-stage microdialysis device. The sample and buffer channels were machined directly into the polycarbonate chip (30 × 30 × 6 mm) using a laser micromachining system. The sample channel was 160 μm wide, 60 μm deep, and 11 cm long. The buffer channel was 500 μm wide, 250 μm deep, and 12 cm long. A Spectra/Por Biotech 1.1 dialysis membrane with a molecular weight cutoff (MWCO) of 8000 was sandwiched between the chips. Figure 6.1b shows a photo of the overlapping sample and buffer channels after alignment of the microchannel dialysis chips.

The microfabricated dialysis device can be used in the online mode for direct ESIMS analysis and has been evaluated for this purpose using the arrangement shown in Fig. 6.2. High voltage was applied to the metal syringe needle used for buffer delivery and provided electrical contact through the buffer and sample solution. In this arrangement, the ESI emitter was a short piece of silica capillary (4–6 cm, 200 μm o.d. × 100 μm i.d.) fixed to the chip through a standard fitting. A 5-μM horse heart myoglobin solution in a

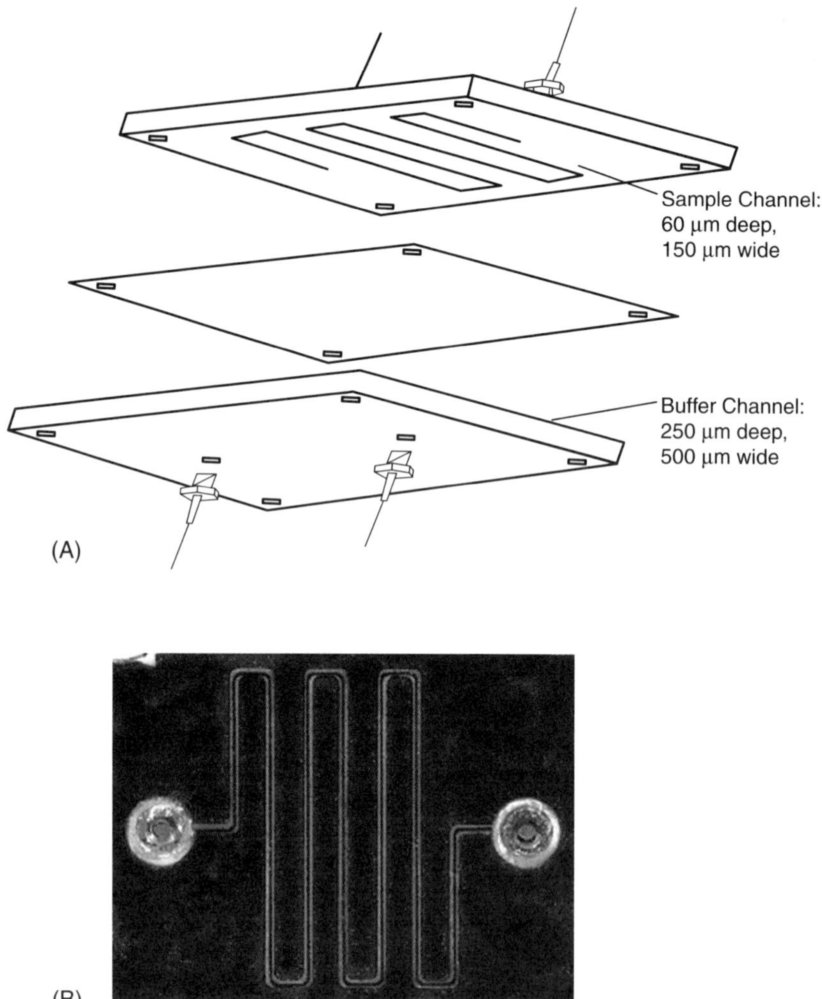

Figure 6.1. Schematic diagram of microfabricated single-stage microdialysis device (a) and photo of overlapping sample and buffer channels after alignment of the microchannel device (b).

complex matrix consisting of 500 mM of NaCl, 100 mM of Tris, and 10 mM of EDTA was studied as an example. Figure 6.3a shows a mass spectrum obtained for this solution by direct infusion with the microspray ESI source. No protein charge state envelope was evident, and the spectrum was complicated by undesired peaks due to Tris and EDTA (commonly used in protein sample preparation). Online microdialysis was carried out using a dialysis buffer solution containing 10 mM of NH_4OAc and 1% (v/v) acetic acid at a flow rate of 100 μL/min, while the sample was introduced at a flow rate of 0.5 μL/min. As

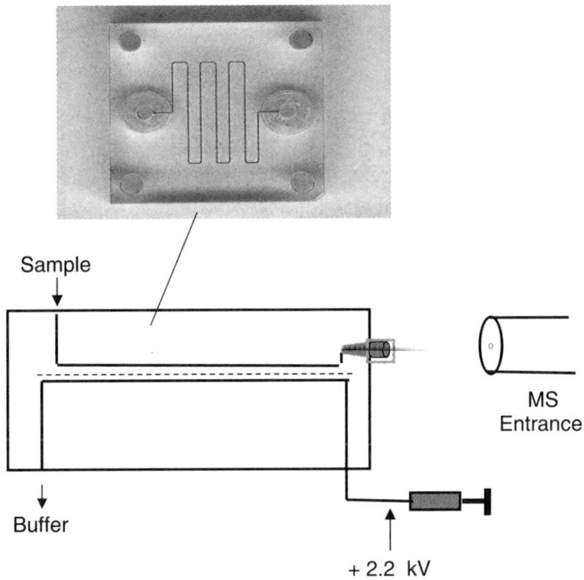

Figure 6.2. Schematic representation of the setup of the microfabricated device for online ESIMS analysis. Top panel shows a photograph of the sample channel chip. (Reprinted with permission from Ref. 17, copyright 1998 American Chemical Society.)

shown in Fig. 6.3b, effective desalting was achieved. Multiple charge states from 19+ to 10+ of denatured myoglobin were clearly resolved, and the interference peaks from the Tris and EDTA were greatly diminished. The signal-to-noise ratio was increased by a factor of more than 40 compared to that obtained without desalting.

With the current design, the total volume of the sample serpentine channel is only 1 μL and the actual dialysis time for the sample, that is, the time the sample spent inside the microdialysis channels was ~2 min at a sample flow rate of 0.5 μL/min. The effectiveness of the cleanup can be ascribed to the size difference between the sample channel and the buffer channel, and the fact that the cleanest sample solution is effectively exposed to the cleanest dialysis buffer, due to its countercurrent flow configuration. For effective desalting, it has been found that air bubbles trapped inside the sample channels need to be forced out before the dialysis occurs. The sample loss during the dialysis is minimal due to the short residence time in the microdialysis channel, and the fact that the analytes are above MWCO of the chosen membrane. The Spectra/Por Biotech 1.1 membrane used for this work has a high tolerance to acidic solvents, such as 90% acetic acid and organic solvents (e.g., 90% methanol), thus providing versatility in application of the microfabricated dialysis device.

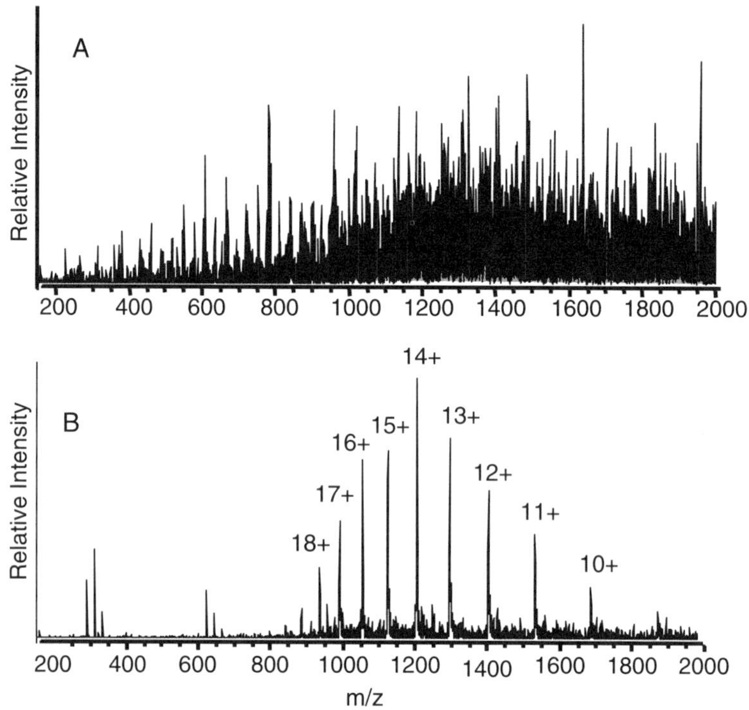

Figure 6.3. (a) ESI mass spectrum of 5-μM horse heart myoglobin in 500 mM of NaCl, 100 mM of Tris, and 10 mM of EDTA by direct infusion; (b) ESI mass spectrum of previous myoglobin sample after online microdialysis using 10 mM of NH_4OAc and 1% acetic acid as a dialysis buffer. (Reprinted with permission from Ref. 17, copyright 1998 American Chemical Society.)

6.4.2. Microfabricated Dual-Stage Microdialysis Device for Rapid Fractionation and Cleanup of Complex Biological Samples

As discussed previously, a single-stage microdialysis device can be used for rapid and efficient desalting of biological samples. More recently, we developed a microfabricated dual-stage microdialysis device for fast and efficient fractionation and cleanup of complex biological samples (i.e., cellular extracts) [24].

Figure 6.4 shows the structure and flow diagram of the dual-stage microdialysis device. The sample channel (60 μm deep, 150 μm wide) was machined directly into the polymer chip (30 × 30 × 6 mm) by using multiple parallel laser passes. A 500-μm-wide buffer channel was machined from 225-μm-thick polyimide with a silicone adhesive on one side. The adhesive was used to attach the buffer channel to a flat polymer chip. Two microdialysis membranes are sandwiched between three polymer chips with microfabricated serpentine channels. This novel design makes the dual-stage microdialysis device highly

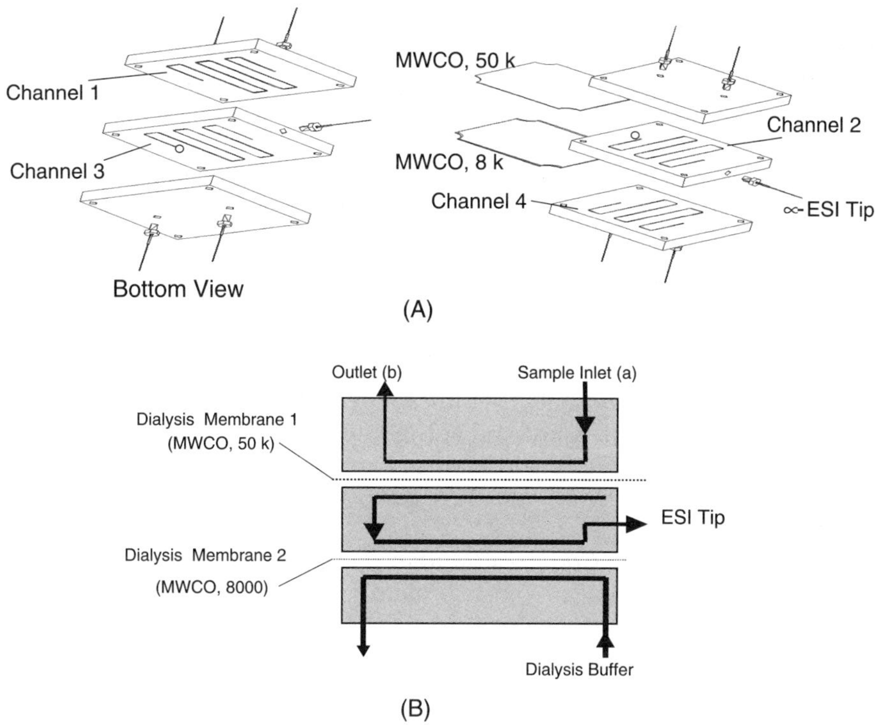

Figure 6.4. (a) Exploded view showing construction of the microfabricated dual-stage microdialysis device; (b) flow diagram of the device showing sample inlet, buffer inlet, and microelectrospray tip. (Reprinted with permission from Ref. 24, copyright 1999 American Chemical Society.)

compact and rugged, eliminating the need for tubing to connect the two stages of microdialysis units. The fluid channels were designed to make the raw sample flow in a microchannel along one side of the first-stage membrane so that a portion of sample flows through the membrane and enters the second stage. This partial flow in the first stage provides a mechanism for sweeping out cellular residues in the sample channel and results in a reduced possibility of membrane clotting. Alternatively, the sample channel outlet (b) can be completely sealed if the lifetime of the membrane is not of concern. The sample flow rate entering the second stage of the device can be adjusted by controlling the flow rate at the outlet (b) by throttling the sample flow through fused-silica tubing of various lengths and inner diameters. The sample entering the second stage of the device is stripped of the small molecules and salts by a dialysis buffer flowing in a microchannel on the opposite side of the second-stage membrane.

A protein mixture consisting of 30 μM of BSA, 8 μM of cytochrome C, and 2.4 μM of ubiquitin in 10-mM PBS was used to evaluate the performance of the dual-stage microdialysis device. A 50-kDa MWCO membrane and a 8-kDa MWCO membrane were used in the first- and the second-stage microdialyses, respectively. The sample was injected at 0.5 μL/min and the cleanup buffer, at 10 μL/min. As shown in Fig. 6.5a, direct infusion of the protein mixture produced no useful ESIMS spectrum. ESIMS performance is well known to be poor for such solutions, presumably due to the excess BSA and NaCl. After the online dual dialysis in the microfabricated device, the spectrum in Fig. 6.5b clearly showed peaks from cytochrome C, ubiquitin, and removal of the unresolved envelope of peaks at higher *m/z*. The improvement in spectral quality using the dual-stage microdialysis device enabled us to effectively assign peaks and accurately determine molecular weights.

For analyzing complex biological samples, initial sample fractionation and cleanup can simplify analytical procedure. Online dual dialysis using the microfabricated device is a fast and efficient means to achieve this. The molecular weight range selected for analysis by dual microdialysis can be controlled by selection of appropriate MWCO membranes at both stages.

6.4.2.1. Application to Complex Cellular Samples. The analysis of cellular extracts and other complex biological samples is of great interest for many purposes, including biotechnology, pathogen detection, medical research, bioremediation, and clinical applications. However, such analysis will require that complex samples be readily separated, fractionated, or "cleaned up" to the point where useful mass spectra can be obtained. Interferences from both the high-molecular-weight and low-molecular-weight components constitute limitations for the direct use of ESIMS for analysis of target biomolecules in such complex systems.

The use of the microfabricated dual-stage microdialysis device for sample processing is attractive because of its ruggedness, its potential for extension to multichannel devices, and its feasibility in greatly increasing speed and care of automation. To evaluate the processing efficiency of the microfabricated dialysis device, an *Escherichia coli* lysate with a total protein concentration of ~1 mg/mL was analyzed by ESIMS before and after the dual dialysis. Direct infusion of the crude cell lysate produced the spectrum shown in Fig. 6.6a. The spectrum is largely uninterpretable because of the complex components, the high concentration of NaCl, and the consequently low signal-to-noise ratio. After the sample was passed through the dual-stage microdialysis device with a 50-kDa (MWCO) membrane in the first stage and a 8-kDa (MWCO) membrane in the second stage, a significant improvement in spectral quality was observed (Fig. 6.6b). A ~20-fold increase of signal-to-noise ratio was obtained. The characteristic peaks from major cellular components are readily discernible in the spectrum and provide the basis for further MS/MS experiments [25]. Lower flow rates with ESIMS provide greater sensitivity [26]. With the current device, online ESIMS can be conducted at a few hundred nanoliters per minute, more than an order of magnitude reduction compared to the previous dual-stage dialysis system [27]. Another feature of the current device is

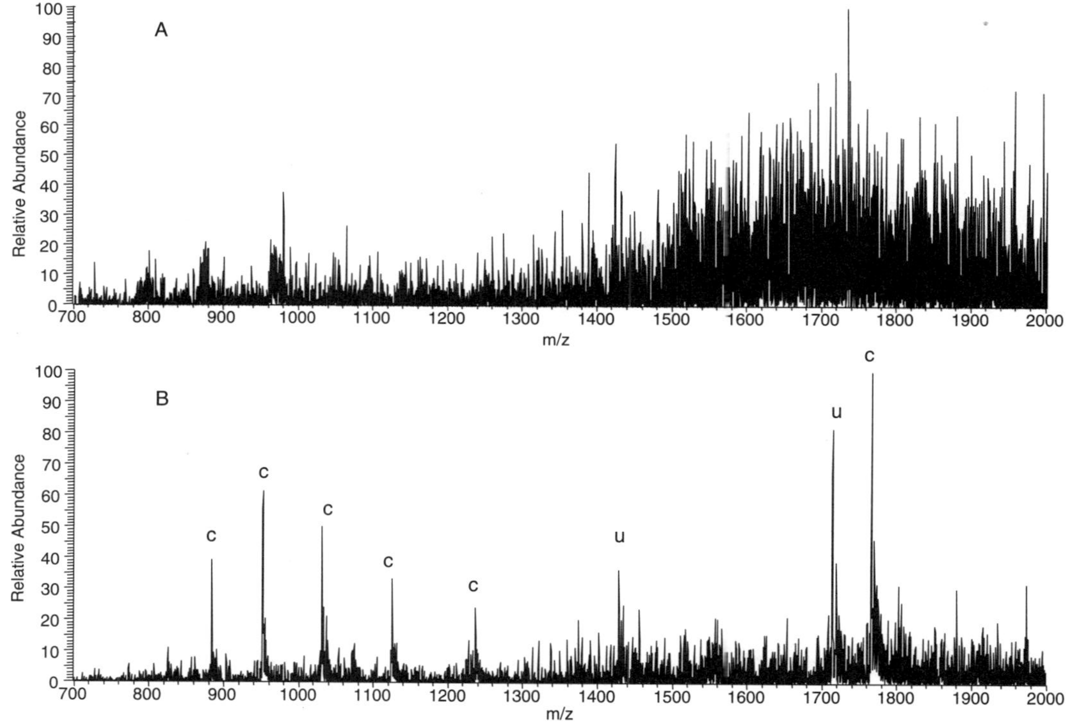

Figure 6.5. ESIMS spectra of a protein mixture consisting of 30 μM of bovine serum albumin, 8 μM of cytochrome C, and 2.4 μM of ubiquitin in 0.01 M of PBS: (a) sample directly infused without passing through the dual-microdialysis device; (b) sample passed by the dual-microdialysis device. Identities of the peaks: u, ubiquitin; c, cytochrome C. (Reprinted with permission from Ref. 24, copyright 1999 American Chemical Society.)

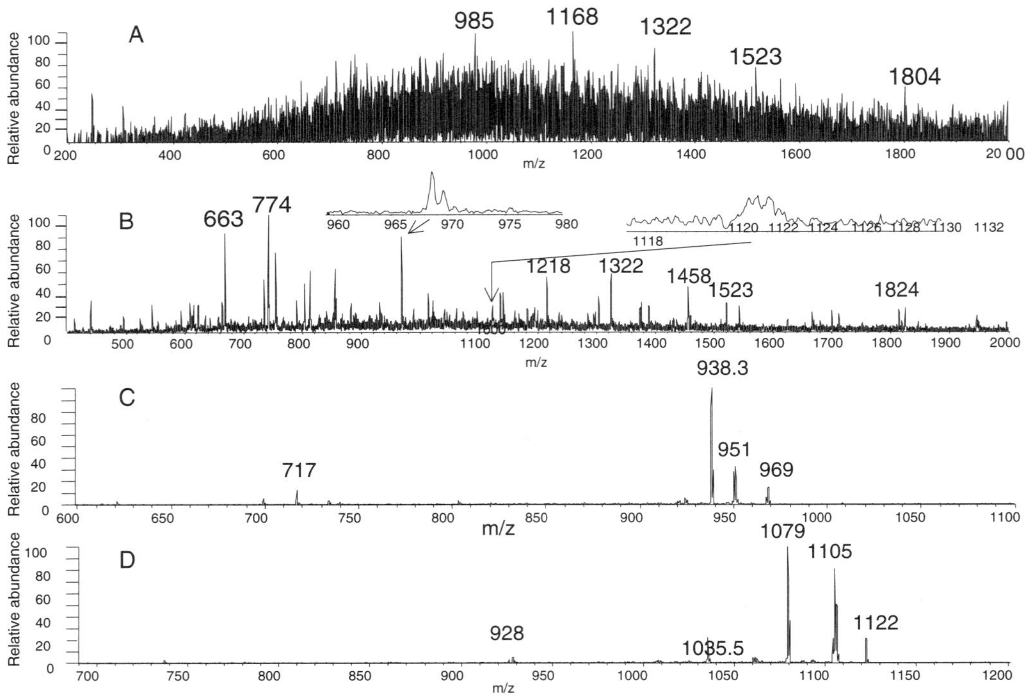

Figure 6.6. ESIMS spectra for an *E. coli* cell lysate: (a) spectrum obtained from the sample directly infused without passing through the dual-microdialysis device; (b) spectrum obtained using online dual-microdialysis sample processing—two expanded views show detail of peaks at *m/z* 969 and 1124; (c) CID spectrum of *m/z* 969 at a relative collision energy of 35%; (d) CID spectrum of *m/z* 1124 at a relative collision energy of 35%. (Reprinted with permission from Ref. 24, copyright 1999 American Chemical Society.)

the reduced dead volume achieve by eliminating connection tubing between the two stages. These features should allow for faster analysis of smaller samples. MS/MS of distinctive ions provides a potential basis for identification (i.e., serving as "biomarkers") of microorganisms. Figures 6.6c,d show two MS/MS spectra of m/z 969 and 1124 ions in the expanded views of Fig. 6.6b. The spectra in Figs. 6.6c,d show an average of 38 scans and present an excellent signal-to-noise ratio. Using a flow rate of 200 nL/min, only 16 μL of sample were consumed for MS/MS data acquisition of 40 cellular components.

We have demonstrated the application of the dual microdialysis/ESI/MS/MS for the identification of microorganisms in our lab [25]. The distinctive MS/MS spectral patterns can be used to identify mass spectrometric-detected species useful as biomarkers, which then provide a basis for confident identification of microorganism. The results presented also demonstrate the application of this for detection of bacteriophage MS2 in the presence of a large excess of *E. coli.*

6.5. MICROFABRICATED ISOELECTRIC FOCUSING (IEF) DEVICE FOR PROTEIN SEPARATIONS

Capillary isoelectric focusing (CIEF) is a technique for high-resolution separations of proteins. However, conventional CIEF using fused-silica capillary tubing commonly suffers from poor reproducibility resulting from inconsistent sample handling and poor hydrolytic stability of the capillary coatings [28–30]. Isoelectric focusing (IEF) devices microfabricated on polymer chips can circumvent some of these limitations and can potentially be integrated with similarly miniaturized injection and interfacing components, enabling much more rapid and automated analyses.

In our most recent work, microfabricated IEF chips on polymer substrates (polycarbonate, polyimide) were used for the separation of proteins [31]. Electroosmotic flows in microchannels directly laser-machined on these polymers are typically lower than those in fused-silica capillary. This suggests that isoelectric focusing experiments may be performed directly in these microfabricated IEF chips without the need for coating of the microchannels, thereby allowing for improved reproducibility in this microchip IEF mode.

To reduce the dead volume in the interface between the microchip and the mass spectrometer and improve both speed and sensitivity, an electrospray emitter was directly fabricated into the end of the device. Extensive experience with ESI at Pacific Northwest National Laboratory has demonstrated that sharper emitter "tips" give higher electric fields at the tip and result in better electrospray performance. Figure 6.7 shows schematic diagram of an integrated IEF microdevice consisting of three pieces: two pieces of 100 × 30 × 5-mm-thick polycarbonate and one piece of polyethylene terephthalate (PET) polymer film with a heat-activated adhesive (25-μm-thick silicone) on one side. The pointed tip of the polycarbonate part was mechanically machined

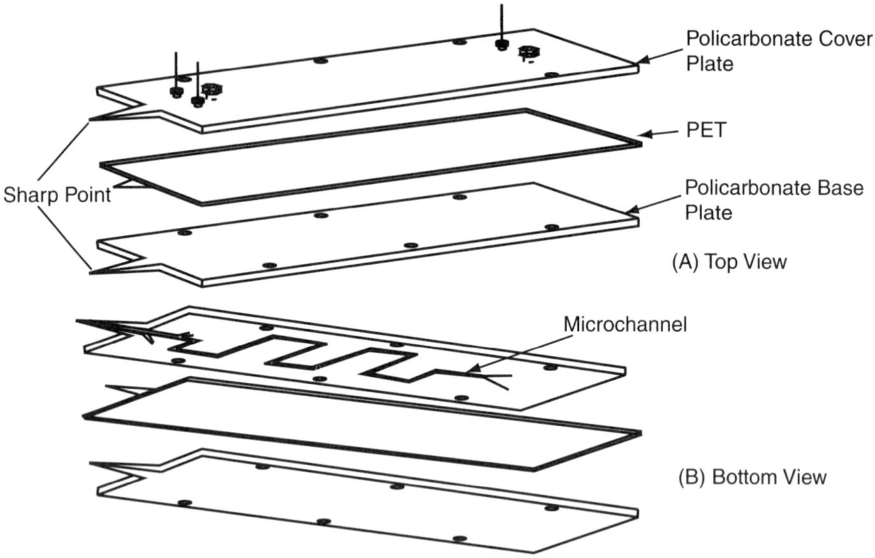

Figure 6.7. Exploded view showing construction of a microfabricated IEF device: (a) top view; (b) bottom view. (Reprinted with permission from Ref. 31, copyright 2000 Wiley-VCH Verlag GmbH.)

with the two polycarbonate pieces pressed together. When separated, each chip contained half of the cone-shaped point. The laser-machined microchannel (50 μm wide, 60 μm deep, and 16 cm long) in the IEF chip (10 × 3 × 0.5 cm) was fabricated directly on the polycarbonate substrate, similar to the sample channel on the microdialysis device described above. Dialysis or ion-exchange membranes were sealed against the bottoms of both the cathode and anode reservoirs by small O-rings (1 mm o.d.) held in place with fingertight fittings, which allowed electric conduction in the microchannel but prevented the fluid from backflowing through the membranes. To form the closed channel network, a PET cover sheet was thermally bonded onto the top of the chip containing the channel and backed with the second polycarbonate piece, as shown in Fig. 6.7. The device included the laser-machined microchannel; three female fittings for sample, sheath gas, and sheath liquid injection; and two buffer reservoirs (2 × 2 mm) machined to accept fingertight fittings. The fittings and reservoirs were mechanically machined into the backside of the chip containing the separation channel prior to sealing the channel with the PET film. Formation of microchannel by direct thermal bonding of two polycarbonate chips was also used for fabrication of IEF microdevices.

IEF separation with mass spectrometric detection was conducted in two steps as shown in Fig. 6.8. In the first step, a sample solution containing proteins and 1.0% pharmalyte (PI 3–10) was injected and filled the entire channel. An electric field was applied across the channel, and isoelectric focusing was

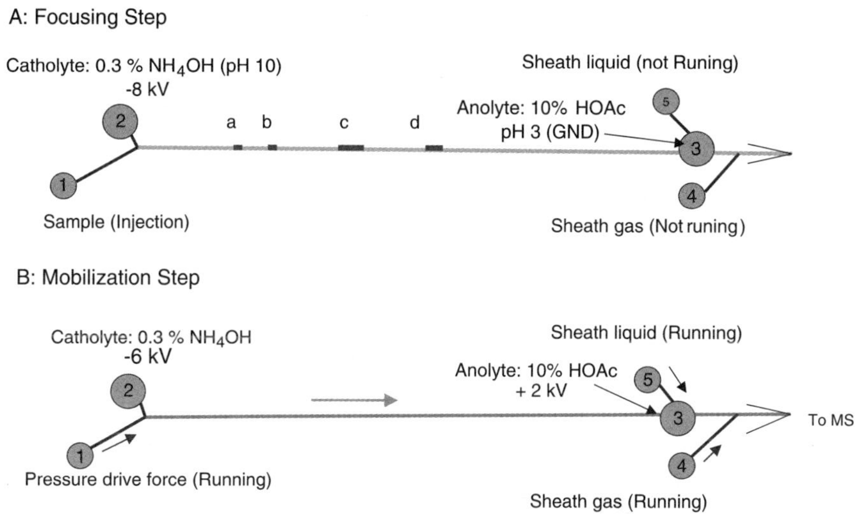

Figure 6.8. A schematic representation of two steps in the separation in a microfabricated IEF/ESIMS. (Reprinted with permission from Ref. 31, copyright 2000 Wiley-VCH Verlag GmbH.)

performed at a constant potential (–8 kV) for approximately 5–10 min. Proteins were focused on different zones according to differences in their isoelectric points (PIs). In the second step, separated proteins were moved into the MS detector by applying pressure using a syringe pump. During operation, the electrospray emitter tip was positioned in front of the sampling orifice of the mass spectrometer. The applied high voltage (2 kV) and nebulizer gas (nitrogen) flowing around the electrospray emitter tip created a fine spray of charged particles for the MS analysis.

We have conducted preliminary experiments to separate a protein mixture with a concentration of 0.05 mg/mL for each protein. Good separation of myoglobin and carbonic anhydrase was achieved without coating of the flow channel. The results indicate that electrospray directly from the sharp-edge microchip is feasible for separation and detection of proteins with microfabricated IEF/ESIMS.

6.6. MULTIPLEXING

Multiplexing is one aspect of microchip MS coupling that offers great potential for high-throughput MS analysis. The ability to handle multiple samples using microchip devices is one of the most significant advantages offered by microfluidic network. The first example of multiplexing combined with microchip MS involved spraying directly from the edge of the microchip, using six

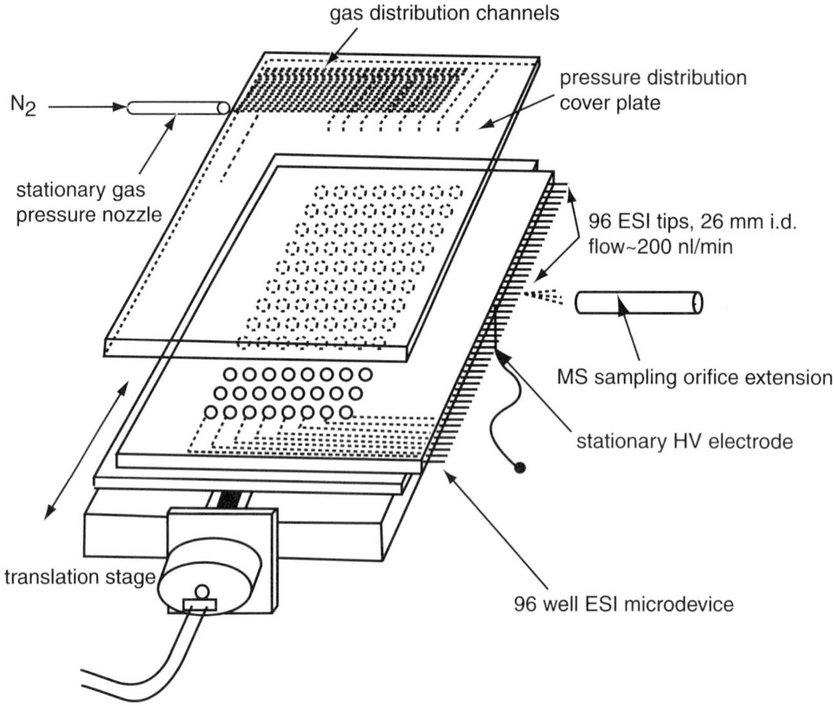

Figure 6.9. Exploded view of the total system design of a 96-ESI channel, 96-well device. The 96-well plate with individual channels and electrospray tips was positioned on a translation stage in front of the extension of the MS sampling orifice. The electrospray analysis of individual samples was activated by sequential pressurization of the sample wells through the pressure distribution coverplate and connection of the ESI high voltage (HV) through the stationary HV electrode positioned under the ESI device. The silicone rubber sealing gasket placed between the ESI device and the pressure distribution coverplate and the aluminum clamping plate were omitted for simplicity. (Reprinted with permission from Ref. 32, copyright 2000 American Chemical Society.)

independent devices on one wafer with six separate spray orifices [14]. The multispray microfluidic device has been extended to 96 microchannels for high-throughput infusion ESIMS of peptide [32]. The prototype plastic devices were fabricated by casting from a solvent-resistant resin. Ninety-six fused-silica capillaries (2.5 cm long, 26 μm i.d., 140 μm o.d.) (electrospray tips) were inserted into the exits of the channels and glued in place. The exploded schematic diagram in Fig. 6.9 shows the total system design. During operation, the 96-channel/96-ESI-tip microdevice was positioned on a computer-controlled translation stage so that ESI tips were aligned with the MS sampling orifice extension. High-throughput ESIMS was demonstrated by analyzing 96 peptide samples in 480 s, corresponding to a potential throughput of 720 samples/h.

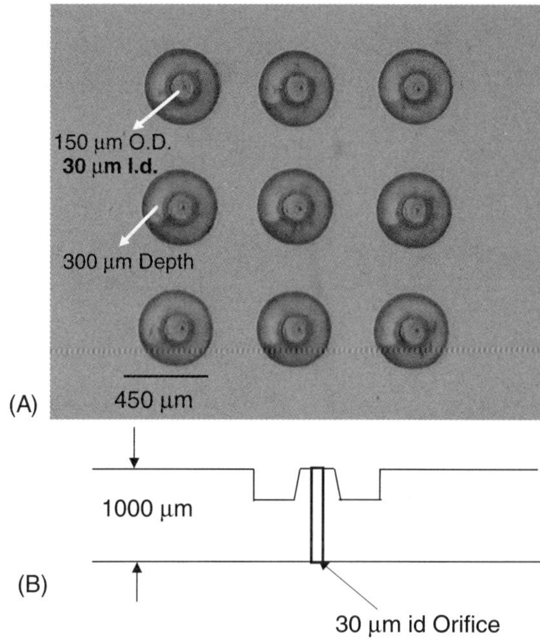

Figure 6.10. Microfabricated electrospray nozzle array: (a) photomicrograph of a nine-electrospray nozzle array arranged in 3×3 configuration; (b) dimensions for each spray nozzle in the array.

Interfacing the multichannel microfluidic device with MS is still a technical challenge. Gluing shortcapillaries to serve as electrospray emitters is a common method [16,18,32]. However, because of the small channel dimensions involved, the assembly of such devices is often time-consuming and difficult to replicate. The need to add a capillary emitter to a microchip can also result in the introduction of "dead volume." Direct electrospray from the flat surface of the edge of microchips was reported by Karger et al. [14] and Ramsey et al. [33]. However, on the basis of our experiment, electrospray directly generated from the flat surface of microchip is unstable. Formation of larger liquid droplets at the flat surface of the chip causes dead volume and sample dilution.

At our lab, we have developed plastic microchips containing arrays of laser-micromachined electrospray nozzles to interface the multichannel microfluidic device with MS [34]. Figure 6.10 shows the photo of a microchip containing 3×3 micro-ESI nozzle arrays. Each nozzle measures 150 μm o.d., 30 μm i.d., and 300 μm deep (high). The number of micronozzles can be easily expanded to 96 if necessary. For different applications, micronozzles with various diameters can be readily fabricated. Nozzles with diameters as small as 50 μm o.d. and 15 μm i.d. for generation of nanospray from chips have been fabricated using this method.

When the laser-machined micronozzle arrays were tested for electrospray formation and stability, we observed that under a variety of conditions the solution exiting the hole rapidly moistened the edge surface of the nozzle. This wetting prevented the formation of a well-focused electric field, which is essential for the generation of a stable electrospray. To inhibit the wetting process, we treated the surfaces of the polycarbonate with CF_4 RF plasma. After this plasma treatment, the hydrogen atoms on the polymer surface were replaced by fluorine atoms, which changed the surface property of the polycarbonate chip from hydrophilic to hydrophobic. The increased hydrophobicity of the treated polycarbonate surface prevented the sample solution from spreading over the edge surfaces of the nozzles and helped focus the electric field at the surface of the liquid exiting the channel. Stable electrospray was directly generated from the microchip without the use of a nebulizing gas. The microelectrospray nozzle array provides a reproducible, controllable, and robust means of producing electrospray of a liquid sample from a plastic microchip. The stable electrospray could be generated at a flow rate as low as 0.1 μL/min. The results indicated that laser-machined micronozzle arrays could be used as an interface between multiple channel microfluidic devices with MS. Microchips with high-density electrospray nozzle arrays can be directly integrated with a multichannel microlfluidic device. Since the fabrication method allows for high-density nozzle arrays of any layout to be made, micronozzle arrays can potentially be fabricated directly from the exits of the microchannels.

6.7. MICRODEVICE WITH HIGH-DENSITY MICRONOZZLE ARRAYS FOR IMPROVING MASS SPECTROMETRY SENSITIVITY

One application of microfabricated multielectrospray arrays is to enhance the sensitivity of mass spectrometry. A microfabricated electrospray array was used as an ionization source and was interfaced with a modified PE-Scex API-3000 triple quadrupole mass spectromter (Fig. 6.11). The standard curtain gas–skimmer interface of the API-3000 was replaced with a heated multicapillary (7 × 500-μm) inlet and an electrodynamic ion funnel interface for improved spray desolvation and ion transmission efficiency, as described in another publication [34]. The inlet was constructed from an array of seven thin-wall stainless-steel tubes soldered into a central hole of a cylindrical heating block. An electrodynamic ion funnel was used in the interface region to more effectively capture, focus, and transmit ions from the multicapillary inlet. The micronozzle array, mounted on a metal block, was positioned about 5 mm away from the multicapillary inlet. A high-voltage DC power supply, connected to the metal block, was used to provide the voltage for electrospray. A syringe pump connected to the metal block through a standard LC fitting was used for the sample infusion.

Figure 6.12 shows photomicrographs of microelectrospray arrays generated from a 3 × 3 micronozzle array chip. As shown in Fig. 6.12a, without surface

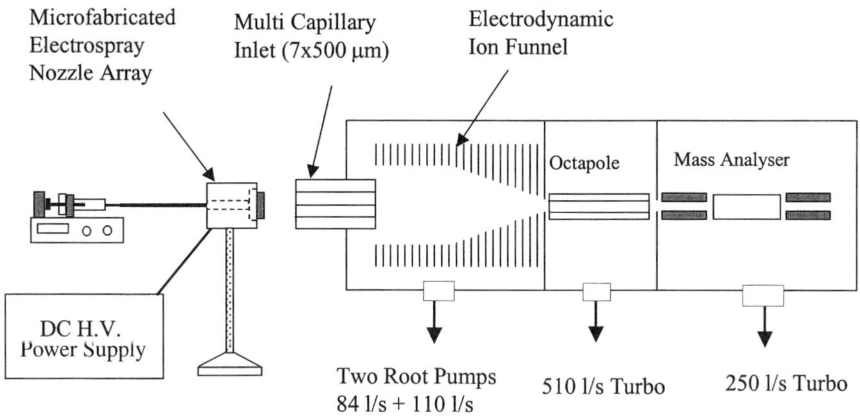

Figure 6.11. Experimental setup for the mass spectrometry evaluation of microfabricated electrospray array as an ionization source [34].

Figure 6.12. Photomicrographs of the Taylor cones and electrosprays generated from the 3 × 3 nozzle array: (a) surface untreated; (b) surface treated with a RF CF_4 plasma.

treatment, liquid tends to wet the outside surfaces of the nozzles and generates larger Taylor cones. After surface treatment with a RF CF_4 plasma, nines stable and well-defined Taylor cones were generated. Sensitivity comparisons between a conventional single electrospray using a fused-silica capillary and multielectrosprays from a microfabricated nozzle array indicated a significant enhancement by using multielectrosprays (up to nine electrosprays) as an ion source for MS. Further sensitivity improvement is possible by further increasing the number of electrosprays and better aligning the spray array and capillary inlet of the mass spectrometer. Even greater increases in the total ESIMS signal and the dynamic range of measurements are expected if each electrospray emitter in the array is operating at a flow rate yielding its maximum ion current. In this manner much greater ion currents should be feasible in situations where higher flow rates are available, and in particular for liquid chromatography where conventionally used flow rates are often up to two orders of magnitude larger (e.g., mL/min) than effectively electrosprayed. The exploitation of such gains will require further advances in ion funnel technology in order to efficiently transmit the much larger ion currents that are anticipated. Success in these efforts would greatly extend the dynamic range of many mass spectrometric applications.

6.8. CONCLUSIONS

Laser-machined microfabricated microdialysis devices were constructed for rapid online sample cleanup/ESIMS analysis of biological samples. Efficient desalting and fractionation of complex biological samples were demonstrated using ESI with an ion trap mass spectrometer. A microfabricated IEF chip was also constructed using laser-machining technology. Concentration and separation of protein mixtures have been successfully demonstrated. Microchip with high-density ESI arrays were fabricated in polymer substrate and the potential for sample multiplexing on a single device has been demonstrated. Because the microfabricated devices developed were readily fabricated on inexpensive polymer substrates, there is a great potential for large-scale production of inexpensive and disposable devices. Integration of these microfabricated devices with other nanoscale manipulations is expected to lead to more powerful, sensitive, less expensive, and rugged instrumentation for biomolecule analysis.

ACKNOWLEDGMENTS

We would like to thank past postdoctoral fellows Drs. Naxing Xu, Jian Y. Wen, and Fan Xiang for their contributions to the original work of microdevice MS. We also thank Dr. Dean W. Matson and Wendy D. Bennett for help in microfabrication. Portions of this work was supported by the National Cancer Institute (Grant CA81654) and the U.S. Department of Energy, Office of Biological and Environmental Research. Pacific Northwest National Laboratory is operated by Battelle Memorial Institute for the U.S. Department of Energy under Contract DE-AC06–76RLO 1830.

REFERENCES

1. D. Schmalzing, A. Adourian, L. Koutny, L. Ziaugra, P. Matsudaira, and D. Ehrlich, *Anal. Chem.* **70**, 2303–2310 (1998).
2. L. C. Waters, S. C. Jacobson, N. Kroutchinina, J. Khandurina, R. S. Foote, and J. M. Ramsey, *Anal. Chem.* **70**, 5172–5176 (1998).
3. J. J. Hoyt and W. G. Wolfer, *Electrophoresis* **19**, 2432–2439 (1998).
4. G. Ocvirk, T. Tang, and D. J. Harrison, *Analyst* **123**, 1429–1434 (1998).
5. D. J. Harrison, K. Fluri, K. Seiler, Z. Fan, C. S. Effenhauser, and A. Manz, *Science* **261**, 895 (1993).
6. J. Cheng, L. C. Waters, P. Fortina, G. Hvichia, S. C. Jacobson, J. M. Ramsey, L. J. Kricka, and P. Wilding, *Anal. BioChem.* **257**, 101–106 (1998).
7. J. N. Vandermoolen, H. Poppe, and H. C. Smit, *Anal. Chem.* **69**, 4220–4225 (1997).
8. S. C. Jacobson, C. T. Culbertson, J. E. Daler, and J. M. Ramsey, *Anal. Chem.* **70**, 3476–3480 (1998).
9. J. C. Fister, S. C. Jacobson, L. M. Davis, and J. M. Ramsey, *Anal. Chem.* **70**, 431–437 (1998).
10. J. C. McDonald, D. C. Duffy, J. R. Anderson, D. T. Chiu, H. Wu, O. J. A. Schueller, and G. M. Whitesides, *Electrophoresis* **21**, 12–26 (2000).
11. J. H. Chan, A. T. Timperman, D. Qin, and R. Aebersold, *Anal. Chem.* **71**, 4437–4444 (1999).
12. H. Becker and C. Gärtner, *Electrophoresis* **21**, 27–40 (2000).
13. C. Henry, *Anal. Chem.* **69**, 359A (1997).
14. Q. F. Xue, F. Foret, Y. M. Dunayevskiy, P. M. Zavracky, N. E. Mcgruer, and B. L. Karger, *Anal. Chem.* **69**, 426–430 (1997).
15. Q. F. Xue, Y. M. Dunayevskiy, F. Foret, and B. L. Karger, *Rapid Commun. Mass Spectrom.* **11**, 1253–1256 (1997).
16. D. Figey, Y. Ning, and R. Aebersold, *Anal. Chem.* **69**, 3153–3160 (1997).
17. N. Xu, Y. Lin, S. A. Hofstadler, D. Matson, C. J. Call, and R. D. Smith, *Anal. Chem.* **70**, 3553–3556 (1998).
18. J. Li, P. Thibault, N. H. Bings, C. D. Skinner, C. Wang, C. Colyer, and J. Harrison, *Anal. Chem.* **71**, 3036–3045 (1999).
19. G. A. Schultz, T. N. Corso, S. J. Prosser, and S. Zhang, *Anal. Chem.* **72**, 4058–4063 (2000).
20. R. D. Olschuk and D. J. Harrison, *Tr. Anal. Chem.* **19**, 379–388 (2000).
21. Q. Wu, C. L. Liu, and R. D. Smith, *Rapid Commun. Mass Spectrom.* **10**, 835–838 (1996).
22. C. Liu, Q. Wu, A. C. Harms, and R. D. Smith, *Anal. Chem.* **68**, 3295–3299 (1996).

23. D. C. Muddiman, D. S. Wunschel, C. Liu, L. Pasa Tolic, K. F. Fox, A. Fox, G. A. Anderson, and R. D. Smith, *Anal. Chem.* **68**, 3705–3712 (1996).
24. F. Xiang, Y. Lin, J. Wen, D. W. Matson, and R. D. Smith, *Anal. Chem.* **71**, 1485–1490 (1999).
25. F. Xiang, G. A. Anderson, T. D. Veenstra, M. S. Lipton, and R. D. Smith, *Anal. Chem.* **72**, 2475–2481 (2000).
26. R. D. Smith, J. H. Wahl, D. R. Goodlett, and S. A. Hofstadler, *Anal. Chem.* **65**, A574–A584 (1993).
27. C. Liu, S. A. Hofstadler, J. A. Bresson, H. R. Udseth, T. Tsukuda, and R. D. Smith, *Anal. Chem.* **70**, 1797–1801 (1998).
28. L. Yang, C. S. Lee, S. A. Hofstadler, and R. D. Smith, *Anal. Chem.* **70**, 4945–4950 (1998).
29. B. Michalke and P. Schramel, *J. Chromatogr. A* **807**, 71–80 (1998).
30. M. H. Lamoree, R. A. M. Vanderhoeven, U. R. Tjaden, and J. Vandergreef, *J. Mass Spectrom.* **33**, 453–460 (1998).
31. J. Wen, Y. Lin, F. Xiang, D. W. Matson, H. R. Udseth, and R. D. Smith, *Electrophoresis*, **21**, 191–197 (2000).
32. H. Liu, C. Felten, Q. Xue, B. Zhang, P. Jedrzejewski, B. L. Karger, and F. Foret, *Anal. Chem.* **72**, 3303–3310 (2000).
33. R. S. Ramsey and J. M. Ramsey, *Anal. Chem.* **69**, 1174–1178 (1997).
34. K. Tang, Y. Lin, D. W. Matson, T. Kim, and R. D. Smith, *Anal. Chem.* **73**, 1658–1663 (2001).

CHAPTER 7

Endogenous Release and Metabolism of Neuropeptides Utilizing In Vivo Microdialysis Microelectrospray Mass Spectrometry

PER E. ANDRÉN, TERRY B. FARMER, and REBECKA KLINTENBERG
Department of Pharmaceutical Biosciences, Uppsala University, Uppsala, Sweden

CONTENTS

7.1. INTRODUCTION

Neuropeptides are present in brain tissue as well as in cerebrospinal fluid (CSF). The evidence that neuropeptides coexist with one or more classic

Mass Spectrometry and Hyphenated Techniques in Neuropeptide Research, Edited by Jerzy Silberring and Rolf Ekman.
ISBN 0-471-35493-7

neurotransmitters suggests that they are involved in modulatory processes and probably in many other types of functions [33]. The synthesis of neuropeptides is a complex process, different from that of conventional neurotransmitters. All neuropeptides are part of a larger molecule precursor, which, after synthesis in the endoplasmatic reticulum, is transferred to the Golgi apparatus for packaging. They are subsequently transported, processed by enzymes (convertases), and released into the extracellular fluid of the brain. Important posttranslational enzymatic modifications of the final neuropeptide may occur, such as glycosylation, *C*-terminal amidation, acetylation, phosphorylation, and sulfation. The presence of the exocytotic release of neuropeptides into the extracellular space can be monitored by the in vivo microdialysis technique [65].

7.2. IN VIVO MICRODIALYSIS

Microdialysis for in vivo sampling has become an important sampling tool in the neurosciences that has provided much information on the complex functions of the intact brain. The microdialysis method, first reported by Ungerstedt and Pycock 1974, is a perfusion technique, by which molecules in the brain extracellular fluid may be monitored [65,66]. It is based on the idea that a microdialysis probe mimics the passive function of a capillary blood vessel by perfusing a thin dialysis tube implanted into the tissue of interest. The microdialysis probe is implanted into a selected brain region. It is then perfused with a physiological solution as close as possible to that of the extracellular fluid of the brain tissue of interest (Fig. 7.1). A syringe pump delivers the perfusion

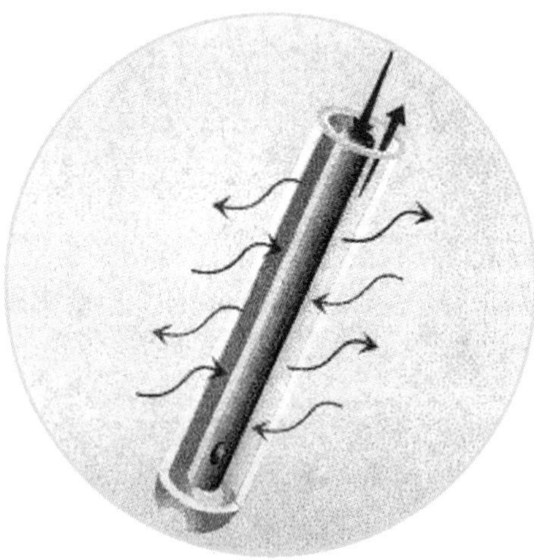

Figure 7.1. Drawing of a microdialysis probe showing the semipermeable membrane over which compounds will diffuse down their concentration gradient into or out of the perfusion medium that flows through the probe.

medium through the microdialysis probe. Analytes outside the probe diffuse into the probe reservoir and are subsequently sampled and their concentration may be determined. Substances can also be administered locally to the region of the brain surrounding the perfused tubular membrane. The compounds are introduced into the extracellular fluids by diffusion from the perfusion medium into the target tissue. The level of substances such as neurotransmitters may be monitored and the response to different treatments may be studied.

Microdialysis sampling provides several advantages over other in vivo methods; specifically, direct measurements in anaesthetized or freely moving animals, analysis of specific target tissues or metabolites from enzymes in tissues, and continuous sampling with no fluid loss [64]. However, a disadvantage of the technique is that the dialysate is relatively dilute (i.e., low analyte recovery) and contains high salt concentrations, conditions that normally give poor sensitivity when using many analytical techniques.

Microdialysis in combination with radioimmunoassay (RIA) has been the common method for studying in vivo release of neuropeptides in the rat brain [9,45,46]. Although RIA provides high sensitivity, it may prove that molecular specificity is limited.

7.3. MICROELECTROSPRAY IONIZATION MASS SPECTROMETRY

Electrospray ionization mass spectrometry (ESIMS) is a continuous device that has the capability of accurately measuring the masses of large proteins as well as small peptides. In order to reduce the volume of liquid flowing into the mass spectrometry source, small inner diameter capillary needles were used, terminating in the spray nozzle. The construction of these spray emitters is briefly described below, but the needles are also available commercially. Fused silica capillary tubing was used as the starting material typically 50–75 μm inner diameter and 200–300 μm outer diameter. The polyimide coating was burned away from the tip (~0.5 cm) using a microtorch, and the tip was carefully ground by hand in order to get a smooth end using a ceramic cutting stone. The tip was then lowered into a 49% solution of hydrofluoric acid (HF), while distilled water was pumped through the capillary at 500 nL/min. The progress of the etching was monitored using a microscope to obtain the desired tapered needle-like tip. A frit was constructed by tapping in a 100-μm length of 10-μm spherical silica particles (SIL-120A-S5, YMC Corp., Morris Plains, NJ) and heating with a microtorch. A slurry of methanol and 10-μm porous spherical C18 particles (ODS-AQ 120A S-10, YMC Corp., Morris Plains, NJ) was loaded into the fused silica column via a Teflon reservoir using a high-performance liquid chromatography (HPLC) pump and a 5-cm-length column was packed (Fig. 7.2). A microscope was used to monitor the progress of the column packing. The opposite end of the needle was then attached to a stainless-steel zero-dead-volume fitting and high voltage was applied to the spray needle through the solvent at this union (Fig. 7.3). Depending on experimental conditions, diameter and length of the capillary, and other variables,

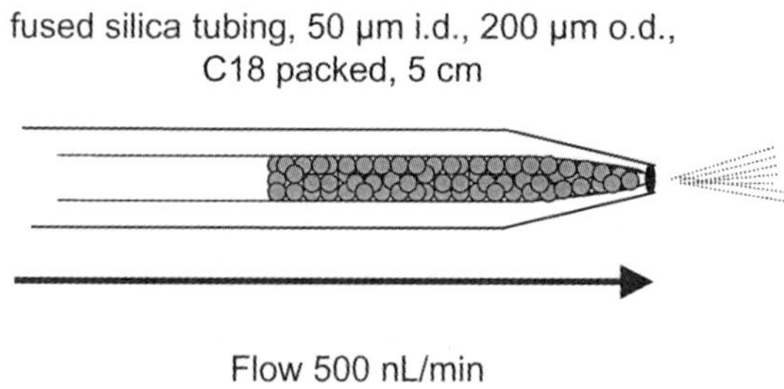

Figure 7.2. Drawing of a microelectrospray emitter. The tip of the fused-silica capillary spray needle of the micro-ESI source is packed with C18 particles to achieve preconcentration and desalting of the sample.

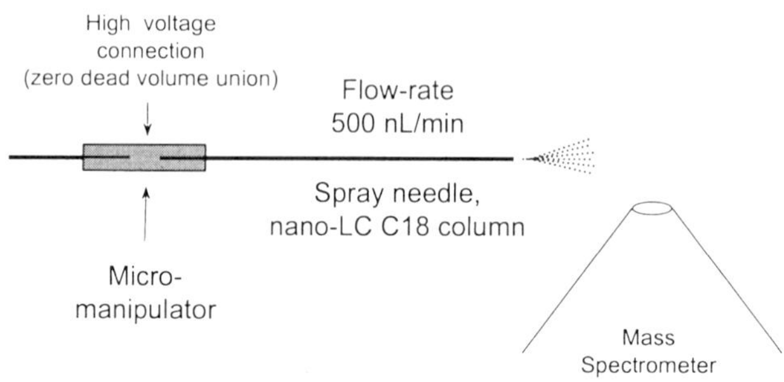

Figure 7.3. Schematic drawing of the microelectrospray ionization source.

the needle voltage varies from 1.0 to 3.5kV. Devices for fabrication of such spray needles for low-flow ES ionization have been reported [14,56]. These devices facilitate a number of manipulations of fused-silica capillary tubing in order to fabricate needles for various uses. The spray needles have been shown to be capable of achieving very high sensitivity, and exhibit high signal stability, making them suitable for multihour usage [56]. The spray needles are durable because there is no external metal coating to wear off, and are vulnerable only to clogging if solvents and liquid samples are not carefully filtered.

7.4. IN VIVO MICRODIALYSIS MICROELECTROSPRAY MASS SPECTROMETRY

The micro-ESI technique was specifically developed for effectively analyzing peptides and proteins compounds in in vivo microdialysis samples [4,19]. Microdialysis probes were placed in target tissues, and in specific locations within target tissues, and endogenous neuropeptides were quantified from microdialysis samples [2,3,20]. The sensitivity and specificity of analysis of small volume microdialysate samples, containing low concentrations of peptides, with the micro-ESI source has been enhanced by coupling this micro-ESI source with different types of mass spectrometers, such as a triple quadrupole, ion trap, quadrupole time-of-flight (Q-TOF), or Fourier transform ion cyclotron resonance (FTICR) mass spectrometers [1–3,5,20,21,27].

The microdialysis samples (10 μl) were loaded onto the reversed phase C18 column/spray emitter using an injection valve with a 10 μl loop. The mobile phase for nanoliter-flow liquid chromatography contained 2% methanol/0.25% acetic acid in distilled water and a splitter was used to achieve a flow rate of 500 nl/min. After 30 minutes of washing, the injection loop was filled with 50–75% methanol/0.25% acetic acid in distilled water and then injected onto the column to elute the neuropeptides.

So far, only a limited number of endogenously released neuropeptides have been identified using in vivo microdialysis and mass spectrometry, namely, neurotensin [3,20] and leucine (Leu)–enkephalin [27] and methionine (Met)–enkephalin [20]. This is due to the very dilute nature of microdialysis samples and the low concentrations of endogenous neuropeptides in the brain, and also the requirement of an ultrasensitive micro-ESI MS device.

In the present experiments, micro-ESIMS or tandem MS (MSMS) was performed on a Finnigan TSQ7000 triple quadrupole (Finnigan, San Jose CA), a Mariner (PerSeptive Biosystems, Framingham MA) time-of-flight (TOF) MS, or a Micromass (Manchester, UK) quadrupole TOF (Q-Tof) mass spectrometer. The micro-ESI device markedly increases sensitivity of the conventional electrospray source [4,19]. The electrospray source was optimized to accommodate nanoliter flow rates spraying directly from a capillary needle that had been packed with C18 particles (nano-LC spray emitter) for the concentration and desalting of the sample prior to its introduction to the mass spectrometer.

7.5. IN VIVO METABOLISM OF NEUROPEPTIDES

Neuropeptides are released from brain cells by exocytotis, but unlike conventional neurotransmitters, no mechanism for their uptake into cells has been described. Instead, they are enzymatically cleaved by membrane-bound or soluble peptidases in the extracellular fluid into peptide fragments that also may have biological activity [33].

The CSF is an enzymatically active environment and mirror events related to the normal homeostasis of the central nervous system (CNS). The neuropeptide content in the CSF may reflect the metabolic balance between different celltypes of the brain and the spinal cord. CSF contains a large number of neuropeptides. In the metabolic modification of extracellular peptides, several enzymes are necessary for each individual processing, conversion, and inactivation. However, analysis of neuropeptides and their metabolism in CSF gives only a "mean" of the events that occur in the brain at specific sites. Therefore, by using the in vivo microdialysis technique to examine the processing of neuropeptdes in specific brain regions and subregions, the extracellular activity in these regions can be studied.

7.5.1. In Vivo Metabolism of Angiotensins in the Brain

Numerous investigators have studied the mechanism of action of angiotensin as well as its metabolism in the body [8,15,16,37,49,52]. Angiotensinogen, the precursor to angiotensin, is cleaved by an aspartyl protease, renin, to form angiotensin I (AngI) (Asp–Arg–Val–Tyr–Ile–His–Pro–Phe–His–Leu). AngI is then cleaved by angiotensin-converting enzyme (ACE) to form angiotensin II (AngII) (Asp–Arg–Val–Tyr–Ile–His–Pro–Phe). Aminopeptidase A then acts to form angiotensin III (AngIII) (Arg–Val–Tyr–Ile–His–Pro–Phe) from AngII and, finally, AngIII is cleaved by aminopeptidase N to form angiotensin IV (AngIV) (Val–Tyr–Ile–His–Pro–Phe) [57].

Circulating AngI, AngII, and AngIII do not readily cross the blood–brain barrier (BBB) [25,55,67]. Because of its importance in the tissue system, researchers looked for the presence of AngII in the brain, where it was localized in the neurons [58,59]. All the proteases involved in the formation of AngII and its metabolites in tissue have been isolated in the brain [43,51,58]. The precursors and enzymes necessary for the formation and degradation of biologically active forms of angiotensins are present in brain tissues, accompanied by at least three specific receptor binding sites and several forms of angiotensin may serve as signaling agents at these sites.

Until 2001, AngII metabolites formed in the brain were thought to be inactive; however, new evidence suggests that these metabolites may be the major effectors in the brain system [57]. AngIII has been shown to be a major effector peptide of the brain renin–angiotensin system, exerting a tonic stimulatory effect in the central control of blood pressure. The finding that AngII and AngIII have similar affinities for the angiotensin receptor type 1 (AT_1) raise the question as to why the conversion of brain AngII to AngIII is required to increase blood pressure [57]. Further, AngIV appears to affect motor activity, inhibit neurite growth, dilate cerebral blood vessels, instigate c-fos expression, and be involved in memory retrieval. This last effect makes it of great interest with regard to drug development.

Previous experiments examining the metabolism of angiotensins in the brain have been performed in vivo and in vitro using radioimmunoassays (RIA) and/or high-performance liquid chromatography (HPLC) for detection of the metabolites. In vivo investigations of AngII and AngIII metabolism were performed by intracerebroventricular injections of AngII and AngIII [26,69]. It was shown that AngIII was metabolized more rapidly than AngII. In vitro metabolism of AngIV showed that only the AngIV fragments 1–5, 1–4, 2–6, 3–6, and 4–6 could be detected [30].

More recent advances in mass spectrometry have allowed for high-resolution detection of biologically relevant concentrations of endogenous compounds. In the present investigation of the metabolism of AngII and AngIV, we utilized in vivo microdialysis and micro-ESI-TOF MS. Analysis of in vivo microdialysis samples obtained from both the marmoset (*Callitrix jacchus*) striatum and the rat frontal cortex allowed us to observe metabolism of AngII and AngIV and to detect the fragments processed by the brain.

7.5.2. Angiotensin II

A microdialysis probe was inserted into the striatum of the marmoset under anesthesia. After a 3-h equilibration period, poststriatal probe implantation, an infusion of AngII (1 μM) in artificial CSF (aCSF) at 0.3 μL/min was performed for 2 h. Several fragments processed from AngII were detected using the nano-LC micro-ESI Q-TOF MS instrument. During the first 30 min of the AngII infusion it was possible to detect AngIV and AngIV 5–8 (Fig. 7.4). In the present study AngIII was detected at very low concentration levels that were near the detection limits of the technique. It is possible that AngIII was processed very rapidly by aminopeptidase cleavage. Further timepoints between 0 and 30 min would have to be examined to determine the in vivo mechanism of processing of AngIII from AngII. Harding et al. [26] have reported that the half-life of intracerebroventricular injected AngII and AngIII were approximately 23 and 8 s, respectively, indicating that AngIII was metabolized more rapidly that was AngII.

7.5.3. Angiotensin IV

The in vivo metabolism of AngIV in rats was performed in freely moving animals. A guide cannula was stereotaxically implanted under anesthesia into the frontal cortex in rats (for experimental procedures, see Refs. 3 and 20). A recovery period of 3–4 days was allowed before a 3-mm microdialysis probe was implanted into the guide cannula, which was located in the frontal cortex. Microdialysis probes were also implanted in anaesthetized marmosets in the striatum. Artificial (aCSF) pH 7.4 was perfused through the probe at 0.4 nL/min in both experiments.

In the marmoset samples were collected 30 and 60 min after a 2-h infusion of 10 μM AngIV had started. The AngIV infusion started after the probes had been in place for at least 2 h. A representative spectrum of infused synthetic AngIV into marmoset striatum is shown in Fig. 7.5. The doubly charged ion (*m/z* 388.2) is the most abundant ion in the spectrum, but a small peak corresponding to the singly charged ion (*m/z* 775.4) can be seen. The possible fragments and their presence or absence in the microdialysate samples at the two timepoints are listed in Table 7.1.

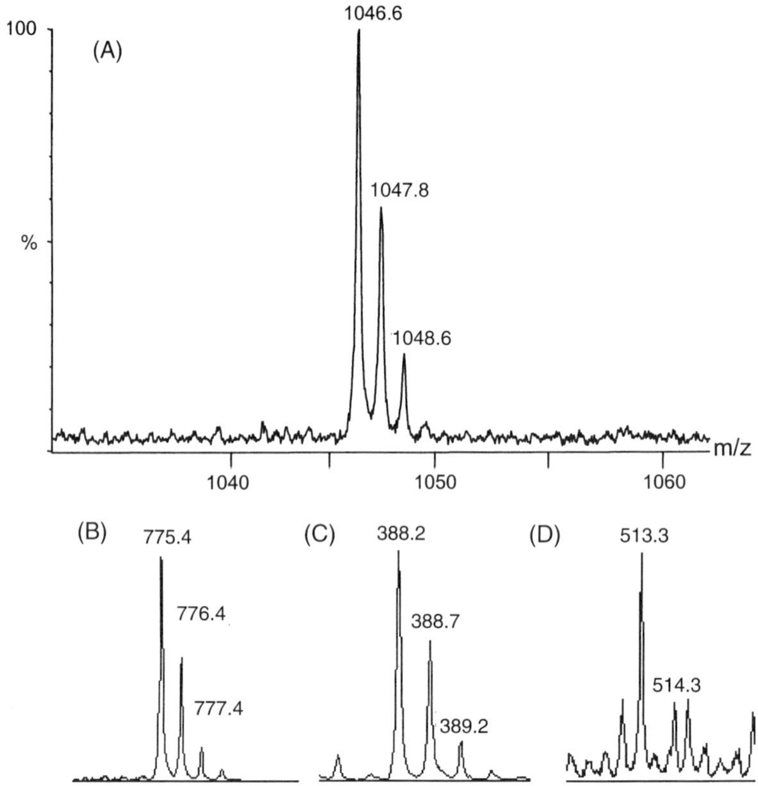

Figure 7.4. Mass spectra of angiotensin II (AngII) and some of the in vivo processed metabolites. AngII was infused for 2 h into the striatum of an anesthetized marmoset monkey 3 h after the microdialysis was implanted. The microdialysis probe was used both to deliver AngII into the brain and to collect the degradation products. The microdialysis probe was perfused with dialysis buffer with a microliter syringe pump at a flow rate of 0.3 µL/min. (a) The mass spectrum of the infused AngII; (b, c) the singly charged $(M + H)^+$, m/z 775.4, and the doubly charged species $(M + 2H)^{2+}$, m/z 388.2, of the AngIV peptide processed in vivo from AngII; and (d) the AngII fragment 5–8, m/z 513.3 $(M + H)^+$.

In the 30-min sample only three *m/z* values were observed corresponding to intact AngIV and two possible cleavage products. Two metabolites, VYI and IHPF, appear to be internal cleavage products. The possibility that they were generated by very rapid exopeptidase cleavage from the *C* terminus for VYI and from the *N* terminus for IHPF cannot be determined from these experiments. Further timepoints between 0 and 30 min would have to be examined to determine the mechanism of generation for the two cleavage products. After 60 min of continuous perfusion of AngIV through the microdialysis

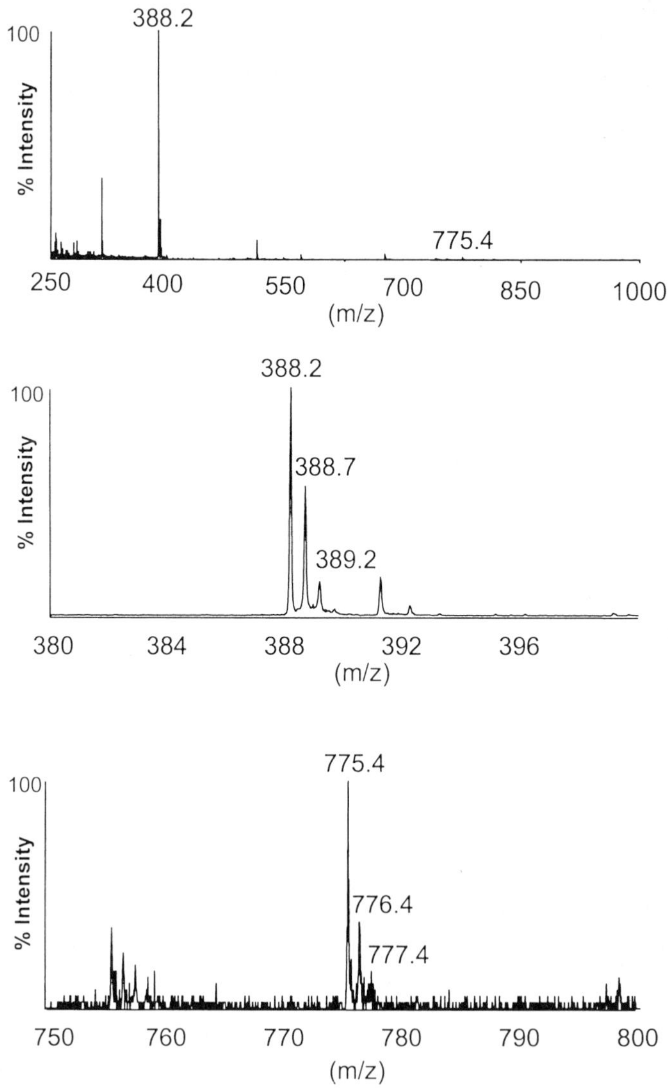

Figure 7.5. Micro-ESI TOF MS spectrum of angiotensin IV (AngIV) obtained after perfusion through the microdialysis probe into the marmoset striatum.

probe to the marmoset striatum, eight additional fragments were detected, while the two fragments observed in the 30-min microdialysate sample are no longer seen in the spectrum.

Processing appears to occur from both the *N* and *C* termini of the intact peptide as well as some internal cleavage denoting possible action by both

TABLE 7.1. Results of Mass Spectrometric Analysis of In Vivo Microdialysis Samples of Infused Angiotensin IV (AngIV) into Marmoset Striatum[a]

AngIV Fragment	Mass (Da)	30 min	60 min
VYIHPF	774.41	*	*
VYIHP	627.34		
VYIH	530.28		*
VYI	393.23	*	
VY	280.14		*
YIHPF	675.34		*
YIHP	528.27		*
YIH	431.22		*
YI	294.16		*
IHPF	512.28	*	
HPF	399.19		*
PF	262.13		
IHP	365.21		*
IH	268.15		
HP	252.12		
VYIHPF (oxidized)	790.58		

[a] The microdialysis probe was used both to administer the AngIV into the brain and to collect the processed fragments in the brain. Metabolites that were detected are indicated by the presence of an asterisk (*) in the appropriate cell.

endopeptidases and exopeptidases in the marmoset striatum. Carboxypeptidases and aminopeptidases are known to act on AngIV in the brain to process the intact peptide. Possibly, carboxypeptidase P removes the *C*-terminal Phe to form AngII(5–7). Both aminopeptidase A and aminopeptidase N have been detected in the rodent brain [28,32,63,70]. There is also known endopeptidase activity that cleaves the 2 *C*-terminal residues from AngIV (VY-) to form AngII(5–8) (Tyr–Ile–His–Pro–Phe) [11].

Four fragments were not detected by mass spectrometric analysis in the two microdialysate samples: VYIHP, PF, IH, and HP. This could be due to rapid degradation of these fragments by the enzymes present in the marmoset striatum or to the fragments not being formed.

The rat frontal cortex is the site of a dense concentration of AngIV receptors, which dictated the location of the microdialysis probe. AngIV was perfused into the rat frontal cortex to examine its metabolism. In our experiments, rats became sleepy and acted sedated on administration of AngIV through the microdialysis probe into the rat frontal cortex. Although the amount of sedation was not rigorously characterized, this indicates that AngIV or one of its metabolites may have action on the behavior of the rat via the frontal cortex.

TABLE 7.2. Angiotensin IV (AngIV) Fragments Detected in the Rat Frontal Cortex Microdialysate Samples at Different Time Periods

Fragment	15 min	30 min	45 min	60 min
VYIHPF (1–6)	*	*	*	*
YIHPF (2–6)	*	*	*	*
IHPF (3–6)	*	*	*	*
HPF (4–6)	*	*	*	*
PF (5–6)	*	*	*	*
VYIHP (1–5)	—	—	—	—
VYIH (1–4)	—	*	*	*
VYI (1–3)	*	*	*	*
VY (1–2)	*	*	*	*
VIHP (2–5)	*	*	*	*
YIH (2–4)			*	*
YI (2–3)	*	*	*	*
IHP (3–5)	—	—	—	—
IH (3–4)	*	*	*	*
HP (4–5)	*	*	*	*
VYIHPF (oxidized)	*	*	*	*

Fragments detected in the rats and the timepoint aliquots in which they are detected are listed in Table 7.2. Cleavage from the *N* terminus occurs within the first 15 min with all possible *N*-terminal fragments seen in the mass spectra. Interestingly, several fragments were preferentially detected when different packing materials in the capillary columns were used. Fragments 1–3 and 2–4 were detected only when using the PepMap (LC Packings, Amsterdam, the Netherlands) packed column, while 1–2, 1–4, 2–3, 2–5, 4–5, and 5–6 were detected only when using the POROS R2 (PerSeptive Biosystems, Framingham MA) packed column. The results show that only two fragments, 1–5 and 3–5, are not detected during the 60 minute time period.

Metabolism of AngIV in the rat frontal cortex appears to proceed rapidly from the *N* terminus. Cleavage of the *C*-terminal Phe may be very slow and fragments may not have been formed in sufficient quantities to detect during the analysis time period of 60 min. The oxidized form of AngIV was also detected in all the microdialysate samples. To determine whether the oxidation was due to sample preparation or the microdialysis probe, experiments were performed by placing a microdialysis probe in aCSF (37°C) and by perfusing AngIV in aCSF through the microdialysis similar to the experiment performed in rat frontal cortex in vivo. No formation of oxidized AngIV was observed by mass spectrometry when the microdialysis probe was not implanted in the rat brain, which suggests that the oxidation may be an in vivo posttranslational modification of AngIV.

7.6. IN VIVO RELEASE OF NEUROPEPTIDES

7.6.1. Endogenous Release of Substance P

Substance P is an 11-amino acid tachykinin peptide and is widely distributed in the nervous system [50]. In the central nervous system, substance P is concentrated in the basal ganglia (striatonigral system), the habenulointerpeduncular system, the brainstem, and the hypothalamus [50]. Neurons containing substance P may also coexist and interact with other neurotransmitters [12,13,18,35,36].

There is evidence suggesting that substance P functions as a neurotransmitter or neuromodulator [50]. Substance P has been suggested to be involved in several processes and functions. Substance P could mediate in peripheral nociceptive processes by its peripheral and central effects [7,17,29], in central regulation of nociceptive impulses (analgesic effect) [41,47,53], in motor regulatory actions as suggested by the large amounts of substance P in the striatonigral tract [24,38,39] and the motor effects induced by injection of substance P into the substantia nigra [44], in control of some learning processes [60–62], in control of growth of some central nerve fibers during development [10], and in anxiolytic activity [42].

In the present experiment microdialysis probes were inserted into the striatum, the hypothalamus, the globus pallidus, or the substantia nigra of rats 4–5 days after a guide cannula was implanted. The guide cannula was implanted under anesthesia [2,3,20]. The microdialysis probe was perfused with artificial cerebrospinal fluid (aCSF) at a flow rate of 0.3 μL/min, and samples were collected every 30 min.

Our results showed that it was possible to detect endogenously released basal extracellular levels of substance P in all the four brain regions studied, the striatum, the hypothalamus, the globus pallidus, and the substantia nigra (Fig. 7.6). The concentration in the samples were not quantified, but appeared to be in the midattomole (amol) range (per 10-μL sample), if compared to different amounts of injected synthetic standards of substance P. The brain region substantia nigra gave the highest intensity of the measured substance P [6]. We are currently developing methods to quantitate endogenous substance P and substance P1–7 using deuterated internal standards.

7.6.2. Quantification of Endogenously Released Neurotensin in Microdialysates

Neurotensin (NT) (pGlu–Leu–Tyr–Glu–Asn–Lys–Pro–Arg–Arg–Pro–Tyr–Ile–Leu) is a tridecapeptide that is localized in neurons and terminals throughout the brain. NT is closely associated with CNS dopamine neurons and is colocalized with dopaminergic neurons in the hypothalamus and midbrain [34]. It exerts hypothermic and analgesic effects when injected into the CNS. From a clinical viewpoint, studies with NT have led to implications of its involvement

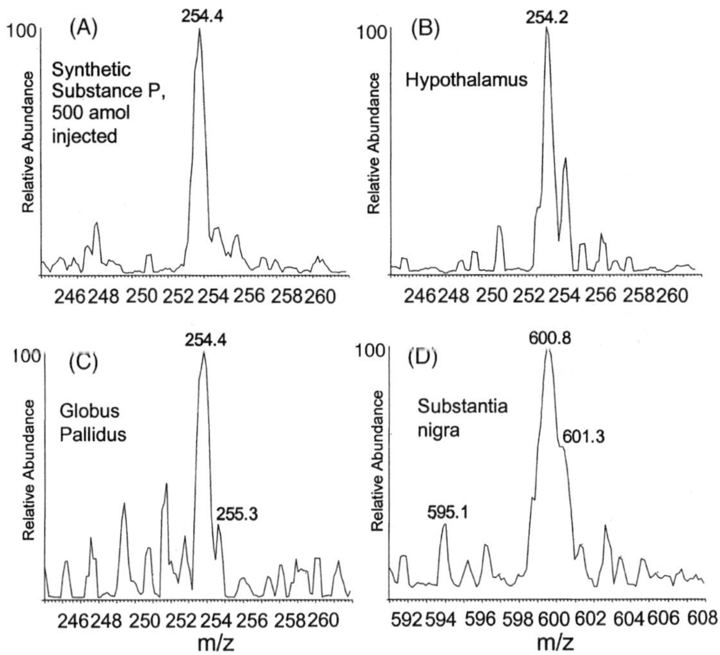

Figure 7.6. Tandem (MSMS) detection of substance P using a triple quadrupole mass spectrometer: (a) injection of synthetic substance P, 500 amol (50 amol/μL in 10 μL); (b–d) endogenously released substance P in different regions of the brain (rats) from in vivo microdialysate samples. The mass spectrometer was set to transmit the doubly charged ion of substance P, m/z 675 $(M + 2H)^{2+}$ in the first analyzer, and the second analyzer was scanned over a narrow mass range (20 m/z units). The singly charged b_2 MSMS ion was detected in (b) and (c) (hypothalamus and globus pallidus), and the doubly charged b_{10} ion was detected in (d) (substantia nigra).

in several disorders, such as schizophrenia, Parkinson's disease, Huntington's chorea, and Alzheimer's disease [23,48,68]. Several of the NT fragments also retain biological activity [31,54].

Two brain areas were chosen for analysis of basal and KCl-induced stimulation of endogenous levels of NT using in vivo microdialysis. It has been reported that the hypothalamus contains high tissue concentration of NT, and globus pallidus/ventral pallidum was reported to have lower tissue concentration of NT [22,40]. For quantitative experiments, synthetic NT in standards made up in artificial CSF and endogenous NT in microdialysis samples were analyzed by using online C18 packed capillaries (nano-LC) and micro-ESI-MS/MS using a triple quadrupole or a Q-Tof mass spectrometer. Standard curves produced from NT (39 amol; 1230 amol in 10 μL aCSF) showed the linearity and interassay variability of the micro-ESI-MS/MS system. The standard curves had an average r^2 linear coefficient of 0.993 and a standard error of the mean (SEM) of <17% ($n = 5$) [3].

Basal levels of NT were established in both the hypothalamus and the globus pallidus/ventral pallidum. The levels were slightly higher in the hypothalamus (~100 amol/10 μL of dialysate) than in the globus pallidus/ventral pallidum region (~74 amol/10 μL of dialysate). Peptidase inhibitors were not included in the perfusate or given systemically because such agents can raise the level of the neuropeptides in the synaptic cleft and otherwise have an effect on regulatory mechanisms. Basal amounts of NT were close to the limit of detection of the mass spectrometric system [3].

Release of NT in the hypothalamus was stimulated by direct depolarization by presenting a (1.0-μL) bolus of artificial CSF supplemented with 100 mM KCl to the microdialysis probe. This perfusion increased the recovered NT to 548 ± 90 amol/10 μL in the hypothalamus (544% increase compared to the baseline). In the globus pallidus/ventral pallidum region KCl stimulation elevated the NT release to 499 ± 99 amol/10 μL during a 30-min sample period (674% increase compared to the baseline). It should be pointed out that there was a large variation in the absolute increase produced by KCl as evidenced by the standard error of the means (SEM) [3].

Endogenous release of NT was also studied in the striatum of the marmoset using nano-LC micro-ESI Q-Tof MS. Injections of synthetic standards to the system showed that the sensitivity was in the low amol range and that the NT peak in the mass chromatogram was approximately 10 s wide (Fig. 7.7). It was possible to detect basal levels (nonstimulated) of NT in the microdialysates

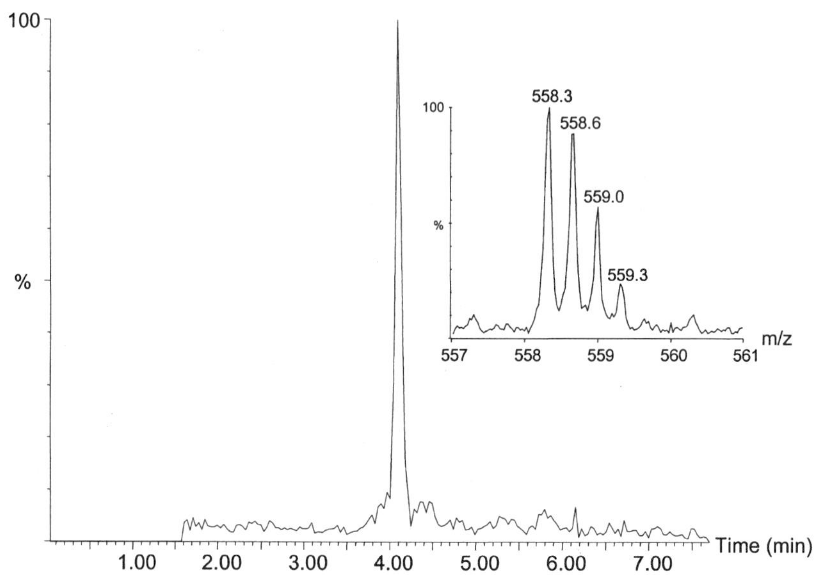

Figure 7.7. Reconstructed mass chromatogram of the triply charged $(M + 3H)^{3+}$ species of synthetic NT and its mass spectrum (inset) at the apex of the chromatography peak; 10 μL of synthetic NT 4.7 amol/μL (47 amol) was injected into the nano-LC column and eluted, ionized, and detected by the Q-Tof mass spectrometer.

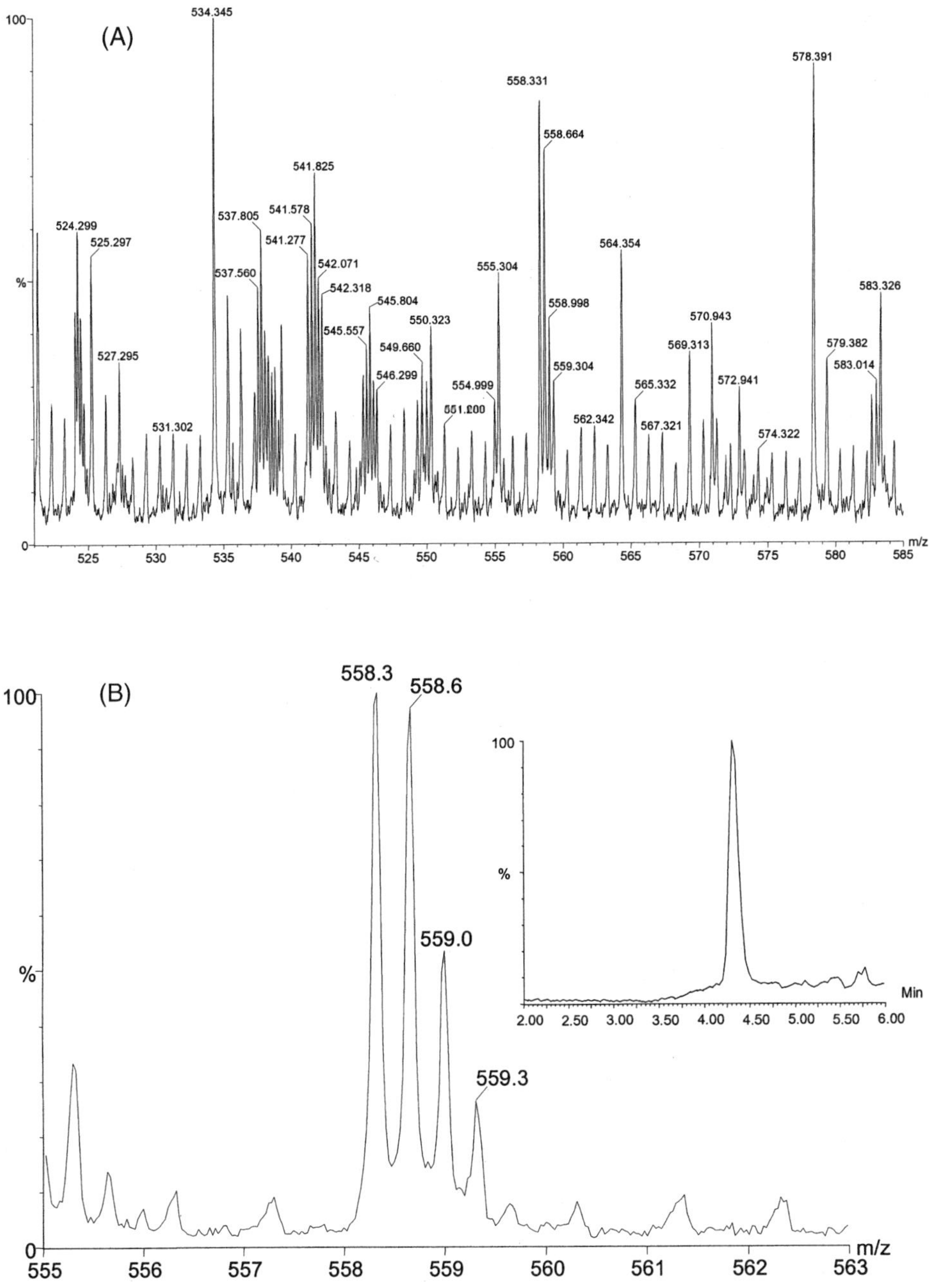

Figure 7.8. (a) Mass spectrum from an in vivo microdialysis sample obtained from the striatum of a marmoset monkey. The spectrum shows the *m*/*z* range 520–585. Endogenous NT was detected in the mass spectrum as the triply charged species $(M + 3H)^{3+}$, *m*/*z* 558.3. The full mass spectrum also showed the doubly charged species of NT $(M + 2H)^{2+}$, *m*/*z* 837.0 (data not shown). As seen in the mass spectrum, several other unidentified multiply charged neuropeptides were detected. (b) Mass spectrum showing the *m*/*z* range 555–563 of the spectrum displayed in (a). The inset shows the reconstructed mass chromatogram of NT, *m*/*z* 558.3 of the microdialysis sample originating from the striatum of the marmoset monkey.

from the striatum of the marmoset monkey, where the $(M + 3H)^{3+}$ ion, m/z 558.3 and the $(M + 2H)^{2+}$, *m/z* 837.0 were detected (Fig. 7.8). These methods are being further developed using deuterated internal standards to get reliable standard curves for quantitation of endogenously released NT.

7.7. CONCLUSION

Microdialysis is widely used for monitoring a number of neuroactive substances in the brain extracellular environment. The attractive simplicity of the technique itself has made it useful to many neuroscientists, but usually the limiting factor is the analytical characterization following sample collection. Provided a sufficiently sensitive assay exists for its measurement, any substance that is able to diffuse across the membrane is capable of being monitored. There are several factors that can negatively affect detection of the substance of interest when using in vivo microdialysis. Diffusion through the membrane depends on the size and hydrophobicity of the neuropeptide and restriction on diffusion through the brain microenvironment can also reduce the amount of the substance recovered for analysis.

The micro-ESI technique provides high sensitivity and is a low-flow-rate version of the basic ESI design. It provides a higher liquid mass transfer and ionization efficiency because of the narrow spray, small droplet diameters, and closeness to the entrance orifice of the mass spectrometer. When coupled to reversed-phase chromatography, the preconcentration, desalting, and fractionation advantages give rise to a very high sensitivity, permitting analyses of peptides in the zeptomole (zmol)/μL concentration range in the best cases [4].

ACKNOWLEDGMENTS

The present studies were supported by the Swedish Medical Research Council grant #11565, The Swedish Natural Science Council Grant I 1179 (special program "Signals of Life"), the Swedish Foundation for International Cooperation in Research and Higher Education (STINT) Grant 723, and the Bank of Sweden Tercentenary Foundation.

REFERENCES

1. P. E. Andren and R. M. Caprioli, Microdialysis/mass spectrometry of neuropeptides, in *Biological Mass Spectrometry: Present and Future*, T. Matsuo, ed., Wiley, New York, 1994, pp. 355–367.
2. P. E. Andren and R. M. Caprioli, In vivo metabolism of substance P in rat striatum utilizing microdialysis/liquid chromatography/micro-electrospray mass spectrometry, *J. Mass Spectrom.* **30**, 817–824 (1995).

3. P. E. Andren and R. M. Caprioli, Determination of extracellular release of neurotensin in discrete rat brain regions utilizing in vivo microdialysis/electrospray mass spectrometry, *Brain Res.* **845**, 123–129 (1999).
4. P. E. Andren, M. R. Emmett, and R. M. Caprioli, Micro-electrospray: Zeptomole/attomole per microliter sensitivity for peptides, *Am. Soc. Mass Spectrom.* **5**, 867–869 (1994).
5. P. E. Andren, S.-N. Lin, and R. M. Caprioli, Microdialysis/mass spectrometry, in *Mass Spectrometry: Clinical and Biomedical Applications*, D. M. Desiderio, ed., Plenum Press, New York, 1994, pp. 237–254.
6. P. E. Andrén, K. Sköld, M. Svensson, U. Eriksson, and F. Nyberg, Detection of endogenous neuropeptides in brain extracellular fluids, *49th ASMS Conf. Mass Spectrometry and Allied Topics*, Vol. 49, American Society for Mass Spectrometry, Chicago, 2001.
7. G. Baranauskas and A. Nistri, Sensitization of pain pathways in the spinal cord: Cellular mechanisms, *Progr. Neurobiol.* **54**, 349–365 (1998).
8. L. A. Belova, Angiotensin II-generating enzymes, **65**, 1337–1345 (2000).
9. E. Brodin, B. Linderoth, B. Gazelius, and U. Ungerstedt, In vivo release of substance P in cat dorsal horn studied with microdialysis, *Neurosci. Lett.* **76**, 357–362 (1987).
10. G. Casini, L. Trasarti, L. Andolfi, and P. Bagnoli, Morphologic maturation of tachykinin peptide-expressing cells in the postnatal rabbit retina, *Brain Res. Devel. Brain Res.* **99**, 131–141 (1997).
11. D. Chansel, S. Czekalski, S. Vandermeersch, E. Ruffet, M. C. Fournie-Zaluski, and R. Ardaillou, Characterization of angiotensin IV-degrading enzymes and receptors on rat mesangial cells, *Am. J. Physiol.* **275**, F535–F542 (1998).
12. I. Christensson-Nylander, M. Herrera-Marschitz, W. Staines, T. Hokfelt, L. Terenius, U. Ungerstedt, C. Cuello, W. H. Oertel, and M. Goldstein, Striato-nigral dynorphin and substance P pathways in the rat. I. Biochemical and immunohistochemical studies, *Exp. Brain Res.* **64**, 169–192 (1986).
13. A. C. Cuello, J. V. Priestley, and M. R. Matthews, Localization of substance P in neuronal pathways, *Ciba Found. Symp.* 55–83 (1982).
14. M. T. Davis, D. C. Stahl, S. A. Hefta, and T. D. Lee, A microscale electrospray interface for online, capillary liquid-chromatography tandem mass-spectrometry of complex peptide mixtures, *Analy. Chem.* **67**, 4549–4556 (1995).
15. D. T. Dinh, A. G. Frauman, C. I. Johnston, and M. E. Fabiani, Angiotensin receptors: Distribution, signalling and function, **100**, 481–492 (2001).
16. D. E. Dostal and K. M. Baker, The cardiac renin-angiotensin system: conceptual, or a regulator of cardiac function? *Circ. Res.* **85**, 643–650 (1999).
17. A. W. Duggan and L. J. Furmidge, Probing the brain and spinal cord with neuropeptides in pathways related to pain and other functions, *Front. Neuroendocrinol.* **15**, 275–300 (1994).
18. R. Elde, M. Schalling, S. Ceccatelli, S. Nakanishi, and T. Hokfelt, Localization of neuropeptide receptor mRNA in rat brain: initial observations using probes for neurotensin and substance P receptors, *Neurosci. Lett.* **120**, 134–138 (1990).
19. M. Emmett and R. M. Caprioli, Micro-electrospray mass spectrometry: Ultra-high-sensitivity analysis of peptides and proteins, *J. Am Soc. Mass Spectrom.* **5**, 605–613 (1994).

20. M. R. Emmett, P. E. Andren, and R. M. Caprioli, Specific molecular mass detection of endogenously released neuropeptides using in vivo microdialysis/mass spectrometry, *J. Neurosci. Meth.* **62**, 141–147 (1995).
21. M. R. Emmett, F. M. White, C. L. Hendrickson, S. D. Shi, and A. G. Marshall, Application of micro-electrospray liquid chromatography techniques to FT-ICR MS to enable high-sensitivity biological analysis, *J. Am. Soc. Mass Spectrom.* **9**, 333–340 (1998).
22. P. C. Emson, M. Goedert, P. Horsfield, F. Rioux, and S. St. Pierre, The regional distribution and chromatographic characterisation of neurotensin-like immunoreactivity in the rat central nervous system. *J. Neurochem.* **38**, 992–999 (1982).
23. A. Fernandez, P. Jenner, C. D. Marsden, and M. L. De Ceballos, Characterization of neurotensin-like immunoreactivity in human basal ganglia: Increased neurotensin levels in substantia nigra in Parkinson's disease, *Peptides* **16**, 339–346 (1995).
24. L. M. Gunne, J. E. Häggstrom, P. Johansson, E. D. Levin, and L. Terenius, Neurobiochemical changes in tardive dyskinesia, *Encephale* **14**(Spec. No.), 167–173 (1988).
25. J. W. Harding, M. J. Sullivan, J. M. Hanesworth, L. L. Cushing, and J. W. Wright, Inability of [125I]Sar1, Ile8-angiotensin II to move between the blood and cerebrospinal fluid compartments, *J. Neurochem.* **50**, 554–557 (1988).
26. J. W. Harding, M. S. Yoshida, R. P. Dilts, T. M. Woods, and J. W. Wright, Cerebroventricular and intravascular metabolism of [125I]angiotensins in rat. *J. Neurochem.* **46**, 1292–1297 (1986).
27. W. E. Haskins, S. R. Witowski, D. H. Powell, and R. T. Kennedy, Ultratrace analysis of neuropeptides, in *48th Conf. Mass Spectrometry and Allied Topics*, Vol. 48, R. M. Caprioli, ed., American Society for Mass Spectrometry, Long Beach, CA, 2000.
28. D. P. Healy and S. Wilk, Localization of immunoreactive glutamyl aminoeptidase in rat brain. II. Distribution and correlation with angiotensin II, *Brain Res.* **606**, 295–303 (1993).
29. J. L. Henry, Relation of substance P to pain transmission: Neurophysiological evidence, *Ciba Found. Symp.* 206–224 (1982).
30. K. Hermann, M. I. Phillips, and M. K. Raizada, Metabolism of angiotensin peptides by neuronal and glial cultures from rat brain, *J. Neurochem.* **52**, 863–868 (1989).
31. D. E. Hernandez, C. M. Richardson, C. B. Nemeroff, R. C. Orlando, S. St.-Pierre, F. Rioux, and A. J. Prange, Jr., Evidence for biological activity of two N-terminal fragments of neurotensin, neurotensin1-8 and neurotensin1-10, *Brain Res.* **301**, 153–156 (1984).
32. L. B. Hersh, N. Aboukhair, and S. Watson, Immunohistochemical localization of aminopeptidase M in rat brain and periphery: Relationship of enzyme localization and enkephalin metabolism, *Peptides* **8**, 523–532 (1897).
33. T. Hokfelt, C. Broberger, Z. Q. Xu, V. Sergeyev, R. Ubink, and M. Diez, Neuropeptides—an overview, *Neuropharmacology* **39**, 1337–1356 (2000).
34. T. Hokfelt, B. J. Everitt, E. Theodorsson-Norheim, and M. Goldstein, Occurrence of neurotensinlike immunoreactivity in subpopulations of hypothalamic, mesencephalic, and medullary catecholamine neurons. *J. Comp. Neurol.* **222**, 543–559 (1984).

35. T. Hokfelt, J. O. Kellerth, G. Nilsson, and B. Pernow, Substance p: Localization in the central nervous system and in some primary sensory neurons, *Science* **190**, 889–890 (1975).

36. T. Hokfelt, S. Vincent, C. J. Dalsgaard, L. Skirboll, O. Johansson, M. Schultzberg, J. M. Lundberg, S. Rosell, B. Pernow, and G. Jancso, Distribution of substance P in brain and periphery and its possible role as a co-transmitter, *Ciba Found. Symp.* 84–106 (1982).

37. T. Inagami, Y. Kambayashi, T. Ichiki, S. Tsuzuki, S. Eguchi, and T. Yamakawa, Angiotensin receptors: Molecular biology and signalling, *Clin. Exp. Pharmacol. Physiol.* **26**, 544–549 (1999).

38. P. E. Johansson, L. Terenius, J. E. Häggstrom, and L. Gunne, Neuropeptide changes in a primate model (Cebus apella) for tardive dyskinesia, *Neuroscience* **37**, 563–567 (1990).

39. Y. Kawada, N. Ogawa, M. Asanuma, and A. Mori, Neuropeptide levels in discrete brain regions in the iminodipropionitrile-induced persistent dyskinesia rat model, *Regul. Pept.* **55**, 103–110 (1995).

40. R. M. Kobayashi, M. Brown, and W. Vale, Regional distribution of neurotensin and somatostatin in rat brain, *Brain Res.* **126**, 584–588 (1977).

41. Y. Kotani, M. Oka, N. Yonehara, T. Kudo, and R. Inoki, Algesiogenic and analgesic activities of synthetic substance P, *Jpn. J. Pharmacol.* **31**, 315–321 (1981).

42. M. S. Kramer, N. Cutler, J. Feighner, R. Shrivastava, J. Carman, J. J. Sramek, S. A. Reines, G. Liu, D. Snavely, E. Wyatt-Knowles, J. J. Hale, S. G. Mills, M. MacCoss, C. J. Swain, T. Harrison, R. G. Hill, F. Hefti, E. M. Scolnick, M. A. Cascieri, G. G. Chicchi, S. Sadowski, A. R. Williams, L. Hewson, D. Smith, and N. M. Rupniak, Distinct mechanism for antidepressant activity by blockade of central substance P receptors, *Science* **281**, 1640–1645 (1998).

43. Z. Lenkei, M. Palkovits, P. Corvol, and C. Llorens-Cortes, Expression of angiotensin type-1 (AT1) and type-2 (AT2) receptor mRNAs in the adult rat brain: A functional neuroanatomical review, *Front Neuroendocrinol.* **18**, 383–439 (1997).

44. U. Liminga, P. E. Johansson, and L. Gunne, Intranigral tachykinin NK3 receptor agonist elicits oral movements in rats, *Pharmacol. Biochem. Behav.* **38**, 617–620 (1991).

45. N. Lindefors, E. Brodin, and U. Ungerstedt, Microdialysis combined with a sensitive radioimmunoassay. A technique for studying in vivo release of neuropeptides, *J. Pharmacol. Meth.* **17**, 305–312 (1987).

46. N. T. Maidment, D. R. Brumbaugh, V. D. Rudolph, E. Erdelyi, and C. J. Evans, Microdialysis of extracellular endogenous opioid peptides from rat brain in vivo, *Neuroscience* **33**, 549–557 (1989).

47. J. R. Naranjo, M. P. Fernandez-Tome, and J. Del Rio, Analgesic activity of substance P in rats: Tolerance and cross-tolerance with enkephalin, *Eur. J. Pharmacol.* **98**, 133–136 (1984).

48. C. B. Nemeroff, W. W. Youngblood, P. J. Manberg, A. J. Prange, Jr., and J. S. Kizer, Regional brain concentrations of neuropeptides in Huntington's chorea and schizophrenia, *Science* **221**, 972–975 (1983).

49. M. G. Nicholls and J. I. Robertson, The renin-angiotensin system in the year 2000, *J. Human Hypertens.* **14**, 649–666 (2000).

50. B. Pernow, Substance P, *Pharmacol. Rev.* **35**, 85–141 (1983).
51. M. I. Phillips and C. Sumners, Angiotensin II in central nervous system physiology, *Regul. Pept.* **78**, 1–11 (1998).
52. R. W. Piepho, Overview of the angiotensin-converting-enzyme inhibitors, *Am. J. Health-Syst. Pharm.* **57**(Suppl. 1), S3–S7 (2000).
53. M. F. Piercey, M. W. Moon, J. R. Blinn, and P. J. Dobry-Schreur, Analgesic activities of spinal cord substance P antagonists implicate substance P as a neurotransmitter of pain sensation, *Brain Res.* **385**, 74–85 (1986).
54. R. D. Pinnock, Neurotensin depolarizes substantia nigra dopamine neurones, *Brain Res.* **338**, 151–154 (1985).
55. M. P. Printz, Regulation of the brain angiotensin system: A thesis of multicellular involvement, *Clin. Exp. Hypertens.* Part A, *Theory Pract.* **10**, 17–35 (1988).
56. J. P. Quinn, M. R. Emmett, and A. G. Marshall, A device for fabrication of emitters for low-flow electrospray ionization, *46th ASMS Confe. Mass Spectrometry and Allied Topics*, Orlando, FL, 1998.
57. A. Reaux, M. C. Fournie-Zaluski, and C. Llorens-Cortes, Angiotensin III: A central regulator of vasopressin release and blood pressure, **12**, 157–162 (2001).
58. J. M. Saavedra, Brain and pituitary angiotensin, *Endocr. Rev.* **13**, 329–380 (1992).
59. J. M. Saavedra, Emerging features of brain angiotensin receptors, *Regul. Pept.* **85**, 31–45 (1999).
60. K. Schlesinger, M. A. Pelleymounter, J. van de Kamp, D. L. Bader, J. M. Stewart, and T. N. Chase, Substance P facilitation of memory: Effects in an appetitively motivated learning task, *Behav. Neural Biol.* **45**, 230–239 (1986).
61. U. Staubli and J. P. Huston, Differential effects on learning by ventromedial vs lateral hypothalamic posttrial injection of substance P, *Pharmacol. Biochem. Behav.* **10**, 783–786 (1979).
62. U. Stäubli and J. P. Huston, Facilitation of learning by post-trial injection of substance P into the medial septal nucleus, *Behav. Brain Res.* **1**, 245–255 (1980).
63. M. Troyanovskaya, G. Jayaraman, L. Song, and D. P. Healy, Aminopeptidase-A. I. CDNA cloning and expression and localization in rat tissues, **278**, R413–R424 (2000).
64. U. Ungerstedt, Introduction to intracerebral microdialysis, in T. E. Robinson and J. B. Justice, eds., *Microdialysis in the Neurosciences*, Vol. 7, Elsevier, Amsterdam, York, 1991, pp. 3–22.
65. U. Ungerstedt, Microdialysis—principles and applications for studies in animals and man, *J. Intern. Med.* **230**, 365–373 (1991).
66. U. Ungerstedt and C. Pycock, Functional correlates of dopamine neurotransmission, *Bull. Schweiz Akad. Med. Wiss.* **30**, 44–55 (1974).
67. M. van-Houten, E. L. Schiffrin, J. F. Mann, B. I. Posner, and R. Boucher, Radioautographic localization of specific binding sites for blood-borne angiotensin II in the rat brain, *Brain Res.* **186**, 480–485 (1980).
68. E. Widerlov, L. H. Lindstrom, G. Besev, P. J. Manberg, C. B. Nemeroff, G. R. Breese, J. S. Kizer, and A. J. Prange, Jr., Subnormal CSF levels of neurotensin in a subgroup of schizophrenic patients: Normalization after neuroleptic treatment, *Am. J. Psychiatry* **139**, 1122–1126 (1982).

69. J. W. Wright, M. J. Sullivan, C. R. Bredl, J. M. Hanesworth, L. L. Cushing, and J. W. Harding, Delayed cerebroventricular metabolism of [125I]angiotensins in the spontaneously hypertensive rat, *J. Neurochem.* **49**, 651–654 (1987).

70. S. Zini, P. Masdehors, Z. Lenkei, M. C. Fournie-Zaluski, B. P. Roques, P. Corvol, and C. Llorens-Cortes, Aminopeptidase A: Distribution in rat brain nuclei and increased activity in spontaneously hypertensive rats, *Neuroscience* **78**, 1187–1193 (1997).

CHAPTER 8

Biomolecular Interaction Analysis Mass Spectrometry (BIA/MS)

DOBRIN NEDELKOV and RANDALL W. NELSON

Intrinsic Bioprobes Inc., Tempe, Arizona, USA

CONTENTS

8.1. INTRODUCTION

With the completion of the human genome sequencing project, public interest is quickly shifting to the extensive scientific effort aimed at characterization of the human proteome. However, the relative ease and pace by which the genomic data have been gathered stands out against the complexity of the proteome world and the efforts required for protein characterization. This complexity stems from the unique three-dimensional structure of each protein, level of protein expression, trafficking and turnover, the existence of

Mass Spectrometry and Hyphenated Techniques in Neuropeptide Research, Edited by Jerzy Silberring and Rolf Ekman.
ISBN 0-471-35493-7

protein variants (point mutations), posttranslational modifications and splice variants, and protein–protein interactions. Even though significant strides have been made in several protein characterization techniques, the enormity of the proteome places much impetus on the development of new technologies.

Since the late 1990s we have been involved in the inception and development of a new multiplexed protein characterization technique termed *biomolecular interaction analysis mass spectrometry* (BIA/MS). BIA/MS is a two-dimensional analytic technique geared toward functional and structural analysis of biomolecules in their native environments [1–3]. In the first (functional) dimension, BIA/MS takes form of a micro-scale affinity interaction analysis performed on a sensor-active surface. Surface plasmon resonance (SPR) is used for detection of biorecognition events that occur at the sensor surface/solution interface [4,5]. One of the interacting partners (the ligand) is immobilized on the surface side that comes in contact with the solution containing the other interactant (the analyte). When binding occurs, the concentration (amount) of biomaterial on the surface increases, resulting in a change of the refractive index at the surface/solution interface. This refractive index change is optically detected via variation in the angle/wavelength at which an incident light is absorbed at the surface (i.e., light photons transformed into surface plasmons). The angle/wavelength difference is proportional to the amount of material present on the surface, as a result of which quantitative assessment of the analyte binding can be made. As SPR sensing is performed in real time, the data are presented in a signal–time format (*sensorgram*); from the *binding* and *dissociation* curves, the kinetic parameters of the biomolecular interactions can be determined. Moreover, the SPR detection is nondestructive, opening the possibilities for further manipulation of the analyte affinity-retrieved and present on the sensor surface.

For the second (structural) dimension, BIA/MS employs matrix-assisted laser desorption/ionization time-of-flight mass spectrometry (MALDI-TOFMS) [6–8]. With minimal physical modifications and appropriate application of MALDI matrices, the SPR-active surfaces can be converted into amenable MALDI targets [2,9]. The MS analysis is performed directly from the sensor surface, yielding masses of the affinity-isolated analyte(s) that are noncovalently bound to the surface-immobilized ligand. Consequently, MALDI-TOFMS serves the purpose of validating the SPR sensing data by providing the molecular weight of the retained analyte. Furthermore, MS analysis yields other qualitative information about the SPR-monitored interaction, such as identification of non-specific binding, binding of analyte variants and fragments and multianalyte binding. Thus, MALDI-TOF MS analysis provides structural information that complements the SPR sensing and overcomes certain shortcoming of the SPR detection stemming from its nondiscriminatory character.

At this stage of development, BIA/MS utilizes a Biacore X biosensor (Biacore AB, Uppsala, Sweden) for the SPR analysis. The SPR-active surface consists of a gold-coated glass slide (*sensor chip*) derivatized with a number

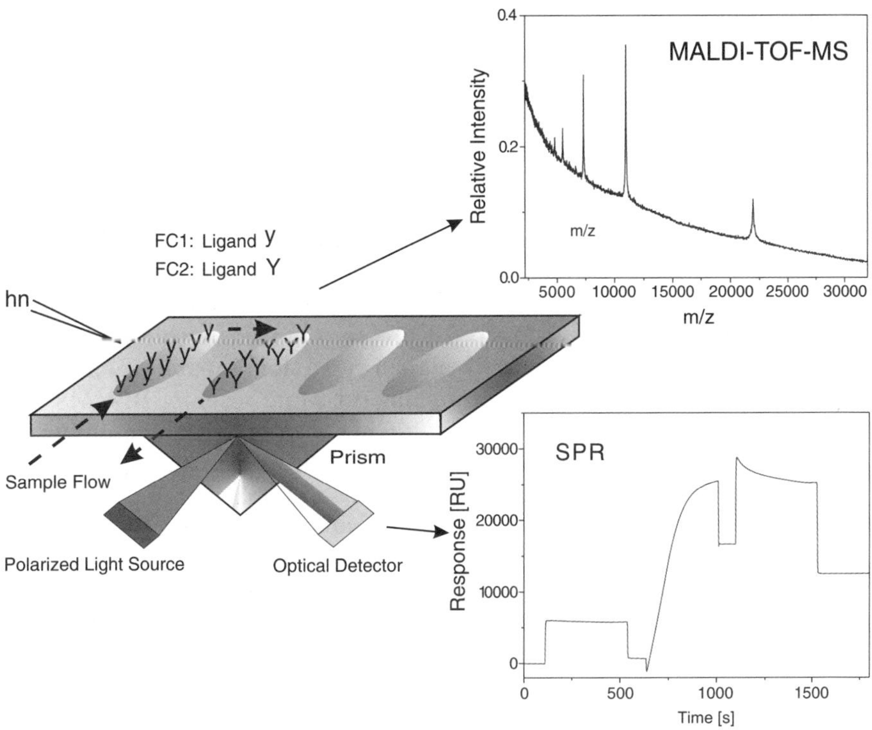

Figure 8.1. Schematics of a BIA/MS analysis. Specific regions [flow cells (FCs)] on the surface of the sensor chip are derivatized with ligands with affinities toward analytes present in a sample solution. The sample solution is directed over the sensor surface via a microfluidics system that forms the flow cells on the chip surface. The interaction between the surface-bound ligand and solution-phase analyte(s) is monitored in real time via surface plasmon resonance (SPR) sensing; SPR is also used to monitor the immobilization of the ligands. A side result of the SPR sample analysis is the selective isolation of analytes within the confines of the flow cells. The retained analytes can then be characterized structurally by MALDI-TOFMS directly from the sensor chip surface.

of functional groups susceptible to covalent ligand attachment (Fig. 8.1). The chip makes contact with the solution via microfluidics system that, when pressed against the glass chip, forms *flow cells* on the chip surface. The surface area of each flow cell is 1 mm^2 (0.5 × 2.0 cm), and the volume above each flow cell is 60 nL. The number of flow cell per sensor chip is either two or four, depending on the instrument configuration. Derivatization of the sensor surface with ligand molecules is performed online, with SPR providing quantitative assessment of the immobilization process. Samples of interest are delivered over each flow cell (individually or serially) via the microfluidics system at low flow rates, typically 1–5 μL/min. Following the SPR analysis of the analyte–ligand binding event, the sensor chips are prepared for MALDI-TOF mass spectrometry via physical modification and MALDI matrix application with in-house fabricated devices [2,9]. Mass spectrometry is then performed using MALDI TOF instrumentation.

In the initial development stages, BIA/MS demonstrated a highly sensitive concerted analysis, with detection limits at or below 20 femtomoles (Fmol) [10]. Dynamic range issues were also investigated and the ability to perform the concerted analysis was demonstrated over a mass range of up to ~150 kDa, with multiple analytes retained from complex mixtures at the 10–100-fmol level [11]. Also demonstrated was the ability to maintain the spatial resolution of multiple interaction sites (flow cells) present on a single chip throughout both SPR and MALDI-TOF, thereby enabling up to four separate interactions to be characterized on a single sensor chip [10]. More recently, BIA/MS was utilized in isolating, detecting, and identifying epitope-tagged recombinant proteins directly from *E. coli* lysate [12]. Also shown was the applicability of BIA/MS in the analysis bacterial toxins in food samples [13]. Furthermore, the basis for a new concerted chip-based proteomics approach (biosensor chip mass spectrometry) for protein capture, processing and digest (all on a single chip) was established [3]. Finally, with an improved matrix deposition approach, the ability of BIA/MS to detect attomole amounts of proteins present in complex biological mixtures was demonstrated [14].

Altogether, BIA/MS represents a multidimensional analytic technique possessing the virtues of quantitatively and qualitatively detecting proteins of interests. The combination of internal reference-free quantitation of SPR sensing and the mass selective detection of MALDI-TOFMS is a valuable asset in protein characterization. Analyses are performed on very small sample quantities, making the BIA/MS a viable approach for studies where the amounts of sample available for proteomics or clinical/diagnostic analyses are limited. This article outlines the BIA/MS protocols and procedures and gives examples illustrating the potential of BIA/MS.

8.2. BIA/MS MODUS OPERANDI

8.2.1. Sensor Surface Preparation

The sensitivity of each BIA/MS analysis is governed by a number of factors that affect the amount of analyte retrieved by the immobilized ligand during SPR analysis. These are: the amount of ligand immobilized on the sensor surface, the affinity of the analyte for the ligand (i.e., the K_D), the analyte concentration in the solution directed over the flow cells, the volume of the sample solution injected, and the flow rate at which the sample solution passes over the active flow cell. Whereas the K_D and the solution concentration of the analyte are beyond the control of the investigator, the other parameters can be varied to achieve maximum binding of the analyte from solution.

To start with, a sufficient amount of ligand should be immobilized on the sensor surface to ensure efficient analyte capture. The ligand surface density should be such to allow not only the initial retrieval of the analyte from the flowing solution but also a recapture of the once-bound and then-dissociated analyte molecules further down in the same flow cell. The ligand density has a direct impact on both the dynamic range of the SPR analysis and the range of analyte concentration that can be detected by SPR. Thus, for each type of ligand, surface derivatization chemistries need to be optimized to achieve high-ligand densities.

For a majority of proteins, EDC/NHS [N-ethyl-*N*′-(dimethylamino-propyl)carbodiimide/*N*-hydroxysuccinimide]-mediated reaction between surface carboxyl and ligand primary amino groups are used for covalent coupling of the ligands to the sensor surface. For low-pI proteins lacking a sufficient number of primary amines, amine surfaces can be created (e.g., via polylysine or 3,3′-iminobispropylamine treatment of the surface carboxyl groups) after which the protein ligands are covalently coupled to the surface amines via EDC-mediated reaction with their carboxyl groups [3]. Alternatively, protein ligand attachment to the sensor surface can be done via streptavidin–biotin or Ni^{2+}/chelate–His–tag noncovalent attachment. These noncovalent attachments are not preferred in BIA/MS because more nonspecific binding is generally introduced during ligand immobilization and BIA/MS analysis. Furthermore, the noncovalently attached ligands are observed in the mass spectrum and can obscure the signal(s) coming from analytes of similar mass.

Different sensor surfaces are utilized for immobilization of small ligands that do not have carboxyl or amino groups. For metal ligands, for example, sensor surfaces are prepared with chelators as functional groups. Amine-functionalized sensor surfaces are derivatized with either EDTA (ethylenediaminetetraacetic acid) or DPTA (diethylenetriaminepentaacetic acid) to create surfaces having three or four chelating carboxyl groups [2]. Solutions containing various metals (i.e., Ca^{2+}, Ni^{2+}, Fe^{3+}, Ga^{3+}) are injected over the flow cells and the metals captured on the chelator surfaces. The metal-loaded sensor surfaces are then used for broad specificity functional screening of various biological fluids.

8.2.2. Sample Analysis

Once the sensor surfaces are prepared, solutions of the biological samples of interest are injected over the flow cells. For efficient analyte capture, low flow rates (<5 μL/min) are generally utilized. In previous experiments, capture efficiencies of 20–50% were achieved using flow rates of 1–5 μL/min [9,14]. At low flow rates, analyte diffusion within the flow cell is maximized, allowing for maximum exposure of the analyte to the surface-immobilized ligand. Thus, these low flow rates de facto maximize the amount of analyte captured in each flow cell, ultimately leading to increased sensitivity of the BIA/MS analysis.

The small surface area and volume of each flow cell dictate that small sample volumes be utilized in each BIA/MS analysis. Typically, the sample volumes are in the 10–100 μL range. However, since dilution of the original sample with running buffer is strongly recommended (to minimize nonspecific binding), the sample volumes needed are, in reality, below 10 μL, turning BIA/MS into a truly microscale proteomics approach.

SPR is used to monitor the interaction between the immobilized ligands and biological material present in the sample solution. The SPR response is measured in resonance units (RUs), with 1 RU equating to 1 picogram of proteinacious material per 1 mm^2 of the flow cell surface. An increase in the SPR response is indicative of a biorecognition event occurring at the sensor interface. Responses below 5 RUs on the BiacoreX biosensor (the instrument that is used at this stage of development) are generally considered to be below its practical and reliable detection limit. Whereas this very low SPR response cannot by itself be a reliable indicator of analyte binding on the sensor surface, the subsequent MALDI-TOFMS analysis can unequivocally confirm the outcome of the affinity interaction. To put this in perspective, a response of 1 RU for a protein (peptide) with molecular mass of 1000 Da would translate into 1-fmol protein on the sensor surface, which, based on our past work and experience, is more than enough for a subsequent MALDI-TOFMS analysis. Therefore, sensor chips for which the SPR readings are below 5 RUs are still subjected to MS analysis.

If the SPR response is to be taken as a true quantitative measure of analyte binding to the surface-immobilized ligand, care must be taken to account for factors such as baseline drift and bulk refractive index change, especially when working in the low SPR response range. In those cases, careful referencing of the response observed in the flow cells must be implemented. When working with pure samples, serial injections over both "active" (ligand-derivatized) and "reference" (derivatized with noninteracting biomolecule) flow cells are commonly performed to correct the bulk refractive index change in real time, via subtraction of the reference flow cell signal from the active-flow cell signal. However, when working with complex biological samples, the "noninteracting" biomolecule immobilized in the reference flow cell can interact with unspecified sample components, leading to SPR responses in the reference flow cell. An alternative to blank flow cell subtraction is baseline referencing from the active flow cell itself. For that purpose, aliquots of the running buffer (in the same volume as the sample aliquots) are injected immediately prior to the sample injection(s). The SPR reading observed at the end of the buffer injection is subtracted from the reading taken at the end of the subsequent sample injection. In that way, the SPR response resulting from binding of proteinacious material to the immobilized ligand can be estimated accurately and used unambiguously for quantitative assessment.

After the sample injection is finished and the SPR reading is completed, sensor chips are removed from the SPR instrument and prepared for MALDI-TOFMS analysis. A general concern during BIA/MS is that the analytes retained during the SPR analysis in the biosensor can be lost during the post-SPR and pre-MALDI experimental steps, especially when the analytes exhibit high dissociation rates. Thus, following binding, the buffer flow over the flow cells is immediately terminated and the sensor chip is rapidly undocked.

Once removed from the biosensor, the chips are prepared for MALDI-TOFMS by application of the MALDI matrix. The MALDI matrix of choice for small (<20-kDa) proteins is an aqueous solution of α-cyano-4-hydrohycinnamic acid, in 33% (v/v) acetonitrile, 0.2% (v/v) trifluoroacetic acid. For larger proteins, an aqueous solution of 3,5-dimethoxy-4-hydroxycinnamic acid (with the same organics composition) is used. The matrix solution is applied by an aerosol delivery system, which we have developed [2,9]. Using this system, the matrix solution is evenly sprayed over the entire chip surface, producing a thin homogeneous matrix layer. The acidified matrix (pH~2) desorbs the affinity-retained analytes from the flow cell/chip surface and on drying (rapidly, in <10s), the matrix/analyte mix is redeposited on the same surface/flow cell area from which the analyte was desorbed. The spatial resolution between the flow cells is preserved, allowing simultaneous BIA/MS analysis in several flow cells. The spraying of the matrix also helps in visualizing the flow cells on the surface of the sensor chip, facilitating the MALDI-TOFMS analysis.

The mass analysis of the retained analytes on the sensor chips can be performed on any MALDI-TOF mass spectrometer. However, only one series of instruments (the Bruker Daltonics Biflex/Reflex series) has a source/target geometry that allows the entire polystyrene support on which the glass sensor chip is mounted to be introduced into the instrument (the dimensions of the support are 2.5 × 10cm; the chip itself is 1 × 1cm). If other mass spectrometers are used, the chip has to be removed from its plastic support. For that purpose, we have designed a circular heated cutter device that reproducibly cuts the sensor chip from its plastic support [9], creating a round chip/support that can be fitted into almost any commercially available mass spectrometer target. Following insertion into the mass spectrometer, each flow cell on the sensor chip is targeted individually with the nitrogen laser, allowing site-specific BIA/MS analysis. If the signals in the mass spectra are poor, the composition of the matrix can be varied and/or more matrices applied and the MALDI-TOFMS analysis repeated.

8.2.3. Data Analysis

The signals present in the mass spectra obtained from the surface of flow cells on the sensor chips will yield a number of important data. First of all, the number of signals is an indication of the number of biomolecules retained specifically (or nonspecifically) on the surface. In the case of functional protein activity screening of complex biological mixtures using a single (protein) immobilized ligand, the appearance of more than one peak in the mass spectrum can be interpreted in several ways. To start with, the immobilized protein might be de facto interacting with more than one protein present in the biological sample screened. Second, the immobilized protein might be interacting with a heterogeneous noncovalently assembled protein complex, which, on laser irradiation, falls apart and yields multiple signals in the mass spectrum. Finally, the presence of multiple signals in the mass spectrum might be an indication of nonspecific binding (NSB) to the surface of the sensor chip. NSB occurs either by binding of sample components to the underivatized (ligand-free) portion of the surface of the flow cell or through other interactions (e.g., charge–charge, nonpolar) with the immobilized ligand. NSB can be minimized by varying the composition of the sample buffer and the character of the sensor surfaces. Nevertheless, when present, the NSB can be straightforwardly revealed (and eliminated from the further data analysis) by the observance of signals that repeatedly appear in mass spectra taken from flow cells derivatized with different protein ligands over which the same sample solution has been directed.

When using the broad-affinity (e.g., chelator/metal) surfaces for screening of biological samples for proteins with particular binding proteins, multiple peaks in the mass spectra are expected and indicate the presence of a number of proteins that have similar functional activity toward the immobilized ligands. Because the dynamic range of MALDI-TOFMS allows for simultaneous analysis of molecules differing by an order of magnitude in concentration, proteins retrieved and present on the sensor surface in various amounts can be detected simultaneously.

The second important information revealed from the mass spectra is the molecular weight (MW) of the bound biomolecules. The molecular weight itself is a valuable parameter as it represents an intrinsic property of each biomolecules. It has been somewhat overlooked in the existing protein identification protocols because of all the potential posttranslational modifications that the current protein (and genomic-based) databases do not account for. Nevertheless, with a good mass accuracy assignment of the peaks observed in the mass spectra obtained from the sensor surface and a knowledge of the sample source (e.g., plasma, urine, tissue type) and the ligand type, one can tentatively identify the bound proteins by doing a Molecular Weight Search at the National Center for Biotechnology Information Website (*http://www.ncbi.nlm.nih.gov/Entrez*). This MW database search allows entering a single MW value, or a MW range, which is particularly helpful having in mind the percent accuracy associated with MW assignment to higher mass proteins. The search can be further limited by specifying the organism (e.g., human) from which the sample was obtained and prepared. Signal peptides are both included and excluded from the calculation of the protein mass in the database, adding to the merit of the search. In one example of using this approach, we were able to identify two peaks present in all of the mass spectra taken from flow cells derivatized with various protein ligands (i.e., antibodies) over which human plasma solution was injected (see Section 8.3 later in the text).

Often *N*-terminal or *C*-terminal amino acids are cleaved from proteins in vivo, producing truncated proteins with varying functional activities. Additionally, posttranslational modifications associated with various diseases are commonly found in a number of proteins. Thus, the knowledge of native molecular weights of intact proteins gains an importance as the extent of the protein modifications vary between individuals and becomes a signature pattern for diagnostic and clinical use. Given that the modifications do not alter the affinity toward the ligand, both the intact and the truncated or modified proteins are affinity retrieved during the SPR analysis and detected in the mass spectra as distinct yet close-in-mass signals. The ability of BIA/MS to distinguish protein variants with varying functional activities clearly sets the technique apart from the rest of the functional screening assays.

8.3. BIA/MS APPLICATIONS

8.3.1. Single-Protein Analysis

Given here is an example of a standard BIA/MS analysis utilizing single protein ligand for single analyte analysis. Human interleukin—1 beta (IL-1β) antibody–antigen system was utilized in this study. The SPR analyses were performed on a Biacore X biosensor, using CM5 research-grade sensor chips (Biacore AB, Sweden) with carboxymethyl dextran-coated gold surface, and HBS-EP (0.01 HEPES, pH 7.4, 0.15 M NaCl, 0.005% (v/v) surfactant P20, 3 mM EDTA) as running buffer at 5 μL/min. Prior to the ligand immobilization, the carboxyl groups of the carboxymethyl matrix were activated (converted to active esters) by an 35-μL injection of EDC/NHS (Fig. 8.2a). Next, monoclonal mouse antihuman IL-1β (anti-IL-1β) was immobilized in the flow cell (FC) by injecting an 80-μL aliquot of the antibody (0.01 mg/mL in 10 mM pH 5.0 acetate buffer). Following the coupling reaction, blocking of the free (unreacted) esters was achieved with a 35-μL injection of ethanolamine (ETA). A total SPR response change of 10,950 RU was observed, indicating immobilization of 10.95 ng of material, which corresponds to ~73 fmol of antibody (MW_{IgG} ~ 150,000) and a maximum possible valence of ~146 fmol.

Figure 8.2b shows the routing of an 80-μL aliquot of IL-1β [10^{-3} mg/mL, with 10 mg/mL of human serum albumin (HSA)] over the antibody-derivatized flow cell, at 5 μL/min. The binding curve levels off toward the end of the injection, suggesting saturation of the antibody binding sites. The total SPR response change of 1770 RU is indicative of retention of ~1.77 ng of material on the surface of the flow cell. Figure 8.2c shows a MALDI-TOF mass spectrum taken from the flow cell surface of this sensor chip. The observed *m*/*z* values of 17,372 and 8684 correspond to the singly and doubly charged ions of IL-1β and are in good agreement with the calculated mass of IL-1β ($MW_{IL\text{-}1\beta}$ = 17,376.9). Neither signals for HSA nor the immobilized antibody are observed in spectra taken from the surface of the flow cell or regions on

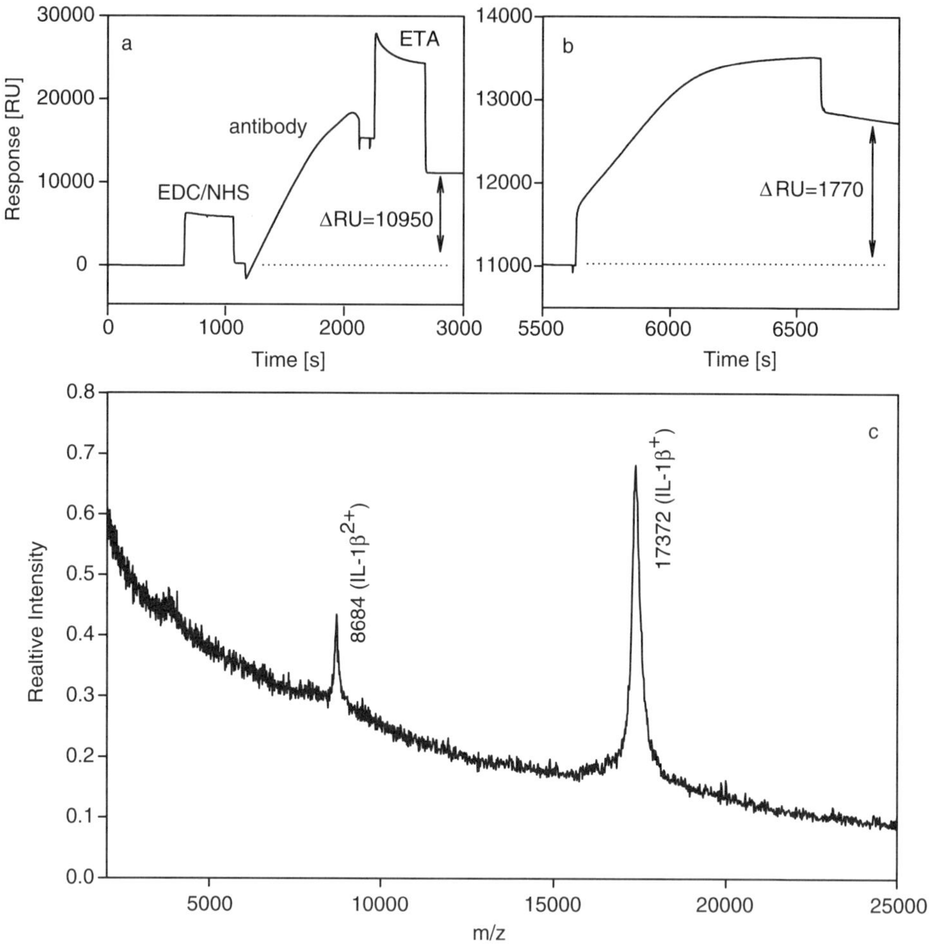

Figure 8.2. BIA/MS analysis of human interleukin-1b (IL-1β). (a) sensorgram showing covalent immobilization of anti-IL-1β; (b) sensorgram resulting from exposure of the flow cell to a solution containing IL-1β; (c) MALDI-TOF mass spectrum taken from the flow cell surface.

the chip bordering the flow cell. Thus, the SPR response of 1770 RU is assumed to be from IL-1β binding and indicates ~100 fmol (1770×10^{-12} g/17,377 g/mol) IL-1β retrieved in the flow cell and detected in the MALDI TOF mass spectrum. Because binding saturation was achieved, the 100 fmol is the antibody valence available for antigen binding, and indicates that 68% of the immobilized antibody has a functional activity. This "activity" percentage is as expected, as some of the antibody coupling to the sensor surface occurs via primary amines located in the Fab region that is responsible for the antigen recognition.

8.3.2. Multiaffinity Surfaces in BIA/MS Analysis

Standalone SPR studies are commonly performed under conditions where only one analyte of interest is targeted by an immobilized receptor on a single flow cell on the sensor chip surface. In this scenario, SPR senses only the amount of analyte retrieved by the surface-immobilized receptor (quantitative assessment), not the number/type of analytes retained on the chip (qualitative assessment). In BIA/MS, however, proteins are ultimately detected during BIA/MS via an intrinsic property of the analyte—the molecular weight. This mass-specific detection allows for the construction of assays where multiple analytes are screened in a single analysis. In this setting, a single flow cell on the sensor surface is derivatized with multiple ligands prior to exposure to a biological fluid. After the initial SPR sensing, retained proteins are detected at precise and characteristic masses using MALDI-TOFMS. Given that the targeted proteins are resolved in the mass spectrum, the absolute specificity (associated with the single flow cell/single ligand spatial arrangement) that is sacrificed on the multiple-ligand configuration is regained during MALDI-TOFMS. Multiaffinity surfaces were explored and their applicability investigated in detecting staphylococcal enterotoxin B (SEB) and toxic shock syndrome toxin-1 (TSST-1), two toxins produced by the major human pathogen *Staphylococcus aureus*.

The surface of the first flow cell (FC1) of a CM5 sensor chip was derivatized with anti-TSST-1 antibody; in the second flow cell (FC2), a 1 : 1 mixture of anti-SEB and anti-TSST-1 antibodies was immobilized. Figure 8.3a illustrates the immobilization of the anti-TSST-1 IgG: following the chip surface activation with 35 μL EDC/NHS, the antibody (50 μL of 0.01 mg/mL anti-TSST-1 in 10 mM acetate buffer, pH 4.8) was injected over FC1, followed by 35 μL ethanolamine wash. A response change of 20,050 RU (~20 ng of immobilized material) was observed, indicating ~130 fmol immobilized antibody. The mixture of two antibodies was immobilized in a similar manner in the second flow cell (Fig. 8.3b), except that a 30-μL mixture of 0.01 mg/mL anti-SEB and 0.01 mg/mL anti-TSST-1 (in acetate buffer) was injected. The response change of 11,800 RU (~11.8 ng) observed is indicative of ~78 fmol immobilized antibodies.

Following the surface derivatization, a 30-μL solution of 10^{-4} mg/mL SEB (with 10 mg/mL HSA, in HBS-EP buffer) was injected over both flow cells (Fig. 8.3c). Binding was observed only in the second flow cell. The 0.23-ng bound material (ΔRU = 230) corresponds to ~8 fmol SEB (MW_{SEB} = 28,366). A subsequent injection of a 35-μL solution of 10^{-3} mg/mL TSST-1 (with 10 mg/mL HSA, in HBS-EP buffer) resulted in response changes of 180 and 30 RU in FC1 and FC2, respectively. These correspond to ~8.2 fmol bound TSST-1 in FC1 and ~1.4 fmol in FC2 ($MW_{TSST\text{-}1}$ = 21,944.6). Thus, the amount of affinity-captured TSST-1 in FC1 was approximately 6 times the amount retrieved in FC2.

Injection of an aliquot (95 μL) of a mixture of both toxins (10^{-4} mg/mL SEB, 10^{-3} mg/mL TSST-1, with 10 mg/mL HSA, in HBS-EP buffer) serially over both

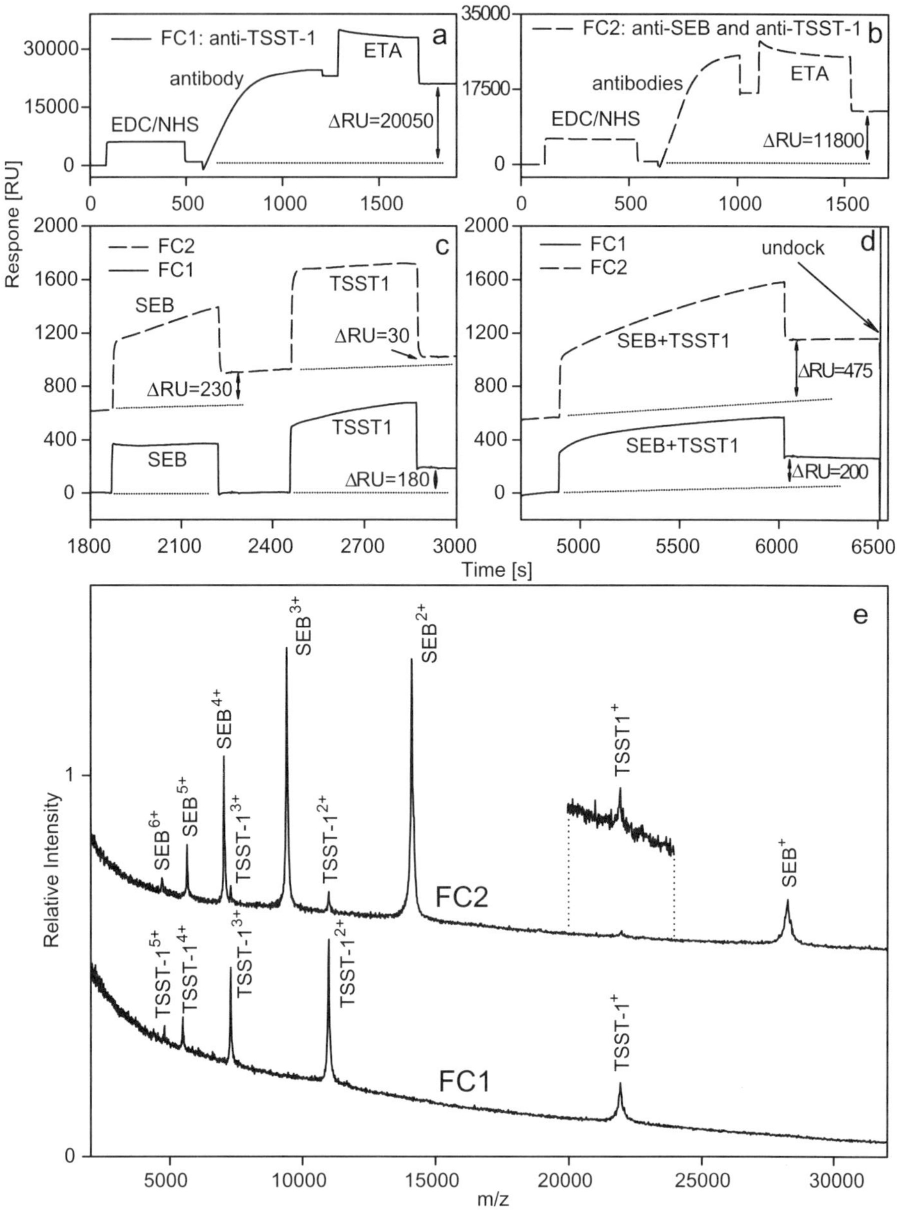

Figure 8.3. Multitoxin BIA/MS analysis: (a) sensorgram showing the covalent immobilization of anti-TSST-1; (b) sensorgram showing immobilization of an equimolar mixture of anti-SEB and anti-TSST-1; (c) sensorgrams resulting from separate injections of SEB and TSST-1 over the flow cells; (d) sensorgram showing the injection of a mixture of SEB and TSST-1 over both flow cells; (e) MALDI-TOF mass spectra taken form the sensor surface following the injection of the SEB/TSST-1 mixture.

flow cells resulted in a response change of ΔRU = 200 in FC1 and ΔRU = 475 in FC2 (Fig. 8.3d). Because the first flow cell was derivatized with anti-TSST-1 IgG only, it is reasonable to assume that only TSST-1 is affinity-retrieved in FC1. Thus, the 0.2 ng of bound material in FC1 is indicative of ~9 fmol captured TSST-1. On the other hand, both SEB and TSST-1 are expected to be affinity-captured in FC1. Because the ratio of captured TSST-1 in FC1 and FC2 derived from Fig. 8.3c was 6:1, it is assumed that ~0.033 ng (~1.5 fmol) out of the 0.475-ng bound material in FC2 is TSST-1; the remaining ~0.44 ng captured material in FC2 was SEB (~15 fmol). Thus, the ratio of SEB and TSST-1 captured in FC2 is ~10:1.

The mass spectra taken from the surface of the two flow cells following the toxin mixture SPR analysis are shown in Fig. 8.3e. Signals corresponding to multiply charged ions of TSST-1 are readily observed in the spectrum taken from the surface of FC1. The mass spectrum taken from the surface of FC2 shows signals corresponding to both SEB and TSST-1. Although no rigorous quantitative estimate can be made based solely on the relative signal intensities for SEB and TSST-1, the signals in Fig. 8.3e support the 10:1 ratio of the affinity-retrieved SEB and TSST-1 in FC1. Neither of the spectra contained signals due to HSA, antibodies, or other nonspecifically retained material.

As shown in this experiment, BIA/MS can simultaneously detect the two *S. aureus* toxins with an order-of-magnitude difference in the amounts retrieved on the sensor chip. This shows the usefulness of BIA/MS in screening food samples for multiple toxic compounds present at various concentrations. Because the maximum number of flow cells per sensor chip is four, these multiaffinity surfaces increase the number of toxins that can be screened simultaneously. In practice, sensor chips with multiaffinity flow cells can be prepared and sample aliquots injected serially over all the flow cells. When (and if) the SPR response indicates toxin(s) binding in any of the flow cells, the chip will be removed from the biosensor and immediately subjected to MALDI-TOF mass spectrometry, identifying the captured toxin(s) via its unique mass. To simplify the MALDI-TOF analysis, each flow cell can be derivatized with antibodies targeted toward toxins with easily distinguishable molecular masses.

8.3.3. Broad-Specificity Functional Screening

The data given in the previous example illustrate the first steps in constructing multianalyte BIA/MS arrays capable of simultaneously determining the absolute and relative amounts of a greater number of proteins present in biological fluids. In a similar approach, sensor surfaces can be derivatized with single ligands that naturally exhibit binding affinities toward a larger number of proteins. Examples of such ligands are metal/chelators (for targeting peptide motifs) or other small molecules (hormones, etc.). Following retrieval on the flow cell surface, MALDI-TOFMS analysis reveals the number and

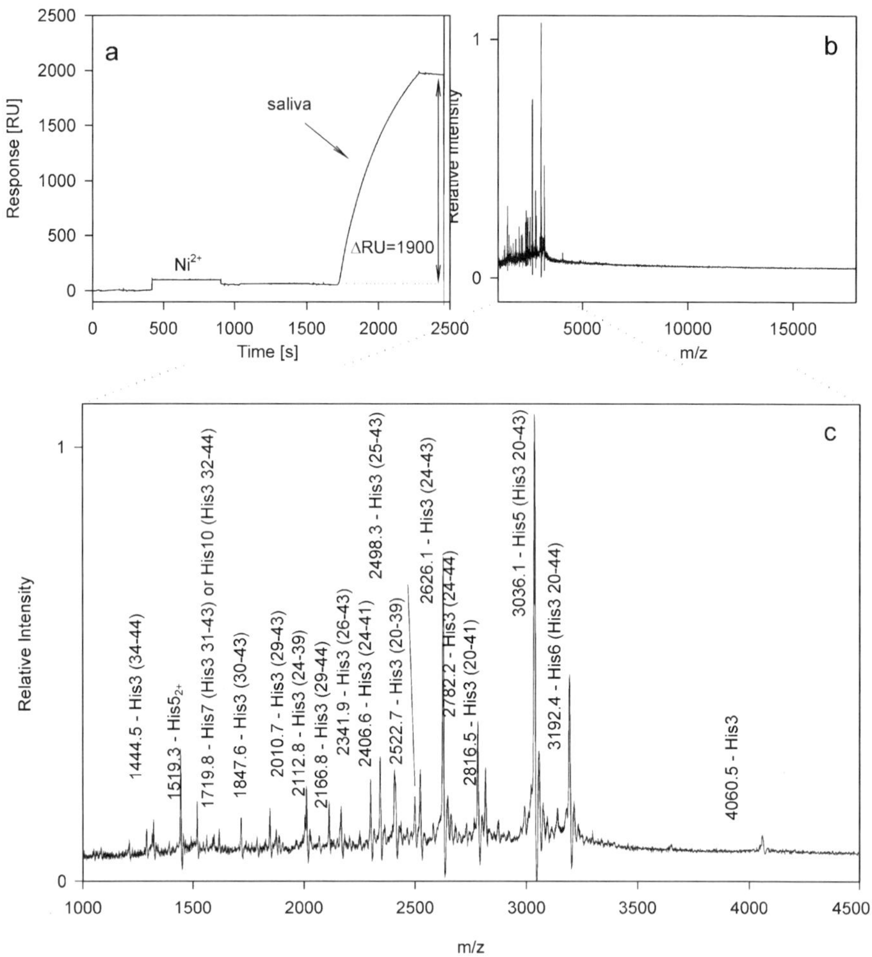

Figure 8.4. Broad-specificity BIA/MS: (a) sensorgram showing the injection of a diluted human saliva aliquot over NTA/Ni^{2+} flow cell surface; (b) MALDI-TOF mass spectrum taken from the flow cell surface; (c) expanded view of the same mass spectrum.

character of the proteins having the same functional activity toward the immobilized ligand.

The data shown in Fig. 8.4 are given as an example of broad-specificity functional screening. Figure 8.4a shows a sensorgram obtained from screening human whole saliva using an NTA/Ni^{2+} chelator sensor surface. The NTA (nitrilotriacetic acid) chelating sensor surface was loaded with Ni^{2+} via small injection of $NiCl_2$, after which an aliquot of a 100-fold diluted (in HBS-P, no EDTA) human saliva was injected at 5 µL/min. The SPR response indicates retention of ~1.9 ng of material on the flow cell surface. A MALDI-TOFMS analysis of the flow cell surface is shown in Fig. 8.4b. Numerous peptide signals in the 1–4kDa range are present; signals in the protein range (>5kDa) were not observed. Figure 8.4c shows an expanded view of the 1000–4500-Da region of the same mass spectrum. Because many salivary proteins have been previously characterized, most of the retained peptides can be tentatively identified by their molecular weight. The peptides retained by the Ni^{2+}-chelate surface are due to Histatin 3 and related enzymatic breakdown products (histatins 4–10). All peptides share the same amino acid sequence (–HEKHHSH–) that is recognized as a Ni^{2+}-binding motif.

This example illustrates the use of BIA/MS in the screening of biological fluids as a function of metal-binding properties. Other broad-range fractionation/screening can be achieved by loading the chelate surfaces with different metals (e.g., Fe^{3+} or Ga^{3+} to address phosphates).

8.3.4. Limit of Detection and Nonspecific Binding

Human plasma and urine samples were screened via BIA/MS for two low-MW biomarkers: urinary protein 1 (UP1, also known as Clara cell protein, CC10 or CC16) and retinol binding protein (RBP). One CM5 sensor chip was utilized for each biological fluid screen. The first flow cell (FC1) on each chip was derivatized with anti-UP1 antibody; in the second flow cell (FC1), anti-RBP was immobilized (not shown). Fresh human plasma was diluted fifty-fold in HBS-EP and a 70-µL aliquot was injected serially over both flow cells on the first sensor chip at 5µL/min (Fig. 8.5a). SPR responses of 5 and 150RU were noted in FC1 and FC2, respectively. The MALDI-TOF spectra taken from the surfaces of these flow cells are shown in Fig. 8.5c. RBP is readily detected in FC2 by the observance of the singly, doubly, and triply charged ion signals at *m/z* of 21,035, 10,515 and 7010, respectively ($MW_{RBP} = 21065$). UP1 is detected (although not as prominently) in the mass spectrum taken from FC1 cell via the singly and doubly charged ions signals at 15,824 and 7906, respectively ($MW_{UP1} = 15{,}826.4$). In addition, two other components (denoted by the signals at *m/z* of 6630 and 6432) were detected in FC1 and can also be scarcely seen in the mass spectrum taken from FC2.

For the second BIA/MS analysis, fresh human urine was diluted 10-fold in HBS-EP and a 45-µL aliquot was injected over the anti-UP1 and anti-RBP derivatized flow cells on the second CM5 sensor chip. The SPR responses at the time of chip undocking were 7RU in FC1 and 15RU in FC2 (Fig. 8.5b). A MALDI-TOF mass spectrum taken from the surface of the first flow cell (Fig. 8.5d) shows signals at *m/z* of 15,820, 7917 and 5278, corresponding to singly, doubly, and triply charged ions of UP1. RBP is barely detected in the second flow cell, as evidenced by the weak singly, doubly, and triply charged signals in the mass spectrum.

The single-digit SPR responses observed in these analyses are very close to the reliable detection limit of the SPR biosensor; thus, careful referencing must be implemented to accurately quantify the amount of affinity-isolated

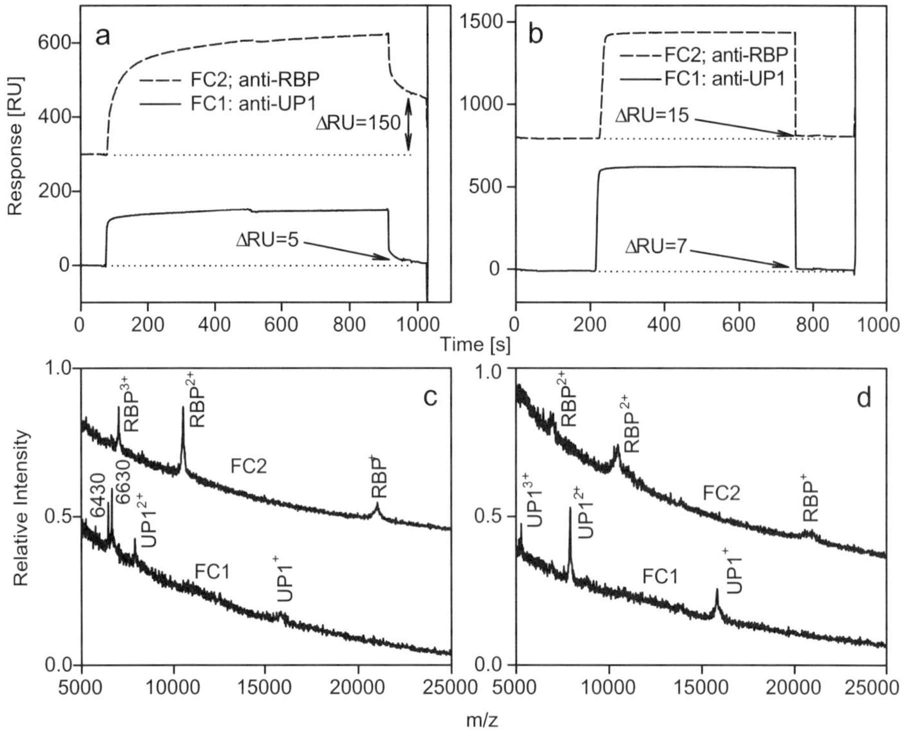

Figure 8.5. Rapid BIA/MS screening for UP1 and RBP: (a) sensorgrams resulting from the injection of a human plasma aliquot over an anti-UP1-deriavized (FC1) and anti-RBP-derivatized (FC2) flow cells; (b) sensorgrams resulting from the injection of a diluted human urine aliquot over FC1 and FC2, derivatized with anti-UP1 and anti-RBP, respectively; (c) MALDI-TOF mass spectra taken from the flow cells over which the human plasma solution was flowed; (d) MALDI-TOF mass spectra taken from the flow cells over which the human plasma solution was flowed.

material (14). Because the object of this BIA/MS experiment was to demonstrate the speed of analysis and simultaneous analysis of two biological markers, both flow cells on the sensor chip were utilized for screening and referencing of the SPR signal was not performed. Nevertheless, bearing in mind the molecular weights of both proteins and the SPR responses observed, it can be unambiguously concluded that the amount of retrieved protein is in the very low femtomole to subfemtomole range. The retrieval and detection of these amounts of UP1 and RBP in both experiments is viewed as a significant achievement because it extends the overall BIA/MS limit of detection beyond the limit of detection of SPR analysis alone. Whereas a low SPR response cannot by itself be a reliable indicator of analyte binding on the sensor surface, subsequent MALDI-TOFMS analysis can unequivocally confirm the outcome of the affinity interaction. As great numbers of proteins exist at intrinsically low concentrations in their native cellular systems, detection of subfemtomole protein amounts shows the applicability of BIA/MS in analyzing small volumes of low-level proteins from their natural milieu.

In addition to the signals from the targeted proteins, two signals at m/z of ~6630 and ~6430 are present in the mass spectra taken from flow cells over which human plasma solution was injected (Fig. 8.5c). In investigating the origin of these peaks, we have noted that, regardless of the type of antibody immobilized on the flow cell surface, these two signals are routinely present whenever human plasma was the source of sample. On the contrary, when urine was the sample source, the two signals were not detected in the mass spectra (Fig. 8.5d). A possible explanation is that they represent two positively charged and closely related high-level plasma proteins or protein fragments that bind to the unreacted carboxyl groups on the sensor surface via charge interactions. A molecular weight search at the National Center for Biotechnology Information Website (*http://www.ncbi.nlm.nih.gov/Entrez* using the MW values obtained from the mass spectra identified the signals at m/z 6630 and 6430 as apolipoprotein C-I (ApoC-I, MW = 6630.6) and a truncated form of apolipoprotein C-1 missing the N-terminal Thr-Pro residues (ApoC-I′, MW = 6432.4), respectively [15]. Their identification was made easier by the fact that two signals were close in the mass spectrum and by the fact that they were abundantly present in human plasma. Thus, the BIA/MS analysis was able to decipher the character of the nonspecific binding, confirming the need for performing MALDI-TOFMS following SPR analysis of complex biological fluids.

8.3.5. Emerging Applications

The examples presented above demonstrate both the compatibility (using the chip-based platform) and complementarity (bimodal detection relating to function and structure) of the BIA/MS approach in protein analysis. Yet, BIA/MS can enter into new arenas of proteome investigations by incorporating additional molecular biology and/or protein chemistry techniques into the analysis. In one approach, BIA/MS can be used with gene tagging techniques for protein identification [12]. Using various tagging approaches, tags are fused or inserted into nominally unknown genes for selectively isolating protein or its tag-containing proteolytic fragment(s) from the expression system. During SPR analysis, a highly selective tag-specific receptor immobilized on the sensor chip surface is used to affinity-isolate, detect, and precisely quantify the tagged polypeptide(s) retrieved (on-chip) from the expression system. The masses of the tagged polypeptides are then accurately determined via MALDI-TOF analysis performed directly from the sensor chip surface. These masses are used in protein structural characterization, such is sequence verification and identification through sequence database search.

In another approach, the BIA/MS technology can be expanded to a full-scale proteomics tool by incorporating postcapture enzymatic digestion of the on-chip retained analytes. In this approach, termed *biosensor chip mass spectrometry* (BCMS) [3], enzymes are immobilized on flow cells downstream of the capture flow cell. Following capture of the analyte by an immobilized ligand in the first flow cell, the analyte is eluted and/or routed into the enzymatically active flow cell(s) where the digestion occurs. With proper experimental design, BCMS is able to yield accurate molecular masses of both the native protein and the proteolytic fragments, and these data are used in the rigorous characterization and/or identification of the protein.

8.4. CONCLUSION

BIA/MS is a comprehensive microscale proteomics approach. The chip platform represents an ideal podium for protein characterization, in parallel to the plethora of DNA/RNA chips that have become the mainstay in the study of the human genome. By using a microfluidics system for sample delivery, and with the help of the small flow cell imprint, the analytes retrieved from the sample solution during the first dimension of the BIA/MS analysis are localized on a very small surface of the chip. Sample loses are kept to a minimum by performing the second dimension of the BIA/MS analysis directly from that same surface. The sample volumes are also minuscule, adding to the microscale attributes of BIA/MS. With this uniqueness, BIA/MS stands to find an increasing role in characterization of a great number of proteins existing at intrinsically low concentrations in their native biological systems.

REFERENCES

1. R. W. Nelson and J. R. Krone, Advances in surface plasmon resonance biomolecular interaction analysis mass spectrometry (BIA/MS), *J. Mol. Recognit.* **12**, 77 (1999).
2. R. W. Nelson, D. Nedelkov, and K. A. Tubbs, Biomolecular Interaction analysis mass spectrometry. BIA/MS can detect and characterize proteins in complex biological fluids at the low- to subfemtomole level, *Anal. Chem.* **72**, 404A (2000).
3. R. W. Nelson, D. Nedelkov, and K. A. Tubbs, Biosensor chip mass spectrometry: A chip-based proteomics approach, *Electrophoresis* **21**, 1155 (2000).
4. E. C. Nice and B. Catimel, Instrumental biosensors: New perspectives for the analysis of biomolecular interactions, *Bioessays* **21**, 339 (1999).
5. J. Homola, S. S. Yee, and G. Gauglitz, Surface plasmon resonance sensors: Review, *Sensors Actuat. B* **54**, 3 (1999).
6. M. Karas and F. Hillenkamp, Laser desorption ionization of proteins with molecular masses exceeding 10,000 daltons, *Anal. Chem.* **60**, 2299 (1988).
7. R. C. Beavis and B. T. Chait, Matrix-assisted laser-desorption mass spectrometry using 355 nm radiation, *Rapid Commun. Mass Spectrom*, **3**, 436 (1989).
8. C. Fenselau, MALDI MS and strategies for protein analysis, *Anal. Chem.* **69**, 661A (1997).

9. D. Nedelkov and R. W. Nelson, Practical considerations in BIA/MS: Optimizing the biosensor-mass spectrometry interface, *J. Mol. Recognit.* **13**, 140 (2000).
10. J. R. Krone, R. W. Nelson, D. Dogruel, P. Williams, and R. Granzow, BIA/MS: Interfacing biomolecular interaction analysis with mass spectrometry, *Anal. Biochem.* **244**, 124 (1997).
11. R. W. Nelson, J. R. Krone, and O. Jansson, Surface plasmon resonance biomolecular interaction analysis mass spectrometry. 1. Chip-based analysis, *Anal. Chem.* **69**, 4363 (1997).
12. R. W. Nelson, J. W. Jarvik, B. E. Taillon, and K. A. Tubbs, BIA/MS of epitope-tagged peptides directly from E. coli lysate: Multiplex detection and protein identification at low-femtomole to subfemtomole levels, *Anal. Chem.* **71**, 2858 (1999).
13. D. Nedelkov, A. Rasooly, and R. W. Nelson, Multitoxin biosensor-mass spectrometry analysis: A new approach for rapid, real-time, sensitive analysis of Staphylococcal toxins in food, *Int. J. Food Microbiol.* **60**, 1 (2000).
14. D. Nedelkov and R. W. Nelson, Exploring the limit of detection in biomolecular interaction analysis mass spectrometry (BIA/MS): Detection of attomole amounts of native proteins present in complex biological mixtures, *Anal. Chim. Acta* **423**, 1 (2000).
15. P. V. Bondarenko, S. L. Cockrill, L. K. Watkins, I. D. Cruzado, and R. D. Macfarlane, Mass spectral study of polymorphism of the apolipoproteins of very low density lipoprotein, *J. Lipid Res*, **40**, 543 (1999).

CHAPTER 9

Ultrasensitive MALDI-TOF-MS with Picoliter Volume Sample Handling

DAVID ERICSSON

Department of Analytical Chemistry, Institute of Chemistry, Lund University, Lund, Sweden

JONAS BERGQUIST

Department of Analytical Chemistry, Institute of Chemistry, Uppsala University and Department of Psychiatry and Neurochemistry, Institute of Clinical Neuroscience, Göteborg University, Sahlgrenska University Hospital, Mölndal, Sweden

CONTENTS

The handling of small sample volumes is imperative for analytic exploration of complex biological systems. In parallel to the nanospray mass spectrometry where extremely low volumes can be reliably analyzed, similar efforts can be achieved using matrix assisted laser desorption/ionization time of flight

Mass Spectrometry and Hyphenated Techniques in Neuropeptide Research, Edited by Jerzy Silberring and Rolf Ekman.
ISBN 0-471-35493-7

(MALDI-TOF) instruments. The strategies to increase the sensitivity in MALDI implicate decreasing the surface where the analyte is deposited. This can be achieved by two strategies: (1) to apply minimal sample aliquots [picoliter–nanoliter (pL–nL)] and (2) to use structured targets, such as nanovials or hydrophilic anchor points. To deposit small amounts of sample, we introduced a small pressure system, a picospritzer, routinely applied in neurobiology and cell biology, for intracellular as well as extracellular ejection. This micropipetting technology was also used when analyzing cells that were grown directly on the MALDI target.

Furthermore, we have shown, together with a number of other groups, the great advantages of using nanovials and prestructured targets for sample handling in direct MALDI and when coupling capillary electrophoresis to MALDI. In combination with picoliter handling using the dispenser technology, these microfabricated devices offer not only the possibility to use the sample extremely effectively but also to perform sample enrichment on target as well as effect chemical modifications. By restricting the limited amount of peptide/protein to a small surface, a higher surface concentration is achieved, which improves the detection sensitivity. The very virgin field of picoliter sample handling in MALDI-MS is presented, with examples, in this chapter. We have based the chapter on our own findings for practical reasons, but have tried to describe solutions presented by other groups where possible. However, it is impossible to cover all areas and applications; therefore, this chapter should be regarded as a small appetizer that hopefully will attract new users into the field of pico/nanoliter sample handling in MALDI.

9.1. REASONS FOR PICO/NANOLITER SAMPLE HANDLING IN MALDI

In MALDI with UV excitation the laser beam is usually focused to a diameter of 50–200 μm. Consequently, the MALDI sample spot should be in the same size in order to maximize the analyte detection in each spectrum. With the standard dried droplet MALDI spot preparation, the spot usually has a diameter between 1–4 mm, implying that only a small fraction of the spot is irradiated in each laser pulse. Another advantage with matched laser beam and sample spot size is that the so-called hot-spot search is obviated. Inhomogeneous sample distribution in the matrix crystals results in large deviations in spectra quality depending on where the laser beam is focused on the spot. This is no problem if the laser beam covers the entire sample spot.

Also, the crystallization behavior for certain matrixes can be controlled; for example, 2,5-dihydroxybenzoic acid (DHB) often crystallizes at the edges of a big sample spot, leaving the center free from crystals. On hydrophilic anchor points the crystals are forced to cover the center of the spot as well [1] and in nanovials the crystals form on the edges and walls [2,3]. This results in higher sample-to-sample uniformity in crystallization and mass spectra characteristics.

9.2. HOW TO APPLY SMALL SAMPLE SPOTS ON THE MALDI TARGET

9.2.1. Pico/Nanoliter Sample Preparation by Capillary Contact Transfer

To achieve as small MALDI spots as possible by dipping the sample onto the target, the capillary diameter needs to be minimized. In most groups this is accomplished by pulling glass micropipettes in the millimeter range down to a few micrometers in diameter. The capillary tips are used in conjunction with a syringe pump or a pressure system, which can eject or aspirate pico- or nanoliter volumes. Positioning of the samples are normally controlled by a 3D micromanipulator, with micrometer increments, and viewed through a microscope.

This concept was utilized by our group taking advantages of a small pressure system, a picospritzer, normally used in the field of neurobiology, cell biology, and biophysics [4,5]. We showed that the micropipetting (*or even picopipetting*) technique also was suitable for the application of small volumes of samples and matrix (down to a few picoliters) onto the MALDI target and that this procedure could increase the utility of MALDI when analyzing specific analytes at low concentrations in complex biological mixtures. In the first report using picoliter volumes in MALDI the main topic was to describe the method and the biological relevance of the samples was secondary, even if the results clearly indicated that the methodology was applicable to a number of different biological as well as pure chemical environments. Since such small sample volumes was used and specifically directed to small spots, the amount of sample needed for MALDI analysis was extremely small with detection limits in the zeptomole (10^{-21} mol) range (Fig. 9.1). The picospritzer technique also allowed for sampling of small sample volumes by using the vacuum-loaded valve box, and single cytosol compartments, single whole cells, or microextracts of cells were analyzed. The method was demonstrated using acutely isolated fetal rat astrocytes and since these cells are very small (Ø~17–20 μm), their total volume is approximately 4 pL and therefor much more difficult to sample than larger cells, such as giant vertebrate neurons. Furthermore, the technique permitted prepurification and enzymatic cleavage to be made at low volume, by packing small amounts of porous reversed-phase material or material with immobilized enzyme (POROS, PerSeptive Biosystem, Framingham, MA) in the very tip of the capillary.

In a second study the picospritzer technology was used for picoliter deposits of matrix on top of individual cells for single-cell analysis [6,7]. In this study MALDI was used to analyze single small mammalian cells (rat adrenal pheochromocytoma PC12 cells) grown directly on the sample target (Fig. 9.2). The cells were screened for peptide and protein contents and analyzed before and after stimulation with a neurotrophic factor, nerve growth factor (NGF). The single cell analysis of on-target-grown cells was also compared with

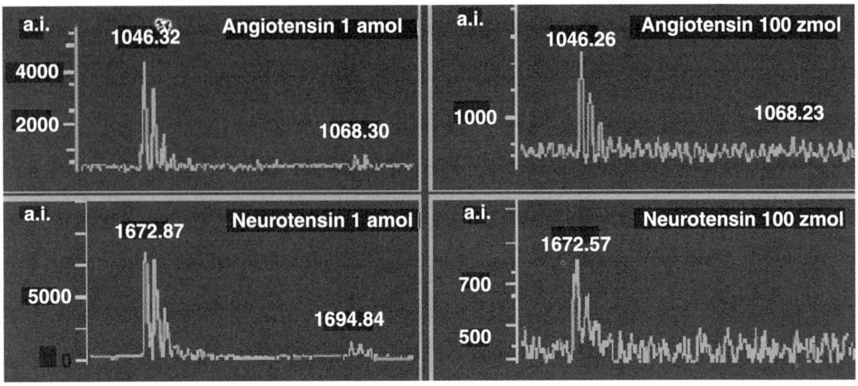

Figure 9.1. The picoliter sampling procedure permits extremely low amounts of sample to be deposited onto a small spot of the target, allowing detection of samples in the zeptomole range to be readily analyzed (100 zmol is ~60.000 molecules). The mass spectra show two examples of synthetic peptides, angiotensin II (1046.54, $M + H^+$ monoisotopic) and neurotensin (1672.92, $M + H^+$ monoisotopic) that have been put in 10 pL volumes on the target. (Reprinted from Bergquist et al. [4,5], copyright 1998, with permission from Elsevier Science.)

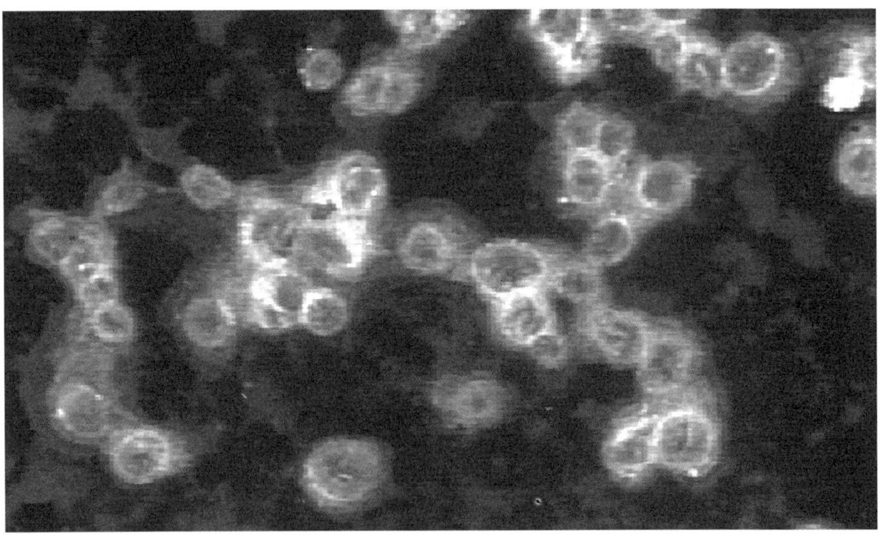

Figure 9.2. An immunofluorescent picture of nonstimulated pheochromocytoma cells (PC12) grown directly on-target for MALDI analysis. The picospritzer was used to deposit picoliter volumes on top of individual cells, in order to allow single-cell analysis. PC12 cells have an approximate diameter of 10 μm.

populations of whole cells put on top of a matrix layer and cellular extracts analyzed using the conventional sample/matrix mixture technique. The analysis of on-target-grown cells resulted in mass spectra with higher intensity, which provided further information of the cellular content of peptides and proteins.

Stimulation with NGF resulted in the detection of over 14 new peptides and proteins, demonstrating the potential of this technique for the investigation of cellular events, direct protein profiling, and cell fingerprinting. This unique and simple strategy provided a possibility to study intact cells, while keeping the cells in contact with other cells and substrates. Furthermore, the complicated sample handling in single-cell analysis was totally eliminated. The report represents a proof of principle, but does not provide identification of peptides or proteins in the mass spectra of the mammalian cells. So far, post-source-decay (PSD) analysis has been hampered by low sensitivity combined with difficult spectra interpretation. New generation MALDI instruments such as MALDI-TOF-TOF and MALDI-Q-Q-TOF will hopefully provide better possibilities to perform sequence analysis and identification online.

Microspot MALDI sample preparations of single subcellular organelles, such as 1–2-μm vesicles from the exocrine atrial gland of *Aplysia californica*, have also been performed by Rubakhin et al. [8]. A micropipette (pulled glass capillary) with <1-μm-tip diameter was used to sample the individual vesicles and subsequently 10 pL of matrix was applied onto the vesicles, resulting in a sample spot diameter of 10–20 μm. The excitation laser spot was focused down to 25 μm, facilitating detection of low to subattomole amounts of peptide within a single vesicle. Several reviews covering direct single-cell peptide profiling by MALDI are available [9,10], and the topic is discussed more thoroughly in another chapter in this book.

In 1995, a microspot MALDI sample preparation technique for peptide purification of brain microdialysate was presented by Zhang et al. [11]. A 50-μm-i.d. fused-silica capillary was packed with reversed-phase (C18) particles with a 20-nL bed volume (Fig. 9.3). It was connected to a multi-port valve together with different solvents for sample loading/washing and elution. Peptides like neurotensin and substance P in phosphate or NaCl buffers were trapped on the reversed phase material and washed with 2% HAc. Finally, the peptides were eluted onto the target in fractions of 5 nL with a solvent containing MALDI matrix. A total of 20 amol of neurotensin from a 20-pM solution in 0.2 M sodium phosphate buffer could be detected. The sensitivity gain of microspot preparation and purification of biological samples was demonstrated in a metabolic study of the neurotensin in the brain of an anesthetized rat by *in vivo* microdialysis.

A simple technique for microspot MALDI preparation as well as purification and enrichment of biological samples and in-gel tryptic digests, separated by 2D gel electrophoresis, has been developed by Gobom et al. [12]. The nanocolumn was made by trapping reversed-phase chromatography medium (30–50 nL) in a narrowed pipette tip (Fig. 9.4). The peptides in the samples were trapped on the reversed-phase material, which was washed. Subsequently

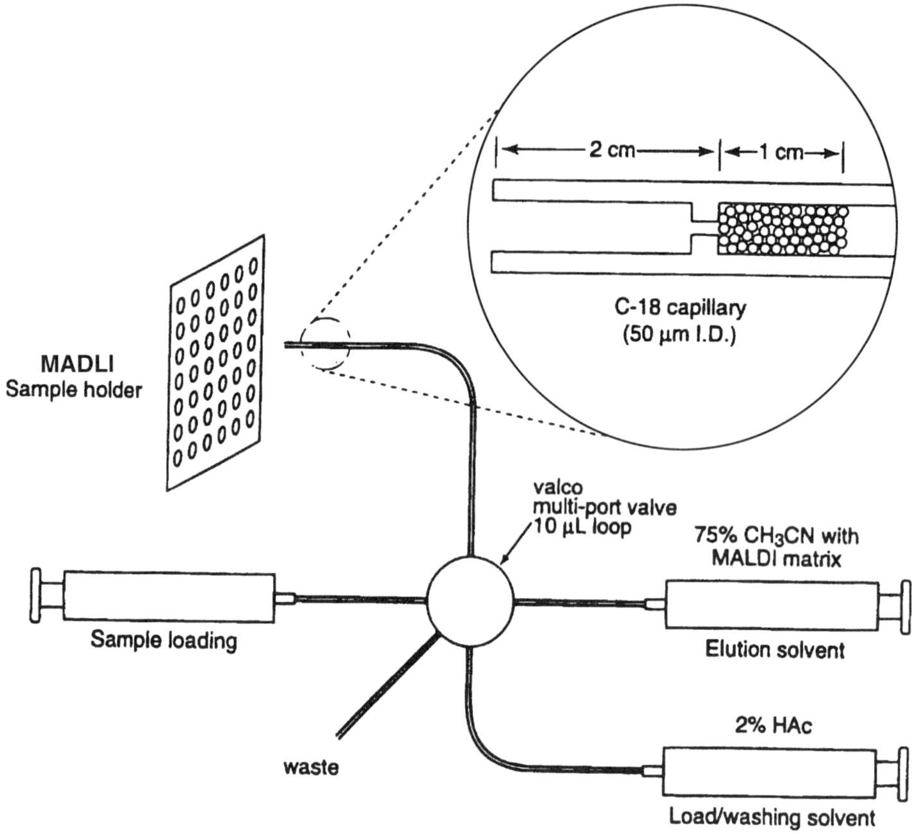

Figure 9.3. Schematic diagram of the MALDI sample microconcentration/desalting/matrix addition device. The loading and washing solvent was 2% acetic acid, the elution matrix was α-cyano-4-hydroxycinnamic acid in 75% acetonitrile with 2% acetic acid, the elution rate was 0.1 μL/min and the sample volume spotted on the target was ~5 nL per fraction. (Reprinted from Zhang et al. [11], copyright 1995, with permission from John Wiley & Sons Limited.)

the peptides were eluted with MALDI matrix and 2–3 droplets (each 50–100 nL) were deposited on the MALDI target. Peptide loads down to 1–10 fmol of tryptic digests were detected and the protein sequence coverage was improved by a factor of 2 compared to a conventional sample preparation method.

In another low-volume sample deposition technique deactivated fused-silica capillaries (10–200 μm i.d.) were used to load and deposit 20 pL–100 nL peptide or protein samples onto precoated MALDI targets [13]. About 100-μm-diameter sample spots was achieved and the sensitivity was in the attomole range for peptides and tens of attomole for proteins. Also 10–30 min tryptic digestions of 100–200 pL samples have been employed in the capil-

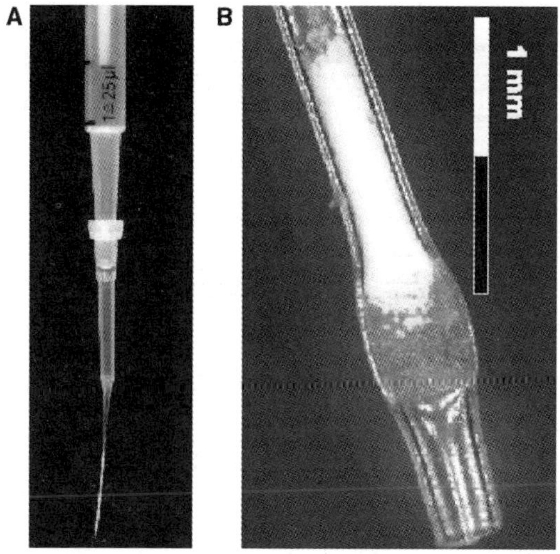

Figure 9.4. (a) A nanocolumn consisting of a long and narrow pipette tip (GELoader, Eppendorf) packed with chromatography media; (b) magnification of the nanocolumn tip with a packed bed volume of 30–50 nL. The flattening near the outlet prevents the chromatography medium from leaking out. (Reprinted from Gobom et al. [12], copyright 1999, with permission from John Wiley & Sons Limited.)

laries for protein identification (Fig. 9.5) [14]. A sample loading of 80-amol cytochrome c resulted in 12 correctly identified peptide fragments. Hemoglobin analysis of single red blood cells (~90 fL) was also presented. The blood cell suspension was loaded into the capillary, let to dry followed by loading of trypsin, which implied cell lysis and digestion of the protein content. Tryptic digest of a single red blood cell followed by microspot sample deposition was sufficient to detect the one amino acid substitution in hemoglobin causing sickle-cell disease.

Sequenom Inc. (*www.sequenom.com*) has an instrument that is commercially available and capable of preparing MALDI microspots in an automatic, high-throughput mode. The spectroJET deposits samples down to 15 nL on a 96- or 384-chip format, but no MALDI-TOF data has been provided.

9.2.2. Pico/nanoliter Sample Preparation by Dispensing

As an alternative to contact deposition, dispensing of pico/nanoliter droplets onto the MALDI target can be employed. The contact-free delivery of samples or matrix obviates carryover effects and dispensing facilitates efficient and precise handling of nanoliter volumes.

Various dispensers based on the inkjet principle have been used for MALDI-spot preparation earlier [15,16], and there are several commercial

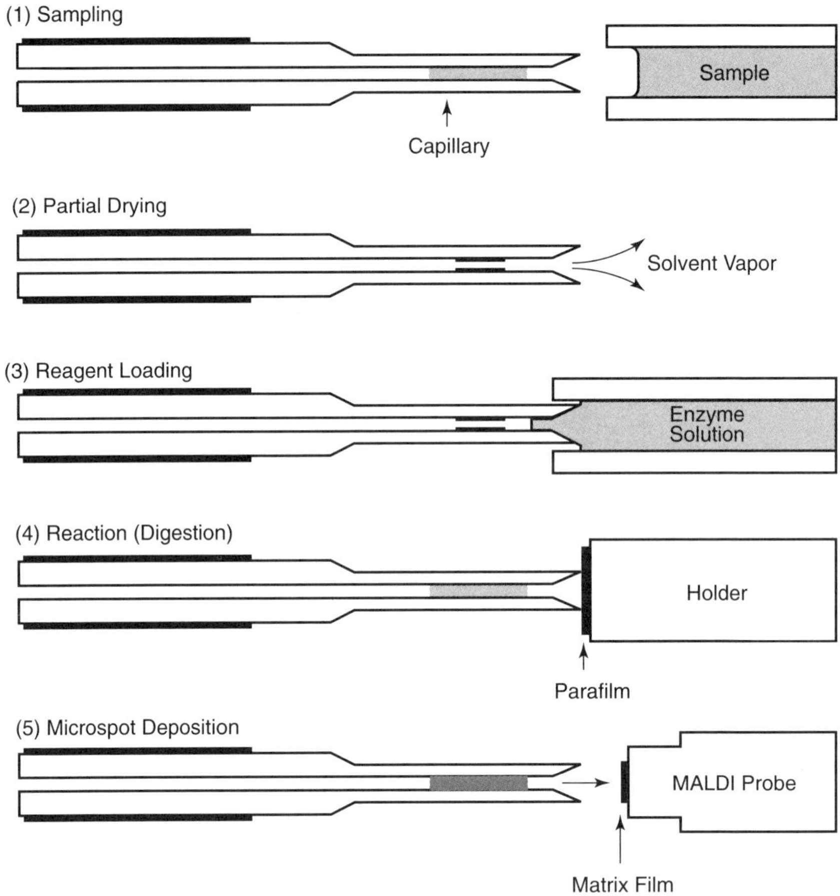

Figure 9.5. Enzymatic digestion and microspot sample deposition with a 20-μm-i.d. fused-silica capillary. (Reprinted from Whittal et al. [14], copyright 1998, with permission from American Chemical Society.)

dispensers available from, for example, Microdrop (*www.microdrop.de*, Norderstedt, Germany) and Microfab (*www.microfab.com*, Plano, TX). However, these dispensers are all single-ended with sample delivery from a stationary reservoir (no flow) or with droplet formation driven by aspiration through the orifice of the dispenser. If interfaced to a chromatographic system, the single-end design can only be used at certain flow rates and dispensing volumes. With this in mind, a piezoelectric silicon flow through microdispenser was developed within our group (Fig. 9.6) [17]. It can be used with a wide span of flow rates as well as with stationary samples. The total internal volume is <1 μL, and the inlet to nozzle volume is only 250 nL.

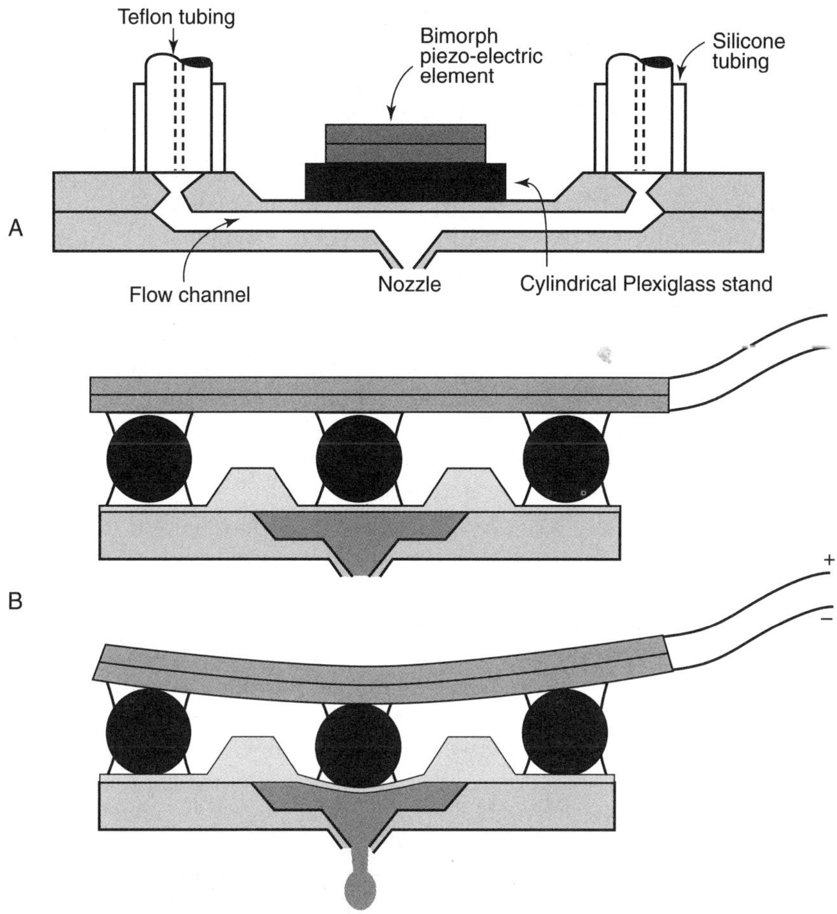

Figure 9.6. (a) A schematic cross section of the dispenser construction; (b) a schematic drawing of the actuation principles for the microdispenser. (Reprinted from Önnerfjord et al. [17], copyright 1998, with permission from American Chemical Society.)

Each dispensed droplet is only 60–100 pL, making the dispenser ideal for handling of small (nL–μL) and precious samples. A conical nozzle was developed to minimize the droplet contact surface, which facilitates accurate and reproducible droplet formation. The microdispenser is able to make precise and rapid sample depositions into nanovials down to 100 × 100 μm in an automatic, computer-controlled mode. The dispenser, as well as the nanovial arrays, is made by etching of monocrystalline silicon, which is widely used in integrated microanalytic systems.

9.2.3. Separation in Conjunction with Pico/Nanoliter Microspot MALDI Preparation

There are numerous applications of interfacing capillary electrophoresis (CE) and microcolumn liquid chromatography (μLC) separation to MALDI-TOF. This is covered mainly in other chapters, but a few examples of separation coupled to microspot MALDI preparation are mentioned here.

In an offline interface between CE and MALDI, a sample track was created by the eluting analytes as the CE capillary was moved over a matrix coated cellulose membrane [18]. Ground potential (cathode) was applied on the pretreated target, implying simultaneous electrophoresis and sample deposition without any need for a sheath flow of electrolyte, which otherwise deteriorates separation and sensitivity. The limit of detection was in the 50–100-amol range for peptides and about 250 amol for proteins. The preserved electrophoretic separation on target, measured by theoretical plate numbers, was as high as 80–90% compared to online UV detection.

Within our group, another CE-MALDI interface has been developed that utilizes MALDI-spot size reduction by a prestructured target, which is described further below [19]. The CE-capillary was inserted into a steel needle functioning as a cathode as well as for sample deposition, which was controlled by a x–y–z axis motion system. To prevent carryover effects, a sheath flow solution was mixed with the CE effluent and a 1.2-μL volume was automatically deposited on each anchor point, which had been precovered with MALDI matrix by dispensing. Subsequently, the CE fractions were concentrated on target by evaporation on the hydrophilic anchor points, a process that is described below. Before MALDI analysis, dispensing of an acidified solvent recrystallized all samples. The number of identified peptides from a tryptic digest of proteins from a human fetal brain cDNA expression library was increased from 47 to 211 with the described technique compared to direct MALDI analysis of the crude mixture. Moreover, the sequence coverage for different proteins was improved 1.3–4.9 times, and peptides in the low femtomole range were detected.

An example of microcolumn LC with online MALDI microspot preparation, Hsieh et al. has separated peptides from single neurons from a freshwater snail [20]. The μLC column was threaded into a glass capillary delivering matrix, which was mixed with the LC eluent and deposited in about 145-nL fractions on a steel MALDI target. A x–y–z-positioning unit controlled sample deposition. Peptide identification was accomplished on the basis of retention order, mass information, and post-source-decay analysis.

Dispensing of HPLC effluent as a contact-free MALDI-spot preparation technique has been performed earlier by using a modified graphics plotter to interface microbore LC to MALDI, but the flow rates and the dispensed volumes (0.5–2 μL) were too large to provide MALDI microspots [16]. In contrast, the flowthrough feature of the micro-dispenser developed in our group, makes the dispensing volume independent of the column flow, facilitating pico/nanoliter sample preparation for MALDI analysis [21,22]. The developed dispenser platform and the interfacing of microcolumn LC and MALDI-MS are described below.

9.3. MALDI TARGET SURFACE STRUCTURES FOR SAMPLE SPOT MINIMIZATION

9.3.1. Silicon Nanovials

Silicon is a favorable material to manufacture small reproducible structures with well-defined dimensions at low costs. Monocrystalline silicon has many mechanical properties comparable to stainless steel and moreover is chemically robust. The vials are created by anisotropic (crystal-plane-dependent) wet etching of ⟨100⟩-oriented monocrystalline silicon, which gives pits of inverted pyramids (Fig. 9.7). Flat-bottomed vials are created when the etching procedure is interrupted prematurely. Pyramidal vials ranging from $100 \times 100 \times 71\,\mu m$ to $400 \times 400 \times 283\,\mu m$ have been used in MALDI-MS, and vials as small as $5 \times 5 \times 4\,\mu m$ have been used in a TOF-SIMS application [23]. As shown

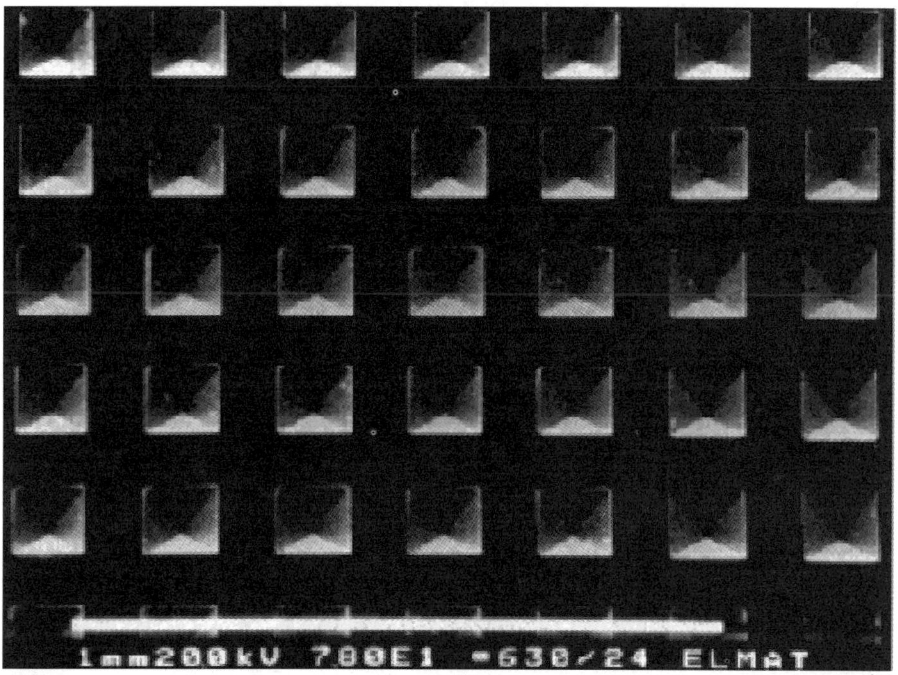

Figure 9.7. A scanning electron microscopy picture of a deep pyramidal nanovial chip, vial size $200 \times 200\,\mu m$, vial depth $140\,\mu m$. (Reprinted from Ekström et al. [3], copyright 2001, with permission from American Chemical Society.)

in Fig. 9.7, the nanovials are arranged in a high-density array form (e.g., 2500 vials/cm^2 for 100 × 100 μm vials), usually in 10 × 10 matrices with well-defined coordinates. The nanovial fabrication steps are described in detail by Ekström et al. [24].

The influence of the vial geometry on the mass accuracy and resolution for peptides and proteins in MALDI-TOFMS has been investigated [3]. It was found that the resolving power ($m/\Delta m$) decreased with increasing vial depth. The resolution was below 1000 for the deepest vial (283 μm), while shallow vials (20–50 μm) provided the same resolution as a flat surface (8000). Mass resolution is decreased with increased vial depth due to spatial distribution of matrix-sample crystals along the vertical axis in the vial, giving inhomogeneous desorption conditions and flight lengths, in turn, resulting in wider distribution of arrival times at the detector for a defined ion species.

Also, the mass accuracy is negatively affected by increasing vial depth, which is expected because of the increased flight length. Although using shallow (20–50 μm) vials and standard peptides for internal mass calibration obviated this problem. High mass accuracy (at least 10–50 ppm) in mass spectral data of peptide maps is very important for correct protein identification by database matching.

9.3.2. Sensitivity Gain by MALDI Target Surface Structures

In 1994 Jespersen et al. presented the first study that used miniaturized sample reservoirs in order to gain sensitivity in MALDI-MS [2]. They used 100 × 100 × 71-μm silicon vials (250 pL) and added the analyte, matrix solution (DHB) (1 : 1 v/v) manually with glass micropipettes having an outer tip diameter <100 μm. Bradykinin (1060 Da) was detected down to 2.5 amol from a 20-nM sample and cytochrom C (12,360 Da) down to 25 amol (200 nM). This was an improvement of the detection limit by three orders of magnitude according to the authors. Our group has performed more thorough investigations and further applications of silicon vials in MALDI-TOFMS. These are described in Section 9.4.

In 1995, Solouki et al. presented an application with microscope-monitored sample deposition onto small MALDI probe-tip indentions for MALDI-Fourier transform ion cyclotron resonance (FTICR) [25]. Indentions (<1 mm^2) were made on a stainless-steel surface in order to gain sensitivity. Substance P was detected at 8 amol; angiotensin II and bradykinin, at 90 amol. This should be compared to prior works where 1–5 fmol was detected by MALDI-FTICR. Multiple depositions of the sample volume (10 aliquots of 0.1 μL) were applied with evaporation in between. By letting the sample evaporate on the MALDI probe, the analyte surface concentration was further enhanced. This strategy was utilized with dispenser addition of samples presented below.

Arrays of hydrophilic sample anchor points (200-μm gold spots) surrounded by a hydrophobic Teflon layer was presented by Schuerenberg et al.

in 2000 [1]. An aqueous solution of a sample/matrix mixture (0.1–4 μL) is placed over the gold spot. The aqueous sample is not wetting the Teflon surface, and as the droplet evaporates, it shrinks toward the gold spot, which keep the droplet in place. Both the analyte and the matrix concentrations increase until the matrix starts to crystallize, incorporating the analytes. The initial matrix concentration must be kept very low (mM-range) compared to normal MALDI-spot preparations (M-range) in order to avoid premature crystallization at a bigger droplet radius. This principle of MALDI target sample enrichment is not working as well for matrixes demanding organic solvents, because the droplet is not repelled as much by the Teflon. Unfortunately, contaminants that interfere in MALDI-analysis, such as salts, are also enriched during evaporation. Therefore only purified samples can be successfully analyzed by this technique. The use of hydrophilic sample anchors give a possible sample up-concentration by more than two orders of magnitude, but the sensitivity gain is less probably due to surface adsorption. Peptides were detected down to 100–200 amol level.

9.4. PLATFORM FOR AUTOMATED NANOLITER SAMPLE HANDLING AND MALDI MICROSPOT PREPARATION BY A PIEZOELECTRIC FLOWTHROUGH MICRODISPENSER

The "Spot-on-a-chip" platform for protein and peptide analysis was presented by Ekström et al. in 2000 [24]. Different investigations and applications of the dispensing platform have been presented in publications since [3,21,22,26] and they are also described in this chapter. The platform consists of the piezoelectric flowthrough microdispenser, described above, and a MALDI target mounted on a movable *x–y* stage (Fig. 9.8). The MALDI-target is either a nanovial silicon chip array or a steel plate, airbrush covered with matrix.

9.4.1. Spot-on-a-Chip Sample Enrichment

The nanovial chip array was used in this application. A similar nanovial/dispensing setup has been designed for DNA analysis by Little et al. [27]. As mentioned above, multiple depositions of sample can be used to enhance the analyte surface density, thus increasing the sensitivity in MALDI-TOF. The spot-on-a-chip technology accomplished this on-spot enrichment efficiently by rapid dispensing and evaporation, facilitated by 45°C heating of the nanovial array. The resulting signal amplification in MALDI-TOF was at least 10–50 times compared to ordinary sample preparation [3]. Most advantageous with the dispensing enrichment is the ability to obtain good quality mass spectra from samples containing low analyte concentrations (low nM range). This was illustrated by the MS signal amplification of a calibration mixture consisting

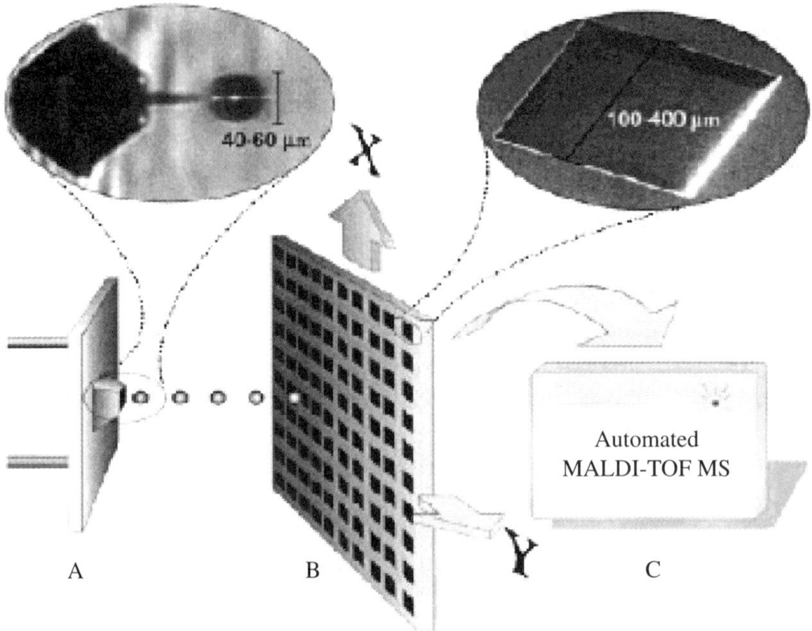

Figure 9.8. Spot-on-a-chip platform: (a) the piezo-actuated microdispenser; (b) heated high-density MALDI target plate; and (c) automated MALDI-TOFMS. (Reprinted from Ekström et al. [3], copyright 2001, with permission from American Chemical Society.)

of four peptides at a 5-nM concentration (Fig. 9.9). Furthermore, signal amplification of biologically relevant material was performed. The cytokine IL-8, involved in inflammatory responses, was not detectable with the standard dried-droplet sample preparation at a 100-nM concentration, but by using the spot-on-a-chip enrichment the IL-8 peaks were evident in the MALDI spectrum.

9.4.2. Automated Protein Identification Using Spot-on-a-Chip Technology

Today, there is a growing demand for automated high throughput technologies in peptide and protein analysis. The arrangement of nanovials or anchorpoints in arrays of well-defined coordinates facilitates automated, nonattended sample deposition and MALDI analysis. This concept was realized by the development of an automated protein identification platform using the microdispenser, nanovial MALDI arrays, and a microchip immobilized enzyme reactor (μ-chip IMER) [24]. Each protein sample was digested within

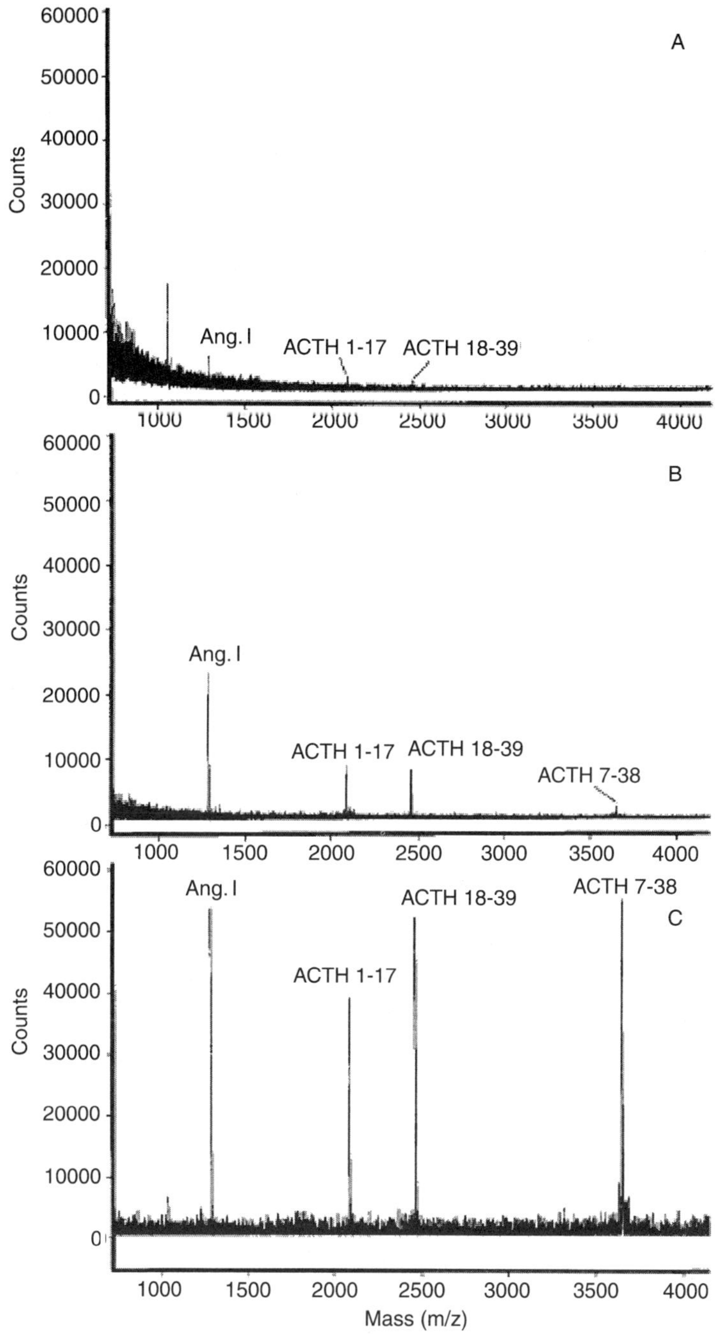

Figure 9.9. Effect of enrichment when analyzing a calibration mixture consisting of four peptides: angiotensin I, ACTH (1–17 clip), ACTH (18–39 clip), and ACTH (7–38 clip) at a concentration of 5 nM. In (a), 100 droplets (50 amol) were deposited and three of the four peptides can be detected. In (b), 2000 droplets (1 fmol) were deposited, and finally, in (c), a total of 8000 droplets corresponding to an absolute amount of 4 fmol were deposited and the saturation level of the detector was reached. Matrix: 2,5-dihydroxybenzoic acid. (Reprinted from Ekström et al. [3], copyright 2001, with permission from American Chemical Society.)

1–3 min in the μ-chip IMER, which was online coupled to the microdispenser, providing picoliter sample preparation in the nanovial MALDI target array. Automated protein digestion, sample preparation and MALDI-TOF analysis for 100 protein samples were accomplished in 3.5 h, demonstrating the high throughput capacity of the technology.

9.4.3. Dispensing on Precoated MALDI Targets

The dispensing platform has been used by Miliotis et al. to interface microcolumn liquid chromatography with automatic MALDI sample preparation [21,22]. To simplify the analysis even more, a homogenous matrix nitrocellulose film was spray-coated on the MALDI target with an airbrush, normally used for painting. Fractions of μLC eluate were dispensed in a precise array pattern onto the precoated target, leaving crystallized spots with a 400–700-μm diameter. Figure 9.10 shows spectra from consecutive spots of a μLC peak, illustrated by the UV chromatogram. The flowthrough feature of the microdispenser enables fraction collection of the μLC eluate. Complementary analysis of the collected fractions can be made by, for instance, electrospray MS or reanalyzed with static mode dispensing, which improves the sensitivity by depositing more of the sample on the target. After tryptic digestion of goat serum albumin, the number of peptide fragments found by MALDI analysis was improved from 22 to 37 with μLC separation. Four synthetic peptides corresponding to phosphorylated/nonphosphorylated tryptic fragments of the protein tyrosine kinase ZAP-70 were detected at 150 amol.

9.4.4. Dispenser-Aided Nanovial Proteolytic Digestion

Silicon vials have been used in earlier applications as compartments for chemical and enzymatic reactions [28,29]. Miniaturized reaction volumes give decreased diffusion distances, facilitating diffusion dependent reactions, such as enzyme reactions. Enzymatic digestion of proteins for mass fingerprint identification by database matching is now a standard technique to identify proteins. We have explored the properties of dispenser-aided proteolytic digestion in nanovial MALDI target arrays [26], based on the spot-on-a-chip platform described above. The dispenser delivers the protein sample to a nanovial precoated with enzyme. Catalytic breakdown at enzyme-specific cleavage sites on the protein begins immediately and continues as long as the dispenser moisturizes the nanovial. By simultaneous dispensing and evaporation, the protein concentration [S] increases in-vial (see Fig. 9.11), implying higher reaction velocity (V), according to the Michaelis–Menten equation:

$$V = \frac{V_{\max}[S]}{[S] + K_M} \tag{9.1}$$

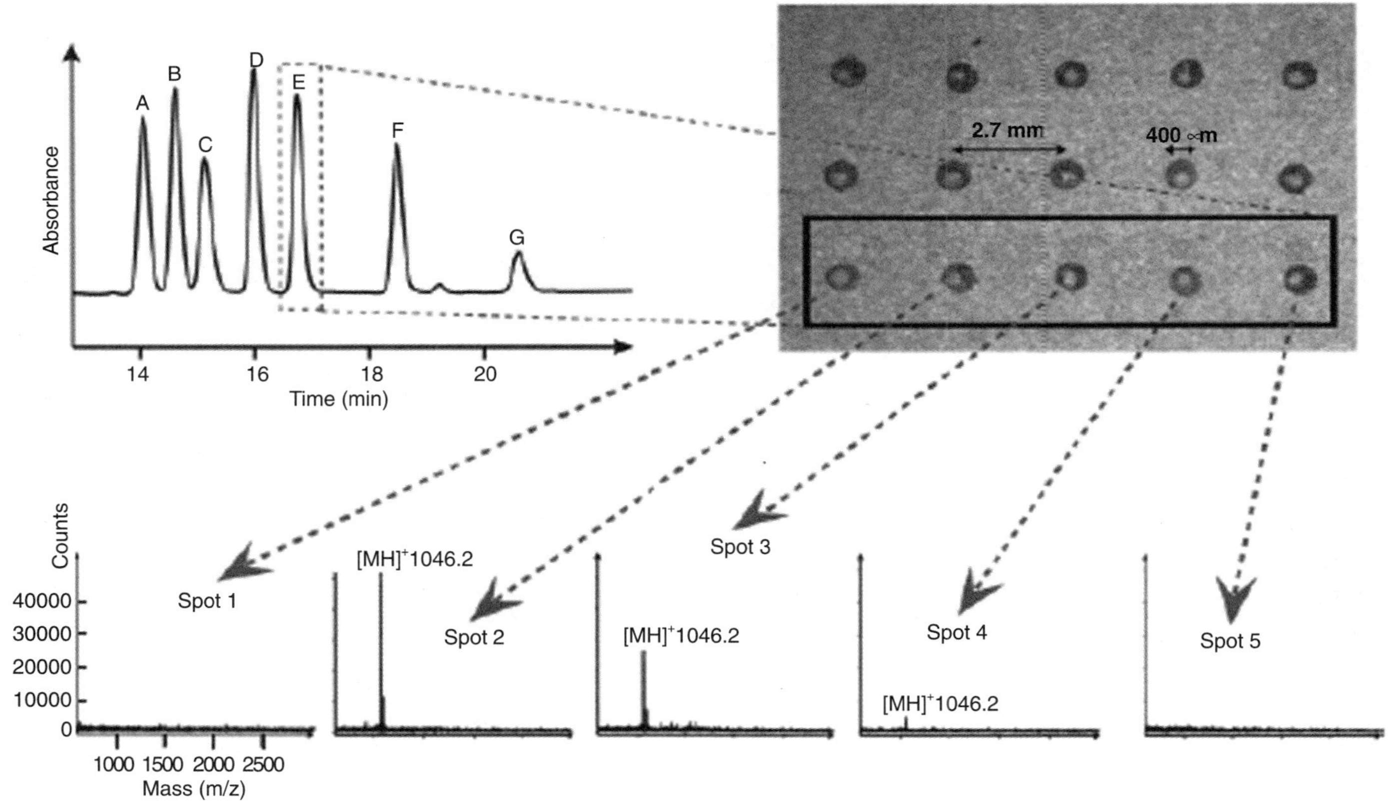

Figure 9.10. UV chromatogram showing reversed-phase capillary column (150 mm × 300 μm i.d.) LC separation of immunoregulating peptides. Peaks: A = ACTH (1–17 clip); B = enkephalin; C = bradykinin; D = angiotensin III; E = angiotensin II; F = angiotensin I; G = ACTH (18–39 clip). A magnified portion of the target plate is also depicted. The five consecutive sample depositions (1–5) correspond to the chromatographic peak enclosed by the broken rectangle (E). The MALDI mass spectra of the five spot positions illustrate the magnitude of the dispersion estimated from the detector cell to the nozzle of the flowthrough dispenser. (Reprinted from Miliotis et al. [21], copyright 2000, with permission from John Wiley & Sons Limited.)

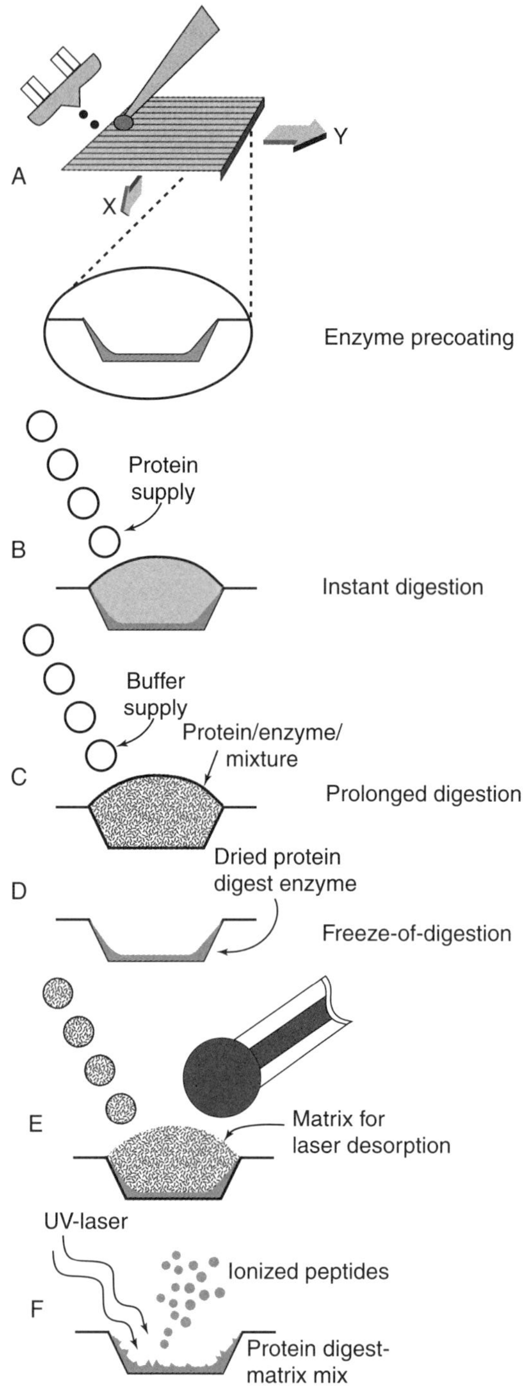

Figure 9.11. Step-by-step illustration of the in-vial digestion protocol: (a) enzyme precoating of the nanovial array by dispenser or pipette; (b) dispensing of protein sample—*instant digestion* begins; (c) dispensing of digestion buffer or water—*prolonged digestion*; (d) digestion stops when let to dry—*freeze of digestion*; (e) MALDI matrix deposition by dispensing (left) or GELoader-tip/capillary-and-pump (right); (f) MALDI-TOFMS analysis. (Modified from Ericsson et al. [26].)

During 10 min of dispensing, the protein concentration was increased 300-fold owing to the dispensing technique. The nano-digestion concept facilitates digestion of low-level proteins (10–500 nM), not possible to digest with standard protocols. Moreover, the digestion time was shortened to minutes or even seconds compared to standard protocols that perform digestion during several hours. As an example of low-level protein identification, a 10-nM solution of lysozyme C was unambiguously identified after 5 min of nanodigestion. In another example, only 30 s of nanodigestion was sufficient to identify hemoglobin (10 μM), which illustrates the fast catalysis of the nanodigestion technique. MALDI analysis can be performed by automation as described earlier.

9.5. CONCLUSION

MALDI has for a rather long time been successfully used for the identification of various substances by precise mass numbers, offering the unique combination of high sensitivity, high mass range, and good compatibility to a variety of biomolecules with fair mass resolution. However, MALDI sample preparation techniques have always been rather mysterious, and only small parts of the matrix-incorporated sample have been consumed in the analysis. In order to fully take advantage of the sensitive MALDI technique, more developmental time must be spent on methods of sampling and application of very small sample volumes onto the MALDI-TOF target. By using the tools described in this chapter, we can reduce the volumes needed for analysis, allow for single-cell analysis, reduce the problems with sample handling and automation, and finally reduce the problems with sample crystallization in MALDI.

ACKNOWLEDGMENTS

We wish to thank all our collaborators for their extensive contribution to the work described in this review. We are also grateful to Professor Per Håkansson for critical reading of the manuscript. This work was supported by the Swedish Foundation for Strategic Research, the Fredrik and Ingrid Thuring Foundation, the Wilhelm and Martina Lundgren Foundation, the Magnus Bergvall Foundation, the Swedish Alzheimer Foundation, the Syskonen Svensson Foundation, the Gamla Tjänarinnor Foundation, the Knut and Alice Wallenberg Foundation, the Swedish Lundbeck Foundation, the Swedish Society for Medical Research, the Swedish Natural Science Research Council (Grant K-AA/KU 12003-300), and the Swedish Medical Research Council (Grant 13123).

REFERENCES

1. M. Schuerenberg, C. Luebbert, H. Eickhoff, M. Kalkum, H. Lehrach, and E. Nordhoff, *Anal. Chem.* **72**, 3436–3442 (2000).
2. S. Jespersen, W. M. A. Niessen, U. R. Tjaden, J. v. d. Greef, E. Litborn, U. Lindberg, and J. Roeraade, *Rapid Commun. Mass Spectrom.* **8**, 581–584 (1994).
3. S. Ekström, D. Ericsson, P. Önnerfjord, M. Bengtsson, J. Nilsson, G. Marko Varga, and T. Laurell, *Anal. Chem.* **73**, 214–219 (2000).
4. J. Bergquist, G. Karlsson, and R. Ekman, *13th Montreux Symp. LC-MS; SFC-MS; CZE-MS; MS-MS*, Montreux, Switzerland, 1996.
5. J. Bergquist, G. Karlsson, and R. Ekman, *Adv. Mass Spectrom.* **14**, 1–15 (1998).
6. J. Bergquist, *Symp. High Performance Liquid Phase Micro Separations*, Lund, Sweden, 1998.
7. J. Bergquist, *Chromatographia* **49**(Suppl. 1), S41–S48 (1999).
8. S. S. Rubakhin, R. W. Garden, R. R. Fuller, and J. V. Sweedler, *Nat. Biotechnol.* **18**, 172–175 (2000).
9. L. Li, R. W. Garden, and J. V. Sweedler, *Trends Biotechnol.* **18**, 151–160 (2000).
10. J. J. Dalluge, *Fres. J. Anal. Chem.* **366**, 701–711 (2000).
11. H. Zhang, P. E. Andrén, and R. M. Caprioli, *J. Mass Spectrom.* **30**, 1768–1771 (1995).
12. J. Gobom, E. Nordhoff, E. Mirgorodskaya, R. Ekman, and P. Roepstorff, *J. Mass Spectrom.* **34**, 105–116 (1999).
13. L. Li, R. E. Golding, and R. M. Whittal, *J. Am. Chem. Soc.* **118**, 11662–11663 (1996).
14. R. M. Whittal, B. O. Keller, and L. Li, *Anal. Chem.* **70**, 5344–5347 (1998).
15. G. Allmaier, *Rapid Commun. Mass Spectrom.* **11**, 1567–1569 (1997).
16. T. I. Stevenson and J. A. Loo, *LC-GC* **16**, 54–60 (1998).
17. P. Önnerfjord, J. Nilsson, L. Wallman, T. Laurell, and G. MarkoVarga, *Anal. Chem.* **70**, 4755–4760 (1998).
18. H. Y. Zhang and R. M. Caprioli, *J. Mass Spectrom.* **31**, 1039–1046 (1996).
19. T. Johnson, J. Bergquist, R. Ekman, E. Nordhoff, M. Schuerenberg, K. D. Kloeppel, M. Muller, H. Lehrach, and J. Gobom, *Anal. Chem.* **73**, 1670–1675 (2001).
20. S. Hsieh, K. Dreisewerd, R. C. van der Schors, C. R. Jimenez, J. R. StahlZeng, F. Hillenkamp, J. W. Jorgenson, W. P. M. Geraerts, and K. W. Li, *Anal. Chem.* **70**, 1847–1852 (1998).
21. T. Miliotis, S. Kjellström, J. Nilsson, T. Laurell, L. E. Edholm, and G. Marko Varga, *J. Mass Spectrom.* **35**, 369–377 (2000).
22. T. Miliotis, S. Kjellström, P. Önnerfjord, J. Nilsson, T. Laurell, L. E. Edholm, and G. Marko Varga, *J. Chromatogr. A* **886**, 99–110 (2000).
23. R. M. Braun, A. Beyder, J. Y. Xu, M. C. Wood, A. G. Ewing, and N. Winograd, *Anal. Chem.* **71**, 3318–3324 (1999).
24. S. Ekström, P. Önnerfjord, J. Nilsson, M. Bengtsson, T. Laurell, and G. Marko Varga, *Anal. Chem.* **72**, 286–293 (2000).
25. T. Solouki, J. A. Marto, F. M. White, S. Guan, and A. G. Marshall, *Anal. Chem.* **67**, 4139–4144 (1995).

26. D. Ericsson, S. Ekström, J. Nilsson, J. Bergquist, G. Marko-Varga, and T. Laurell, *Proteomics* **1**, 1072–1081 (2001).
27. D. P. Little, T. J. Cornish, M.J. O'Donnell, A. Braun, R. J. Cotter, and H. Koster, *Anal. Chem.* **69**, 4540–4546 (1997).
28. C. A. Clark and A. G. Ewing, *Chemtech* **28**, 20–25 (1998).
29. E. Litborn, A. Emmer, and J. Roeraade, *Anal. Chim. Acta* **401**, 11–19 (1999).

PART 3

SELECTED APPLICATIONS IN PEPTIDE RESEARCH

CHAPTER 10

Peptide Mapping Using MALDI-TOFMS

ULF HELLMAN

Ludwig Institute for Cancer Research, Uppsala, Sweden

CONTENTS

10.1. INTRODUCTION

The technique by which proteolytic products of a given protein is analyzed (i.e., peptide mapping) has undergone a dramatic change since the early 1990s. Earlier, the protein of interest had to be isolated in a homogeneous form, and digested in solution and the resulting peptides purified by liquid chromatography. The peptide pattern in the chromatogram could be regarded as a fingerprint from that particular protein, and when patterns from two proteins were compared, subtle differences between related proteins could be identified [1]. In other applications, selected peptides were subjected to automated amino acid sequence analysis. The protein could then be identified by scanning a sequence database using the peptide's primary structure as probe. Often

Mass Spectrometry and Hyphenated Techniques in Neuropeptide Research, Edited by Jerzy Silberring and Rolf Ekman.
ISBN 0-471-35493-7

several peptide sequences had to be used as probes for a confident identification. If no secure match was found, it could be concluded that the protein was uncharacterized and the information was used to construct an oligonucleotide probe by which the corresponding gene could be isolated and sequenced.

The sensitivity of the analytical instruments used for amino acid sequencing was such that a large quantity of pure protein was required, namely, in the microgram–milligram range for an average sized protein.

In the early 1990s, two revolutionary techniques appeared; one in the preparation of samples and one in the analysis of processed samples. For sample preparation, the use of polyacrylamide gel electrophoresis for isolation of proteins [2,3] was an extremely important improvement. Now it became possible to handle practically any protein since they were in a detergent environment and the sample could be relatively heterogeneous; it was sufficient that the target protein appeared homogeneous in a gel band or spot. The yield after in-gel digestion, when compared to digestion of identical amount in solution is 60–70% [3].

For high-sensitivity analysis, the introduction of peptide mass fingerprinting (PMF) using MALDI-TOFMS was a tremendous breakthrough [4]. Identification of proteins by PMF can be done in the low fmol range—at least a 100-fold increase in sensitivity compared to Edman degradation. Other important advantages are speed of analysis (a few minutes compared to overnight runs) and running cost (a few cents for one analysis compared to about 150 USD per peptide).

The improvements in preparation and analysis were well timed with the steadily growing and now almost finished characterization of complete genomes from several species. It is no longer necessary to put much effort into sequence work, whether it is on the protein or cDNA level, since almost every peptide mass fingerprinting leads to the identification of known proteins.

10.2. ISOLATION OF PROTEINS BY GEL ELECTROPHORESIS FOLLOWED BY IN-GEL DIGESTION

The mixture containing the target protein must not be too complex, especially if SDS-PAGE is the electrophoresis to be done. Some kind of enrichment has to be performed in order to obtain a homogeneous protein band. Preferred techniques are affinity chromatography on an immobilized ligand with as few interaction points as possible. For example, a synthetic peptide bound to AffiGel (Bio-Rad, Hercules, CA) will pull out distinct interacting proteins from a total-cell lysate [5], whereas an immobilized antibody or a GST fusion ligand will retard a large number of proteins from a whole-cell lysate.

It is important that cysteine-containing proteins be reduced and alkylated. This should be done prior to the electrophoresis and digestion since proteins with several SS bonds will be sterically hindered for proteolysis. Furthermore, when peptide mass fingerprinting is performed, the search will be more stringent if status of cysteines is well defined. Depending on the quantity applied to the gel, different staining methods can be chosen. When a protein can be visualized by Coomassie staining, there is plenty of material, and a good signal should be obtained by MALDI-TOFMS without any special workup procedures. If, on the other hand, silver or fluorescent dyes were needed to "see" the protein, then it may be necessary to desalt and concentrate the sample prior to analysis. The following description relies on several years of experience and may be considered robust methods.

Before electrophoresis, make sure that the target protein is so much enriched that it is realistic to find it homogeneous in a gel band.

1. Run the reduced and alkylated sample in a standard SDS-PAGE or a two-dimensional gel. Be extremely careful not to contaminate the preparation. (It is however almost impossible to avoid traces of in particular keratins!) Make sure that all solutions and equipment used for electrophoresis are clean. As a general rule, avoid manipulations as much as possible. Bands and spots from several lanes/2D gels may be combined, but if the sample is dilute, it is better to concentrate by an appropriate technique [6].
2. Stain and destain by standard Coomassie techniques, using the modified silver-staining procedure [7] or other techniques, such as Sypro Ruby [8]. If the quantity is unknown, it might be a good idea to first stain by Coomassie, and if that fails to detect protein bands, the same gel can be restained by silver.
3. Alternatively, a dried and archived gel may be rehydrated using water and the procedure followed from step 4 below. Gels at least 5 years old have been treated successfully. If the gel for some reason needs to be dried (e.g., for autoradiography or archiving), it is preferable to dry between plastic wrap, and not on filter paper.
4. Cut out the band(s) of interest. It may be a good practice to run one empty lane for a negative control and one lane with a known amount of a standard protein. A molecular weight marker protein can be used as a positive control.
5. *For Coomassie-stained proteins*, wash the gel piece(s) twice with buffer/50% acetonitrile—enough volume to cover the gels. For trypsin, use 0.2 M ammonium bicarbonate; for LysC, use 0.5 M Tris–HCl pH 9.25. Incubate with some shaking for at least 30 min at 30°C. Discard the washings. *For silver-stained proteins*, if the stain was very weak, the silver should be removed following the procedure by Gharahdaghi et al. [9], that is, treat with potassium ferricyanide in sodium thiosulfate until the gel piece is transparent (this takes a few minutes). Then rinse with water, incubate in digestion buffer, and dehydrate in neat acetonitrile. *For proteins stained with a fluorescent dye*, such as Sypro Ruby

(Molecular Probes, Inc. Eugene, OR), Rinse with water, incubate in digestion buffer, and dehydrate with neat acetonitrile. Washing time can for convenience be prolonged without any problem. It is very difficult to wash out intact proteins!

6. Dry the gels completely under a stream of nitrogen or in a rotary evaporator. The completely dry gel will, on reswelling, force the protease solution to absorb into the gel matrix and meet the protein.
7. Start the rehydration/reswelling of the gels by adding the protease at a concentration of 10–20 ng/μL in the appropriate digestion buffer. (*For trypsin*, 0.1 M ammonium bicarbonate; *for LysC*, 0.1 M Tris–HCl pH 9.25.) Add the protease in the volume that will bring back the dried gel to its original size. It has been suggested (Darryl Pappin, personal communication) to let the gel reswell on ice for about 45 min in order to allow the gel to absorb the protease before it autodigests. Add more buffer if necessary. We use "modified trypsin, sequence grade" from Promega, (Madison, WI), a preparation that is almost resistant to autodigestion, and LysC is from WAKO Chemicals, Neuss, Germany. Those are stored at −20°C in aliquots in 1 mM HCl (trypsin) or in 0.1 M Tris–HCl pH 9.25, (LysC).
8. Incubate at 30°C overnight. We use an air-heated column oven, equipped with a small shaker.
9. To stop the reaction, acidify by adding 10% trifluoroacetic acid (TFA) to a final concentration of 1%, vortex and centrifuge briefly.
10. Save the supernatant. Add 0.1% TFA/60% acetonitrile (enough volume to cover the gel) and incubate for >30 min at 30°C. Combine extract with the previous supernatant and repeat extraction. Use the same pipette tip to minimize losses due to surface adsorption.
11. Reduce volume by vacuum centrifugation to about a third; this removes most of the organic solvent, which is a prerequisite for the following miniature reversed-phase-based desalting/concentration step. This may not be necessary for strong samples, such as Coomassie- or dark silver-stained spots, but is needed for weak silver or fluorescent spots.
12. Desalt, and if necessary concentrate, the peptide mixture on a "nanocolumn." Once the sample is loaded and salts washed off, it can be eluted in different ways:

    a. Elute with 60% acetonitrile in 0.1% TFA into a small tube
    b. Elute in steps of increasing concentrations of acetonitrile into small tubes. This creates a group separation of the peptides according to hydrophobicity. Analysis of each group individually will give more peptide mass information compared to analysis of the whole mixture because the ionic suppression is reduced.
    c. Elute with 60% acetonitrile containing matrix in smallest possible droplets directly onto the target.

d. Elute as in substep b, but include matrix in the eluent and collect each step onto target.

We prepare columns by packing about 0.1 µL of POROS R2 20 (Perceptive Biosystems, Foster City, CA) in GelLoader pipette tips (Eppendorf, Hamburg, Germany) [10]. The same functionality can be found in commercially available ZipTip (Millipore Corporation, Bedford, MA).

10.3. PEPTIDE MAPPING BY MALDI-TOFMS

There are numerous ways to prepare the crystalline sample for MALDI-MS analysis [11]. In essence, the peptide mixture is mixed with the same volume of a saturated matrix solution, which is then placed onto the surface of the target and allowed to dry. The two most common matrices for peptides are α-cyano-4-hydroxycinnamic acid (ACCA) and dihydroxy benzoic acid (DHB). These are made up about once a week as saturated solutions in, for instance, 0.1% TFA : acetonitrile (2 : 1, v : v). We use a piece of Parafilm on which we mix a 0.3-µL sample with the same volume of the matrix solution. Half of the mix is applied onto the target and allowed to dry. The target is then inserted into the vacuum system of the instrument for analysis.

The most crucial considerations are the sample quality and concentration. For example, when a protein was detected by Coomassie, and the extracted peptides are in a volume of 50–100 µL, even containing up to 50 mM ammonium bicarbonate, one can take a 0.3-µL sample directly from the extract and usually get a clean spectrum. If, on the other hand, the extract was from a weak silver-stained spot, the low concentration of peptides will, together with the presence of buffer salts, not be sufficient for direct analysis. Such samples must be concentrated and desalted, a procedure that is elegantly done by reversed phase in a diminutive scale (see text above).

The spectrum is taken in an instrument specific way; some brands are equipped with a camera that displays an enlarged view of the sample crystal which helps in finding a "sweet spot," whereas other makes rely on an exact positioning of the sample spot. In either case, several shots of the laser are fired and spectra recorded. When a realistic number of peptide masses with a sufficient signal to noise ratio is achieved, the spectrum is stored for further evaluation.

One of the most important issues in MALDI peptide mass fingerprinting is to obtain the data with highest possible accuracy, or at least should the accuracy be known. High-accuracy mass determinations require internal calibration, which often can be done using autodigestion products from the protease. For the commonly used porcine trypsin, one should look for the following monoisotopic MH^+ masses: 842.510, 1045.564, and 2211.105. It is also common to find peptides from contaminating proteins, such as keratins of various forms, whose peptide masses soon become familiar to the mass spectrometrist, and that also can be used for internal calibration. An accuracy of ±0.1 Da or better should be obtained with a correct calibration.

When searching with the peptide mass list against the sequence databases via any of the search engines over the Internet, it is important first to make an unfiltered scan, specifically, without specifying species, size, or pI. With these settings it will be possible to detect contaminating proteins from unexpected sources and of other molecular weights and pI values than the target protein's migration indicate. Also allow one miscut, since endoproteinases are poor exopeptidases; thus, for trypsin, when there are two Arg or Lys near each other, only one of the two will be cleaved, leading to one acceptable miscut. Most search engines offer the possibility of allowing common modifications in the search, such as cysteine derivatizations, phosphorylations, and oxidations. Some search engines enable the user to construct any modification and include that in the search; examples of the latter can be fatty acids or ubiquitinylation. A list of four common search engines, including their URLs (universal resource locators), is given here: ProFound (http://*www.proteometrics.com*), Mascot (http://*www.matrixscience.com*), PeptideSearch (http://*www.mann.embl-heidelberg.de*), MS-Fit (http://*prospector.ucsf.edu*).

10.4. ATTEMPTS TO FIND POSTTRANSLATIONAL MODIFICATIONS BY MALDI-TOFMS

Identifying proteins by peptide mass fingerprinting is becoming a routine task, now that the primary structure of a vast number of proteins is available in public sequence databases. The new challenge will be to analyze the various modifications that native proteins carry, and that probably give rise to many unmatched mass numbers in a search. This is because the majority of the sequence databases are constructed from translated DNA sequences, and therefore very little information on modifications is given. Unfortunately MALDI-TOFMS has a few inherent limitations that make interpretations insecure. The most serious one is the fact that all analytes in a peptide mixture are not detected. Furthermore, some modifications, including phosphorylations, make that modified peptide less sensitive. Therefore, what is positively detected by MALDI-MS is real, but what is not detected may still be present in the sample. Another drawback is that a positive finding of a modification (e.g., an addition of 42 Da, indicating an acetylated residue or *N* terminus) is not absolute. There is some probability that this particular peptide mass comes from a nonmodified peptide in the target protein or a contaminant, and that it coincides with the mass of a modified peptide. Therefore these discoveries must be confirmed by a direct identification, such as MS/MS analysis and perhaps ultimately a mutation of the suspected modified residue.

The following are some drawbacks in peptide mass fingerprinting by MALDI-TOFMS:

All peptides do not fly	Probably an inherent property; change matrix?
Ionic suppression	Separate peptides prior to MALDI
Amount and signal nonlinear	Label also very small signals
Salt interference	Desalt or wash on-target
Protease miscuts	Use high-quality enzymes
Contaminating proteins	Prefractionate before electrophoresis; work clean
Several unmatched masses	Probably due to unknown modifications
Modifications	Are seldom detected

10.5. APPLICATIONS

10.5.1. Identification of a Large Protein from a Coomassie-Stained SDS-PAGE

In an attempt to isolate PDGF-beta receptors from mouse kidneys by immunoprecipitation using protein-A-bound antibodies, five discrete bands, seen clearly after Coomassie staining, were obtained on analysis by SDS-PAGE on a 6% Mini-PROTEAN 3 gel (Bio-Rad, Hercules, CA) (gel not shown). The bands were from around 100 to 280 kDa, and all were subjected to identification analysis by PMF. The analysis of the band at 190 kDa is shown as example. After tryptic in-gel digestion, acidification, and extraction, a 0.3-μL sample was taken from the 60-μL extract and mixed with the same volume of a saturated solution of the matrix (ACCA). The sample was pipetted and allowed to dry on the target surface, and the spectrum was taken using instrument settings optimized for peptides. The instrument, a Bruker Biflex III, was equipped with delayed extraction and reflector. A total of 80 shots were combined and the monoisotopic peaks were labeled first, since only one known peptide could be identified, using external calibration (Fig. 10.1). The peptide mass list from the spectrum was used in ProFound to search for matching proteins of sizes up to 3000 kDa and pI values from 3 to 10 in all taxa, that is, a nonfiltered search. Two miscuts and partial oxidation of methionines was allowed. The status of Cys (alkylated with iodoacetamide) was indicated in the search. As a convincing match was obtained, the spectrum was recalibrated internally using three of the matched protein's peptide masses (Table 10.1). Of 76 peptides used for the search, 63 fit into the sequence of rat Clathrin heavy chain. Apparently the mouse sequence database does not contain this protein. Nine of the peptide masses represent peptides with oxidized methionines, and 20 are results of noncleaved Arg or Lys residues. This may seem

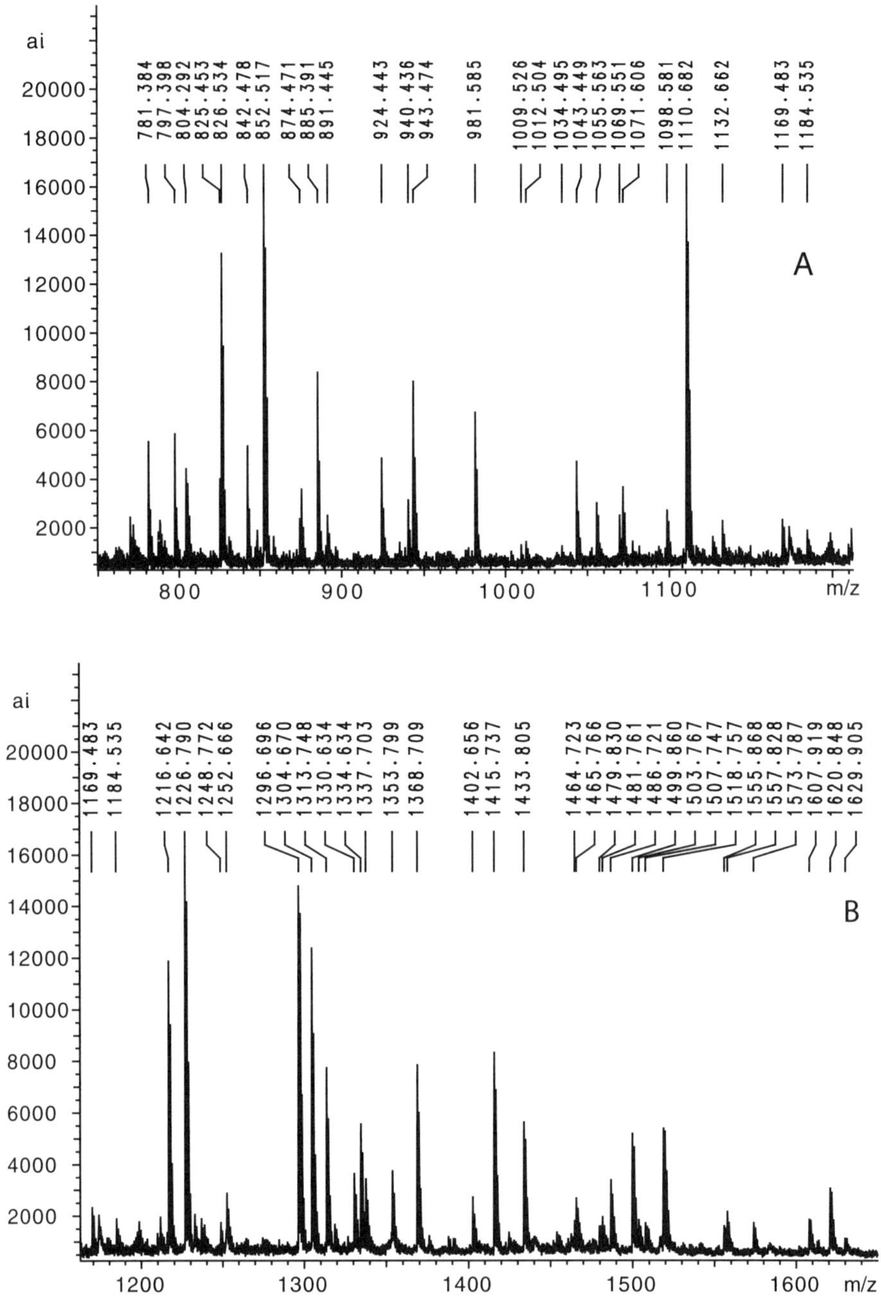

Figure 10.1. MALDI-TOF spectrum from the in-gel digest of a Coomassie-stained 190-kDa protein. Charged (MH^+) monoisotopic peaks are labeled. For clarity, the spectrum has been divided into four sections; *A–D*.

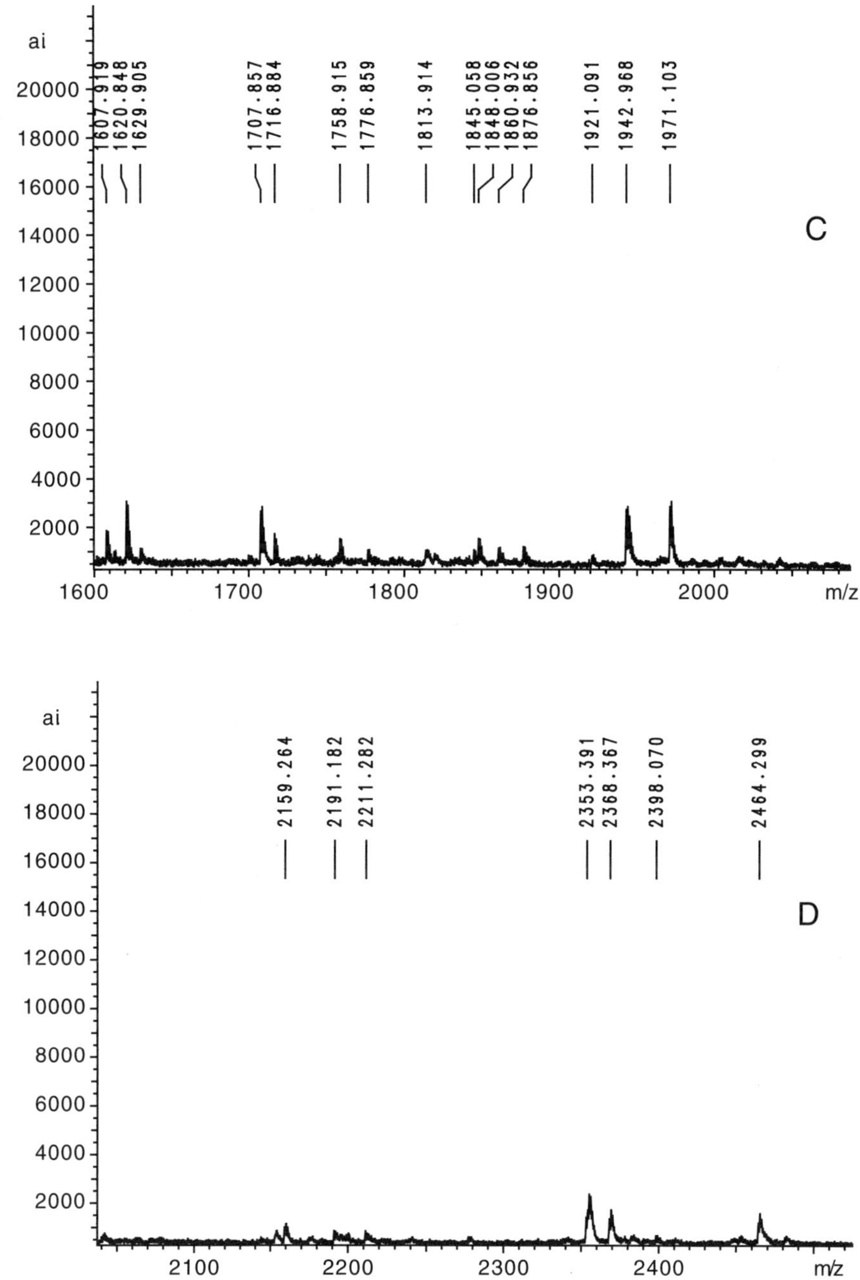

Figure 10.1. *Continued.*

TABLE 10.1. List of Peptides Matching Rat Clathrin Heavy Chain[a]

Monoisotopic Masses		Error (Da)	Residues		Missed Cut	Peptide Sequence	
Measured	Computed		Start	To			
780.377	780.395	–0.018	1156	1161	0	YLQMAR	
796.390	796.390	0.000	1156	1161	0	YLQMAR	OxMet
824.445	824.479	–0.034	1510	1516	0	IAAYLFK	
825.526	825.543	–0.018	831	837	0	NLILVVR	
884.383	884.402	–0.020	1095	1101	0	AYEFAER	
890.437	890.464	–0.027	669	676	0	AMLSANIR	OxMet
923.435	923.453	–0.018	1040	1046	0	VMEYINR	
939.428	939.452	–0.023	1454	1461	0	SVQNHNNK	
939.428	939.448	–0.020	1040	1046	0	VMEYINR	OxMet
942.466	942.492	–0.026	1502	1508	0	HELIEFR	
980.578	980.580	–0.003	1509	1516	1	RIAAYLFK	
1042.441	1042.435	0.006	896	903	0	ENPYYDSR	
1054.555	1054.569	–0.014	799	806	0	YIEIYVQK	
1054.555	1054.581	–0.026	1435	1443	1	AVNYFSKVK	
1068.543	1068.538	0.005	611	619	0	AHIAQLCEK	
1070.598	1070.612	–0.014	270	278	0	HDVVFLITK	
1097.573	1097.575	–0.002	1398	1406	0	VANVELYYK	
1109.675	1109.659	0.015	857	865	0	LLLPWLEAR	
1168.475	1168.445	0.031	750	758	0	ESNCYDPER	
1183.527	1183.500	0.028	1255	1264	0	EVCFACVDGK	
1215.635	1215.603	0.031	1334	1342	0	EHLELFWSR	
1225.782	1225.754	0.028	1444	1453	0	QLPLVKPYLR	
1251.658	1251.653	0.006	87	96	0	TLQIFNIEMK	OxMet

1295.688	1295.662	0.026	1216	1226	0	LLYNNVSNFGR	
1303.662	1303.651	0.010	355	366	0	NNLAGAEELFAR	
1312.740	1312.713	0.027	1499	1508	1	LEKHELIEFR	
1329.626	1329.613	0.013	913	923	0	DPHLACVAYER	
1333.626	1333.626	0.001	882	892	0	IYIDSNNNPER	
1336.695	1336.680	0.015	177	188	0	VVGAMQLYSVDR	
1352.791	1352.675	0.116	177	188	0	VVGAMQLYSVDR	OxMet
1352.791	1352.700	0.091	1119	1130	1	GMVKEAIDSYIK	
1352.791	1352.838	–0.047	1023	1034	0	NLQNLLILTAIK	
1367.701	1367.690	0.011	572	583	0	NNRPSEGPLQTR	
1401.648	1401.670	–0.023	1102	1113	0	CNEPAVWSQLAK	
1414.729	1414.720	0.009	1011	1022	0	IVLDNSVFSEHR	
1432.797	1432.792	0.005	469	481	0	SVDPTLALSVYLR	
1463.715	1463.729	–0.014	626	637	0	ALEHFTDLYDIK	
1464.758	1464.775	–0.017	177	189	1	VVGAMQLYSVDRK	
1478.822	1478.792	0.031	508	519	0	VGYTPDWIFLLR	
1480.753	1480.770	–0.017	177	189	1	VVGAMQLYSVDRK	OxMet
1485.713	1485.714	–0.001	912	923	1	RDPHLACVAYER	
1498.852	1498.825	0.027	164	176	0	WLLLTGISAQQNR	
1502.759	1502.745	0.015	1332	1342	1	MREHLELFWSR	
1506.739	1506.749	–0.010	488	500	0	VIQCFAETGQVQK	
1517.749	1517.740	0.009	145	157	0	HSSLAGCQIINYR	
1554.860	1554.843	0.017	768	780	0	LTDQLPLIIVCDR	
1556.821	1556.797	0.023	64	78	0	RPISADSAIMNPASK	
1572.779	1572.792	–0.013	64	78	0	RPISADSAIMNPASK	OxMet
1606.911	1606.886	0.024	507	519	1	KVGYTPDWIFLLR	
1619.840	1619.830	0.010	626	638	1	ALEHFTDLYDIKR	

(*Continued*)

TABLE 10.1. ***Continued.***

Monoisotopic Masses		Error (Da)	Residues		Missed Cut	Peptide Sequence	
Measured	Computed		Start	To			
1706.849	1706.924	–0.075	611	625	1	AHIAQLCEKAGLLQR	
1706.849	1706.938	–0.089	84	98	2	AGKTLQIFNIEMKSK	
1706.849	1706.847	0.002	838	852	1	GQFSTDELVAEVEKR	
1715.876	1715.862	0.014	190	205	0	VSQPIEGHAASFAQFK	
1757.907	1757.873	0.034	367	382	1	KFNALFAQGNYSEAAK	
1775.851	1775.825	0.026	866	881	0	IHEGCEEPATHNALAK	
1844.050	1843.957	0.093	189	205	1	KVSQPIEGHAASFAQFK	
1846.998	1846.935	0.063	430	444	0	YESLELCRPVLQQGR	
1859.924	1859.817	0.107	206	221	0	MEGNAEESTLFCFAVR	
1875.848	1875.812	0.036	206	221	0	MEGNAEESTLFCFAVR	OxMet
1941.960	1941.906	0.054	1482	1498	0	TSIDAYDNFDNISLAQR	
1970.095	1970.003	0.092	1102	1118	1	CNEPAVWSQLAKAQLQK	
1970.095	1970.021	0.074	1227	1245	0	LASTLVHLGEYQAAVDGAR	
2158.174	2158.116	0.058	950	967	2	RKDPELWGSVLLESNPYR	
2190.175	2190.086	0.088	451	468	2	WLKEDKLECSEELGDLVK	
2210.274	2210.168	0.105	1164	1182	2	ARESYVETELIFALAKTNR	
2352.383	2352.242	0.141	298	320	0	ISGETIFVTAPHEATAGIIGVNR	
2367.359	2367.254	0.105	1074	1094	1	KFDVNTSAVQVLIEHIGNLDR	
2397.063	2397.076	–0.013	279	297	0	YGYIHLYDLETGTCIYMNR	OxMet
2463.199	2463.213	–0.014	1183	1204	0	LAELEEFINGPNNAHIQQVGDR	

[a] Monoisotopic mass numbers shown are unharged; peptides with an oxidized methionine residue are labeled "OxMet." Unmatched monoisotopic masses (M): 803.316 841.502 851.540 873.494 1008.544 1011.522 1033.512 1131.675 1247.780 1628.895 1812.894 1920.065 2879.968.

to be an indication of a noncomplete cleavage; however, almost all of the miscuts are located *N*- or *C*-terminal in the respective peptide. When a miscut is found in the middle of a peptide, that particular cleavage site is often near an acidic residue, which makes tryptic digestion slower. Although as many as 63 peptides matched into the protein, only 35% of its sequence was covered. This may seem a low figure, but one seldom reaches higher coverage of such a large protein. Several of the matching peptides do overlap, because of miscuts and because the same Met residue may or may not be oxidized. The 13 unmatched peptide masses may represent unknown modifications, miscuts other than those discussed, or, maybe most probably, contaminating proteins. The other four proteins were positively identified with similar coverage.

10.5.2. Analysis of Acetylated Lysine Residues

Now that the genomes from many species are fully characterized, much effort will be directed toward protein function, protein–protein interactions, regulation, secondary and tertiary structures, and not least, posttranslational modifications (PTMs). The use of MALDI-TOFMS to discover PTMs has its limitation; the main one is that all analytes in a complex sample are not always detected; another one is the risk that a modification makes the altered peptide less detectable. This is, for example, the case with phosphorylations on Ser, Thr, or Tyr. For such analyses specialized techniques have been developed, for example, dephosphorylation on target by alkaline phosphatase [12] or enrichment of phosphopeptides on IMAC supports [13].

The following example shows a case where at least one PTM was positively detected, and thus demonstrates that MALDI-TOFMS analysis can be useful for acetylated residues. The model molecule used was mouse P53, which is known to become aceylated on one or two Lys residues near its *C* terminus [14]. P53 was expressed in fusion with GST and after treatment of the P53-GST complex with the acetylating enzyme P300 and after SDS-PAGE and Coomassie staining, the target molecule was excised and in-gel-digested with trypsin. After MALDI-TOFMS analysis, using untreated P53-GST as a negative control, the peptide mass lists were loaded into ProFound. A potential addition to Lys of C_2H_2O (i.e., 42 Da for acetylation) was allowed for both sets. A modified lysine sidechain makes that Lys resistant to trypsin clevage, so one must remember to allow one or more miscuts during database searching. The unfiltered search revealed both GST and P53, with a coverage of about 42–50%. In the P300-treated sample, a unique mass of 1033.625 (M) was found, which most likely corresponds to the peptide TKKGQSTSR with one acetyl group on the first or the second Lys (theoretical mass 1033.551). The nontreated sample did not show this mass, and although it can be argued that the 1033.625 mass is a pure coincidence, it is highly likely that MALDI-TOFMS detected an acetylated lysine residue. Unfortunately one cannot at this stage identify which of the two Lys was modified; an MS/MS analysis would probably reveal the exact position. We have investigated several other samples for acetylation, and it seems that this modification is relatively easy to detect by MALDI-TOFMS.

10.5.3. Analysis of a Protein from a Silver-Stained Two-Dimensional Gel

Two-dimensional gel electrophoresis has become the sample preparation technique of choice in proteomics projects. Although it has clear limitations in terms of resolution and dynamic range, for example, it is very useful to compare expression patterns between for example two cell lines, where one has been exposed to a stimulant of some kind. Unfortunately, one will detect mainly those proteins that are abundant, unless an efficient specific enrichment has been performed. Still, analysis of differences in expression of those proteins as well will be useful for diagnosis and other purposes.

The following example is from an experiment where we wanted to identify proteins isolated by 2D electrophoresis for later use as markers. The cells were from a human breast cancer line (MCF-7), and about 50 μg of total cell lysate was separated per gel, using equipment and standard protocol from Amersham-Pharmacia Biotech. After silver-staining using the modified protocol [7], prominent spots were excised and identical spots from three gels were combined (gel not shown). First the silver was destained [9], gel pieces dried, and a trypsin solution was added and absorbed into the gel. After incubation and extraction, the organic phase was removed by rotary evaporation, and the sample was desalted and concentrated on a POROS R2 20 column packed in a GelLoader tip. Peptides were eluted from the tip using half-saturated ACCA matrix in about 65% acetonitrile directly onto the MALDI target in drops of about 100 nL. MALDI spectra were taken (65 shots collected) and searches performed. From 20 2D gel spots, 13 were positively identified, 6 were probably not homogeneous and could not be unambiguously identified, and 1 spot was negative. The spectrum from spot 19, which migrated on the gel at pI ~ 5, and a molecular mass of about 75 kDa is shown in Fig. 10.2. Although relatively few peptide masses were recorded, with the precise internal calibration obtained using two autolytic fragments from trypsin and one common keratin fragment, an unambiguous match was found in human BiP (immunoglobulin heavy-chain binding protein) also named Grp 78 (78 kD glucose regulated protein precursor). Of 34 peptide masses used for the search, 13 match BiP with a coverage of 22% with a mean error of about 0.06 Da. Of the remaining 21 peptides, 2 belong to trypsin autodigestion fragments, and at least 5 to human keratin 1. As seen in the Table 10.2, the top ranked protein is given a Z value of 2.23, which indicates that this match is very far from being coincidental. Although an unfiltered search was done among 643,065 sequences in the NCBInr database, the best matching protein fits with species, pI, and molecular weight. To our experience, we would not have been able to get this convincing result without the destaining of silver and concentration and desalting on the nanocolumn with elution directly onto target.

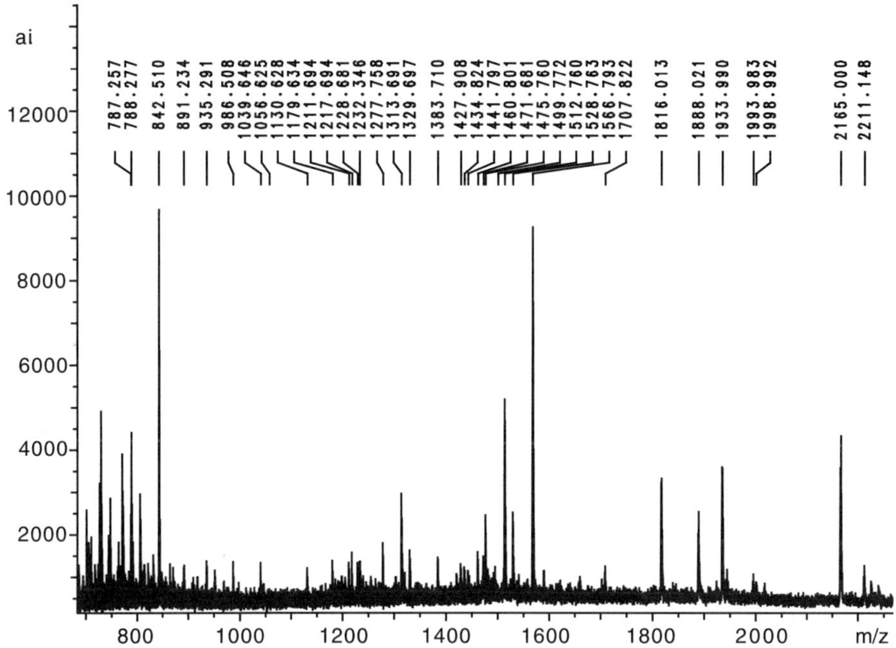

Figure 10.2. MALDI-TOF spectrum from in-gel digest of a 71-kDa protein from a silver-stained 2D gel obtained after silver destaining and concentration/desalting on a nanocolumn. Charged (MH^+) monoisotopic peaks are labeled.

TABLE 10.2. Search Result Summary from MALDI-TOF Analysis of the 71-kDa Protein from a Silver-Stained 2D Gel[a]

Rank	Probability	*Z* Value	Matching Protein	% Coverage	pI	kDa
1	1.0 e^{00}	2.23	BiP *Homo sapiens*	22	5.2	70.91
2	5.5 e^{-10}	—	L1905.9 *Leishmania major*	3	6.0	732.70
3	1.9 e^{-11}	—	ORF N0646 *S. cerevisiae*	8	6.5	174.48
4	2.7 e^{-12}	—	Hypothetical protein o761 *E. coli*	13	5.0	82.31
5	2.2 e^{-12}	—	Signal transducer 5B *Gallus gallus*	10	5.9	90.06

[a] The first five matching proteins from a ProFound search is shown.

10.6. CONCLUDING REMARKS

The development of PMF using MALDI-TOFMS for protein identification has appeared very well timed with the completion of the characterization of many species' genomes. The ease and precision by which proteins are identified by PMF is making direct sequencing of proteins for identification obsolete. With this comes higher sensitivity, higher speed of analysis, and extremely low running cost. Furthermore, for laboratories that require high-throughput analyses, the technique is possible to automize throughout the whole line using robots, starting from the stained 2D gel with an image analysis, cutting spots and transfering those to microtiter plates where proteolytic digestion and peptide extraction is done. Extracts are then mixed with matrix and samples applied onto target. After MALDI measurements, spectra are labeled and searches performed either over the Internet or via local sequence databases. Robots for high throughput in proteomic laboratories are being designed to carry out over 10,000 samples unattended, and yet the postgenome era has just begun.

ACKNOWLEDGMENTS

I thank Rainer Heuchel, Johan Ericsson, and Serhiy Souchelnytskyi from our institute for their kindness in allowing me to use unpublished material that has been generated in collaboration.

REFERENCES

1. J.-H. Ha, U. Hellman, E. Johnson, L. Li, D. McKay, M. Sousa, S. Takeda, C. Wernstedt, and S. Wilbanks, Destabilization of peptide binding and inter-domain communication by an E543K mutation in the seventy kilodalton bovine heat shock cognate protein, a molecular chaperon, *J. Biol. Chem.* **272**, 27796–27803 (1997).
2. J. Rosenfeld, J. Capdeville, J. C. Guillemot, and P. Ferrara, In gel digestion of proteins for internal sequence analysis after one- or two-dimensional gel electrophoresis, *Anal. Biochem.* **203**, 173–179 (1992).
3. U. Hellman, C. Wernstedt, J. Goñez, and C. H. Heldin, Improvement of an "in-gel" digestion procedure for the micropurification of internal protein fragments for amino acid sequencing, *Anal. Biochem.* **224**, 451–455 (1995).
4. D. J. C. Pappin, P. Hojrup, and A. J. Bleasby, Protein identification by peptide mass fingerprinting, *Curr. Biol.* **3**, 327–332 (1993).
5. L. Rönnstrand, A. K. Arvidsson, A. Kallin, C. Rorsman, U. Hellman, U. Engström, C. Wernstedt, and C. H. Heldin, SHP-2 binds to Tyr763 and Tyr1009 in the PDGF β-receptor and mediates PDGF-induced activation of the Ras/MAP kinase pathway and chemotaxis, *Oncogene* **18**, 3696–3702 (1999).
6. D. Wessel and U. I. Flügge, A method for the quantitative recovery of proteins in dilute solution in the presence of detergents and lipids, *Anal. Biochem.* **138**, 141–143 (1984).
7. A. Shevchenko, M. Wilm, O. Vorm, and M. Mann, Mass spectrometric sequencing of proteins silver-stained polyacrylamide gels, *Anal. Chem.* **68**, 850–858 (1996).
8. M. F. Lopez, K. Berggren, E. Chernokalskaya, A. Lazarev, M. Robinson, and W. F. Patton, A comparison of silver stain and SYPRO Ruby protein gel stain with respect to protein detection in two-dimensional gels and identification by peptide mass profiling, *Electrophoresis* **21**, 673–3683 (2000).

9. F. Gharahdaghi, C. R. Weinberg, D. A. Meagher, B. S. Imai, and S. M. Mische, Mass spectrometric identification of proteins from silver-stained polyacrylamide gel: A method for the removal of silver ions to enhance sensitivity, *Electrophoresis* **20**, 601–605 (1999).
10. J. Gobom, E. Nordhoff, E. Mirgorodskaya, R. Ekman, and P. Roepstorff, Sample purification and preparation technique based on nano-scale reversed-phase columns for the sensitive analysis of complex peptide mixtures by matrix-assisted laser desorption/ionization mass spectrometry, *J. Mass Spectrom.* **34**, 105–116 (1999).
11. M. Kussmann, E. Nordhoff, H. Rahbek-Nielsen, S. Haebel, M. Rossel-Larsen, L. Jacobsem, J. Gobom, E. Mirgorodskaya, A. Kroll-Kristensen, L. Palm, and P. Roepstorff, Matrix-assisted laser desorption/ionization mass spectrometry sample preparation techniques designed for various peptide and protein analytes, *J. Mass Spectrom.* **32**, 593–601 (1997).
12. O. N. Jensen, M. R. Larsen, and P. Roepstorff, Mass spectrometric identification and microcharacterization of proteins from electrophoretic gels: strategies and applications, *Proteins* (Suppl. 2), 74–89 (1998).
13. D. C. Neville, C. R. Rozanas, E. M. Price, D. B. Gruis, A. S. Verkman, and R. R. Townsend, Evidence for phosphorylation of serine 753 in CFTR using a novel metal-ion affinity resin and matrix-assisted laser desorption mass spectrometry, *Protein Sci.* **6**, 2436–2445 (1997).
14. W. Gu and R. G. Roeder, Activation of p53 sequence-specific DNA binding by acetylation of the p53 *C*-terminal domain, *Cell* **90**, 595–606 (1997).

CHAPTER 11

Electron-Capture Dissociation of Peptides

ROMAN A. ZUBAREV

Department of Chemistry, University of Southern Denmark, Odense, Denmark

CONTENTS

11.1. INTRODUCTION

Electron-capture dissociation (ECD) stands alone among different MS/MS approaches. It is the first truly "mild" fragmentation technique, a perfect match to the mild ionization technique electrospray ionization (ESI) that produces multiply protonated polypeptide molecules. Conventional approaches, such as

Mass Spectrometry and Hyphenated Techniques in Neuropeptide Research, Edited by Jerzy Silberring and Rolf Ekman.
ISBN 0-471-35493-7

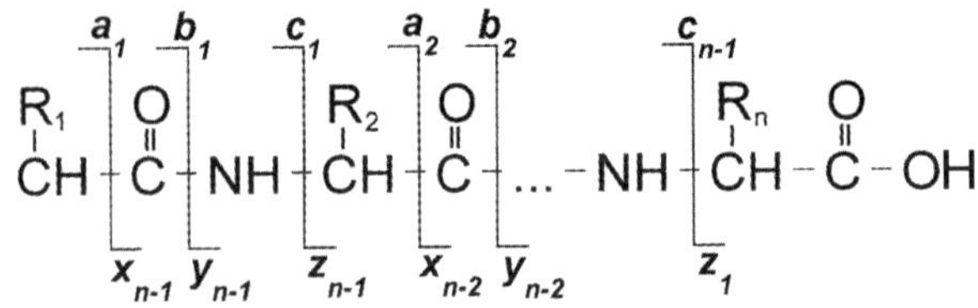

Figure 11.1. Nomenclature of the polypeptide chain fragmentation proposed by Roepstorff and Fohlman.

low-energy collisional activation dissociation (CAD), infrared multiphoton dissociation, and surface-induced dissociation, are based on excitation of vibrational modes. In these vibrational excitation (VE) techniques, where intramolecular energy redistribution (IVR) takes place prior to bond dissociation, the resulting fragmentation pattern depends on the total amount E of the deposited energy but not on the way this energy was imparted:

$$[M+n\text{H}]^{n+} + \text{E} \rightarrow \left([M+n\text{H}]^{n+}\right)_{\text{hot}} \rightarrow \text{fragments} \tag{11.1a}$$

Consequently, MS/MS spectra obtained with different VE techniques look rather similar. For example, the major fragmentation channel of the polypeptide molecular ions is the heterolytic cleavage of the peptide bond, yielding *b*,*y* fragments (Fig. 11.1) in the commonly used notation [1]. Other low-energy channels produce losses of water, ammonia, carbon dioxide, and labile groups. With some degree of simplification, one can say that weaker bonds fragment in VE techniques first.

In ECD, the bond strength is less most important parameter; often, strong bonds are cleaved while the weak bonds remain intact. A different type of cleavage is achieved—mainly $c,z\cdot$ backbone fragmentation. The fragment abundances are more homogeneous and generally bear little correlation with those in VE techniques. The sites that undergo preferential cleavage in the latter can be resistant to ECD and vice versa.

These differences are determined by the very different mechanisms of excitation and fragmentation. In VE techniques, the fragmenting ion is the same species (although with a different energy content) as the parent ion. In ECD, ion–electron recombination changes converts even-electron n-cations into $(n - 1)+$ radical ions, with different bond strengths and different fragmentation pathways:

$$[M+n\text{H}]^{n+} + \text{e}^- \rightarrow \left([M+n\text{H}]^{(n-1)+.*}\right)_{\text{transient}} \rightarrow \text{fragments} \tag{11.1b}$$

ECD bond cleavage is also more abundant and less specific than in VE techniques. This makes ECD suitable for de novo sequencing of larger polypeptides and whole proteins, as has been nicely demonstrated by McLafferty et al. [2].

Substantial accumulated evidence suggests that the ECD cleavage is faster than IVR, that is, *nonergodic*. Because of this nonergodicity, strong bonds are cleaved in the presence of weak bonds. This observation is now being extensively applied for determination of the positions of labile modifications in polypeptides. Yet another peculiarity is that disulfide bonds that do not readily fragment in CAD are preferentially cleaved in ECD. This feature provides a means to avoid wet chemistry and do direct MS/MS analysis of disulfide-containing peptides.

The interest toward ECD in the mass spectrometric community is rapidly growing, not the least due to the wider spread of Fourier transform mass spectrometry (FTMS).

11.2. HISTORY

In mass spectrometry, ion–electron recombination processes were extensively used in neutralization–reionization (NR) experiments performed in the 1970s–1980s [3]. In NR MS, cations produced by electron ionization (EI) or chemical ionization (CI) are first *m/z*-selected in a MS analyzer, then neutralized by collisions with gas-phase species exhibiting low ionization energies (usually alkali-metal atoms or mercury), and subsequently reionized by EI for the second-stage MS analysis. This technique allowed for studying short-lived unusual species, such as hypervalent radicals. Porter and co-workers were the first to demonstrate that protonated ammonium NH_4^+ fragments on a very short ($\ll 10^{-7}$ s) timescale on neutralization [4]. This fragmentation of the $\cdot NH_4$ hypervalent radical was found to be nonergodic, which inspired McLafferty in 1986 to make a daring prediction that one-electron neutralization of multiply protonated polypeptides should produce nonergodic backbone cleavage [5]. This prediction turned out to be amazingly accurate, taking into account how poorly gas-phase protonation of polypeptide chains was understood at that time. It took more than a decade to realize this reaction in practice. Neutralization of protein polycations by a metal surface gave no nonergodic fragmentation [6]; neither did gas-phase polycation–anion reactions [7,8].

The first truly ECD-type mass spectra were observed in UV photodissociation experiments, where melittin (2.8 kDa) 4+ and bovine ubiquitin (8.6 kDa) 10+ ions trapped in a FTMS cell were irradiated by 193-nm laser pulses [9]. Surprisingly, charge reduction of the parent ions was the main process, along with the unusual *c*,*z* fragmentation. When the FT cell was later modified to trap not only cations but also electrons, the effect increased greatly. The conclusion was made that the responsible particles were not the UV photons but the secondary electrons photoemitted from the copper cell walls. The UV laser was then replaced by a standard EI source (filament-base electron gun), and the ECD technique was finally born [10]. The cell modification for simultaneous trapping of both polarities has proved helpful but not essential, and commercial FTMS instruments produced by the major manufacturers Bruker [11] and IonSpec [12] were possible to adopt for ECD experiments with only software modifications.

11.3. INSTRUMENTATION

11.3.1. MS Analyzer

So far, controlled ECD has been implemented only in FTMS, mainly due to the lack of truly effective sources of slow electrons. In an FTMS cell, the typical reaction time between polycations and low-energy electrons is seconds, which is too long for ECD to be implemented in time-of-flight and quadrupole instrumentation. Dynamic (RF) ion traps allow for storing of ions even of opposite polarities for such a period of time, but not electrons and cations simultaneously because of the vastly different m/z values.

The basics of the Fourier transform mass spectrometry can be found in this volume or elsewhere [13]. A diagram of the ECD FTMS experiment is shown in Fig. 11.2 [14]. The FT cell can be either of the conventional type, or nested, allowing for trapping of species of both polarities simultaneously. The cations of interest reside in the center of the cell, where the potential is slightly positive (the magnitude depends on the cell construction and the trapping potentials). The magnitude of the trapping potentials (V_2 and V_3 in Fig. 11.2) for cations is not very important; good ECD spectra have been observed at 0.5–5 V. One should, however, avoid too high trapping potentials in order to prevent uncontrolled expansion of the magnetron radius.

11.3.2. Electron Gun

The low-energy electrons are produced by an electron gun, which in all reported cases was just a conventional heated filament. Since the magnetic field lines converge in the magnet center, which is also the center of the cell, the gun may be positioned off-axis at a certain distance from the cell. The electrons emitted by the gun will follow the magnetic lines to the center of the cell. This arrangement allows the infrared laser beam to pass along the axis, thus permitting both ECD and infrared activation on the same setup [14]. The ability to excite ions with IR radiation is valuable for breaking the multiple intramolecular bonds in large biomolecules, which may otherwise prevent separation of ECD fragments.

The filament is usually made of tungsten or rhenium. The current through the filament and thus the temperature are chosen empirically; hotter filaments produce more electrons but also give larger energy spread, so the best results are often obtained with a red-glowing filament.

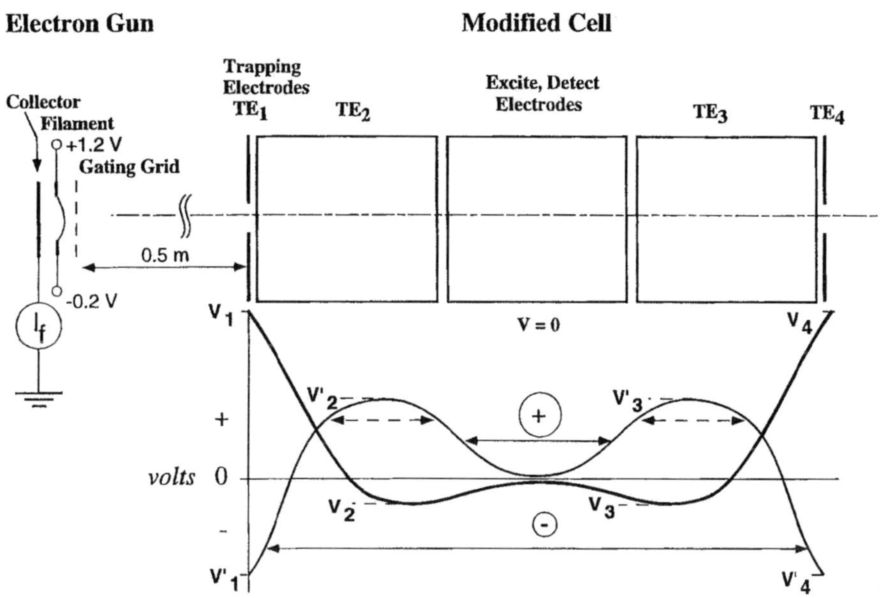

Figure 11.2. The diagram of the ECD FTMS experiment. Schematic of the nested ion cell geometry used for ECD. The flat-plate trapping electrodes TE_1 and TE_4 have been added outside the conventional cylindrical electrodes. The solid double-headed arrows correspond to the axial oscillation amplitudes of positive ions and freshly captured electrons; after cooling, the latter occupy (dashed arrows) potential wells on either side of the positive ion well. (Reproduced from Ref. 14 with permission.)

11.3.3. Optimal Regime

The optimal regime of ECD depends on the particular instrumentation. When the nested cell is used, the outer potentials are kept slightly negative (–1.0, . . . ,–0.5 V) to trap the electrons [14]. Besides these potentials, there are rather few adjusting parameters in ECD, namely, the filament potential, the current through the filament, and the duration of the irradiation event. Optimization of ECD begins with maximizing the charge reduction, since the reduced peaks are usually the most abundant products. Before turning the filament on, it is advisable to make sure that a single charge state $n+$ is isolated at a sufficient abundance (signal/noise > 20). The current through the filament can be set to a value corresponding to an orange-red glow. The time of irradiation is selected between 3 and 10 s, depending on the charge state (longer time for lower charge). If no previous setting of the filament potential is available, +0.5 V is a good starting point. By scanning the filament potential, the abundance of the reduced $(n - 1)+\cdot$ peak is maximized. If little or no charge reduction is observed, increasing the pressure in the cell during irradiation by letting in small portions of neutral gas (nitrogen or argon) can be performed. If no charge reduction is still observed, one should check the alignment of the fila-

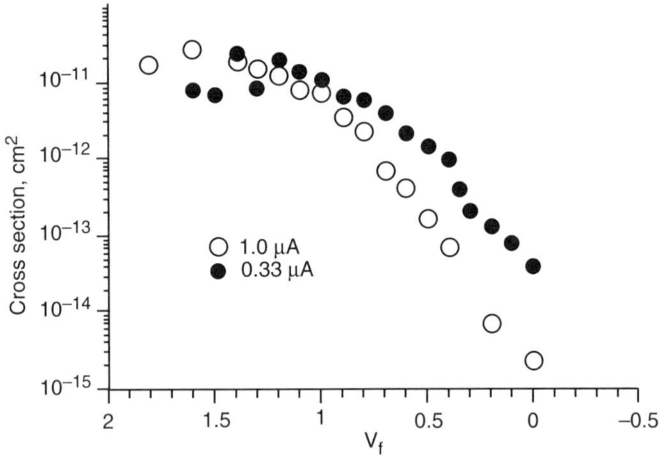

Figure 11.3. Dependence of the capture cross section of cytochrome C 15+ ions on the potential on the filament for two electron-beam currents. Electron energy of <0.1 eV corresponds to $V_f \approx 1.2$ V. (Reproduced from Ref. 14 with permission.)

ment. After the charge reduction, the signal intensity of the ECD fragments is maximized by varying the chirp excitation parameters.

11.4. MECHANISMS OF ELECTRON-CAPTURE DETECTION (ECD)

11.4.1. Electron Capture

The cross sections for electron capture were studied in Ref. 14. Figure 11.3, where the capture cross section of cytochrome C 15+ ions is shown, demonstrates a strong dependence on the electron energy. The measured capture rate increases 10^3 times when the electron energy is reduced from 1 to 0.1 eV; this result emphasizes the importance of "cold" electrons.

The electron-capture cross section of cytochrome C 15+ ions measured at typical ECD conditions exceeds the ion-neutral collisional cross section by two orders of magnitude [14]. That means that electrons are captured at distances significantly exceeding the physical dimension of the molecular ion and that the capture is primarily Coulombic. At large capture distances, the point-charge model is valid, in which the capture distance (Thomson radius) is proportional to the ionic charge, and thus the cross section is proportional to the charge squared. The experimental data are in a rather good agreement with this prediction [14]. In the point-charge model, the shape of the molecule, its mass, the strength of intramolecular bonds, and other factors are unimportant. From the experimental experience, it is, however, known that the efficiency of electron capture is highest for linear, fully stretched ions.

11.4.2. Recombination Energy

The recombination RE energy can be estimated as

$$RE = 13.6\,\text{eV} - PA[(n-1)+] + HA[(n-1)+] \tag{11.2}$$

where PA and HA are the proton and hydrogen atom affinities, respectively, and $(n-1)+$ denotes a molecular with $(n-1)$ sites protonated [15]. The gas-phase proton affinity of peptides can be as high as 10.6 eV for $n = 1$ but it decreases with n [16]. For electrospray-produced ions, the proton affinity cannot be much smaller than that of the solvent, ~8 eV. The hydrogen atom affinity is typically rather small, <1 eV. Therefore, the recombination energy is somewhere between 4 and 7 eV depending on the ionic charge state. Although this figure is larger than the typical bond strength (3–4 eV), it is much smaller than the recombination energy of a free proton, 13.6 eV. Being distributed over a large number of degrees of freedom, recombination energy is causing just a moderate temperature increase. For example, in a 3-kDa molecule, this increase is estimated to be between 50 and 100 K, which is far less than the typical value for VE techniques 300 to 600 K. This explains why, in ECD, labile groups are rarely lost.

11.4.3. Hydrogen Atom Capture Model

Although the exact mechanism of $c,z\cdot$ cleavage is yet to be fully understood, it appears to be a fast, radical-site-initiated process that can be qualitatively described by the hot hydrogen atom model [17]. The original capture of the electron to a high Rydberg state is followed by either autoemission of the captured electron back to the continuum, or its "landing," resulting in charge neutralization. Since the charge is carried by a proton, neutralization creates an excessive hydrogen atom. This atom carries away some recombination energy in the form of kinetic energy.

For example, for a protonated lysine

$$R{-}N^{+}H_3 + e^{-} \rightarrow R{-}NH_2 + (H\cdot)_{hot} \tag{11.3}$$

In gas-phase peptides, all protons are primarily located at the sidechains of the basic residues Arg, Lys, and His. If these sidechains were isolated, ECD would give losses only of small groups (e.g., NH_3). Since, in reality, the charges are shared (solvated) with the backbone functionalities, the hot hydrogen atom becomes mobile along the backbone. Exploring the vicinity of the solvation site, it can be captured by a group with a sufficient affinity, such as carbonyl

$$\begin{aligned} &\ldots{-}CHR{-}CO{-}NH{-}CHR'{-}\ldots + H\cdot \rightarrow \\ &\ldots{-}CHR{-}C(OH){-}NH{-}CHR'{-}\ldots \end{aligned} \tag{11.4}$$

or disulfide

$$\mathrm{R{-}S{-}S{-}R' + H\cdot \rightarrow R{-}S{-}\overset{H\cdot}{S}{-}R'} \tag{11.5}$$

The exothermicity of the charge neutralization and the H· capture is used for bond cleavage, initiated by the presence of a radical site:

$$\mathrm{\ldots{-}CHR{-}C(OH){-}NH{-}CHR'{-}\ldots \rightarrow \ldots{-}CHR{-}C(OH) = NH + \cdot CHR'{-}\ldots} \tag{11.6}$$

or

$$\mathrm{R{-}\overset{H\cdot}{S}{-}S{-}R' + H\cdot \rightarrow R{-}S\cdot + HS{-}R'} \tag{11.7}$$

The even-electron product in reaction (11.6) corresponds to a *c* fragment in peptide nomenclature [1], while the radical species is the *z*· fragment. This homolytic *amine* bond cleavage is dissimilar to the heterolytic *amide* bond (peptide bond) *b,y* fragmentation in collisional excitation.

The hot hydrogen atom model predicts that other groups with high affinity toward H· (e.g., the indole group in the tryptophan residue) must also be reactive in electron-capture dissociation [17]. Indeed, the ion peaks corresponding to backbone cleavage around tryptophan residues are often prominent in ECD spectra [18]. The only interresidue bond that is 100% immune to ECD is that on the *N* side of the proline residue, due to the presence of the tertiary nitrogen there.

11.4.3.1. Nonergodic Cleavage. Nonergodic fragmentation is the most debatable feature of ECD, which can be explained by the success of the RRKM theory in accounting for the observed cleavage rates in VE techniques. Since RRKM explicitly assumes IVR prior to bond dissociation (ergodic behavior), the nonergodic ECD mechanism means that RRKM is no longer applicable. Indeed, RRKM modeling could account for the observed ECD cleavage only at the condition of releasing the whole recombination energy in a small molecular region containing just a few atoms, that is, without energy redistribution over the whole molecule [17]. Moreover, the cleavage time was $<10^{-12}$ s, that is, faster than IVR that takes at least 10^{-11} s. In other words, the RRKM modeling yielded essentially non-RRKM results, consistent with a non-ergodic ECD mechanism. More insight into the mechanism could be obtained from high-level ab initio modeling of relevant polypeptide chains, which has so far been difficult to perform because of the time and memory limitations.

The accumulated experimental evidence is clearly in favor of the nonergodic mechanism. Perhaps the best example is ECD of supramolecular aggregates, the most labile species observed in biological mass spectrometry. VE techniques produce separation of the cluster constituents, and only occasion-

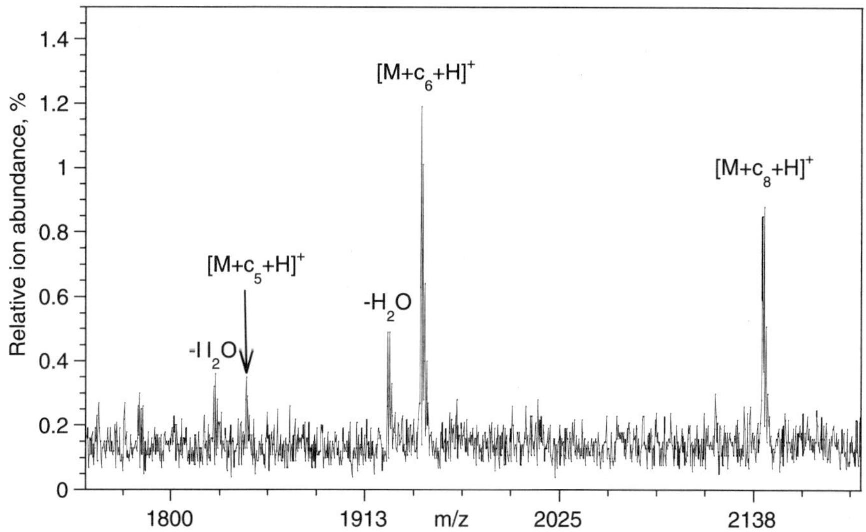

Figure 11.4. Part of the ECD mass spectrum of the trication $[2M + 3H]^{3+}$ of a non-specific peptide dimmer (M = TTTDSTTPAPTTK). CAD of the same ion lead exclusively to monomer separation. (Reproduced from Ref. 19 with permission.)

ally covalent bond fragmentation, mostly trivial. With the non-ergodic ECD cleavage, one expected to achieve fragmentation of strong covalent bond without dissociation of the weakly bound constituents. The successful result of such an experiment is shown in Fig. 11.4. ESI-generated trications of a non-specific head-to-toe dimer of a 13-residue peptide have produced, on electron capture, three weakly bound fragments of the composition $[M + c_x + H]^+$, $x = 5,6,8$ [19]. The c_7 fragment was absent because of the proline residue at the eighth position. CAD of the $[M + c_6 + H]^+$ ion produced $[M + H]^+$ ions, as was expected for noncovalent bonding.

The importance of the nonergodic ECD cleavage for structure characterization of supramolecular complexes by MS/MS will soon be clear, as more studies are currently under way.

11.4.3.2. Minor Fragmentation Channels. More than 80% of all cleavages observed in ECD are of the $c,z\cdot$ type. Other channels of fragmentation include losses of small molecules and radicals from the reduced species, such as −17, −44, and −59 Da, and constitute ~6% of all product ions [10]. The analytical importance of these losses is yet to be determined. Another contribution to the total fragmentation is the rather unusual $a\cdot,y$ backbone cleavage that becomes especially prominent if the parent ions are vibrationally excited. It has been proposed that this cleavage is a result of neutralization of the protonated backbone nitrogen [17]:

$$\begin{gathered}\ldots\text{—CHR—CO—N}^{+}\text{H}_2\text{—CHR}'\text{—}\ldots+e^- \rightarrow \\ \underset{a\cdot}{\ldots\text{—CHR}\cdot}+\text{C}\equiv\text{O}+\underset{y}{\text{NH}_2\text{—CHR}'\text{—}\ldots}\end{gathered} \tag{11.8}$$

The $a\cdot,y$ fragmentation, unlike other ECD cleavages, is a breakage of two bonds, and therefore the masses of the complementary fragments do not add up to the molecular mass. Yet another unusual cleavage is $c\cdot,z$ fragmentation [14]. This type of cleavage is characteristic mostly for low-charge-state ions in which the charge is located at the *C* terminus; it appears that H· is attracted, due to its high polarizability, by the charge gradient. The same cleavages at higher charge states where this gradient is smaller are likely to be the $c,z\cdot$ type or a mixture of the two types.

11.4.3.3. Efficiency. Efficiency is determined as a sum of the total signal from the product ions divided by the intensity of the parent ion peak. While low-energy CAD can be almost 100% efficient, the figure drops to <80% for infrared multiphoton dissociation and to <50% for ECD, although in some experiments up to 80% efficiency has been registered [14]. Since the cross section for electron capture increases as z^2 as the ionic charge z increases, fragments capture far less electrons than the parent ions under extended irradiation. Excitation is thus rather selective towards parent ions, and one would expect to obtain higher ECD efficiency. However, one of the biggest advantages of the technique—extensive backbone cleavage—leads to a large amount of low-abundance fragment ion signals, which could be difficult to detect. Integration of multiple scans brings up the low-abundance fragments as the signal/noise improving, at the expense of the longer times of data accumulation [19].

11.5. APPLICATIONS

While the issue of non-ergodicity of ECD is a matter of great theoretical interest, but for practical purposes it is important that the process produces abundant cleavage of just a few types of bonds and that the bonds of other types, however labile, remain largely intact. This feature is used for deriving primary structure information on polypeptides, including those with posttranslational modifications.

11.5.1. De Novo Sequencing

Conditions for de novo sequencing of long polypeptide chains are discussed Ref. 2. One of the basic requirements is the backbone cleavage between each pair of the adjacent amino acids. This necessary condition is, however, not sufficient for unequivocal sequence determination. Since the identity of the fragment ions (i.e., their *C*- or *N*-terminal origin) is seldom known a priori, confusion can arise when the mass difference between two fragments, one of which is *N* and another *C* terminal, is accidentally equal to zero or to the mass of an amino acid residue. Simulations show that this situation is rather typical for peptides >15 residues long [20]. Application of both vibrational excitation and ECD techniques resolves the confusion, since the mass difference between *c* and *b* fragments is 17.03 Da (mass of NH_3), while that between *y* and $z\cdot$ fragments is 16.02 Da (mass of NH_2) [2]. For the bond cleavage after the *x*th residue in an *n* peptide with the mass *M*, the following relationships are valid:

$$
\begin{aligned}
m(b_x) + m(y_{n-x}) &= M \\
m(c_x) + m(z_{n-x}\cdot) &= M + 1.008 \\
m(b_x) + m(z_{n-x}\cdot) &= M - 16.02 \\
m(c_x) + m(y_{n-x}) &= M + 17.03
\end{aligned}
\tag{11.9}
$$

where *m* is the neutral mass of the corresponding fragment, determined through its mass-to-charge ratio *m/z* as

$$
m = z[m/z - 1.007277] \tag{11.10}
$$

If at least one fragment from each dissociation technique is present in the spectrum, the fragment identity can be readily determined. Such a set of fragments is known as a *golden set* [2]. Golden sets greatly improve the sequencing performance, which urges the complementary use of CAD and ECD techniques.

11.5.2. MS^n

The presence of golden sets for all pairs of adjacent residues is the ideal but rather unrealistic situation, since VE rarely produces full cleavage for chains >15 residues long. When complementary use of the two fragmentation techniques proves insufficient, multiple-stage tandem mass spectrometry is the solution. Although ECD cleaves many more interresidue bonds than does CAD [10,18,21], the intensities of ECD fragments are generally too low to be suitable for further fragmentation. VE, on the contrary, is capable of producing a few intense fragments composing most of the fragment ion current, which is a perfect situation for MS^3. ECD of *y* ions is easy [14], as they are just shorter peptides. Accumulation at sufficient abundances of the less stable [22] *b* ions has only relatively recently become routinely available as the external accumulation technique progressed [23]. Figure 11.5 shows an example where bombesin molecular ions were first collisionally dissociated in the storage hexapole of the ESI interface, and then the intense b_{13}^{2+} fragments were subjected to ECD [19]. The CO loss from the reduced *b* ion confirmed their identity, while the observed *c* fragments revealed the ion identities in the ECD spectrum of $[M + 2H]^{2+}$. The full *c*-ion series from c_3 to c_{12} were detected, with

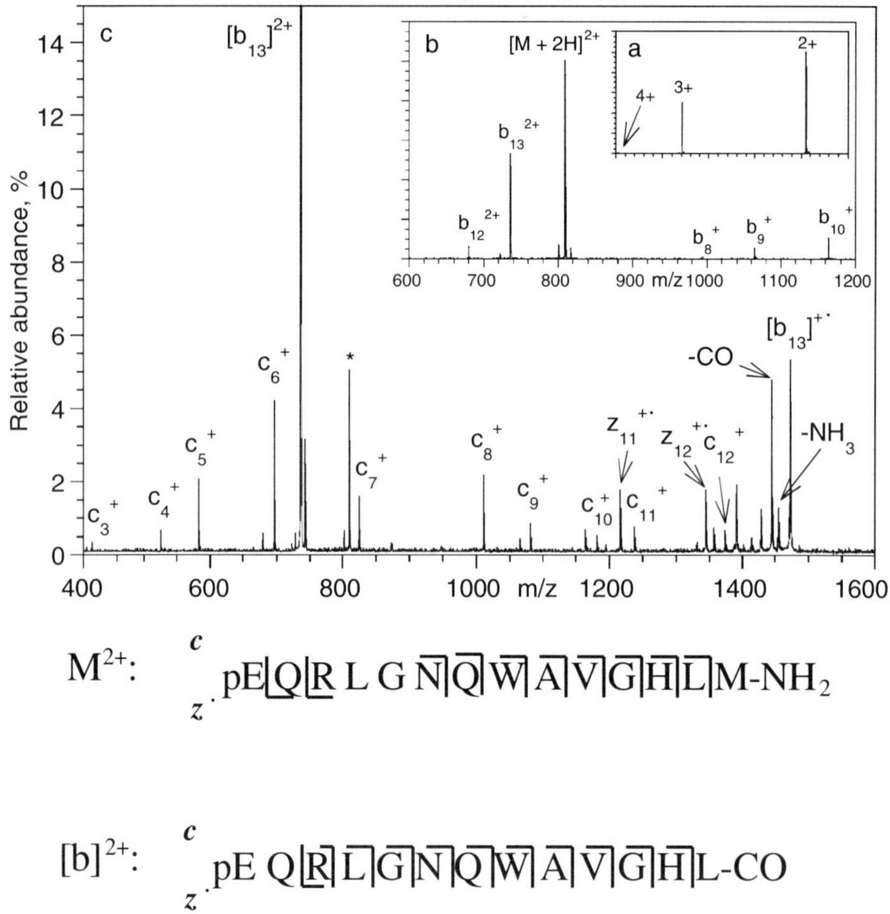

Figure 11.5. Upper panel, mass spectra of bombesin: (a) parent ions; (b) CAD of parent ions; (c) ECD of b_{13}^{2+} ($m/z = 736.8\,\text{Th}$). Lower panel: ECD cleavages in the molecular ion and the b_{13}^{2+} ion. (Reproduced from Ref. 19 with permission.)

only c_1 and c_2 absent due to the detection window cutoff at m/z 400. Compared to ECD of $[M + 2H]^{2+}$ ions (Fig. 11.5), the bonds between R_3 and N_6 have now been broken. Since all interresidue bonds were finally fragmented, and the fragment ion identities were established, full sequence could easily be determined.

McLafferty et al. have demonstrated the principle possibility of de novo sequencing of 9-kDa proteins and of obtaining extensive fragmentation information for proteins up to 42 kDa [2,24]. The key innovation was breaking prior to ECD the extensive intramolecular hydrogen bonding by gentle collisional excitation, so-called activated ion ECD.

11.5.3. S—S Bonded Peptides

Disulfide bond cleavage in ECD is discussed in detail in Ref. 17. Briefly, if two chains are connected by an S—S bridge, their separation upon electron capture occurs at dominant rates. In the presence of the internal S—S bridge, fragments from the internal cycle are often observed, although at lower intensities than the fragments from the linear region. The internal-cycle fragments originate from the secondary reactions with the ·S radical [see Eq. (11.7)]. Molecules with multiple disulfide bonding require capture of more than one electron for fragments to be observed, unless collisional activation prior or after electron capture causes intramolecular S—S bond scrambling. For example, "normal" ECD of insulin 5+ ions gives only charge reduced ions; activated ion ECD ("in beam") produces chain separation, with B-chain 3+ and A-chain 1+ molecular ions (in insulin, linear A and B chains are linked by two disulfide bonds) [17].

11.5.4. Determination of Labile Posttranslational Modifications

This is the area where electron-capture dissociation provides arguably the greatest advantage over VE techniques. Because of its nonergodic nature, ECD backbone cleavage is "mild" with respect to labile modifications that are usually preserved on fragmentation. This is in a stark contrast to VE techniques, where labile bonds are cleaved first. To avoid undesirable losses from molecular ions, collisional excitation has to be prevented, which often means that the activated ion technique cannot be used. External accumulation of ions in a *rf*-only hexapole prior to injection to the FTMS cell can provide intact labile molecular ions at high abundances [19]. During electron irradiation event, a gas pulse that normally increases backbone fragmentation must be used with caution, since it can also cause collisional excitation and undesirable losses [19]. In many cases, *c* fragments, as even-electron species, survive better and provide more reliable information than radical $z\cdot$ ions.

Phosphorylated peptides are rather stable in the gas phase and produce meaningful spectra in VE techniques, despite abundant losses of the phosphate and small groups (water, ammonia, etc.) from fragments. ECD of phosphorylated and multiple phosphorylated molecules gives more extensive fragmentation and no losses. The peak intensities in ECD decrease with the relative amount of phosphorylation increased. Therefore, integration of many scans is required for obtaining reasonable signal/noise values. The ECD achievements include identification of the positions of up to four phosphate groups in proteolytic peptides of β-casein (Fig. 11.6) [25]. Phosphopeptides were isolated in this work from the proteolytic mixture by metal-ion affinity chromatography (Fe(III)-IMAC technique). For the same 24-kDa phosphoprotein, direct ECD of the intact molecular ions identified one phosphorylation site, restricted positions of three other groups to five possible sites, and that of the last phospho group to four other sites [26]. While proteolysis-based approach provided

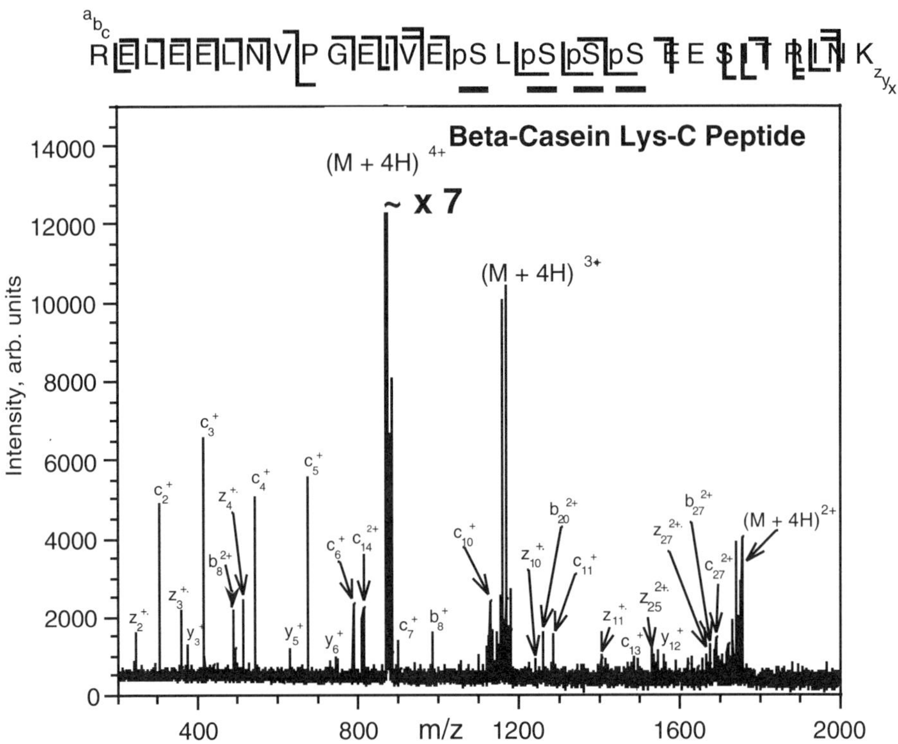

Figure 11.6. ECD of quadruply phosphorylated proteolytic peptides of β-casein, with identification of all four positions. (Reproduced from Ref. 25 with permission.)

superior specificity and compatibility with the standard 2D gel/MS protein identification approach, direct ECD of phosphoproteins is potentially faster and more sensitive.

O-glycosylation possessed, until 1999, significant difficulties for MS/MS analysis. With the advent of ECD, it has been shown that up to five *O*-linked glycans (GalNAc) could be localized in <4-kDa peptides (Fig. 11.7) [12]. The use of the external accumulation technique allowed to extend the complexity range two-fold: in a 8 kDa 60-residue peptide, positions of five out of six GalNAc groups were determined with the sixth position restricted to two neighboring sites (Fig. 11.8). Also, positions of up to five sialic acid groups were identified in a 25-residue molecule already containing six GalNAc groups [19]. The analysis was based on almost exclusively *c* ions that showed minimal losses of glycans. No scrambling of the *O*-linked groups was found, suggesting that ECD could determine the degree of heterogeneity of the glycosylation sites [19].

Sulfation of cysteine was historically the first labile modification where ECD advantages over VE techniques have been demonstrated (the idea of

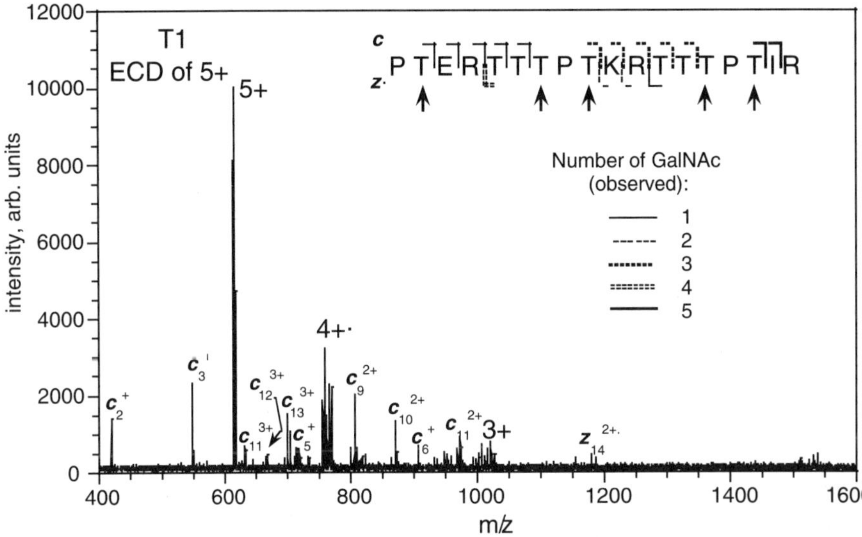

Figure 11.7. ECD of an 18-residue peptide, with identification of the positions of five *O*-linked glycans (GalNAc groups). (Reproduced from Ref. 12 with permission.)

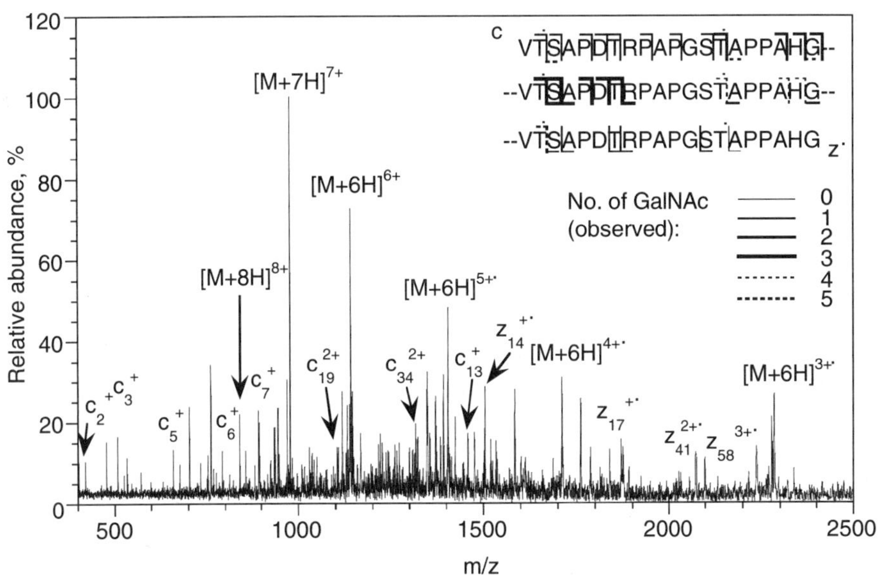

Figure 11.8. ECD of a 60-residue peptide, with determination of the positions of five out of six GalNAc groups and restricting the position of the sixth group to two adjacent residues. (Reproduced from Ref. 19 with permission.)

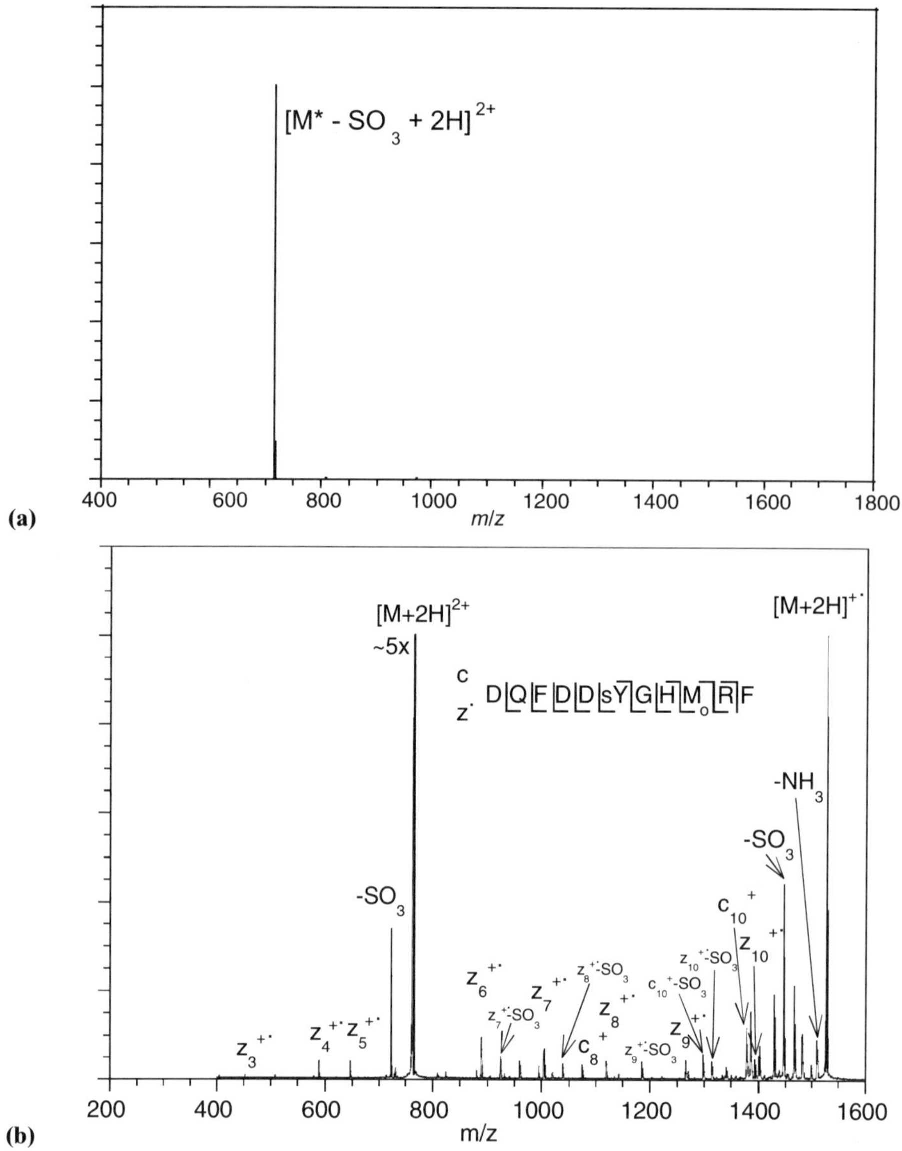

Figure 11.9. Mass spectra of drusulfakinin 2+ ions. (a): mild CAD produces 100% sulfate loss; (b): ECD gives complete cleavage with minimal losses. (Reproduced from Ref. 19 (modified) with permission.)

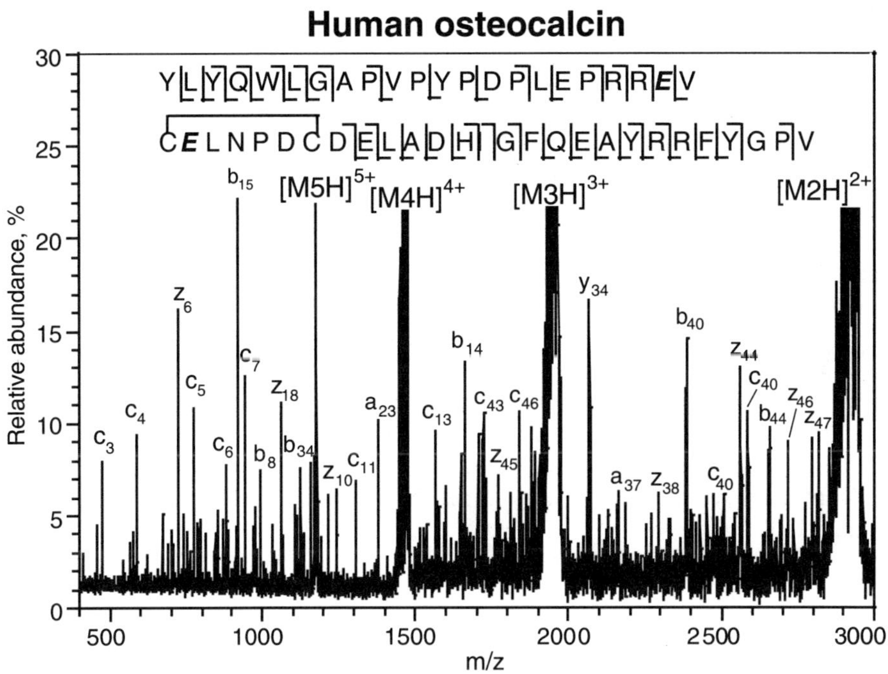

Figure 11.10. Determination of two sites of naturally occurring γ-carboxylation in a sample of human osteocalcin. (The sample was provided courtesy of S. Auriola.)

the experiment and in vitro preparation of a Cys-sulfated peptide was due to N. L. Kelleher) [27]. While even slightest collisional excitation leads to facile loss of the SO_3 group, backbone ECD fragments all retained the sulfate [27]. Sulfotyrosine-containing peptides are more abundant in nature, but their MS/MS analysis has only recently become possible. Figure 11.9 shows mild CAD and ECD spectra of drusulfakinin 2+ ions [19]. Quantitative desulfation is achieved under mild CAD conditions (Fig. 11.9, upper panel), consistent with the extreme lability of sulfotyrosine. At higher collisional energies, all peptide bonds were broken, but only y_9^+ ions of all fragments retained some amount (ca. 25%) of the sulfate. At the same time, all ECD fragments retained >75% of the labile modification (Fig. 11.9, lower panel).

γ-Carboxylation of glutamic acid residues belongs to one of the most labile modifications. Early ECD experiments showed the possibility of dealing with synthetic peptides containing up to two residues, while an attempt with natural products was unsuccessful [27]. More recently, the sites of the naturally occurring γ-carboxylation of two Glu residues in human osteocalcin has been confirmed by ECD (Fig. 11.10) [28]. Although one of the glutamic acids was located inside the internal cycle created by a S—S bond, its modification could be deduced from the cleavages before and after this cycle.

11.6. FUTURE DEVELOPMENT

ECD is still in the early stage of its development, but an increasing number of groups become comfortable with this technique. At the moment, certain instrumental difficulties need to be overcome before ECD becomes a robust routine procedure. The prediction is that in designing the new source of low-energy electrons lies the greatest potential. With a bright source, the FTMS script will become very short, with the corresponding sensitivity improvement and the increase of the complexity of molecules amenable to analysis. Implementing ECD on types of instrumentation other than FTMS is a more distant future, and ion traps are the first candidate. Another task is to incorporate ECD into the sequencing algorithms for routine applications; the first step toward this goal has already been taken [29]. And finally, novel applications of ECD beyond those mentioned in this chapter are bound to appear. These applications, which are difficult to foresee at the moment, will be driven by the adventurous scientific spirit and helped by serendipity, "the mother of inventions" [30].

REFERENCES

1. P. Roepstorff and J. Fohlman, *Biomed. Mass Spectrom.* **11**, 601 (1984).
2. D. M. Horn, R. A. Zubarev, and F. W. McLafferty, *Proc. Natl. Acad. Sci.* (USA) **97**, 10313 (2000).
3. C. Wesdemiotis and F. W. McLafferty, *Chem. Rev.* **87**, 485 (1987).
4. G. I. Gellene and R. F. Porter, *J. Phys. Chem.* **88**, 6680 (1984).
5. F. W. McLafferty, in *Mass Spectrometry in the Analysis of Large Molecules*, C. J. McNeal, ed., J Wiley, New York, 1986, pp. 107–120.
6. R. A. Chorush, I. Vidavsky, and F. W. McLafferty, *Org. Mass Spectrom.* **28**, 1016 (1993).
7. R. R. Ogorzalek Loo, U. R. Udseth, and R. D. Smith, *J. Am. Soc. Mass Spectrom.* **3**, 695 (1992).
8. J. L. Stephenson, G. J. Van Berkel, and S. A. McLuckey, *J. Am. Soc. Mass Spectrom.*, **8**, 637 (1997).
9. Z. Guan, N. L. Kelleher, P. B. O'Connor, D. J. Aaserud, D. P. Little, and F. W. McLafferty, *Int. J. Mass Spectrom. Ion Process.* **157/158**, 357 (1996).
10. R. A. Zubarev, N. L. Kelleher, and F. W. McLafferty, *J. Am. Chem. Soc.* **120**, 3265 (1998).
11. J. Axelsson, M. Palmblad, K. Håkansson, and P. Håkansson, *Rapid Commun. Mass Spectrom.* **13**, 474 (1999).
12. E. Mirgorodskaya, P. Roepstorff, and R. A. Zubarev, *Anal. Chem.* **71**, 4431 (1999).
13. I. J. Amster, *J. Mass Spectrom.* **31**, 1325 (1996).
14. R. A. Zubarev, E. K. Fridriksson, D. M. Horn, N. L. Kelleher, N. A. Kruger, B. K. Carpenter, and F. W. McLafferty, *Anal. Chem.* **72**, 563 (2000).

15. R. A. Zubarev, in *Dissociative Recombination, Theory, Experiments and Applications*, Vol. IV, M. Larsson, J. B. A. Mitchell, and I. F. Schneider, World, Scientific, Singapore, 2000, pp. 214–217.
16. I. A. Kaltashov, D. Fabris, and C. C. Fenselau, *J. Phys. Chem.* **99**, 10046 (1995).
17. R. A. Zubarev, N. A. Kruger, E. K. Fridriksson, M. A. Lewis, D. M. Horn, B. K. Carpenter, and F. W. McLafferty, *J. Am. Chem. Soc.* **121**, 2857 (1999).
18. N. A. Kruger, R. A. Zubarev, B. K. Carpenter, N. L. Kelleher, D. M. Horn, and F. W. McLafferty, *Int. J. Mass Spectrom. Ion Process.*, **182–183**, 1 (1999).
19. K. F. Haselmann, B. A. Budnik, J. V. Olsen, M. L. Nielsen, and R. A. Zubarev, *Anal. Chem.* (in press).
20. R. A. Zubarev, unpublished results.
21. N. A. Kruger, R. A. Zubarev, D. M. Horn, and F. W. McLafferty, *Int. J. Mass Spectrom. Ion Process.*, **185/186/187**, 787 (1999).
22. D. J. Aaserud, D. P. Little, P. B. O'Connor, and F. W. McLafferty, *Rapid Commun. Mass Spectrom.* **9**, 871 (1995).
23. K. F. Haselmann, B. A. Budnik, and R. A. Zubarev, *Rapid Commun. Mass Spectrom.* **14**, 2242 (2000).
24. D. M. Horn, Y. Ge, and F. W. McLafferty, *Anal. Chem.* **72**, 4778 (2000).
25. A. Stensballe, O. N. Jensen, J. V. Olsen, K. F. Haselmann, and R. A. Zubarev, *Rapid Commun. Mass Spectrom.* **14**, 1793 (2000).
26. S. D.-H. Shi, M. E. Hemling, S. A. Carr, D. M. Horn, I. Lindh, and F. W. McLafferty, *Anal. Chem.* **73**, 19, (2000).
27. N. L. Kelleher, R. A. Zubarev, K. Bush, B. Furie, B. C. Furie, F. W. McLafferty, and C. T. Walsh, *Anal. Chem.* **71**, 4250 (1999).
28. S. Auriola, B. A. Budnik, K. F. Haselmann, and R. A. Zubarev, unpublished results.
29. D. M. Horn, R. A. Zubarev, and F. W. McLafferty, *J. Am. Soc. Mass Spectrom.* **11**, 320 (2000).
30. D. C. Blanchard, *Bull. Am. Meteorol. Soc.* **77**, 1279 (1996).

CHAPTER 12

Synthesis of Combinatorial Peptide Libraries

KRZYSZTOF ROLKA

Faculty of Chemistry, University of Gdansk, Gdansk, Poland

CONTENTS

Mass Spectrometry and Hyphenated Techniques in Neuropeptide Research, Edited by Jerzy Silberring and Rolf Ekman.
ISBN 0-471-35493-7

12.1. INTRODUCTION

The first synthetic peptides were obtained by Curtius at the end of nineteenth century. During the synthesis of hippuric acid he obtained Bz–Gly–Gly as a byproduct, and an unidentified γ acid, which, 20 years later, appeared to be the *N*-benzoylhexaglycine. Nevertheless, Fisher is considered to be a father of peptide chemistry. At the beginning of twentieth century he turned his interest to the synthesis of peptides. In his pioneering work he established the principles of peptide bond formation. The introduction of a benzyloxycarbonyl group as a reversible protection of amino function by Zervas and Bergmenn was a milestone in the modern peptide chemistry. Further improvement in this field was connected with the development of analytic techniques that facilitated purification and identification of peptides. A significant impact on the biology and pharmacology of peptides, reported in 1953 by du Vigneaud, was the complete chemical synthesis of the first peptidergic hormone—oxytocin. His achievement was awarded 2 years later by the Nobel Prize.

The turning point in peptide chemistry, biochemistry, pharmacology, and also other fields dealing with biologically active peptides, was the introduction by Merrifield in 1962 of the method of peptide synthesis on the insoluble polymers. This method, called solid-phase peptide synthesis (SPPS) is nowadays the main approach in the chemical synthesis of peptides. The recognition of this genius invention was the Nobel Prize conferred on Merrifield in 1984. Many modifications of SPPS allow the synthesis of polypeptides with the length of 30–40 amino acid residues in a short time, utilizing routine protocols. The availability of peptides for biological investigations yielded enormous progress in understanding the molecular mechanism involving endogenous peptides and brought several peptidergic drugs into the market. The increasing demand for synthetic peptides stimulates many research group to improve SPPS protocols.

One of the main targets of SPPS modifications was to synthesize more than one peptide simultaneously. Two concepts were developed. One is the parallel peptide synthesis, where the number of compounds synthesized during the synthesis is constant. The second concept, proposed in late 1980s by Furka, was the controlled synthesis of a peptide mixture with the number of peptides increasing with the progress of the synthesis. The development achieved by other groups allowed for selection from such structurally related peptides mixture, only those with the designed activity, and such strategy opened an extremely fast growing field, especially attractive for the drug discovery programme. This approach, called combinatorial chemistry, originally designed for peptides, has also been successfully applied for identification and structure elucidation of other biologically active compounds, such as oligonucleotides, benzodiazepines, peptidomimetics, receptor ligands, and other organic molecules. The compound mixtures are called *combinatorial libraries*. The construction and biological screening of such libraries in relation to the peptides will be discussed. Other techniques for creating combinatorial libraries such as phage display method, will not be described in this chapter.

12.2. SOLID-PHASE PEPTIDE SYNTHESIS (SPPS)

In the 1970s a "classical" peptide synthesis in solution was a dominant method for obtaining peptides. It took more than a decade since Merrifield published his first papers [1,2], where he described the synthesis of a simple tetrapaptide, Leu–Ala–Gly–Val, on insoluble polymer to convince most of peptide chemists to the advantages of this method. Nevertheless, the "classical" technique is still used, mainly by pharmaceutical industry to obtain large quantities of peptidergic drugs. It is also useful to synthesize peptides containing certain modifications, including peptidomimetics. It should also be noted that this strategy is successfully developed and utilized for many years by Sakakibara and his research group. Perhaps one of his most impressive works was the total solution to the chemical synthesis of a 238-residue protein, the *green fluorescent protein* [3].

12.2.1. Supports for SPPS

The key idea of SPPS is that the entire synthesis is carried out on an insoluble polymer. Originally Merrifield's support was chloromethylpolystyrene, crosslinked with 2% divinylbenzene (DVB) [1,2]. With years of experience, many types of resins have been developed. Most of them consist of the crosslinked polystyrene and polyacrylamides. In more recent SPPS procedures, copolymers containing polyethylene glycol (PEG) are especially recommended for the synthesis of peptide libraries. Modern SPPS polymers are equipped with functional groups attached to the polymer by special linkers. They enable one to synthesize peptides with *C*-terminal carboxyl or amide groups, which can be cleaved from the resin at basic or acidic conditions. According to the procedure selected, either free peptides or those with functional groups protection can be obtained. It is also possible to prepare fully deprotected peptides attached to the polymeric support via their *C*-terminal carboxyl groups. This creates the opportunity for the direct analysis of peptide libraries (discussed further in this chapter). Manufacturers supplying reagents for peptide synthesis offer many kinds of resins that are suitable for combinatorial chemistry. To select the optimal type of resin for a particular synthesis, several excellent reviews [4–6] can be recommended. Very useful information comes also from the suppliers (e.g., Novabiochem GmbH, Bachem AG, Rapp Polymere GmbH, Fluka, Advanced ChemTech).

12.2.2. Principles of SPPS

The general concept of SPPS in shown in Fig. 12.1. In the first step, the *C*-terminal amino acid of the synthesized peptide is attached to the appropriate functionalized resin (step a). Obviously, the α-amino group and functional side-chain group (if present) of this amino acid must be protected. The reaction is usually carried out outside the reaction vessel. In order to save time,

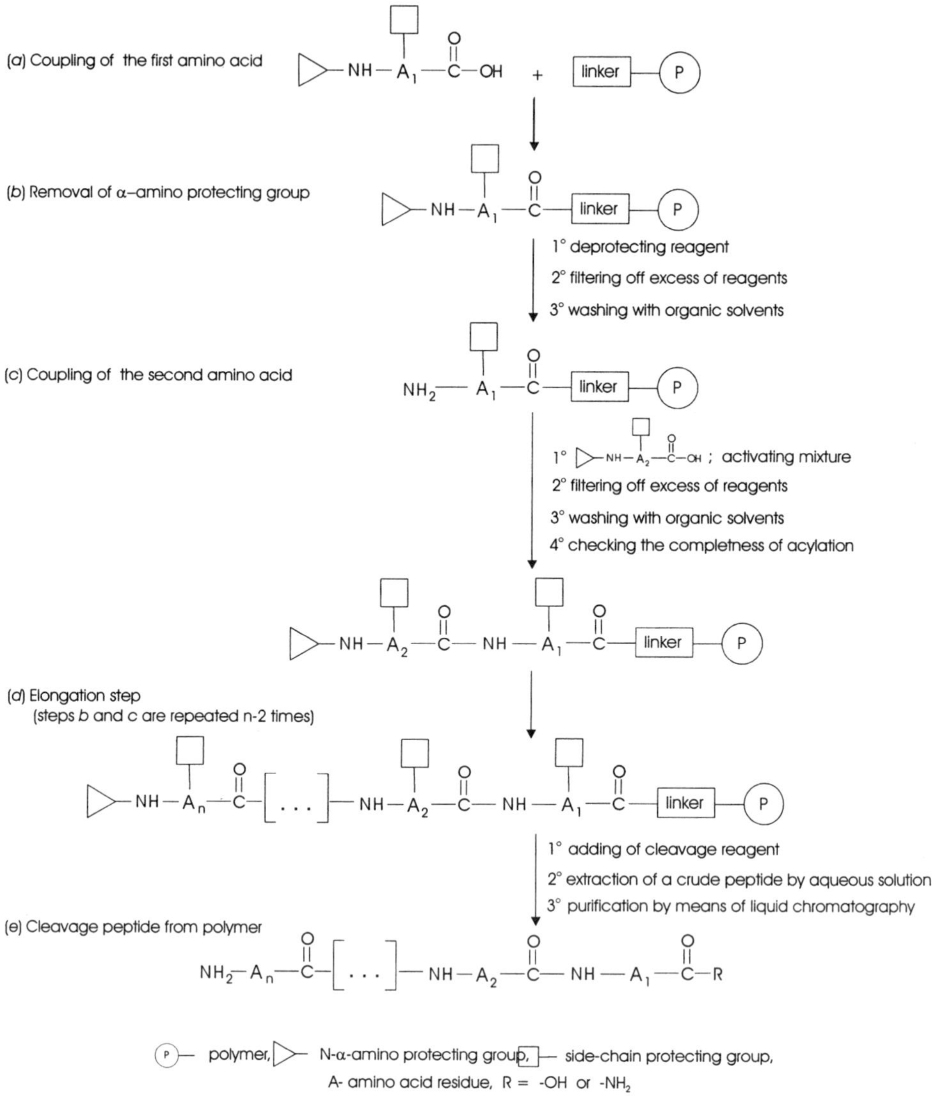

Figure 12.1. General scheme of solid-phase peptide synthesis.

the resin with the first amino acid residue attached can also be purchased from the manufacturers. Starting from that point, the synthesis proceeds in the reaction vessel constructed in such a way that all soluble reagents (except the resin and the peptide attached to it) and solvents can be washed out. The solvents commonly utilized for SPPS are dimethylformamide, *N*-methylpyrrolidinone, and dichloromethane. The coupling of the next amino acid derivative (step c) is preceded by the selective splitting of the α-amino protecting group (step b). Nowadays, two removable α-amino protecting groups are used in SPPS: *tert-*

butoxycarbonyl (Boc) and 9-fluorenylmetoxycarbonyl (Fmoc). For standard procedures all protecting amino acids used for the synthesis contain only one of these groups. If the acid-labile Boc group is used, trifluoroacetic acid (TFA) solution is usually applied as a deprotecting reagent throughout the entire synthesis. The second choice is the base-labile Fmoc group, which can be removed at basic conditions, such as by piperidine solution. The deprotecting step is followed by the removal of all the soluble reagents that might react with the incoming amino acid residue. Then, the appropriately protected amino acid and coupling reagents dissolved in organic solvent(s) are added. Carbodiimides, and more recently uronium salts based on 1-hydroxy-7-azabenzotriazole [7] are the most popular activating reagents in peptide synthesis. When the Boc chemistry is applied for the synthesis, the coupling step is preceded by the neutralization of the amino group with the tertiary amine in an organic solvent. The coupling usually takes a few hours, after which the liquid phase is filtered off and the presence of free amino group is inspected by one of the tests developed. The negative result of such test indicates that the amino acid derivative was quantitatively incorporated. The cycle constituting steps b and c is than repeated in order to couple the next protected amino acid. If the peptide consists of n amino acid residues, $n - 1$ such cycles have to be performed in the elongation step (step d). Finally, after removing the protection from the *N*-terminal amino group, the resin with the attached peptide having all side chains protected, is treated by the cleavage reagent (step e). In the case of Fmoc chemistry, TFA solution is applied, which simultaneously also splits all sidechain protections. Boc chemistry is used under more harsh conditions such as treatment with liquid hydrogen fluoride or trifuoromethanesulfonic acid. In order to minimize side reactions, special scavengers are also added. The crude free peptide is then extracted by aqueous solutions, and purified using liquid chromatography techniques, usually HPLC on C18 columns in the reversed-phase solvent systems. Solid supports developed for Fmoc chemistry allow to remove the peptide with protected functional groups from the resin. This makes it possible to synthesize e.g., large polypeptides, applying fragment condensation strategy. Orthogonal protections of side chain groups make possible the selective cleavage of some of them and to prepare cyclic peptides by the formation of disulfide bridges or amide bonds. Another option is to remove all the protecting groups and leave a peptide covalently attached to the solid support. This approach is applied in the combinatorial chemistry methods.

SPPS considerable speeds up the synthesis of peptides. Since the building blocks consist of repetition steps, the method can easily be fully automated. Therefore, this method has become very attractive for investigators who are interested in the use of synthetic peptides for biological or conformational studies. Several companies offer peptide synthesizers. The most advanced instruments set the optimal method of synthesis based on the given sequence. The entire procedure is fully automated. However, for the proper use of such an automate, at least a basic knowledge of the SPPS method is still recom-

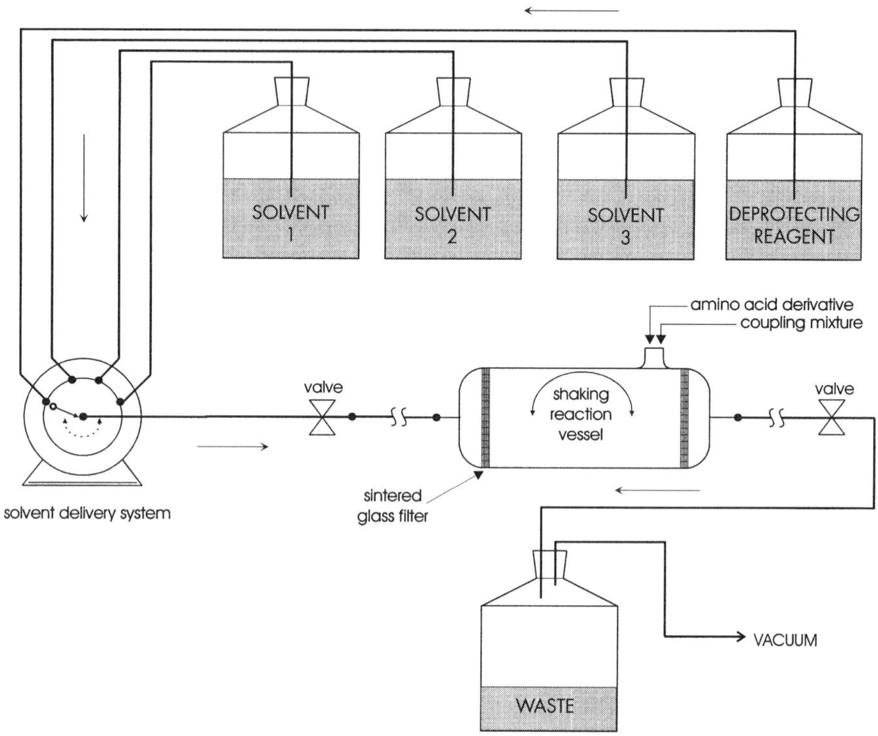

Figure 12.2. Simple manual equipment for solid-phase peptide synthesis.

mended. Obviously, SPPS can also be carried out manually. Figure 12.2 shows a very simple equipment, which can be set up in most laboratories. The reaction vessel can be gently shaken or twisted by any laboratory shaker. Solvents and eventually other liquid reagents are delivered into the reaction vessel through a simple device. The disconnection of solvent lines when the reaction vessel is in the vertical position allows to empty the reactor. Coupling reagents and amino acid derivatives are manually added directly into the reaction vessel.

12.2.3. Multiple-Peptide Synthesis

The growing demand for synthetic peptides, mainly from the pharmaceutical industry, gave a strong impetus for the development of new methods and techniques in SPPS. More efficient reagents, in conjunction with modern analytic methods, especially HPLC on the reversed-phase columns, allowed the synthesis of peptides in a shorter time with a higher purity. The "classical" approach is based on the synthesis of one peptide in one reaction vessel. In order to make SPPS more efficient, the obvious direction was to develop a method, which would result in obtaining more than one peptide in one run. It can be achieved by designing a multivessel reactor. It was shown [8] that a simple modification of the Merrifield-type reactor was efficient to carry out a simultaneous synthesis of four peptides. The introduction of polymeric support with proper physicochemical properties enabled the synthesis in a reaction vessel similar to the chromatographic column. The method in which reagent solutions and solvents pass continuously through the reaction vessel is called *continuous-flow SPPS*. Placing more than one of such reactors into the system allows for parallel synthesis of the peptide in each reactor [9].

Another very interesting method of simultaneous SPPS, called *multipin peptide synthesis*, was described by Geysen et al. [10]. Instead of the reaction vessel, they used a block of polyethylene rods (pins) constructed in such a way so that the protected amino acids can be attached to the pin through its α-carboxyl group. Each pin can fit into the well of ELISA plate. The coupling is achieved by immersing the pins into the well where solutions of the protected amino acids and coupling reagents are present. Washings and deprotections are performed in the appropriate bath solutions. The order of amino acid derivatives introduced into the wells determines the amino acid sequence of the peptide synthesized on each pin. In the original procedure, after removing the side-chain protecting groups, the peptides were tested by ELISA while still covalently bound to the pins. It is also possible to remove the peptides from the rods and analyze them separately in solution.

Several modifications of SPPS that are worth mentioning in this chapter concern various solid supports. Frank et al., on the basis of his early experiments on the simultaneous synthesis of oligonucleotides [11], proposed the simultaneous SPPS on cellulose disks [12]. This approach is known as "SPOT synthesis" [13,14]. A sheet of cellulose paper 55 × 105 mm was used for such a purpose. By pipetting the respective reagent solutions onto 49 spots, followed by adequate washing, 49 different peptides were simultaneously obtained and successfully used for the antibody binding assay, while they remained bound to the cellulose support. Lebl et al. [15,16] used another inexpensive support—cotton strips. Because of their mechanical stability, the synthesis can be carried out continuously when the strips pass through the baths with the appropriate reagents. Both methods described above are easily handled manually or may be automated.

Another, very elegant approach of simultaneous SPPS was described by Fodor et al. [17], who carried out peptide synthesis on an aminated glass plate. α-Amino groups were protected by photoremovable nitroveratryloxycarbonyl moiety. By masking some sectors of the plate, and illuminating the rest of surface with a mercury lamp, spatially addressable photodeprotection of the α-amino groups in unmasked sectors was achieved. Except for the deprotection, the synthesis was carried out by the standard SPPS method. The combination of spatially localized photodeprotection allowed for synthesis of the different peptides on one glass plate. The usefulness of the method called "a light-directed, spatially addressable parallel synthesis" was proved by the synthesis of 1024 peptides in ten steps on a 1.6-cm^2 glass plate. The peptides were localized by their interaction with a fluorescent-labeled antibody [17].

In contrast to the method utilizing the photolabile protecting group, that require special equipment for the synthesis and peptide detection, the approach described by Houghten [18] was very simple, and in addition, allowed to synthesize peptides on a preparative scale. Resin "packets" (3 × 4 cm) resembling tea-bags in the polypropylene mesh were prepared. After placing each "bag" in a single reaction vessel (e.g., polypropylene scintillation vials), all common steps—washing, neutralization, and deprotection—can be carried out. The coupling step can be performed separately with each individual "bag," or several "bags" can be treated by the same coupling mixture. With the help of the "classical" Boc chemistry it was possible to synthesize 248 different 13-residue peptides in 4 weeks in a yield of 10–20 mg [18]. This method can be used even in a very simple manual system such as that shown in Fig. 12.2. Later, this group considerably improved the efficiency of the method by the introduction of an HF apparatus for simultaneous cleavage of 24 peptides from the resin [19].

12.3. LIQUID-PHASE PEPTIDE SYNTHESIS

Four general methods for chemical synthesis of peptides are currently available: SPPS, the "classical" method in solution, liquid-phase peptide synthesis and semisynthesis. The principle of SPPS was briefly described above. Traditionally, the oldest method—solution-phase peptide synthesis, and also semisynthesis where peptide bonds are formed by the combination of chemical and enzymatic methods—are not commonly used for the synthesis of peptide libraries and, therefore, are not discussed here. Liquid-phase peptide synthesis was for the first time described by Bayer and Mutter [20]. The scheme of the process is very similar to that of the SPPS, but instead of the insoluble polymer, the *C*-terminal amino acid residue is covalently linked to the polyethylene glycol monomethyl ether. This monofunctional homopolymer shows good solubility in a variety of organic and aqueous solvents. The building blocks are then carried out in the solution and the protected product after each step of the elongation process is precipitated by diethyl ether. The product can be purified and used in the next elongation step. Despite some limitations, this method also combines some advantages of both solution-, and solid-phase peptide synthesis. It was also successfully applied in combinatorial chemistry approach.

12.4. CHEMICAL METHODS USED FOR CONSTRUCTION OF PEPTIDE LIBRARIES

As opposed to the "classical" organic chemistry approach, when compounds are synthesized one by one, the parallel SPPS considerably improved the efficiency of peptide synthesis. Parallel SPPS can be considered as a starting point of combinatorial chemistry. Using both approaches, compound libraries can be synthesized, but in contrast to the first strategy, where the number of peptides is constant during the synthesis, combinatorial synthesis produces the increasing number of compounds synthesized during each coupling step.

Peptide libraries can also be obtained by the methods of cloning synthetic DNA that encodes random peptide sequence [21–23]. The best example of this approach is the phage display library. The synthetic oligonucleotide library is inserted into the DNA strand of phages. The DNA replication is achieved by infection of bacteria with those transformed phages. Each formed phage clone carries a copy of a different coat protein with the sequence corresponding to the inserted DNA segment. This type of libraries called biological libraries, although widely studied, are outside the scope of this chapter.

12.4.1. Portioning Mixing Method

The first concept of the synthesis of combinatorial peptide libraries was presented by Furka et al. in 1988 at two international symposia, and in 1991 this idea was published as an original paper [24]. The principle of the method described as portioning–mixing synthesis of combinatorial libraries is illustrated in Fig. 12.3. In the first step, a solid support is divided into equal portions (in the example shown in Fig. 12.3—into three portions). To each portion, three different protected amino acids are coupled (coupling step 1) by applying any of the methods developed for SPPS. Then, all portions of the resin with attached amino acid derivatives are mixed, and again divided into three portions. After deprotection of the α-amino groups, the second coupling step is performed, followed by pooling of these portions of the resin with attached protected dipeptides. By such a simple combination, nine different dipeptides consisting of three different amino acids can be synthesized, and the procedure requires only two coupling steps. Finally, after the cleavage step, a library of nine dipeptides is ready for further analysis. The number of peptides present in the library depends on their length and a number of amino acids variations in each position. Let us assume that the peptide library consists of n-amino acid residue peptides (e.g., pentapeptides), and in each position k different amino acid residues are present. This means that if all naturally occurring amino acids will be used, the k will be equal to 20; however, any lower number can be applied for the reaction.

$$(X_k)^1 — (X_k)^2 — (X_k)^3 — \cdots — (X_k)^n$$

in this case the number (N) of peptides present in the library is equal to $N = k^n$.

If a dipeptide library consists of all 20 naturally occurring amino acids, $20^2 = 400$ peptides are present. Obviously, with the increasing length of peptides, the number of peptides in the library is increasing exponentially. In the

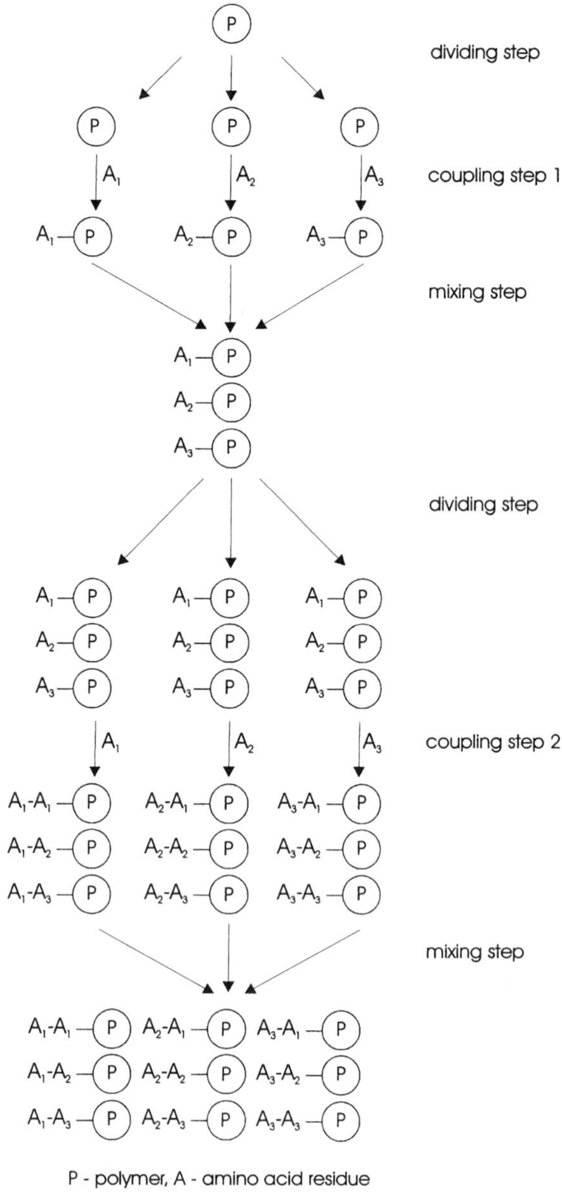

Figure 12.3. The principle of portioning-mixing synthesis of combinatorial library. The method is also known as the split method or divide–couple–recombine.

pentapeptide library, $20^5 = 3.20 \times 10^6$ compounds are present, whereas a decapeptide library consists of $20^{10} = 1.02 \times 10^{13}$ peptides. In his pioneering work Furka et al. [24] applied a routine analytic methods (HPLC and paper electrophoresis) for the identification of library members. Therefore, only libraries of the small size were analyzed. In the same year (1991) two other groups, Lam et al. [25] and Houghten et al. [26], described the peptide libraries of millions of peptides, and demonstrated their practical use for the selection of biologically active sequences. The principle of the construction of peptide libraries was the same as that described by Furka et al. [24]. Both groups seemed to be unaware of Furka's reports in 1988 at symposium [27]. Leaving a question open who was the father of combinatorial chemistry, it should be stressed that the idea of the use of peptide mixtures in biological investigations gave a great impetus for many research groups. On the other hand, the idea of the use of a peptide mixture for biological tests was criticized by many researches. The impressive first examples [25,26] on how to select biologically active peptides from a mixture of million other components, threw new light on this problem. The selection and identification of active peptides in the library (deconvolution of peptide library) is, therefore, a key problem in combinatorial chemistry.

12.4.2. Tea-Bag Method

The "tea-bag method" described by Houghten [18] was very efficient for synthesis of a large number of peptides in a short time and on the preparative scale. It appeared to be also a very convenient way for the preparation of the peptide library. By simply pooling the peptidyl-resin portions from the appropriate mesh "bags" followed by redistribution and further coupling with defined amino acid derivatives, a peptide library was obtained. The method called "divide-couple-and recombine" was conceptually the same as that described by Furka et al. [24] and Lam et al. [25]. Further modification proposed by the same group was a chemical mixture approach in which mixtures of protected amino acids at the predetermined ratio were coupled to the resin portion [28]. In this method, Boc chemistry was first applied, but obviously Fmoc chemistry can also be used. Such a method of peptide library preparation opened new possibilities for deconvolution (to be discussed further in the text).

12.4.3. Multipin Peptide Synthesis

The method of peptide synthesis on polyethylene rods [10] can be considered as one of the SPPS variants. With the use of cleavable linker, the synthesized peptides can be released from the pins and screened individually. Wang et al. [29] prepared a library of 512 substance P analogs using one pin/one peptide strategy. The individual quantitative analysis of these peptides for their competition assay showed this method to be efficient for the synthesis of peptide libraries. Multipin synthesis was also used in the combinatorial approach to synthesize a mixture of peptides on one pin. In this case, the application of a mixture of protected amino acids is the only way to prepare a peptide combinatorial library [30]. The coupling was performed by immersing pins into the solution containing the mixture of protected amino acids. To obtain the ratio of peptides concentrations close to l, the protected amino acid derivatives were arranged in several groups, according to coupling efficiencies of the amino acid derivatives. The advantages of this method were shown with the use of peptide libraries for epitope mapping. The method is especially attractive for fast screening, but its limitation is rather small quantity of the libraries obtained.

12.4.4. SPOT Synthesis

Similar to the method described above, also the synthesis of a combinatorial peptide library with the SPOT approach requires the use of a mixture of protected amino acid derivatives for coupling. As shown by Frank et al. [31], the subequimolar amino acid mixtures had a homogenous representation of synthesized peptides in the library. A cellulose sheet used for the peptide synthesis can be considered as an open reactor. In contrast to the conventional reaction vessels, in spot synthesis the concentration of the reagents increases with time, due to the evaporation of the solvents. This might be the reason why the coupling efficiency of different protected amino acid derivatives was similar. The development of the appropriate linkers, novel polymeric surfaces, and the modified cleavage procedures, allows screening of peptide libraries both in solution and while attached to the support. The synthesis of a peptide library by the SPOT method was significantly improved by the application of automated pipetting system. Hence, possibilities are created to use a variety of methods for the library deconvolution. The SPOT method appeared to be a valuable tool for the epitope mapping, identification of sequences bound to a double-stranded DNA and silver [32], as well as for the determination of substrate specificity of protein kinases [33]. More recent inventions of the method are summarized in the literature [34,35].

12.4.5. Liquid-Phase Peptide Synthesis

All the peptide library methods described above utilized solid supports. As discussed briefly in the previous paragraphs, the advantages of this approach are obvious. It should be stressed that most of peptide libraries are synthesized using this concept. Significantly, less common is the use of a liquid-phase method in combinatorial chemistry. Janda and Han [36] described the synthesis of peptapeptide libraries on the polyethylene glycol monoether (MeO–PEG) using Boc chemistry. The libraries were constructed by the split method. After each coupling, the peptidyl–MeO–PEG portions were individually precipitated with ether, and than mixed together. ELISA was used to identify the active peptides. The usefulness of the method was proved by selection from the peptapeptide libraries the enkephalin sequence, which was the strongest inhibitor of the binding of leucine enkephalin to anti-β-endorphin monoclonal antibody. More recently, Janda et al. [37,38] also reported the liquid-phase synthesis of nonapeptide libraries.

12.4.6. Coupling with Single Amino Acid Derivative versus the Use of a Mixture of Protected Amino Acids

It took several years for most of the researchers to accept the suitability of compound mixtures (libraries) for biological screening. The same "resistance" was observed among chemists with respect to the application of a mixture of derivatives in a single coupling step. In the synthesis of the first combinatorial peptide libraries [24–26], each coupling step was carried out with a single amino acid derivative, as in the "traditional" one-by-one organic synthesis. Deconvolution of such libraries, and its advantages and limitations, will be described in the next paragraph. In the second method for preparation of synthetic libraries, the mixture of protected amino acids is utilized. This approach is called "a premix method." In order to obtain equimolar representation of peptides in the library, the rate constants of coupling for each amino acid should be the same. This can be achieved by the optimization of coupling conditions. One of the important parameters to be considered is the molar ratio of amino acid derivatives used in the mixture. Some of the research groups reported that equimolar amino acid derivatives gave reasonable results in terms of equimolarity of peptides of the library [39]. On the other hand, the use of amino acid mixtures at predefined ratios was also reported [39,40]. The reasons why, in some cases, the equimolar ratio of protected amino acids yielded nearly equimolar peptide representation in the library but not in others are coupling conditions such as: method of synthesis, type of amino acid derivatives (Boc, Fmoc, etc.), activating reagents, solvents, concentrations of reagents, and time of coupling. Therefore, if someone plans to use a premix method, it is advisable to perform a separate test for equimolarity of peptides in the library, before a biological screening will be performed. As was already mentioned, the use of the premix method is the only choice to prepare combinatorial libraries in, for example, multipin and spot methods. This also facilitates the synthesis of peptide libraries screened in solution (to be discussed later in the text).

12.5. DECONVOLUTION OF PEPTIDE LIBRARIES

Since the first reports [25,26] where the biological screening of synthetic peptide libraries was described, a vast number of papers (recently about 1000 every year) concerning chemistry and biology of peptide libraries have been published. There are several journals (e.g., *Molecular Diversity*, *Journal of Combinatorial Chemistry*, *Combinatorial Chemistry & High Throughput*

Screening) devoted to combinatorial chemistry. Despite the synthetic method used for the library construction, two approaches of their biological screening are utilized. In the first one, biological testing is performed when the peptide library is still attached to the polymeric support used for the synthesis. In the second method, the peptide library is released from the resin and then subjected to biological investigations in solution. Both methods have advantages and limitations, which will be discussed in this paragraph.

12.5.1. One Bead–One Compound Concept

This method, one of the oldest, was invented by Lam et al. [25]. The authors constructed a peptide library by means of a split method. As stressed by the authors, the interesting feature of this synthesis is that every bead of resin contains only multiple copies of one sequence. Based on the so-called "one bead-one peptide concept," a deconvolution known as the "selectide process" was proposed [41–43]. Figure 12.4 illustrates the principle of this method. The synthesis of the peptide library is carried out by Fmoc chemistry. On completion of the building blocks, the Fmoc group from the *N*-terminal α-amino group is removed by piperidine solution, followed by the removal of all side chain protecting groups (acid-labile) by means of trifluoroacetic acid (TFA) solution in a mixture of scavengers [44]. Under the conditions applied, the synthesized peptides are still covalently attached to the resin, and are further used for the determination of bioactive sequences present in the library. The general idea is that the peptide molecules attached to the bead specifically interact with the target molecules that are tagged by a chromophore or a fluorescence probe. Therefore, the beads carrying out the active sequences can be directly distinguished by color changes, or after illumination with the light [43]. Chen et al. [42] also developed a method for selection of thrombin inhibitors where the target (thrombin) molecules were modified by biotinylation, and the formed thrombin complexes with high-affinity sequences were subjected to interact with streptavinin-alkaline phosphatase, which deposited a color derivative of indigo on the beads. In the method developed for determination of protein kinase substrate specificity [42], the peptide library was incubated with the enzyme in the presence of [γ-^{32}P]ATP and the phosphorylated, radiolabeled beads were identified by autoradiography. In the final step, beads containing active peptides were collected, followed by the sequential analysis of the these peptides. The microsequencing of peptide molecules attached to one bead is, at present, within the capability of modern mass spectrometry. It considerably increases the efficiency of the deconvolutoin process. There are several factors that have to be considered when the peptide library is screened while attached to the solid support. One is the type of resin used for the experiment. SPPS is carried out in organic solvents, whereas the biological tests are performed in aqueous solutions. "Classical" Merrifield supports are based on the crosslinked polystyrene, which swells well in organic solvents but not in water and, therefore, such polymer is not suitable for the deconvolution of peptide library

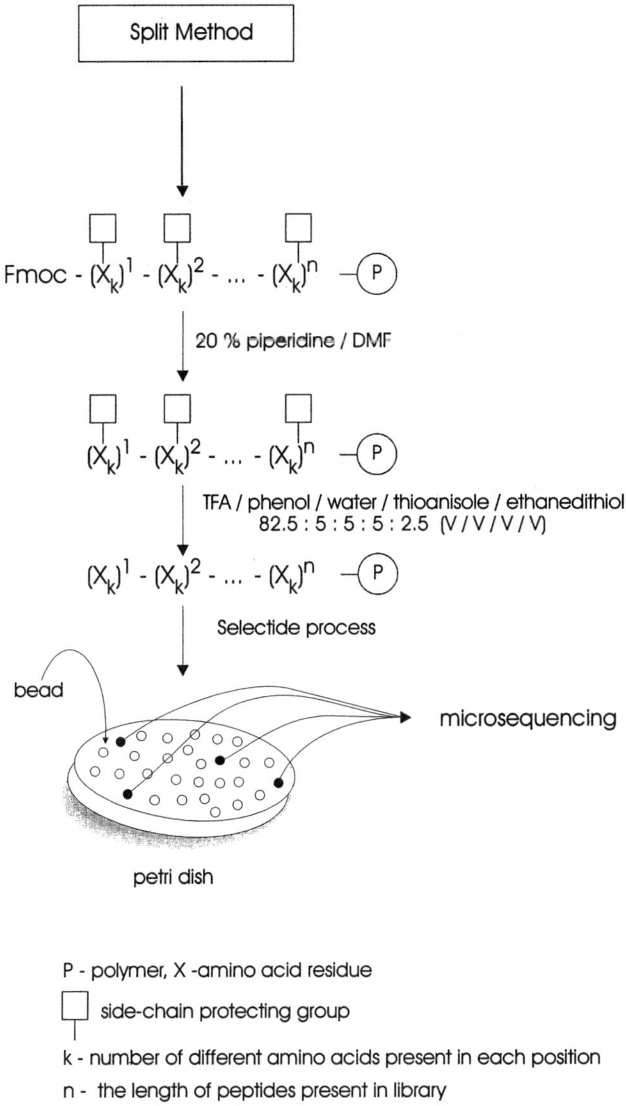

Figure 12.4. Deconvolution using the one bead–one compound concept.

attached to the solid support. For this purpose, commercially available polydimethylacrylamide resin (Pepsyn) [45], and TentaGel resins (Rapp Polymere) [46] were found satisfactory [42,43]. The second support—a polystyrene poly(ethyleneglycol)-tentacle resin—displays the swelling properties almost equal in water and organic solvents used for SPPS, and are probably the most frequently used in the synthesis of peptide libraries. Another advantage is the possibility of applying the orthogonal protecting groups for the reactive sites on the surface and inside the beads [47]. More recent developments in the solid support preparation for combinatorial chemistry are discussed in the literature [5,6,48].

For the screening of support-bound libraries, the amount and the bead size of resin used for the experiment should be considered. When applying this approach, the number of beads should be at least the same as the number of peptides in the library synthesized. In a library consisting of pentapeptides prepared from all 20 naturally occurring amino acids, $20^5 = 3.2$ million different peptides are present. To obtain a full representation of this library, at least 3.2 million beads should be taken for the experiment, but in practice at least 10 times more should be recommended. The bead size of the commercially available TentaGel S resins varies from 10 to 750 µm. They contain from 1.95×10^9 to 4.62×10^3 beads per gramm resin, respectively. Therefore 6.9 kg of 750-µm bead-size resin or 16 mg of the 10 µm bead size resin is necessary. Since the amount of peptide required for microsequencing should also be kept in mind, very small bead size might not be suitable for deconvolution using one-bead-one compound concept. Although the current analytical techniques (HPLC, NMR, IR, and particularly MS) allow to perform unambiguous structural analysis of a peptide using minute amounts of a sample released even from a single bead [49], thus, the bead size and, in consequence, the amount of resin, are still the critical factors for this type of deconvolution. One of the method to overcome this problem is synthesis of a series of smaller libraries instead of the full representation of every species in the library. The concept is called "the library of libraries." Each sublibrary contains all randomized structures within one motif and all sublibraries representing all possible motifs [50]. The construction of "library of libraries" considerably reduced the amount of resin used for the experiment. When a peptapeptide library was constructed in the "library of libraries" approach, approximately 40 times less resin was required than would be needed for the entire representation [43].

12.5.2. Encoded Libraries

The final step of the library screening described above was the sequencing of selected peptide(s). The chemical synthesis of libraries was at the primarilly designed for peptides and oligonucleotides—compounds for which sequential analysis can be considered as a routine. As already mentioned, combinatorial chemistry became especially attractive for the drug development process. From that point of view, peptides and oligonucleotides are not the best candidates as potential drugs. Peptidomimetics and small organic molecules seem to be more suitable for such purposes. On the other hand, in contrast to the oligomeric ligands, the structure elucidation of these compounds is sometimes much more complicated. One approach to improve the structure elucidation of compounds with potential activity, selected from the library, is the synthesis of encoded libraries. Such libraries contain two types of compounds. One

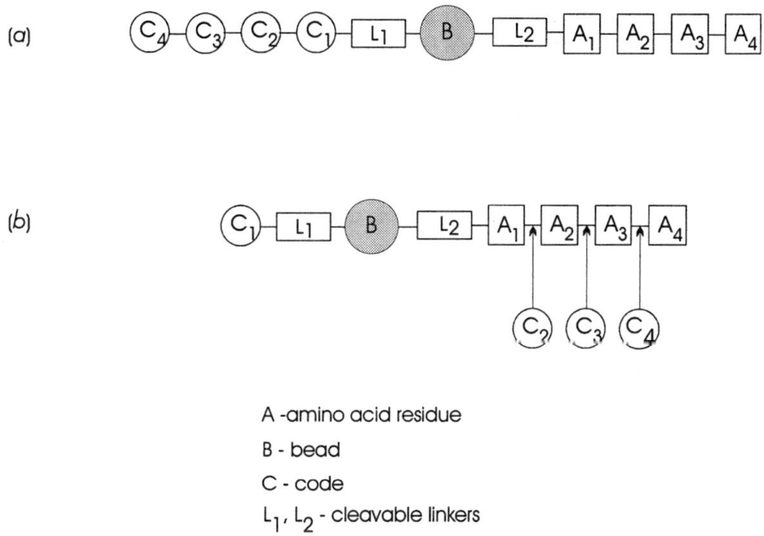

Figure 12.5. Encoded library beads obtained by (a) parallel synthesis of encoded sequence and (b) binary code method.

includes members of library screened in biological tests and the second, structurally different compounds considered as chemical tags, that encode the chronology of the synthesis. Two general methods for the synthesis of encoded library were described.

Brenner and Lerner [51] and later Needels et al. [52] carried out the parallel combinatorial synthesis of peptides and nucleotides attached to the different reactive sites of the same beads. In each building block, coupling of an amino acid derivative to the synthesized peptide library was followed by the addition of an oligonucleotide sequence which encodes this amino acid derivative (Fig. 12.5a). Finally, after the ligand screening, beads with the active peptides are selected and the oligonucleotide sequences encoding the peptide sequences are removed from the beads, multiplied by the PCR and sequenced. This method, although applied for the deconvolution of peptide library, can be adapted for the synthesis of other organic compounds. The use of PCR to multiply copies of oligonucleotide tags significantly increased the sensitivity of the method as compared with the direct sequential analysis of peptides. Nikolaiev et al. [53] synthesized an encoded library of nonsequencable compounds using peptides as molecular tags. After selection of positive beads, sequential analysis of the peptide attached gives information about the history of the ligand synthesis. A different method for encoding libraries was described by Ohlmeyer et al. [54], who used a split method to prepare a peptide library. During each step of synthesis, a tag molecules that encode in a binary fashion both the step and the chemical reagent used in that step, were attached. Using a binary code composed of 18 tags, encoded combinatorial library of 117,649 peptides was obtained. Amino acid derivatives and tag molecules (alcohols differentiated by the length of hydrocarbon chain and electrophoric halobenzenes) were coupled to the same free amino groups (Fig. 12.5b). In this way, the elongating peptide chain was in ~0.5% terminated by each tag added. Tags were attached through a photocleavable linker. After biological screening of the peptide library tethered to the beads, the tag molecules were released from the positive beads by an UV irradiation, derivatized and analyzed by electron capture capillary gas chromatography. The tag composition unambiguously encoded the peptide sequence. In contrast to the polipeptide and oligonucleotide tags, this type of tags do not form sequencable compounds. A similar approach was also successfully used for the deconvolution of small organic molecules [55]. Ni et al. [56] described the synthesis and deconvolution of the encoded library that combines both approaches. Ligand molecules and tag molecules were attached to the different sites of the same beads via different cleavable linkers. Ninety percent of available sites were occupied by ligand (e.g., peptide) molecules and to the remaining 10% tags (secondary amines) organized in a binary code, were covalently linked. After completion of the parallel synthesis of both, the ligand library and tagging entities, single beads were distributed into individual wells of the microtiter plates. Ligand molecules were photocleaved from the resin and assayed in biological testing. Positive beads were recovered and tag molecules were removed from the resin by acid hydrolysis, dansylated, and analyzed by HPLC. The amines (tags) identified along the chromatogram unambiguously described the structure of ligand.

Among many other encoded libraries, the one published by Vaino and Janda [57] is worth noting because of its simplicity. The polymer support was synthesized in different Euclidean shapes. In this case, the polymeric support itself (shapes of polymeric monoliths) served as an encoding element during the synthesis. Another possibility is to place the solid support for the SPPS in units with different shapes, and arrange them spatially. The appropriate redistribution of these units during the library synthesis encodes the structures of the synthesized compound in each solid support unit [58]. Technically more advanced approach is to enclose electronic chips into the capsules, where the information about the history of synthesis is stored [59].

12.5.3. Iterative Deconvolution

This method of screening of the peptide libraries was described, for the first time, by Houghten et al. [26]. The authors used their original tea-bag method, and synthesized library by the "divide-couple-and recombine" approach. Iterative deconvolution allows one to identify active sequences even in a large library containing millions compounds. Because the biological assay is carried out using the peptide library, which is not attached to the solid support, any of the standard biological tests can be adapted for this approach. Figure 12.6 illustrates schematic representation of the iterative deconvolution process of

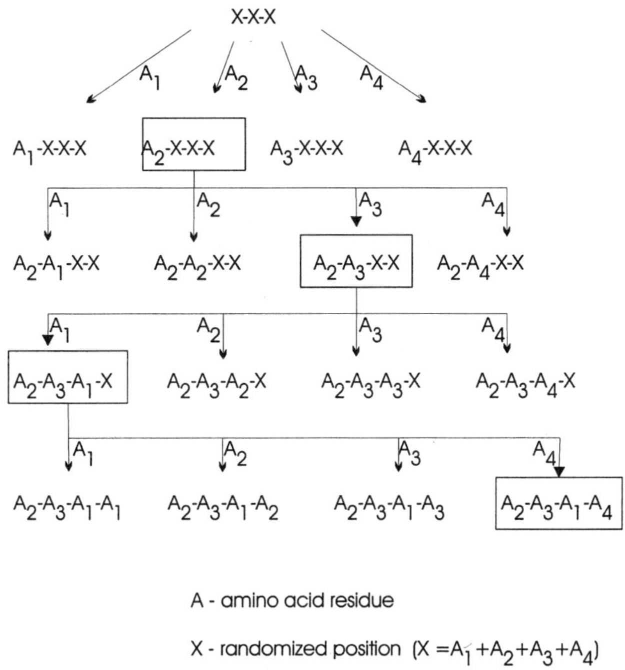

Figure 12.6. Iterative deconvolution method.

a tetrapeptide library consisting of four amino acids. In the first step, a tripeptide library is synthesized by "divide-couple-and recombine" approach. At every coupling step, portion of the resin (X-polymer, XX-polymer) is set aside. Then the peptidyl-resin(XXX-polymer) is divided into four portions. To each portion, one of the four amino acids is coupled. After removal of the protecting groups, and splitting peptide libraries from the support, each sublibrary is individually screened in biological test. A positive result obtained for peptide library with amino acid residue A_2 at the *N* terminus, defined this position. In the next step, XX-polymer is divided into four portions, and to each portion one of the four appropriate protected amino acid is coupled, followed by the coupling of protected A_2 derivative. The biological screening and the synthetic steps are repeated and finally the active A_2–A_3–A_1–A_4 sequence is selected. The peptide library presented in this example is very small. In practice, much bigger libraries (varied in seven positions) were synthesized and screened by this method [60].

Multistep fractionation of a library by HPLC and biological screening [61] can also be considered as an iterative approach. Fractions of the compound library obtained after HPLC separation were analyzed. The one with the appropriate activity was further fractionated until a single compound was selected.

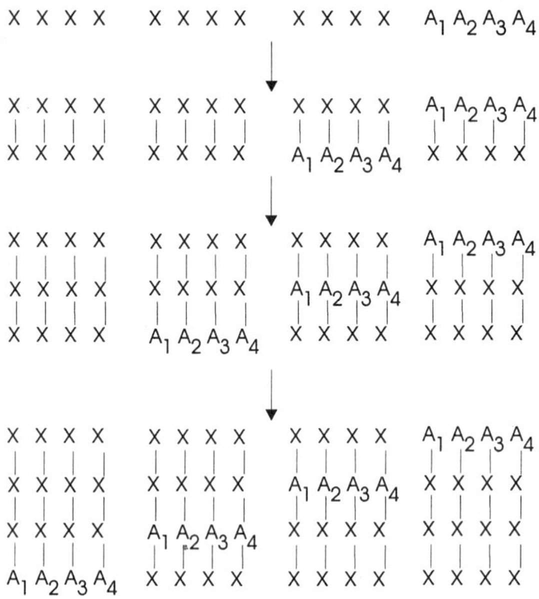

A - amino acid residue

X - randomized position

Figure 12.7. Synthesis of terapeptide library for positional scanning deconvolution.

12.5.4. Positional Scanning Deconvolution

Iterative process is applied for the libraries described in Section 12.4.1; therefore, this deconvolution method (although flexible) is limited by the nature of the synthetic method used for the library construction (see also discussion in Section 12.5.1), and is impractical for large libraries preparation. Libraries prepared using a premix method are free from such problem. The term: "positional scanning" of peptide libraries was introduced by the group of Houghten [62]. Similar deconvolution process was also proposed by Sebestyén et al. [63]. The scheme of the synthesis of a simple tetrapeptide library built up of four amino acids is illustrated in Fig. 12.7. To prepare such a library, the polymeric support is divided into 16 portions. To the first twelve, a predefined mixture of all four protected amino acids is coupled (randomized positions). To the remaining four, each of the four amino acids is added. The coupling step is repeated 3 times, changing every time the portions to which single amino acid derivative is coupled. Finally, from all sixteen sublibraries, peptides are cleaved from the resin simultaneously with the deprotection of functional side chain

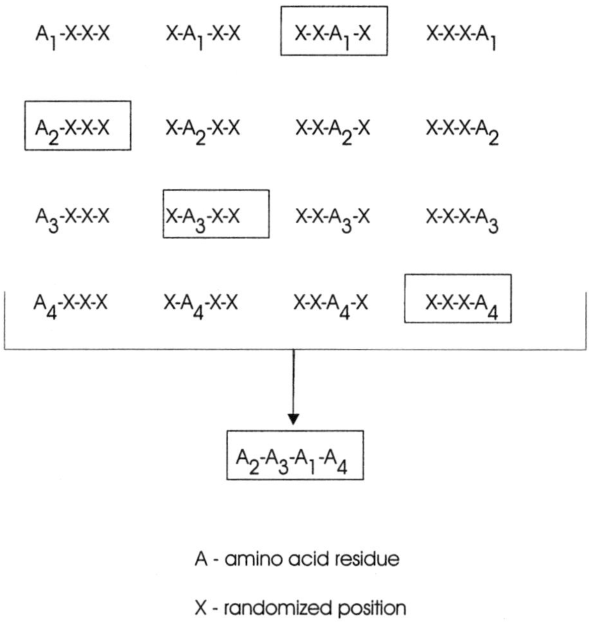

Figure 12.8. Positional scanning deconvolution method.

groups, and used for biological tests. Alternatively, only side chain protections can be removed, and libraries covalently attached to the polymeric support can be used for further screening. Exemplification of the deconvolution process is shown in Fig. 12.8. All 16 sublibraries are screened separately. Among all four sublibraries with fixed position 1, the activity was found only when A_2 was present. By combining the results concerning the remaining positions, a single amino acid sequence was selected. In practice, more than one sublibrary might display similar biological activity. In this case, additional sublibraries could be synthesized. The final verification of the results obtained would be synthesis of the selected sequences and to test them individually.

Both iterative and positional scanning methods were successfully used [64,65] for the selection of peptides exhibiting varied activity: agonists and antagonists of different receptors, proteinases inhibitors, peptides with antigenic activity against different antibodies, substrates for protein kinases, and peptides that bind to the double-stranded DNA. These deconvolution methods were widely used for peptide libraries obtained by SPOT synthesis [31].

The group of Houghten [66–68] extended their investigations also on peptidomimetics by *N*-alkylation of all peptide bonds in the synthesized libraries. This modification was carried out when the libraries were still attached to the solid support. The chemically transformed libraries were further screened as described above. This concept was described as "libraries from libraries."

12.5.5. Multiuse Peptide Libraries

Biological screening of compound libraries is a critical part of the combinatorial methods. In addition to the question to what extent the activity of one compound is influenced by other members of a library, there is also a problem caused by the low concentration of individual compounds in the screened library. Some improvements in splitting the peptide library from the support were published. Lebl et al. [41,69,70] attached peptide libraries to the support via three different linkers, and synthesized multicleavable libraries. The sequential release of two portions of the peptide library offers a possibility to screen the same sequences of the library in a two-step procedure for selection of the active peptides. The third portion of peptide, covalently linked to the support, was used for microsequencing, after selection of "positive" beads. The approach of multiuse peptide library, described by Lerner's group [71,72], was much simpler from the chemical point of view. The peptide libraries were synthesized by the Fmoc chemistry using the split method, followed by the removal of the side chain protecting groups. For the biological screening, portions of peptide library were released from the beads by gaseous TFA. In each step, only few percent of the library was released. The repetition of the releasing process performed on the screened beads allows one to perform the biological test iteratively.

12.6. CONCLUSIONS

The development of combinatorial chemistry, although established only in 1991, is extremely fast. It is impossible, in one chapter, to give even a very brief description of most of the published methods and strategies. Therefore, the concepts summarized here may be considered as representative examples. On the other hand, recent inventions in this field were based conceptually on these "classical" approaches. Combinatorial chemistry was originally established for peptides. Considering its potentially high efficiency for drug discovery it also attracted industrial laboratories. For most of the methods described above, the manual, semiautomated, automated, or robotic systems were designed and became commercially available. Because application of peptides as therapeutic agents is rather limited, many pharmaceutical and biotechnology companies have now shifted their attention to combinatorial libraries of small organic molecules. Therefore, only ~20% of more recent publications concerned peptide libraries [73]. As mentioned above, the synthetic peptide libraries are almost exclusively synthesized on a solid support. The same tendency is also observed for nonpeptidic libraries. Since 1996 ~80% of all libraries were synthesized by this method. During 1992–1997 two-third of papers were affiliated by industry, but more recently this number has shifted in favor of academia [73]. One of the main objectives raised by skeptics against combinatorial chemistry was that in contrast to the efforts and investments in this field, none of the drugs was selected by this approach. Reports from pharmaceutical laboratories have shown that using a combinatorial chemistry approach, a potent orally active thrombin inhibitor [74], and an irreversible human rhinovirus 3C protease inhibitor [75] were designed. These are just first examples of the compounds with potential clinical use. Regularly updated information including published papers and books, as well as patents in the field of combinatorial chemistry, is available on the Internet [76].

ACKNOWLEDGMENTS

This work was supported by the Polish State Committee for Scientific Research (KBN), Grant PB 0158/T09/2000/18.

REFERENCES

1. R. B. Merrifield, *Fed. Proc., Fed. Am. Soc. Exp. Biol.* **21**, 412 (1962).
2. R. B. Merrifield, *J. Am. Chem. Soc.* **85**, 2149–2154 (1963).
3. S. Sakakibara, in *Peptides 1998, Proc. 25th Eur. Peptide Symp.*, S. Bajusz and F. Hudecz, eds., Budapest, Hungary, Aug. 30–Sept. 4, 1998, Akadémiai Kiadó, pp. 1–13.
4. M. Winter, in *Combinatorial Peptide and Nonpeptide Libraries*, G. Jung, ed., VCH, Weinheim, Germany, 1996, pp. 465–510.
5. M. Meldal, *Meth. Enzymol.* **87**, 289–283 (1997).
6. I. Sucholeiki, *Mol. Divers.* **4**, 25–30 (1998).
7. F. Albericio, M. Cases, J. Alsina, S. A. Triolo, L. A. Carpino, and S. A. Kates, *Tetrahedron Lett.* **38**, 4853–4856 (1997).
8. J. J. Gorman, *Anal. Biochem.* **136**, 397–406 (1984).
9. V. Krchňák, J. Vagner, and O. Mach, *Int. J. Peptide Protein Res.* **3**, 209–213 (1989).
10. H. M. Geysen, R. H. Meloen, and S. J. Barteling, *Proc. Natl. Acad. Sci.* (USA) **81**, 3998–4002 (1984).
11. R. Frank, W. Heikens, G. Heisterberg-Moutses, and H. Blöcker, *Nucl. Acids Res.* **11**, 4365–4377 (1983).
12. R. Frank and R. Döring, *Tetrahedron* **44**, 6031–6040 (1988).
13. R. Frank, S. Güler, S. Krause, and W. Lindenmaier, in *Peptides 1990, Proc. 21st Eur. Peptide Symp.*, E. Giralt and D. Andreu, eds., Platja d'Aro, Spain, Sept. 2–8, 1990, ESCOM, Leiden, pp. 151–152.
14. R. Frank, *Tetrahedron* **48**, 9217–9232 (1992).
15. M. Lebl, A. Stieranová, J. Eichler, and M. Bienert, in Ref. 13, pp. 153–155.
16. J. Eichler, M. Binert, A. Stieranová, and M. Lebl, *Pept. Res.* **4**, 296–307 (1991).
17. S. P. A. Fodor, J. L. Read, M. C. Pirrung, L. Stryer, A. T. Lu, and D. Solas, *Science* **251**, 767–773 (1991).
18. R. A. Houghten, *Proc. Natl. Acad. Sci.* (USA) **82**, 5131–5135 (1985).

19. R. A. Houghten, M. K. Brey, S. T. DeGraw, and C. J. Kirby, *Int. J. Peptide Protein Res.* **27**, 673–678 (1986).
20. E. Bayer and M. Mutter, *Nature* **237**, 512–513 (1972).
21. J. K. Scot and G. P. Smith, *Science* **249**, 386–390 (1990).
22. J. J. Devlin, L. C. Panganiban, and P. C. Devlin, *Science* **249**, 404–406 (1990).
23. G. P. Smith and J. K. Scott, *Meth. Enzymol.* **217**, 228–257 (1993).
24. Á. Furka, F. Sebestyén, M. Asgedom, and G. Dibó, *Int. J. Peptide Protein Res.* **37**, 487–493 (1991).
25. K. S. Lam, S. E. Salomin, E. M. Hersh, V. J. Hruby, W. M. Kazimierski, and R. J. Knapp, *Nature* **354**, 82–84 (1991).
26. R. A. Houghten, C. Pinilla, S. E. Blondelle, J. R. Appel, C. T. Dooley, and J. H. Cuervo, *Nature* **354**, 84–86 (1991).
27. M. Lebl, *J. Comb. Chem.* **1**, 3–24 (1999).
28. J. M. Ostresh, J. H. Winkle, V. T. Hamashin, and R. A. Houghten, *Biopolymers* **34**, 1681–1689 (1994).
29. J.-X. Wang, A. J. Dipasquale, A. M. Bray, N. J. Maeji, D. C. Spellmeyer, and H. M. Geysen, *Int. J. Pept. Protein Res.* **42**, 392–399 (1993).
30. S. Rodda, G. Tribbick, and M. Geysen, in Ref. 4, pp. 303–326.
31. R. Frank, S. Hoffman, M. Kieß, H. Lahmenn, W. Tegge, C. Behn, and H. Gausepohl, in Ref. 4, pp. 363–386.
32. A. Kramer, R. Volkmer-Engert, R. Malin, U. Reineke, and J. Schneider-Mergener, *Pept. Res.* **6**, 314–319 (1993).
33. W. R. Dostmann, C. Nickel, S. Thiel, I. Tsigelny, R. Frank, and W. J. Tegge, *Pharmacol. Ther.* **82**, 373–387 (1999).
34. H. Wenschuh, R. Volkmer-Engert, M. Schmidt, M. Schulz, J. Schneider-Mergener, and U. Reineke, *Biopolymers* **55**, 188–206 (2000).
35. A. Kramer, U. Reineke, L. Dong, B. Hoffmann, U. Hoffmuller, D. Winkler, R. Volkmer-Engert, and J. Schneider-Mergener, *J. Pept. Res.* **54**, 319–327 (1999).
36. K. D. Janda and H. Han, *Meth. Enzymol.* **267**, 234–247 (1996).
37. D. J. Gravert and K. D. Janda, *Curr. Opin. Chem. Biol.* **1**, 107–113 (1997).
38. P. H. Toy and K. D. Janda, *Acc. Chem. Res.* **33,** 546–554 (2000).
39. K. H. Wiesmüller, S. Feiertag, B. Fleckenstein, S. Kienle, D. Stoll, M. Herrmann, and G. Jung, in Ref. 4, pp. 203–246.
40. K. M. Ivanetich and D. V. Santi, *Meth. Enzymol.* **267**, 247–260 (1996).
41. M. Lebl, K. S. Lam, P. Kocis, V. Krchňák, M. Patek, S. E. Salomon, and V. J. Hruby, *Peptides 1992, Proc. 22nd Eur. Peptide Symp.*, C. H. Schneider and A. N. Eberle, eds., Interlaken, Switzerland, Sept. 13–19, 1992, ESCOM, Leiden, pp. 67–69.
42. C. L. Chen, P. Strop, M. Lebl, and K. S. Lam, *Meth. Enzymol.* **267**, 211–219 (1996).
43. K. S. Lam and M. Lebl, in Ref. 4, pp. 173–201.
44. D. S. King, C. G. Fields, and G. B. Fields, *Int. J. Peptide Protein Res.* **36**, 255–266 (1990).
45. R. Arshady, E. Aetherton, D. L. J. Clive, and R. C. Sheppard, *J. Chem. Soc. Perkin* **2**, 529–537 (1981).

46. E. Bayer and W. Rapp, in *Poly (Ethylene Glycol) Chemistry: Biotechnical and Biomedical Applications*, J. M. Harris, ed., Plenum Press, New York, 1992, pp. 325–345.
47. W. Rapp, in Ref. 4, pp. 425–464.
48. M. Grotli, C. H. Gotfredsen, J. Rademann, J. Buchardt, A. J. Clark, J. O. Duus, and M. Meldal, *J. Comb. Chem.* **2**, 108–119 (2000).
49. N. J. Wells, M. Devies, and M. Bradley, *J. Org. Chem.* **63**, 6430–6431 (1998).
50. N. F. Stepenov, V. Krchňák, M. Staňková, K. S. Lam, S. Wade, and M. Lebl, *Peptides 1994, Proc. 23rd Eur. Peptide Symp.*, H. L. S. Maia, ed., Braga, Portugal, Sept. 4–10, 1994, ESCOM, Leiden, pp. 467–468.
51. S. Brenner and R. A. Lerner, *Proc. Natl. Acad. Sci.* (USA) **89**, 5381–5383 (1992).
52. M. C. Needels, D. G. Jones, E. H. Tate, G. L. Heinkel, L. M. Kochersperger, W. J. Dower, R. W. Barret, and M. A. Gallop, *Proc. Natl. Acad. Sci.* (USA) **90**, 10700–10704 (1993).
53. V. Nikolaiev, A. Stierandova, V. Krchňák, B. Seligman, K. S. Lam, S. E. Salomon, and M. Lebl, *Pept. Res.* **6**, 161–170 (1993).
54. M. H. J. Ohlmeyer, R. N. Swanson, L. W. Dillard, J. C. Reader, G. Asouline, R. Kobayashi, M. Wigler, and W. C. Still, *Proc. Natl. Acad. Sci.* (USA) **90**, 10922–10926 (1993).
55. J. J. Burbaum, M. H. J. Ohlmeyer, J. C. Reader, I. Henderson, L. W. Dillard, G. Li, T. L. Randle, N. H. Sigal, D. Chelsky, and J. J. Baldwin, *Proc. Natl. Acad. Sci.* (USA) **92**, 6027–6031 (1995).
56. Z.-J. Ni, D. Maclean, C. P. Holmes, and M. A. Gallop, *Meth. Enzymol.* **267**, 261–272 (1996).
57. A. R. Vaino and K. D. Janda, *Proc. Natl. Acad. Sci.* (USA) **97**, 7692–7696 (2000).
58. A. Furka, *Comb. Chem. High Throughput Screening* **3**, 197–209 (2000).
59. E. J. Moran, S. Sarshar, J. F. Cargill, M. Shahbaz, A. Lao, A. M. M. Mjalli, and R. W. Armstrong, *J. Am. Chem. Soc.* **117**, 10787–10788 (1995).
60. J. M. Ostresh, S. E. Blondelle, B. Dörner, and R. A. Houghten, *Meth. Enzymol.* **267**, 220–234 (1996).
61. R. H. Griffey, H. An, L. L. Cummins, H. J. Gaus, B. Haly, R. Hermann, and P. D. Cook, *Tetrahedron* **54**, 4067–4076 (1998).
62. C. Pinilla, J. R. Appel, and R. H. Houghten, in Ref. 42, pp. 65–66.
63. F. Sebestyén, G. Dibó, and Á. Furka, in Ref. 42, pp. 63–64.
64. C. Pinilla, J. Appel, C. Dooley, S. Blondelle, J. Eichler, B. Dörner, J. Ostresh, and R. H. Houghten, in Ref. 4, pp. 139–171.
65. E. Erb, K. D. Janda, and S. Brenner, *Proc. Natl. Acad. Sci.* (USA) **91**, 11422–11426 (1994).
66. R. H. Houghten, J. M. Ostresh, G. M. Husar, B. Dörner, and S. E. Blondelle, in Ref. 50, pp. 459–460.
67. B. Dörner, J. M. Ostresh, G. M. Husar, and R. H. Houghten, in Ref. 50, pp. 463–464.
68. J. M. Ostresh, G. M. Husar, S. E. Blondelle, B. Dörner, P. A. Weber, and R. H. Houghten, *Proc. Natl. Acad. Sci.* (USA) **91**, 11138–11142 (1994).
69. M. Lebl, M. Patek, P. Kocis, V. Krchňák, V. J. Hruby, S. E. Salomon, and K. S. Lam, *Int. J. Pept. Protein Res.* **41**, 201–203 (1993).

70. S. E. Salomon, K. S. Lam, M. Lebl, A. Kandola, P. S. Khattri, S. Wade, M. Patek, P. Kocis, V. Krchňák, D. Thorpe, and S. Felder, *Proc. Natl. Acad. Sci.* (USA) **90**, 11708–11712 (1993).

71. C. K. Jayawickreme, G. F. Graminski, J. M. Quillan, and M. R. Lerner, *Proc. Natl. Acad. Sci.* (USA) **91**, 1614–1618 (1994).

72. J. M. Quillan, C. K. Jayawickreme, and M. R. Lerner, *Proc. Natl. Acad. Sci.* (USA) **92**, 2894–2898 (1995).

73. R. E. Dolle, *J. Comb. Chem.* **2**, 383–433 (2000).

74. S. F. Brady, K. J. Stauffer, W. C. Lumma, G. M. Smith, H. G. Ramjit, S. D. Lewis, B. J. Lukas, S. J. Gardell, E. A. Lyle, S. D. Appleby, J. J. Cook, M. A. Holahan, M. T. Stranieri, J. J. Jr. Lynch, J. H. Lin, I. W. Chen, K. Vastag, A. M. Naylor-Olsen, and J. P. Vacca, *J. Med. Chem.* **41**, 401–406 (1998).

75. P. S. Dragovich, R. Zhou, D. J. Skalitzky, S. A. Fuhrman, A. K. Patick, C. E. Ford, J. W. Meador, and S. T. Worland, *Bioorg. Med. Chem.* **7**, 589–598 (1999).

76. M. Lebl and Z. Leblova, Internet: *http://www.5z.com*

CHAPTER 13

Fourier Transform Ion Cyclotron Resonance Mass Spectrometry for Peptide Analysis

KRISTINA HÅKANSSON, TERRI L. QUENZER, ALAN G. MARSHALL, and MARK R. EMMETT

Center for Interdisciplinary Magnetic Resonance, National High Magnetic Field Laboratory, Florida State University, Tallahassee, USA

CONTENTS

13.1. INTRODUCTION

The ultimate goal in many laboratories is to analyze biological compounds from biological fluids at endogenous levels. Endogenous levels of biological analytes

Mass Spectrometry and Hyphenated Techniques in Neuropeptide Research, Edited by Jerzy Silberring and Rolf Ekman.
ISBN 0-471-35493-7

are extremely low, typically only 50–500 attomoles (amol) of total sample. Furthermore, biological analytes present themselves in complex solvents at a salt concentration greater than 100 mM [1]. The high salt concentration can inhibit analyte signal by adduct formation, and by introducing impurities that can interfere with the analyte signal. Analysis of endogenous biological analytes requires high sensitivity in order to detect such small amounts.

Fourier transform ion cyclotron resonance mass spectrometry (FT-ICR MS) [2–8] offers several advantages for the analysis of biological samples, including high mass resolution, ultrahigh mass accuracy, and wide mass range. High resolution enhances sensitivity by making it possible to distinguish between analyte and background species at or near the detection limit. Ultrahigh mass accuracy offers an alternative to tandem mass spectrometry (MS/MS) for identification, an advantage if the amount of sample is limited. Spectra are collected in a single scan over a wide mass range without loss of sensitivity. FT-ICR sensitivity is ultimately limited by broadband image current detection, which typically requires approximately 100 charges to generate a measurable signal at a given *m/z* ratio. Although ion-counting mass spectrometers can detect a single charge and are thus more sensitive, the detection limit for nano-LC microelectrospray ionization (micro-ESI) FT-ICR MS can nevertheless be extremely low. Thus, all the inherent advantages of FT-ICR provide a unique platform for the analysis of biological samples [9].

13.2. INSTRUMENTATION

The heart of the FT-ICR mass analyzer is a cell, located in a strong, homogeneous, static magnetic field. Ions are trapped inside the cell by a combination of the magnetic field and applied electric fields. Superconducting magnets of 3–11.5 T (tesla) and spatial homogeneity of a few ppm over a 10-mm sphere are usually employed. Advantages of higher magnetic fields include improved mass resolving power, signal-to-noise ratio, dynamic range, and mass accuracy [10]. However, the obvious tradeoff is that a larger magnet increases the cost of the instrument.

The most common peptide/protein ion source used for FT-ICR biological analysis is low-flow electrospray ionization (ESI) [11–16], i.e., nanospray [17], or microelectrospray [18]. The first coupling of ESI and FT-ICR was reported in 1989 by Henry et al. [19]. The FT-ICR mass analyzer benefits from the multiple charging produced by ESI [20,21], which usually results in *m/z* values below 2000. This *m/z* range is compatible with FT-ICR MS because its performance is higher for low-*m/z* ions [2]. The resolving power decreases linearly with increasing *m/z* [8], and, because of the limited size of the ICR cell, the presence of a trapping potential (see below), and the presence of other ions, the technique has an upper *m/z* limit [2]. Ions produced by matrix-assisted laser desorption ionization (MALDI) in most cases are singly charged and therefore have higher *m/z* values [22]. Also, the high initial kinetic energies of ions formed by MALDI makes it more difficult to trap the ions efficiently in the cell [23]. However, high trapping potential in combination with collisional cooling can partially solve the latter problem [24].

The electrospray ionization source is usually located outside the magnetic field, with some exceptions [25,26], and the ions are transported along the field lines into the magnet. Thus, a large magnetic field gradient must be overcome. For ions with radial velocity components, such as those produced by ESI, traversing the field gradient is difficult because the ion magnetic moment increases as the magnetic flux density becomes stronger [27]. As a result, ion radial velocity increases, so that the axial velocity must decrease in order to keep the kinetic energy constant. If the initial ratio between radial and axial velocity components is sufficiently large, the magnetic mirror effect occurs in which the ions will stop their forward motion and reverse direction without ever entering the ICR cell. In order to avoid the magnetic mirror effect, focusing optics such as radiofrequency (RF)—quadrupoles [28], hexapoles [29], or octapoles [30]—can be used for transferring the ions into the cell. An alternative solution is to accelerate the ions to high axial velocity through high-voltage electrostatic lenses [31].

A schematic diagram of a 9.4-T FT-ICR mass spectrometer, located at the National High Magnetic Field Laboratory in Tallahassee, Florida, is shown in Fig. 13.1. This instrument is equipped with a microelectrospray ion source, an octapole ion trap for external accumulation [33] of the electrosprayed ions,

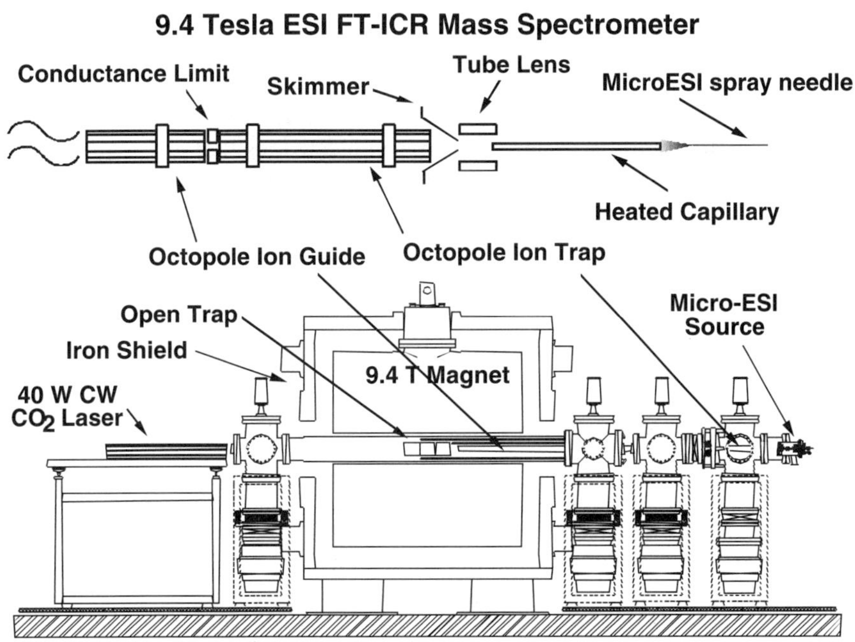

Figure 13.1. A schematic diagram of a homebuilt, passively shielded, 9.4-T FTICR mass spectrometer located at the National High Magnetic Field Laboratory in Tallahassee, Florida. (The figure is reproduced from Ref. 32.)

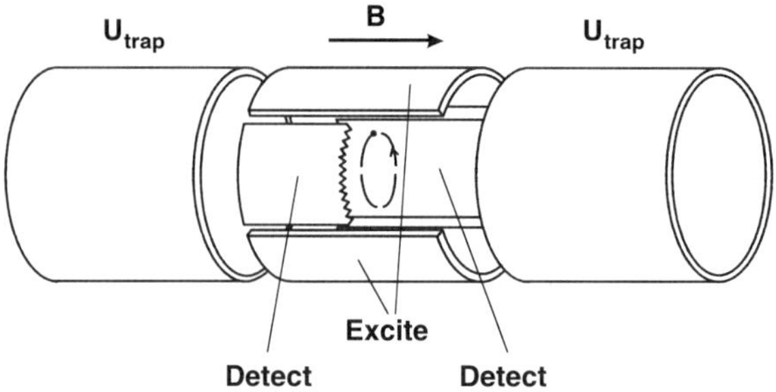

Figure 13.2. The geometry of an open cylindrical ICR trapped ion cell. The ions are trapped axially by applying a small DC voltage (U_{trap}) to each end cylinder of the cell. The cyclotron motion of an ion is indicated.

and an octapole ion guide for transferring the ions into the magnet. External accumulation increases the duty cycle of the experiment; thus a larger portion (close to 100%) of the gas-phase ions are detected since no ions are thrown away during the detection event. The CO_2 laser shown on the backside of the instrument is used to fragment ions by infrared multiphoton dissociation (IRMPD) [34]. Fragmentation can also be accomplished by sustained off-resonance irradiation (SORI) [35]. Both fragmentation techniques are described in more detail below.

The cell in which the ions are trapped can be any of many different geometries [36]. Figure 13.2 shows a cylindrical cell, a common design. The axis of the cylinder is aligned with the magnetic field, **B**, and the central envelope surface is divided into two pairs of opposed electrodes that are used to introduce RF electric excitation, and for detection (see text below). The cell shown in Fig. 13.2 has open ends. Another alternative is a closed cell in which the two end cylinders are replaced by two flat circular endplates, each with a small (~5-mm-diameter) central aperture. However, an open cell has the great advantage of improved access to the cell interior, thereby simplifying external ion injection and manipulation [37].

13.3. OPERATING PRINCIPLE

Ions entering the magnetic field of the FT-ICR mass analyzer experience a Lorentz force. As a consequence of this force, the ions move in circular orbits in the plane perpendicular to the magnetic field (the *xy* plane) (see Fig. 13.2). The angular velocity, v/*r* or ω, is given by

$$\mathrm{v}/r = \frac{qB}{m} \tag{13.1}$$

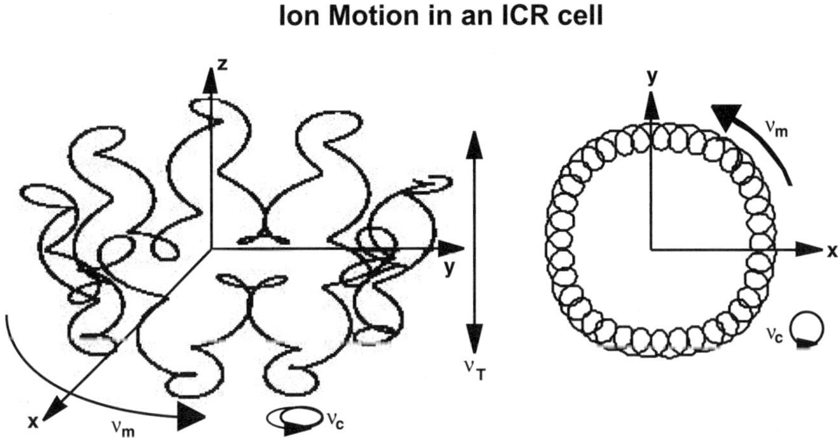

Figure 13.3. Ion motion in a 2-in. cubic ICR cell in a perfectly homogeneous magnetic field of 3T for an ion of m/z 2300, for 10V trapping voltage. (The figure is reproduced from Ref. 2.)

in which v is the ion speed, r is the radius of the circular trajectory, q is the charge of the ion, B is the magnetic flux density, and m is the mass of the ion, all in SI units. Equation (13.1) is the basic cyclotron equation [38]. The frequency, ν, in hertz is obtained by dividing Eq. (13.1) by 2π.

To prevent the ions from leaving the cell along the axial direction, a trapping potential (U_{trap}) of 1–10V is applied to the two end cylinders/endplates of the cell (Fig. 13.2). As a result, the ions undergo harmonic oscillation parallel to the z axis. The ions can be trapped in this manner for long time periods (minutes or even hours). The presence of the electric field also influences the radial motion of the ions. The shape of the electric potential results in a radially outward directed force, which opposes the inward-directed Lorentz magnetic force. As a consequence, the center of the cyclotron motion will also orbit the axis of the cell in a circular path (see Fig. 13.3). This extra orbital motion introduced by the electric field is called *magnetron motion*. The magnetron motion leads to radial diffusion of ions and sidebands, and reduces the cyclotron frequency [2,8]. The latter effect introduces a nonlinearity in the frequency-to-m/z calibration equation [39], so that ions of at least two m/z ratios are needed for calibration, rather than the single m/z value expected from Eq. (13.1). Moreover, the cyclotron motion is also affected by the space charge built up when more than one ion is present in the cell, so that an internal calibration mixture containing ions of several m/z values is optimal for calibration [2].

To obtain a mass spectrum, a radiofrequency voltage containing a band of frequencies covering the desired m/z range is applied differentially between one pair of cell plates. A common excitation waveform is a frequency sweep ("chirp"), namely, a constant-amplitude sinusoidal waveform whose

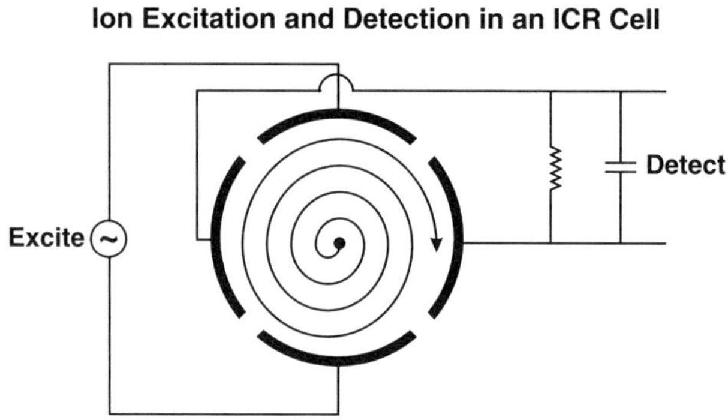

Figure 13.4. Excitation of an ion trapped inside the ICR cell. An RF voltage waveform containing the resonance frequency of the ion is applied differentially between one pair of cell plates. The other pair of plates detects an image current of the orbiting coherent ion packets.

instantaneous frequency increases linearly with time [40]. Ions undergoing cyclotron motion at these particular frequencies absorb energy and are excited coherently to higher kinetic energies so that their cyclotron radii increase [2] (see Fig. 13.4). In this manner, clouds of ions move alternately toward and away from the cell walls and a second pair of plates can be used to detect an image current from the orbiting ion clouds (Fig. 13.4). This time-domain signal contains the frequencies of all ions present in the cell. The time domain signal is converted to the frequency domain by applying a Fourier transform [40]. Figure 13.5a shows the time-domain transient detected from microelectrosprayed bovine ubiquitin with a 7-T FT-ICR instrument, and Fig. 13.5b shows the same data after Fourier transformation. The frequency scale is easily converted to an m/z scale by applying Eq. (13.1). The frequencies of the most abundant ions correspond to the m/z values of the different nuclides of the isotopic clusters of the 10^+, 11^+, 12^+, and 13^+ charge states of ubiquitin.

13.4. FT-ICR MS PERFORMANCE

13.4.1. Mass Accuracy and Resolving Power

As described above, the physical quantity measured in FT-ICR MS is frequency. Because frequency can be measured very accurately, unprecedented mass accuracy is offered. Mass measurements as accurate as 1 ppm can be achieved on a routine basis for large biomolecular ions [41,42], and sub-ppm accuracy is obtained for low-m/z (<300) ions [43].

Besides yielding unparalleled mass accuracy, FT-ICR MS also provides mass resolving power far superior to other mass analyzers. In the low-pressure limit, the resolving power is directly proportional to the duration of the time-

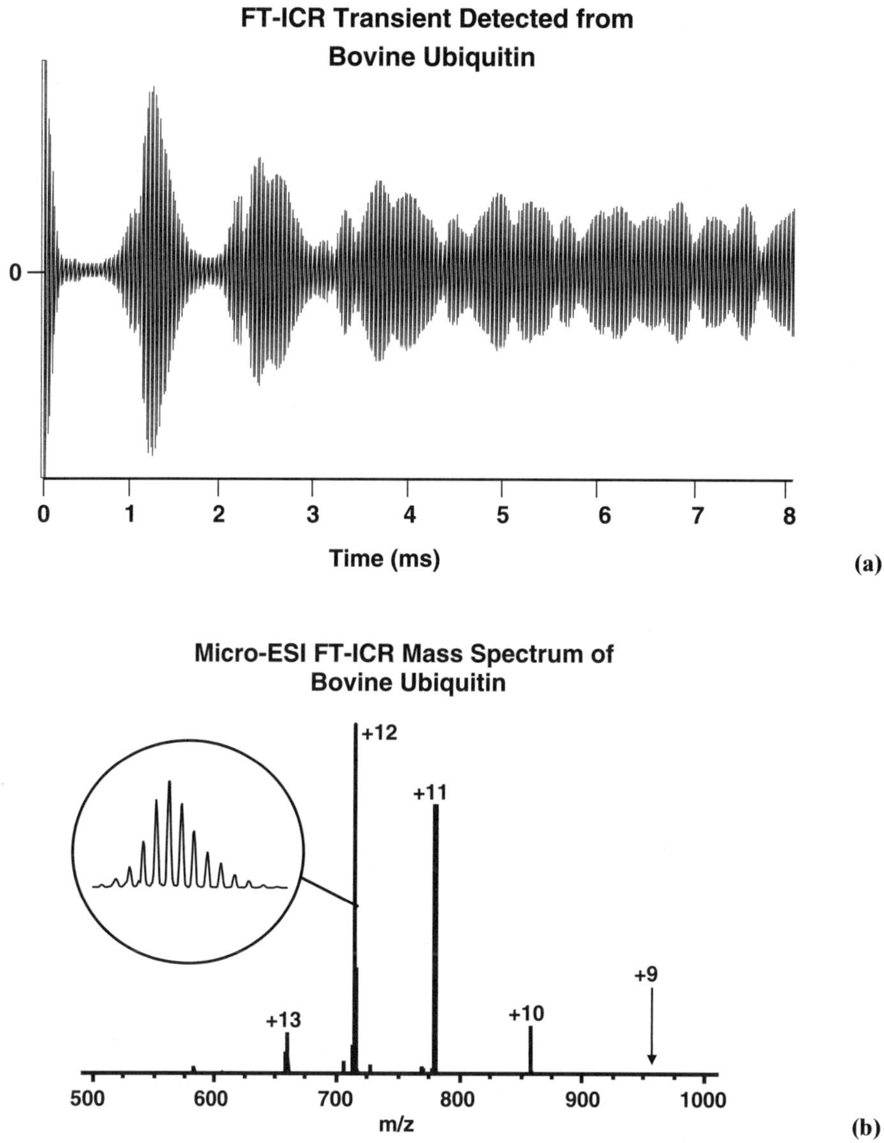

Figure 13.5. The first 8 ms of the time-domain FT-ICR transient detected from bovine ubiquitin with a 7-T FT-ICR mass spectrometer (a). The same signal is also shown after Fourier transformation and calibration to convert from time-domain to *m/z*-domain display (b). Note the high resolving power obtained with FT-ICR, manifested by the baseline isotopic resolution.

domain signal [44]. In the ultra-high-vacuum environment of the analyzer, the detection signal often lasts for several seconds and resolving power exceeding 10^6 can be obtained for large proteins. For example, Kelleher et al. have detected a 112 kDa protein with unit mass resolution [45], that is, with separation of the nuclides differing in the number of heavy stable isotopes, such as ^{13}C, ^{15}N, ^{17}O, ^{18}O, ^{2}H, ^{33}S, ^{34}S, and ^{36}S. Thus, the monoisotopic mass of a peptide or protein, specifically, the mass corresponding to the unique elemental composition containing only ^{12}C, ^{14}N, ^{16}O, ^{1}H, and ^{32}S, can be determined. Another advantage of isotopic resolution is that it allows a direct determination of the charge state, z, of an ion from the isotopic distribution [46]. The m/z difference between adjacent nuclides separated by 1 Da in mass is simply $1/z$. It is also possible to resolve isotopic fine structure with FT-ICR, such as the difference in mass that results when two ^{12}C are replaced by ^{13}C versus replacement of one ^{32}S by ^{34}S. The separation of such peaks enables the counting of individual sulfur atoms in a protein, as demonstrated by Shi et al. [47].

In addition to the long lasting transients, FT-ICR has superior resolving power over mass analyzers that rely on differences in kinetic energy for separating the ions, such as time-of-flight (TOF) or sector instruments. That is because the ion cyclotron frequency [Eq. (13.1)] is independent of the ion velocity. Stated most simply, FT-ICR MS resolving power is the number of cycles of cyclotron frequency during the detection period. In TOF and sector instruments, the spread in ion initial position, direction, and speed all limit performance.

High mass accuracy and resolving power are of crucial importance in many applications. For example, mass spectrometry is often employed in order to identify posttranslational modifications, such as phosphorylation, glycosylation, and disulfide bond formation [48–51], as described in detail elsewhere in this book. In such applications, accurate mass determination is a prerequisite to pinpoint the exact modification, whereas high resolving power is necessary to distinguish between related forms of the peptide or protein with different degrees of modification. One example in which high mass accuracy is crucial is determination of the presence or absence of a disulfide bond in a large protein, corresponding to a mass difference of only 2 Da. Another example is the determination of metal atom oxidation state in heme proteins, in which the reduced and oxidized forms of the same charge state of the protein differ by only the mass of one proton [52]. For peptide sequencing, the mass accuracy of FT-ICR MS makes it possible to directly distinguish between the amino acids glutamine and lysine, which differ by only 36 mDa. In fact, two peptides differing in mass by less than 0.0005 Da (i.e., by less than the mass of one electron) have been resolved [53].

Another area in which high mass accuracy and resolving power greatly improve the performance is the identification of proteins from complex mixtures. A common strategy, also described in detail elsewhere in this book, is gel separation (one- or two-dimensional) of the protein mixture, in-gel enzymatic digestion, peptide mass fingerprinting by mass spectrometry, and database searches based on the peptide masses [54–58]. FT-ICR greatly limits the number of candidate proteins for a given peptide, based on its high mass accuracy and resolution.

13.4.2. Sensitivity and Detection Limit

It is worth noting that the true test of sensitivity is the one in which the researcher is presented with a given total amount of a compound or mixture and is able to provide high resolution and accurate mass spectra from that limited supply of sample. Such sensitivity results are reported in terms of the *total* amount of sample presented in the experiment. Sensitivity is often reported as the amount of sample consumed from a higher initial concentration and volume during the fraction of a second that the ions are detected. However, that measure does not represent the smallest total amount of sample that can be handled. Analytes present at low concentration are a challenge in ESI FT-ICR MS because electrospray is inherently a concentration sensitive technique [59], and the limited number of gas-phase analyte ions created from a dilute sample is transported a significant distance over a 10-order-of-magnitude pressure-gradient prior to detection. Detection of 400 amol/μL (400 pM) of horse cytochrome C and horse myoglobin by constant infusion has been demonstrated by Belov et al. with an instrument equipped with an electrodynamic ion funnel for improved ion transmission [60,61].

The coupling of capillary electrophoresis (CE) to FT-ICR MS has produced impressive sensitivity in terms of total sample required. For example, Hofstadler et al. [62] detected hemoglobin from a single red blood cell containing ~450 amol of hemoglobin. Valaskovic et al. [63] achieved an even lower CE FT-ICR MS detection limit of 7 amol of human carbonic anhydrase from a single red blood cell and only 0.7 amol of equine cytochrome C. However, a disadvantage of CE is that the analyte must be relatively concentrated, and then only a subnanoliter volume is needed to obtain a low detection limit.

Liquid chromatography (LC) is appropriate for the analysis of samples that are too dilute for CE. Kennedy and Jorgenson opened a new pathway to high-sensitivity bioanalysis with their pioneering work in nano-LC [64], enabling analyses at biological concentrations. Nano-LC also improves sensitivity with its ability to preconcentrate the sample and desalt online [65] prior to sample introduction into the mass spectrometer.

Microelectrospray ionization (micro-ESI), first described by Emmett and Caprioli in 1993 [66], coupled nano-LC with ESI. Micro-ESI is more compatible with nano-LC than conventional ESI due to the lower (nL/min) flow rate. Increased sensitivity of micro-ESI relative to conventional ESI is attributed to a narrower spray dispersion of smaller droplets, thereby increasing both sampling and desolvation efficiencies [18]. Thus, more analyte (and less solvent) enters the mass spectrometer.

The sensitivity of FT-ICR MS improved by interfacing nano-LC and micro-ESI [67]. Although that performance represented a 2×10^6-fold improvement over previously reported sensitivity [68], it still ranked approximately one order of magnitude below the sensitivity required to analyze biological samples at endogenous levels in biological solvents. Several modifications to a nano-LC micro-ESI FT-ICR MS system have further improved the sensitivity [29]. Shortening the external accumulation octapole allows for a shorter ion accumulation period, thereby improving the efficiency of ion transfer to the cell. Accumulating one batch of ions in the octopole while detecting the prior batch in the ICR cell provides a duty cycle of nearly 100% [33]. Improvements in the nano-LC micro-ESI emitter design [69] have increased durability and minimized plugging of the nano-LC column. Finally, incorporating new reversed-phase resins that require no ion-pairing agent has reduced the number of background peaks due to impurities associated with the ion pairing agent.

These modifications have resulted in the detection of subfemtomolar levels of peptides with FT-ICR MS. Quenzer et al. [29] demonstrated the reproducible detection of peptides in water with total amounts loaded onto the column as low as 500 amol for peptide mixtures and only 100 amol for a single peptide (Fig. 13.6). The peptide peaks were resolved from other background peaks and were identified on the basis of exact mass and exact spacing of the isotope peaks. Quenzer et al. [29] also demonstrated the reproducible detection of peptides in artificial cerebrospinal fluid (aCSF) with total amounts loaded onto the column as low as 500 amol for peptide mixtures and 300 attomoles for a single peptide (Fig. 13.7). Analysis in aCSF, chosen to represent a true biological solvent, lowers the signal-to-noise ratio relative to water because the high salt concentration (>100 mM) and impurities present in aCSF interfere with the analyte signal, especially at low analyte concentration. Even at a lower signal-to-noise ratio, the peptide peaks are resolved from other background peaks, enabling the peptides to be identified on the basis of accurate mass and spacing between adjacent nuclides in the isotopic distribution. Martin et al. [70] used nanoelectrospray to detect a total of 10 amol of a cytochrome *c* tryptic peptide from a protein mixture. Again, FT-ICR MS enabled identification based on accurate mass and peak spacing.

The attomole sensitivity presented by Quenzer et al. [29] represents the total amount of sample loop-injected onto the nano-LC column, eluted, and subsequently detected. Thus the sensitivity is sufficient to detect biological analytes at endogenous levels in biological solvents. High sensitivity, coupled with high resolution, exact mass measurement, and wide mass range, make FT-ICR MS an attractive tool for the analysis of biological analytes.

13.5. TANDEM MASS SPECTROMETRY

13.5.1. Precursor Ion Isolation

Isolation of one or several precursor ions for tandem mass spectrometry (MS/MS) is achieved by resonant excitation of the unwanted ions in a manner

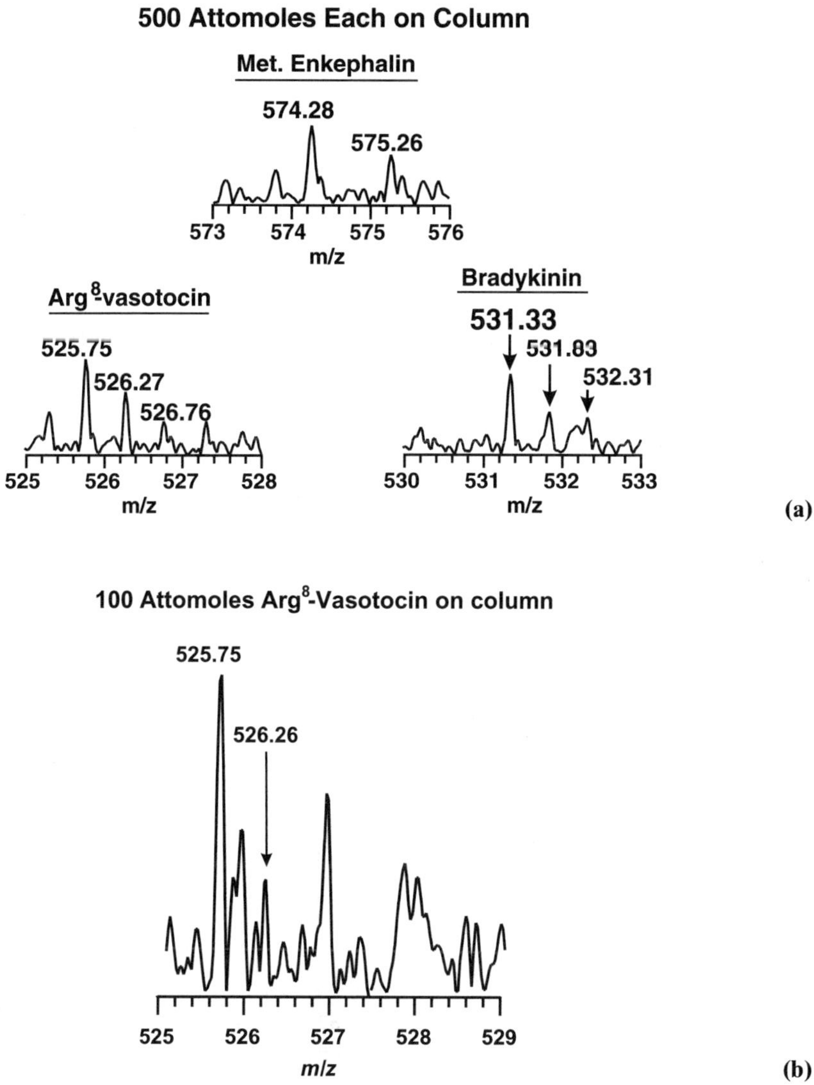

Figure 13.6. Isocratic elution of a total of 500 amol each of an equimolar mixture of the peptides Arg^8–vasotocin, methionine enkephalin, and bradykinin (50 amol/μL in water, 10 μL) (a) and a total of only 100 amol Arg^8–vasotocin (100 amol/μL in water, 1 μL) (b). The samples were loaded onto the column (Pharmacia Source 15RPC packed micro-ESI spray emitter), washed with 2% methanol, then eluted with 75% methanol, 0.25% acetic acid. (These diagrams are reproduced from Ref. 29.)

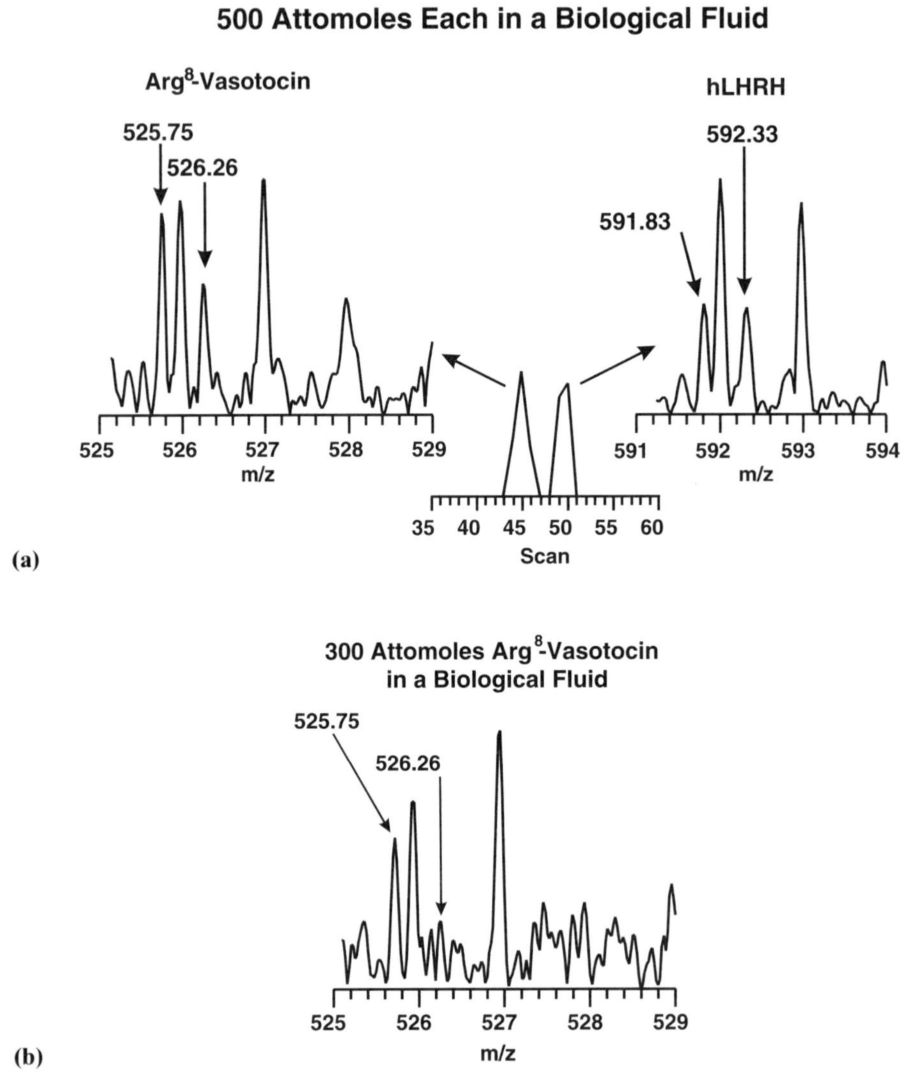

Figure 13.7. Isocratic elution of 500 amol each of an equimolar mixture of the peptides, Arg^8–vasotocin, and hLHRH ([500 amol/μL in artificial cerebrospinal fluid (aCSF)], 1 μL) (a) and a total of 300 amol Arg^8–vasotocin in aCSF (300 amol/μL, 1 μL) (b). The samples were loaded onto the column (Pharmacia Source 15RPC packed micro-ESI spray emitter), washed with 2% methanol, then eluted with 75% methanol, 0.25% acetic acid. The reconstructed ion chromatogram is included in (a). (These diagrams are reproduced from Ref. 29.)

very similar to the excitation of the trapped ions preceding image current detection. The difference is that in the case of ion isolation, ions of all *m/z* values except those to be kept for further fragmentation are excited; thus the excitation signal should contain all frequencies *except* those of the ions to be isolated. Also, the excitation signal amplitude is higher so that the unwanted ions are excited all the way to the cell walls (i.e., the ions are removed from the cell). One efficient method for defining such a signal is stored waveform inverse Fourier transform (SWIFT), introduced by Marshall et al. in 1985 [71]. The desired SWIFT excitation magnitude profile is defined in the frequency (*m/z*) domain and that waveform is then converted to the time-domain by applying an inverse Fourier transform [72].

13.5.2. Sustained Off-Resonance Irradiation Collisionally Induced Dissociation

A common strategy for ion activation in mass spectrometry is by collisions with a nonreactive gas [73]. Sustained off-resonance irradiation collisionally induced dissociation (SORI-CID) [35] is the preferred tandem FT-ICR technique for collisionally activated mass-selective fragmentation. The isolated parent ions are subjected to low-energy (<10-eV) collisions with an inert gas leaked into the ICR cell. The translational energy of the parent ions is increased by exciting them slightly off-resonance. As a result of off-resonance excitation, the ions will undergo acceleration–deceleration cycles throughout the irradiation period, so that the ion cyclotron radius increases and decreases periodically. Gas collisions during the acceleration cycles increase the ion internal energy. Advantages of off-resonance compared to on-resonance irradiation are (1) a longer timescale (several milliseconds instead of microseconds) is allowed for excitation because the repeated decrease in ion cyclotron radius will ensure that the ions do not leave the cell, (2) a larger number of collisions can occur, and (3) the fragment ions are created close to the center of the cell simplifying excitation and detection of the fragment ions.

SORI-CID is a rather "soft" fragmentation technique. In peptides, cleavage of the peptide amide bonds, or loss of small neutrals, such as water and ammonia, are typically observed. An example of a SORI-CID spectrum of melittin from bee venom obtained on a commercial 9.4-T FT-ICR instrument [74] is shown in Fig. 13.8. Ten doubly protonated *y* ions [75] could be identified, giving two peptide sequence tags of two and six amino acids, respectively.

13.5.3. Infrared Multiphoton Dissociation

An alternative to SORI-CID is infrared multiphoton dissociation (IRMPD) [34]. In IRMPD, the isolated precursor ions are irradiated with infrared photons that slowly (millisecond timescale) heat them to their dissociation threshold. A 10.6-μm CO_2 laser is usually employed to provide the photons. The main advantage of IRMPD compared to SORI-CID is that there is no

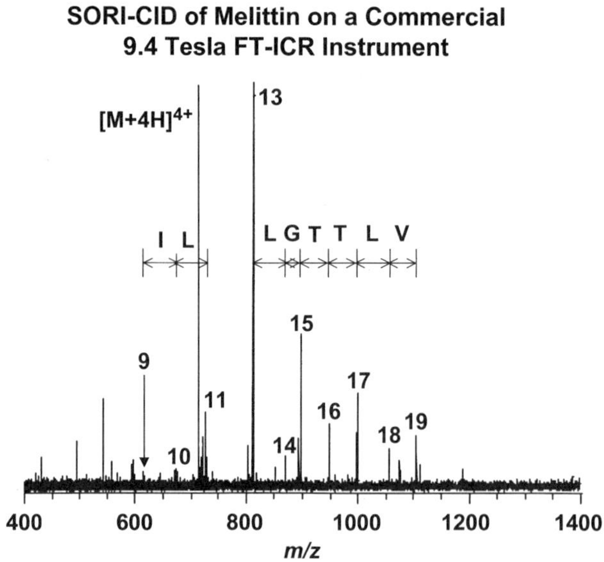

Figure 13.8. Sustained off-resonance irradiation fragment spectrum of melittin obtained on a commercial 9.4-T FT-ICR mass spectrometer; 10 doubly protonated *y* ions can be identified, marked with their corresponding numbers in the figure. The fragment ions give two peptide sequence tags of two and six amino acids, respectively. (Data kindly provided by Per Håkansson.)

need to pulse gas into the mass analyzer, thereby degrading the high vacuum and thus reducing analyzer performance. IRMPD also permits faster fragmentation analysis because no time is required for pumpdown before excitation and detection of the fragment ions. Finally, IRMPD has no mass discrimination, whereas SORI requires irradiation near the ICR frequency of the *m/z* of interest, and can thus be problematic if the spectrum is congested.

IRMPD usually provides fragment spectra that are similar to the corresponding SORI-CID spectra. An IRMPD fragment ion spectrum of human luteinizing hormone releasing hormone obtained on a 7-T FT-ICR mass spectrometer is shown in Fig. 13.9. All peptide amide bonds, except the ones closest to the two peptide termina, are cleaved, resulting in *b*- and *y*-ion series. A sequence tag of six amino acid residues is obtained from the fragment spectrum.

13.5.4. Electron-Capture Dissociation

Electron-capture dissociation (ECD) [76–78] is a fairly new fragmentation technique offering several unique features and advantages [79–83]. The characteristics of ECD are described in detail elsewhere in this book.

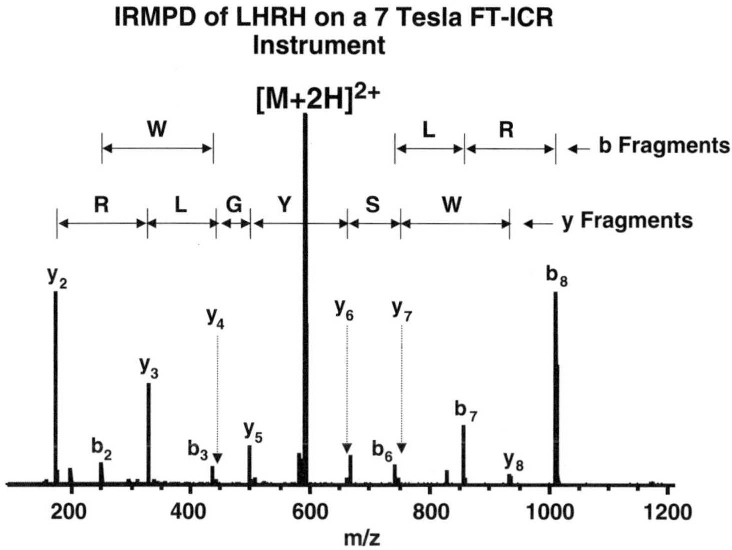

Figure 13.9. Infrared multiphoton dissociation fragment ion spectrum of human luteinizing hormone-releasing hormone obtained on a 7-T FT-ICR mass spectrometer. Both *b*- and *y*-ion series [75] are observed, giving a sequence tag of six amino acids.

Until 1999 or so, the main drawback of ECD has been that only moderate sensitivity could be obtained. Sample concentrations of 2–20 μM were needed [82,84] and the low efficiency of ECD in its simplest experimental configuration to convert parent ions into product ions has been suggested to be "a problem for very limited sample sizes" [85]. However, more recent work by Håkansson et al. has shown that reproducible ECD fragment ion spectra of peptides can be achieved from solutions microelectrosprayed at a concentration as low as 10 nM (10 fmol/μL), an improvement by a factor of 200–2000 [86]. Examples of such spectra, obtained from substance P and neurotensin on a 7-T FT-ICR mass spectrometer, are shown in Fig. 13.10. All expected peptide backbone amine bond cleavages are observed, except on the *N*-terminal side of proline. Such cleavages are not expected because of the cyclic structure of proline [76].

13.5.5. Other Fragmentation Techniques

Surface-induced dissociation (SID) [87], blackbody infrared radiative dissociation (BIRD) [88], and ultraviolet (UV) photodissociaton [89–90] are other fragmentation techniques available with FT-ICR MS. Although those techniques will not be discussed further in this text, further information can be found in the aforementioned references.

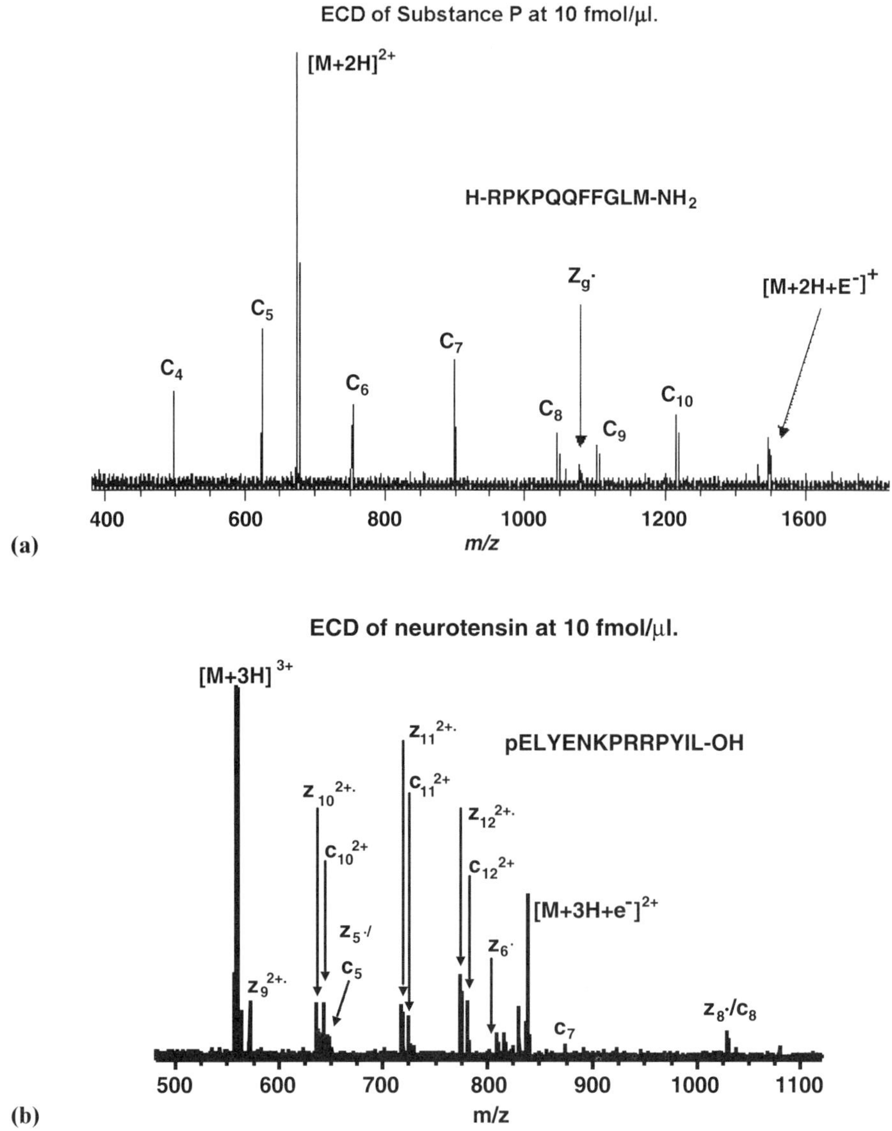

Figure 13.10. Electron-capture dissociation fragment FT-ICR mass spectra obtained from 10-fmol/μL samples of microelectrosprayed substance P (a) and neurotensin (b). The observed fragment ions correspond to cleavages at all peptide backbone amine bonds, except for the *N*-terminal side of proline. (The figure is reproduced from Ref. 86.)

13.5.6. MS^n Analysis

The number of steps of precursor ion isolation and fragmentation is, in theory, unlimited with FT-ICR MS, allowing for MS^n analysis. In practice, the magnitude of the product ions compared to the parent ion is usually not higher than 10–50%; thus, a limit will be reached at which the desired precursor ion magnitude is too low to produce any observable fragment ions. However, the seemingly small increase of the number of MS cycles from two (MS/MS) to three (MS^3) can drastically increase the amount of structural information that can be achieved from MS analysis [91].

13.6. SUMMARY

Fourier transform ion cyclotron resonance mass spectrometry has several advantages over other mass spectrometric techniques, including ultrahigh resolving power (>10^6 for large biomolecular ions) and mass accuracy (1 ppm or better). In addition, attomole sensitivity has been demonstrated with nano-LC micro-ESI FT-ICR MS. FT-ICR MS also offers several fragmentation techniques for tandem mass spectrometry. In particular, electron-capture dissociation has been shown to provide extensive peptide sequence coverage from only femtomoles of sample. The combination of all these features makes FT-ICR MS a unique platform for biological analysis.

ACKNOWLEDGMENTS

This work was supported by the NSF National High Field FT-ICR Facility (CHE-99-09502), Florida State University, the National High Magnetic Field Laboratory, and the Swedish Foundation for International Cooperation in Research and Higher Education (STINT). We thank Per Håkansson for providing the SORI-CID spectrum shown in Fig. 13.8.

REFERENCES

1. M. R. Emmett, P. E. Andren, and R. M. Caprioli, *J. Neurosci. Meth.* **62**, 141–147 (1995).
2. A. G. Marshall, C. L. Hendrickson, and G. S. Jackson, *Mass Spectrom. Rev.* **17**, 1–35 (1998).
3. M. B. Comisarow and A. G. Marshall, *Chem. Phys. Lett.* **25**, 282–283 (1974).
4. K.-P. Wanczek, *Int. J. Mass Spectrom. Ion Process.* **95**, 1–38 (1989).
5. N. M. M. Nibbering, *Analyst* **117**, 289–293 (1992).
6. A. G. Marshall and L. Schweikhard, *Int. J. Mass Spectrom. Ion Process.* **118/119**, 37–70 (1992).

7. T. Dienes, S. J. Pastor, S. Schurch, J. R. Scott, J. Yao, S. Cui, and C. L. Wilkins, *Mass Spectrom. Rev.* **15**, 163–211 (1996).
8. I. J. Amster, *J. Mass Spectrom.* **31**, 1325–1337 (1996).
9. A. G. Marshall, *Int. J. Mass Spectrom.* **200**, 331–356 (2000).
10. A. G. Marshall and S. Guan, *Rapid Commun. Mass Spectrom.* **10**, 1819–1823 (1996).
11. C. K. Meng, M. Mann, and J. B. Fenn, *Z. Phys. D* **10**, 361–368 (1988).
12. J. B. Fenn, M. Mann, C. K. Meng, S. F. Wong, and C. M. Whitehouse, *Science* **246**, 64–71 (1989).
13. J. B. Fenn, M. Mann, C. K. Meng, S. F. Wong, and C. M. Whitehouse, *Mass Spectrom. Rev.* **9**, 37–70 (1990).
14. R. D. Smith, J. A. Loo, C. G. Edmonds, C. J. Barinaga, and H. R. Udseth, *Anal. Chem.* **62**, 882–899 (1990).
15. R. D. Smith, J. A. Loo, R. R. Ogorzalek-Loo, M. Busman, and H. R. Udseth, *Mass Spectrom. Rev.* **10**, 359–451 (1991).
16. S. J. Gaskell, *J. Mass Spectrom.* **32**, 677–688 (1997).
17. M. Wilm and M. Mann, *Int. J. Mass Spectrom. Ion Process.* **136**, 167–180 (1994).
18. M. R. Emmett and R. M. Caprioli. *J. Am. Soc. Mass Spectrom.* **5**, 605–613 (1994).
19. K. D. Henry, E. R. Williams, B.-H. Wang, F. W. McLafferty, J. Shabanowitz, and D. F. Hunt, *Proc. Natl. Acad. Sci.* (USA) **86**, 9075–9078 (1989).
20. C. L. Hendrickson and M. R. Emmett, *Annu. Rev. Phys. Chem.* **50**, 517–536 (1999).
21. S. A. Lorentz, E. P. Maziarz, and T. D. Wood, *Appl. Spectrosc.* **53**. 18A–36A (1999).
22. M. Karas and F. Hillenkamp, *Anal. Chem.* **60**, 2299–2301 (1988).
23. J. A. Castro, C. Koster, and C. Wilkins, *Rapid Commun. Mass Spectrom.* **6**, 239–241 (1992).
24. R. T. McIver, Y. Z. Li, and R. L. Hunter, *Int. J. Mass Spectrom. Ion Process.* **132**, L1–L7 (1994).
25. S. A. Hofstadler and D. A. Laude, *J. Am. Soc. Mass Spectrom.* **3**, 615–623 (1992).
26. Z. Q. Guan, V. L. Campell, J. J. Drader, C. L. Hendrickson, and D. A. Laude, *Rev. Sci. Instrum.* **66**, 4507–4515 (1995).
27. D. A. Laude, E. Stevenson, and J. M. Robinson, in *Electrospray Ionization Mass Spectrometry: Fundamentals, Instrumentation & Applications*, R. B. Cole, ed., Wiley, New York, 1997, pp. 291–319.
28. D. F. Hunt, J. Shabanowitz, R. T. McIver, R. L. Hunter, and J. E. P. Syka, *Anal. Chem.* **57**, 765–768 (1985).
29. T. L. Quenzer, M. R. Emmett, C. L. Hendrickson, P. H. Kelly, and A. G. Marshall, *Anal. Chem.* **73**, 1721–1725 (2001).
30. A. G. Marshall and S. H. Guan, *Nucl. Instrum. Meth. Phys. Res. A* **363**, 397–405 (1995).
31. P. Kofel, M. Allemann, H. P. Kellerhals, and K. P. Wanczek, *Int. J. Mass Spectrom. Ion Process.* **65**, 97–103 (1985).
32. M. A. Freitas, C. L. Hendrickson, M. R. Emmett, and A. G. Marshall, *Int. J. Mass Spectrom.* **185/186/187**, 565–575 (1999).
33. M. W. Senko, C. L. Hendrickson, M. R. Emmett, S. D. H. Shi, and A. G. Marshall, *J. Am. Soc. Mass Spectrom.* **8**, 970–976 (1997).

34. D. P. Little, J. P. Speir, M. W. Senko, P. B. O'Connor, and F. W. McLafferty, *Anal. Chem.* **66**, 2809–2815 (1994).
35. J. W. Gauthier, T. R. Trautman, and D. B. Jacobson, *Anal. Chim. Acta* **246**, 211–225 (1991).
36. S. H. Guan and A. G. Marshall, *Int. J. Mass Spectrom. Ion Process.* **146**, 261–296 (1995).
37. S. C. Beu and D. A. Laude, *Int. J. Mass Spectrom. Ion Process.* **112**, 215–230 (1992).
38. E. O. Lawrence and M. S. Livingston, *Phys. Rev.* **40**, 19–35 (1932).
39. E. B. Ledford Jr., D. L. Rempel, and M. L. Gross, *Anal. Chem.* **56**, 2744–2748 (1984).
40. A. G. Marshall and F. R. Verdun, *Fourier Transforms in NMR, Optical, and Mass Spectrometry: A User's Handbook*, Elsevier, Amsterdam, 1990.
41. M. K. Green, M. M. Vestling, M. V. Johnston, and B. S. Larsen, *Anal. Biochem.* **260**, 204–211 (1998).
42. J. E. Bruce, G. A. Anderson, J. Wen, R. Harkewicz, and R. D. Smith, *Anal. Chem.* **71**, 2595–2599 (1999).
43. R. P. Rodgers, E. N. Blumer, M. A. Freitas, and A. G. Marshall, *Environ. Sci. Technol.* **34**, 1671–1678 (2000).
44. A. G. Marshall, M. B. Comisarow, and G. Parisod, *J. Chem. Phys.* **71**, 3343–4444 (1979).
45. N. L. Kelleher, M. W. Senko, M. M. Siegel, and F. W. McLafferty, *J. Am. Chem. Soc.* **8**, 380–383 (1997).
46. K. D. Henry and F. W. McLafferty, *Org. Mass Spectrom.* **25**, 490–492 (1990).
47. S. D. H. Shi, C. L. Hendrickson, and A. G. Marshall, *Proc. Natl. Acad. Sci.* (USA) **95**, 11532–11537 (1998).
48. K. A. Resing and N. G. Ahn, *Meth. Enzymol.* **283**, 29–44 (1997).
49. J. Qin and B. T. Chait, *Anal. Chem.* **69**, 4002–4009 (1997).
50. B. Kuster and M. Mann, *Curr. Opin. Struct. Biol.* **8**, 393–400 (1998).
51. D. L. Smith and Z. R. Zhou, *Meth. Enzymol.* **193**, 374–389 (1990).
52. F. He, C. L. Hendrickson, and A. G. Marshall, *J. Am. Soc. Mass Spectrom.* **11**, 120–126 (2000).
53. F. He, C. L. Hendrickson, and A. G. Marshall, *Anal. Chem.* **73**, 647–650 (2001).
54. M. Mann, P. Hojrup, and P. Roepstorff, *Biol. Mass Spectrom.* **22**, 338–345 (1993).
55. J. S. Cottrell, *Pept. Res.* **7**, 115–124 (1994).
56. S. D. Patterson and R. Aebersold, *Electrophoresis* **16**, 1791–1814 (1995).
57. J. Qin, D. Fenyö, Y. M. Zhao, W. W. Hall, D. M. Chao, C. J. Wilson, R. A. Young, and B. T. Chait, *Anal. Chem.* **69**, 3995–4001 (1997).
58. G. Li, M. Waltham, N. L. Anderson, E. Unsworth, A. Treston, and J. N. Weinstein, *Electrophoresis* **18** (1997) 391–402.
59. P. Kebarle and L. Tang, *Anal. Chem.* **65**, A972–A986 (1993).
60. M. E. Belov, M. V. Gorshkov, H. R. Udseth, G. A. Anderson, and R. D. Smith, *Anal. Chem.* **72**, 2271–2279 (2000).
61. M. E. Belov, M. V. Gorshkov, H. R. Udseth, G. A. Anderson, A. V. Tolmachev, D. C. Prior, R. Harkewicz, and R. D. Smith, *J. Am. Soc. Mass Spectrom.* **11**, 19–23 (2000).

62. S. A. Hofstadler, J. C. Severs, R. D. Smith, F. D. Swanek, and A. G. Ewing, *Rapid Comm. Mass Spectrom.* **10**, 919–922 (1996).
63. G. A. Valaskovic, N. L. Kelleher, and F. W. McLafferty, *Science* **273**, 1199–1202 (1996).
64. R. T. Kennedy and J. W. Jorgenson, *Anal. Chem.* **61**, 1128–1135 (1989).
65. D. C. Moon, and J. A. Kelly, *Biomed. Mass Spectrom.* **17**, 229–237 (1988).
66. M. R. Emmett and R. M. Caprioli, *Proc. 41st ASMS Conf. Mass Spectrometry and Allied Topics,* San Francisco, CA, 1993, p. 47.
67. M. R. Emmett, F. M. White, C. L. Hendrickson, S. D.-H. Shi, and A. G. Marshall, *J. Am. Soc. Mass Spectrom.* **9**, 333–340 (1998).
68. C. C. Stacey, G. H. Kruppa, C. H. Watson, J. Wronka, F. H. Laukien, J. F. Banks, and C. M. Whitehouse, *Rapid Commun. Mass Spectrom.* **8**, 513–516 (1994).
69. J. P. Quinn, M. R. Emmett, and A. G. Marshall, *Proc. 46th ASMS Conf. Mass Spectrometry and Allied Topics*, Orlando, FL, 1998, p. 1388.
70. S. E. Martin, J. Shabanowitz, D. F. Hunt, and J. A. Marto, *Anal. Chem.* **72**, 4266–4274 (2000).
71. A. G. Marshall, T.-C. L. Wang, and T. L. Ricca, *J. Am. Chem. Soc.* **107**, 7893–7897 (1985).
72. S. Guan and A. G. Marshall, *Int. J. Mass Spectrom. Ion Process.* **157/158**, 5–37 (1996).
73. R. N. Hayes and M. L. Gross, *Meth. Enzymol.* **193**, 237–263 (1990).
74. K. Hakansson, J. Axelsson, M. Palmblad, and P. Hakansson, *J. Am. Soc. Mass Spectrom.* **11**, 210–217 (2000).
75. P. Roepstorff and J. Fohlman, *Biomed. Mass Spectrom.* **11**, 601–601 (1984).
76. R. A. Zubarev, N. L. Kelleher, and F. W. McLafferty, *J. Am. Chem. Soc.* **120**, 3265–3266 (1998).
77. J. Axelsson, M. Palmblad, K. Hakansson, and P. Hakansson, *Rapid Commun. Mass Spectrom.* **13**, 474–477 (1999).
78. R. A. Zubarev, D. M. Horn, E. K. Fridriksson, N. L. Kelleher, N. A. Kruger, M. A. Lewis, B. K. Carpenter, and F. W. McLafferty, *Anal. Chem.* **72**, 563–573 (2000).
79. R. A. Zubarev, N. A. Kruger, E. K. Fridriksson, M. A. Lewis, D. M. Horn, B. K. Carpenter, and F. W. McLafferty, *J. Am. Chem. Soc.* **121**, 2857–2862 (1999).
80. N. A. Kruger, R. A. Zubarev, B. K. Carpenter, N. L. Kelleher, D. M. Horn, and F. W. McLafferty, *Int. J. Mass Spectrom.* **182/183**, 1–5 (1999).
81. N. L. Kelleher, R. A. Zubarev, K. Bush, B. Furie, B. C. Furie, F. W. McLafferty, and C. T. Walsh, *Anal. Chem.* **71**, 4250–4253 (1999).
82. E. Mirgorodskaya, P. Roepstorff, and R. A. Zubarev, *Anal. Chem.* **71**, 4431–4436 (1999).
83. A. Stensballe, O. Norregaard-Jensen, J. V. Olsen, K. F. Hasselman, and R. A. Zubarev, *Rapid Commun. Mass Spectrom.* **14**, 1793–1800 (2000).
84. D. M. Horn, Y. Ge, and F. W. McLafferty, *Anal. Chem.* **72**, 4778–4784 (2000).
85. N. A. Kruger, R. A. Zubarev, D. M. Horn, and F. W. McLafferty, *Int. J. Mass Spectrom.* **185/186/187**, 787–793 (1999).
86. K. Hakansson, M. R. Emmett, C. L. Hendrickson, and A. G. Marshall *Anal. Chem.* **73**, 3605–3610 (2001).

87. G. Tsaprailis, H. Nair, A. Somogyi, V. H. Wysocki, W. Zhong, J. H. Futrell, S. G. Summerfield, and S. J. Gaskell, *J. Am. Chem. Soc.* **121**, 5142–5154 (1999).

88. W. D. Price, P. D. Schnier, and E. R. Williams, *Anal. Chem.* **68**, 859–866 (1996).

89. W. D. Bowers, S.–S. Delbert, R. L. Hunter, and R. T. McIver, *J. Am. Chem Soc.* **106**, 7288–7289 (1984).

90. E. R. Williams, J. J. P. Furlong, and F. W. McLaffery, *J. Am. Soc. Mass Spectrom.* **1**, 288–294 (1990).

91. T. Solouki, B. B. Reinhold, C. E. Costello, M. O'Malley, S. H. Guan, and A. G. Marshall, *Anal. Chem.* **70**, 857–864 (1998).

CHAPTER 14

Peptide Sequencing Promoted by *N*-Terminal Derivatization

DOUGLAS F. BAROFSKY

Department of Chemistry, Environmental Health Sciences Center, Oregon State University, Corvallis, Oregon, USA

TSE-FANG CHEN

Department of Microbiology, Oregon State University, Corvallis, Oregon, USA

CONTENTS

14.1. INTRODUCTION

Knowledge of a protein's primary structure is a prerequisite for understanding its higher levels of organization and, ultimately, its biological function [1]. The earliest attempt to sequence peptides by mass spectrometric means apparently dates back to 1959 [2]. After that beginning however, progress remained limited until fast-atom bombardment (FAB) was introduced in the early 1980s [3–6] as a method for ionizing peptides, and tandem mass spectrometric techniques became available in the late 1980s [7–10] for analyzing peptides ionized by FAB. Development moved so rapidly from that point that by the mid-1990s spectrometric analyses using electrospray ionization (ESI) [11–13] and matrix-assisted laser desorption/ionization (MALDI) [14–18] in combination with a variety of chemical and biochemical methods had become essential tools for

Mass Spectrometry and Hyphenated Techniques in Neuropeptide Research, Edited by Jerzy Silberring and Rolf Ekman.
ISBN 0-471-35493-7

directly probing the sequence and structural fidelity of the amino acid residues that compose a protein. The sequencing information can be specific enough to identify the protein's accession number in a genomic database, and the structural characterization of individual amino acids can reveal the presence, location, and identity of mutations or post-translational modifications in the protein's primary structure.

By and large, peptides remain intact when they are ionized by electrospray or by laser desorption from a matrix. In order to obtain sequencing or other structural information from a particular protonated peptide, it must be separated according to mass from the other charged species produced in the ionization process and then, by some mechanism that increases the ion's internal energy, induced to dissociate into charged fragments that can be identified in a succeeding, tandem mass spectrometric analysis. Currently, the most common way to cause an ionized peptide to fragment into sequence-specific product ions is to subject it to low- or high-energy, gas-phase, collision-induced dissociation (CID) [10,19–21]. Electrosprayed ions are routinely analyzed by low-energy CID on triple quadrupole [19], ion trap [21], or most recently, hybrid quadrupole time-of-flight (Q-TOF) [22–27] mass spectrometers. Laser desorbed ions are usually analyzed at present by postsource decay (PSD) [28–33] on time-of-flight (TOF) mass spectrometers, but analysis by high-energy CID on tandem time-of-flight (TOF/TOF) [34–42] or by low-energy CID on Q-TOF [43–47] instruments has also become possible.

Almost immediately after it became possible to perform high-energy CID analyses on peptides ionized by FAB, derivatization methods aimed at increasing sensitivity or simplifying data interpretation began to appear. Derivatization procedures intended to improve sensitivity were based for the most part either on increasing surface activities [48–50] or generating preformed ions [51] in FAB matrices. Derivatization procedures developed to facilitate interpretation focused on creating structures capable of channeling the fragmentation of peptides into structurally diagnostic pathways. The preponderance of these studies devoted to FAB and related desorption/ionization phenomena were performed between the mid-1980s and mid-1990s. For the purposes of this chapter, it will suffice to summarize the major findings of this era; readers interested in a full account of this large body of work are referred to the comprehensive review published recently by Roth et al. [52].

Early in the FAB era of peptide analysis, it was recognized that the positive charge in protonated peptides is generally localized on basic amino acids and, furthermore, that the extent of charge localization influences the types of fragment ions produced [53]. Specifically, it was found that a basic amino acid located close to the *N* terminus or the *C* terminus of a peptide tends to promote formation of a series of sequence-specific fragment ions in which the charge is retained respectively on the N-termini (i.e., a_n, b_n, c_n, and d_n ions) or the *C* termini (i.e., v_n, w_n, x_n, y_n, and z_n ions) of the fragments (conventional nomenclature [54,55] is used throughout this paper to designate ion fragments from nonderivatized peptides). It was subsequently inferred that these structurally diagnostic fragmentation patterns could by and large be rationalized either on the basis of a charge-directed [9] or a charge-remote [56–58] mechanism and, furthermore, that derivatives with a fixed charge direct fragmentation more effectively than do derivatives with elevated proton affinities [52]. Most of the endeavors during this period were devoted to *N*-terminal derivatives, but a few efforts to develop *C*-terminal derivatives were also made [59–62]. In general, *N*-terminal derivatization has proved to be more practicable than *C*-terminal derivatization because (1) the difference in p*K* between the *N* terminus and the ε-amino group on a lysine is large enough to allow selective derivatization whereas the differences in p*K* between the *C* terminus and acidic sidechains are not and (2) fewer types of fragment ions and, consequently, more easily interpretable mass spectra, result from *N*-terminal derivatives than *C*-terminal derivatives. Roth et al. summed up the ideal method for charge derivatization as one in which a high-yielding chemistry rapidly proceeds, without modifying peptide side-chains, to a derivative with a fixed charge that (1) does not require purification prior to analysis, (2) is compatible with all mass spectrometric ionization and induced fragmentation techniques, (3) has a lower detection limit than the underivatized peptide, and (4) can be energetically activated to dissociate, with the added charged moiety intact, into a single series of sequence-specific fragments [52].

During the developmental phases of ESI and MALDI, little need was perceived for using derivatization in conjunction with either of these modes of ionization. There was almost no incentive to use derivatization as a means for increasing ESI or MALDI signal intensities of analytes because the detection limits for electrosprayed and laser-desorbed peptides and proteins are several orders of magnitude lower than those possible with FAB. Furthermore, it was (and still is) generally felt that, with ESI, MALDI, or a combination of both, it should be possible by purely instrumental means to generate enough sequence information from a given protein to identify or characterize it. Over the ensuing years, however, numerous examples have accumulated from ESI and MALDI analyses in which CID of key peptides derived chemically or enzymatically from a protein yields fragmentation that is too sparse for sequencing or, conversely, too complex for facile interpretation. ESI of a peptide that possesses a strongly basic residue, such as the arginine or lysine on the *C* terminus of a tryptic peptide, will protonate the molecule to varying degrees, and double protonation usually is the most likely in the distribution. A doubly protonated peptide containing a strongly basic residue can yield extensive amounts of sequence information when it is subjected to the low-energy CID conditions used with electrospray instruments. The mobility of one of the ionizing protons in such a peptide is now generally accepted as a requirement for the formation of most of the structurally diagnostic fragments [63–66]. Tryptic peptides, for example, tend in most instances to fragment readily into y_n ions and, to a far lesser degree, b_n ions. Fragmentation of doubly protonated peptides becomes considerably less predictable and the mass spectra correspondingly more complicated when two or more highly basic residues are located along a tryptic peptide's chain. Competing processes, such as sidechain rearrangement and cyclization reactions, form ion fragments that further confound mass spectral interpretation. MALDI of a peptide that has a highly basic residue will, in essence, only singly protonate the molecule. In contrast to doubly protonated ions, the presence of a strongly basic residue in a singly protonated ion greatly impedes the formation of sequence-specific fragments under low-energy CID conditions. When the same singly protonated peptide precursor is subjected to high-energy CID conditions, which have become possible for MALDI with the advent of Q-TOF and TOF/TOF instruments, charge-remote fragmentation reactions [56–58] that lead to structurally diagnostic fragment ions can readily occur. In general however, high-energy CID mass spectra of singly protonated peptides are complicated because of the overlapping presence of the different possible types of amino-terminal fragments (i.e., a_n, b_n, c_n, and d_n ions), carboxy-terminal fragments (i.e., v_n, w_n, x_n, y_n, and z_n ions), and internal fragments [10]. MALDI-PSD of singly protonated peptides seems to induce fragmentation that is characteristic of both low- and high-energy CID [67]. Consequently, fragmentation patterns resulting from MALDI-PSD are complex and difficult to interpret even with the aid of computer algorithms [32]. Moreover, the use of delayed extraction (DE) in conjunction with MALDI [68], a technique that significantly improves mass resolution and mass accuracy for precursor ions [35], reduces the rate of PSD fragmentation by at least an order of magnitude with the consequence that PSD fragmentation of many low-intensity precursor ions is not sufficient to yield even short segments of sequence [69].

The complexities outlined in the preceding paragraph have slowly kindled an interest in methods that produce predictable and complete fragment ion series from electrosprayed and laser-desorbed precursors that are independent of the individual characteristics of any given peptide. In the mid-1990s, a relatively small number of studies were conducted by both ESI [66,70–73] and MALDI [74–82] to elucidate and exploit charge-driven and charge-remote fragmentation processes in the analysis of protonated peptides. Whereas these studies did reveal much about the fragmentation mechanisms, they have not yet led to widespread analytic applications. Quite recently, the use of negatively charged derivatives in connection with ESI and MALDI has been investigated. These latter studies suggest that negatively charge-derivatized peptides might more closely match the definition of an ideal procedure for charge derivatization that was defined by Roth et al. [52] than do positively charge-derivatized peptides.

The focus of this chapter is on making the reader aware of derivatives that localize a charge on the peptide in order to control its fragmentation during single or tandem stages of mass spectrometry. Readers interested in the use of charge derivatization to increase ionization efficiencies or simplify mass spectra in FAB and related forms of particle-induced desorption/ionization mass spectrometry are referred to the review by Roth et al. [52]. Fragmentation mechanisms will be summarized in the next section of this chapter but not explained in detail. In the same vein, only brief outlines rather than detailed descriptions of actual derivatization procedures will be presented in the last two sections of this chapter. The emphasis in the last two sections will be on illustrating the potential analytical advantages of using charged derivatives to direct the fragmentation to one end of a peptide and thereby facilitate interpretation of the resulting sequence data.

14.2. STRUCTURES AND MECHANISMS

The structures of peptide fragment ions and the mechanisms that produce those ions have been investigated and debated for many years. Intensive analyses of numerous high-energy CID experiments conducted on tandem mass spectrometers have, in particular, shown that fragmentation of a singly protonated, linear peptide can be rationalized in terms of one of the following four generalized mechanisms: (1) cleavage of a single bond along the peptide backbone to form a sequence-diagnostic ion specific to either the *N* or the *C* terminus; (2) cleavage of a single bond along the peptide backbone plus a second bond within the side chain of one of the amino acid residues adjacent to the broken backbone bond to form a sequence-diagnostic ion specific to either the *N* or *C* terminus; (3) cleavage of two backbone bonds to form a nonsequence-specific internal acylium, immonium, or amino acid immonium ion; and (4) loss of an amino acid sidechain (R) from one of the residues in the peptide to form a nonsequence-specific $(M + \mathrm{H} - \mathrm{R})^+$ ion [8–10,83]. The interested reader can consult the literature at large for both conventional and alternative structural diagrams that correspond formally to the generic peptide fragment ions formed by these four processes. The nomenclature conventionally used to name the peptide fragment ions was originally proposed by Roepstorff and Fohlman [54] and subsequently refined and extended by Biemann and co-workers [8,9,55].

With so many possible fragmentation pathways open to a collisionally energized peptide, one would expect every high-energy CID spectrum to be complicated. Early investigators soon learned, however, that the high-energy CID spectra of a relatively large number of peptides are strikingly simple. Pursuant to this observation, they discovered that the appearance of a singly protonated peptide's high-energy CID fragmentation pattern is determined primarily by the number and location of amino acids with basic functionalities strong enough to localize positive charge [9,84]. Specifically, charge-driven fragmentation mechanisms, in which cleavage adjacent to the charge site leads chiefly to b_n and y_n ions [9], prevail in the absence of highly basic moieties whereas charge-remote mechanisms, in which cleavage remote from the charge site results mainly in a_n, d_n, and y_n ions [58], predominate in the presence of strongly basic moieties. An opportunity was perceived in this behavior for exercising some degree of control over the appearance of high-energy CID spectra [84], and many studies ensued to find derivatives that would fix a positive charge at either the *N* terminus or the *C* terminus and thereby might facilitate the charge-remote decomposition of singly protonated peptides into diagnostic ions belonging to a single type of fragment series. In reviewing this body of work [52], Roth et al. discuss several of the structures that have been proposed for the fragment ions of a charge-derivatized peptide, and they introduce a nomenclature in which the presence of the charge moiety on a fragment ion is simply denoted by placing an asterisk in front of the conventional letter symbol used for that type of ion when it originates from a nonderivatized peptide.

The effect of low-energy activation on a charge-derivatized peptide is markedly different from that of high-energy activation. In the former case, energy is incrementally imparted to the precursor through a succession of collisions that occur on the average every few microseconds or so while, in the latter case, energy is transferred to the precursor ion in a single, fast, low-scattering-angle collision [85]. It is not surprising, therefore, that the mode used to activate protonated peptides strongly influences the pathway through which fragments are formed. In particular, these two dissimilar activation mechanisms cause peptides with basic residues to dissociate in strikingly different ways. The presence of one highly basic residue in a singly protonated peptide will promote facile charge-remote fragmentation following high-energy activation, whereas it will severely impede decomposition of an identical peptide in a low-energy experiment [86]. If that same peptide were doubly protonated, the presence of the basic residue would lead to a complex combination of charge-remote and charge-directed processes after high-energy activation [87] but to almost pure, relatively uncomplicated, charge-proximal fragmentation subsequent to low-energy activation [65,86]. While high-energy activation of peptides that contain weakly basic amino acids, such as histidine or lysine, results in various combinations of charge-directed and charge-remote fragmentation, low-energy activation of these same structures induces varying degrees of charge-proximal fragmentation [52].

The difference between the high- and low-energy experiments in the case of the singly protonated peptide can be rationalized to a large degree in terms of the mobile-proton model proposed by Wysocki and co-workers [66]. Results from this group's systematic surface-induced dissociation (SID) experiments on electrosprayed peptide ions clearly showed that the degree to which a singly protonated peptide fragments is markedly dependent on the absence or presence of basic residues and that shifts in the onset energy for fragmentation to higher collision energies correlates with increases in the peptide's gas-phase basicity (e.g., peptides without a basic residue < lysine-containing peptides < arginine-containing peptides) [66,88]. Wysocki and company reasoned from these results that, for a singly charged peptide possessing a strongly basic moiety, the ion population leaving the source is dominated by one stable form in which the peptide is protonated at the site of its basic moiety [66]. In like manner, experiments with doubly charged peptides containing two arginine residues demonstrated that both protons were preferentially localized at the basic arginine residues. Ab initio plus MNDO bond order and energy partitioning calculations by Wysocki's group indicated that a proton localized on a basic residue can have only a local effect on fragmentation and, furthermore, that cleavages more than three bonds from the charge site could only result from charge-remote processes [63,64]. The quantum mechanical calculations also revealed that protonation of an amide nitrogen atom in the backbone of a peptide dramatically weakens the amide bond, whereas protonation of the amide oxygen atom strengthens the amide bond. Drawing on findings and ideas published earlier by several other researchers [9,86,88–94] as well as their own experimental and theoretical results for singly charged peptides, Wysocki and co-workers concluded that, regardless of its initial site of attachment, the ionizing proton can be mobilized via an activation process and eventually transferred to a less basic site in the peptide. By this means, protonated forms with higher internal energies than that of the most stable structure can become populated, and in particular, those structures in which an amide nitrogen is protonated can become sites for facile charge-proximal fragmentation. In this picture, the proton's initial site of localization depends on the type and number of basic residues present in a given peptide and, in turn, determines the activation energy necessary to mobilize the proton and induce dissociation. The correlation between the gas-phase basicity of a peptide and the onset-energy for its dissociation is a strong indication that the activation energy required for mobilization of the ionizing proton at a reasonable rate is directly influenced by the difference between the internal energy of the most stable protonated form and the internal energies of the other less stable protonated forms.

The mobile proton model provides a basis for gaining insight into many of the features observed in high- and low-energy fragment ion spectra of peptides. For example, it accounts well for why high-energy CID mass spectra of singly protonated peptides with a strongly basic moiety, such as Arg or a cationic derivative, located near the *N* or *C* terminus can exhibit signals corresponding to nearly complete series of charge-remote fragment ions respectively from the *N* terminus (e.g., a_n and d_n ions) or the *C* terminus (e.g., y_n ions) while low-energy CID mass spectra of the same peptides show only a paucity of peaks corresponding to diagnostically useful fragment ions. It also provides an explanation for why a doubly protonated peptide precursor possessing a single, strongly basic residue will fragment so much more efficiently than will its singly protonated analog—one of the protons in the most stable form of the doubly charged molecule would be sequestered on the highly basic moiety, and the second proton would reside on a much less basic site, where, with only a relatively small increase in internal energy, it could be mobilized to facilitate charge-proximal fragmentation. The extent to which the proton transfer competes with charge remote fragmentation or other pathways, such as the loss of NH_3 from a protonated arginine side chain, varies with amino acid composition and is beyond the mobile proton model's scope [66]. Nevertheless, the model can be gainfully used as a guide for designing charge derivatives that channel the dissociation of peptides into diagnostic pathways and, thus, simplify the interpretation of their fragment ion mass spectra.

14.3. POSITIVELY CHARGED *N*-TERMINAL DERIVATIVES

The use of charged derivatives to promote peptide fragmentation that can be more easily interpreted originated in the middle eighties with fixed, positively charged groups. Not unexpectedly, therefore, the literature since that time has become replete with reports on techniques for coupling positively charged functionalities to the *N* termini of peptides. The reader interested in a thorough introduction to the development, chemistry, and application of these derivatization procedures is again referred to the review by Roth et al. [52]. From this large body of work, only three methods come close to meeting the ideal for charge derivatization defined by Roth et al. Of these, two are for attaching quaternary ammonium functionalities—dimethylalkylammonium acetyl (DMAA) [70,84,95] and C5Q [74,82]—and one is for coupling a phosphonium group, namely, tris(2,4,6-trimethoxyphenyl)phosphonium acetate (TMPP–Ac) [77], to the *N* termini of peptides. These three derivatization reactions require little or no purification prior to analysis. They can be performed in a relatively short time, on relatively small amounts of peptide, and at high yields. All can be made selective for the *N* terminus by adjusting the pH of the reaction mixture. The three derivatives are apparently compatible with FAB, ESI, and MALDI, and all can be detected in the low-picomole–subpicomole range. It is not within the scope of this chapter to describe these three approaches in detail, but it is in order to give a synopsis of the information that Roth and his co-authors compiled on them so that the reader can appreciate the state of the art to which this methodology has developed.

DMAA derivatization of a peptide is carried out with commercial reagents in two steps by adding first a halide acetyl group at pH 6 to the *N* terminus and second a dimethylalkyl amine (Scheme 14.1) [95]. The entire procedure takes about 2 h and requires around 100 pmol of peptide. It is usually necessary to purify DMAA derivatives by HPLC before mass spectrometric analysis. To date, DMAA derivatives have only been analyzed by FAB [84,95,96] and ESI [66,70], but there is no intrinsic reason why they should not be compatible with MALDI. The detection limits for DMAA derivatives are apparently not much different, that is, only slightly lower or higher, than for their nonderivatized analogs. High-energy CID mass spectra of nonderivatized peptides generally exhibit a complex superposition of weak peaks corresponding to partial series of a_n, b_n, v_n, w_n, x_n, y_n, and z_n fragment ions, such as those seen in the spectrum shown in Fig. 14.1a. By contrast, high-energy CID spectra of DMAA derivatives contain strong peaks corresponding to nearly complete series of $^*a_n$ and $^*d_n$ ions plus a few $^*b_n$ and $^*c_n$ ions. A peak due the loss of the alkyl group is also usually present at the high-mass end of the spectrum. These characteristics are readily apparent in the mass spectrum

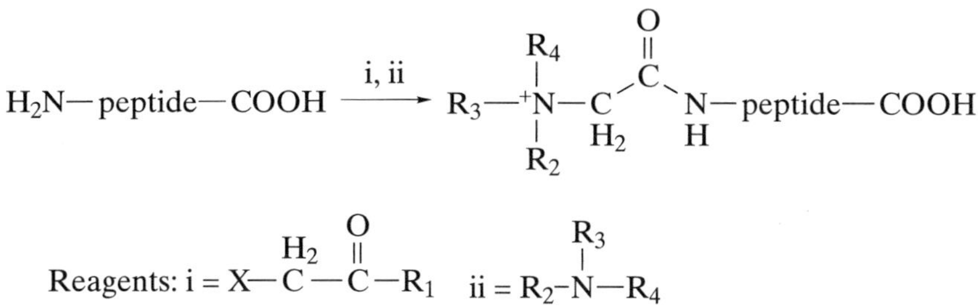

(A) X = Cl, R_1 = Cl, $R_2 = R_3 = R_4$ = Et
(B) X = Cl, R_1 = Cl, $R_2 = R_3 = R_4 = CH_3$
(C) X = I, $R_1 = OC(O)CH_2I$, $R_2 = R_3 = CH_3$, $R_4 = CH_3, C_8H_{13}, C_8H_{17}$

Scheme 14.1

shown in Fig. 14.1b. Also evident in this spectrum is a significant improvement in the fragment ions' signal strengths, which is attributed to the fact that the derivatized precursor ion dissociates into fewer fragments than its nonderivatized analog [95]. Low-energy CID mass spectra of DMAA derivatives contain complete or nearly complete series of $*a_n$ and $*b_n$ ions and a smattering of $*c_n$ and $*d_n$ ions.

Formation of the C5Q derivative of a peptide can be performed in a single 20-min step (Scheme 14.2) on as little as 100 fmol of peptide; if additional steps are performed to remove an *N*-terminal residue or to derivatize the sidechains of lysine and arginine, the procedure can take up to 4 h and require peptide amounts in the low-picomole range [82]. The active reagent, trimethylammonium alkyl *N*-hydroxysuccinimidyl ester (Scheme 14.2), for C5Q derivatizations must be synthesized, but it can survive for months. As with DMAA derivatives, C5Q derivatives usually require purification by HPLC to remove excess reagent. Although C5Q derivatives have only been analyzed by MALDI, it should be possible to analyze them by FAB and ESI as well. No loss of signal intensity or mass resolution has been observed to accompany C5Q derivatization. The first reported MALDI-PSD spectrum of a C5Q-derivatized peptide exhibited a few $*a_n$ peaks and one $*b_n$ peak, but most of the fragment-ion peaks corresponded to immonium, $*a_n$—59, and $*b_n$—59 ions (i.e., $*a_n$ and $*b_n$ ions less the triethylamine moiety) [76]. By contrast, the peaks in MALDI-PSD mass spectra of C5Q derivatives published subsequently by another group of researchers [82,97] correspond mostly to $*a_n$ and $*b_n$ ions and occasionally to $*c_n$ and $*d_n$ ions as seen, for example, in Figure 14.2. The distinction between the fragmentation patterns in the mass spectra produced by these two laboratories may be due to a difference in the laser power used or in some other experimental parameter as was shown in a case involving the analysis of another type of charged derivative by MALDI-PSD [78]. The MALDI-PSD mass spectra shown in Fig. 14.2 illustrate the fragmentation of

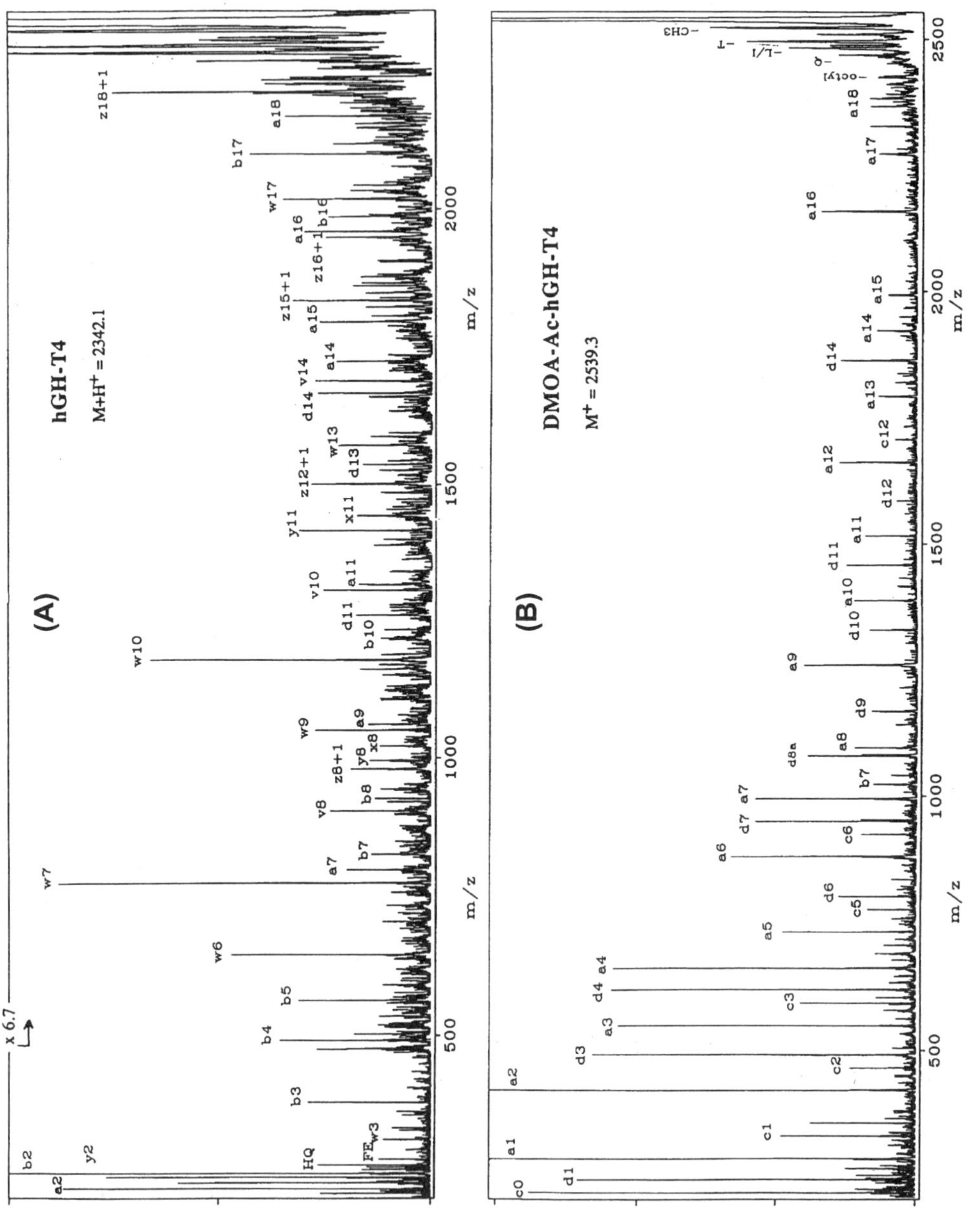

(A)
hGH-T4
M+H+ = 2342.1
x 6.7
a2 b2 y2 HQ FE w3 b3 b4 b5 w6 w7 a7 b7 v8 b8 z8+1 y8 x8 w9 a9 w10 b10 d11 v10 a11 y11 x11 z12+1 d13 w13 d14 v14 a14 a15 z15+1 z16+1 a16 b16 w17 b17 a18 z18+1
m/z
500 1000 1500 2000
(B)
DMOA-Ac-hGH-T4
M+ = 2539.3
c0 d1 a1 c1 a2 c2 d3 a3 c3 d4 a4 a5 c5 d6 a6 c6 d7 a7 b7 d8a a8 d9 a9 d10 a10 d11 a11 d12 a12 c12 a13 d14 a14 a15 a16 a17 a18
-CH3
m/z
500 1000 1500 2000 2500

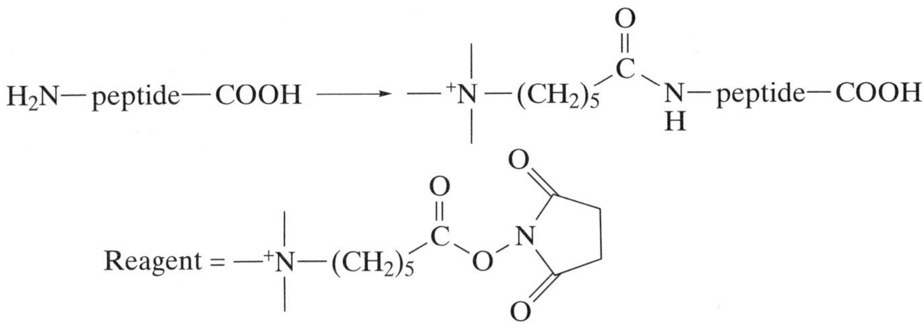

Scheme 14.2

a peptide, which does not possess any strong basic amino acids, before and after derivatization with C5Q. The native peptide (Fig. 4.2a) dissociates into a confusing mixture of mostly not very abundant *C*-terminal and *N*-terminal fragment ions, whereas the charge-derivatized peptide (Fig. 4.2b) decomposes almost entirely into relatively abundant $*a_n$ fragment ions. The uniform signal intensities and nearly complete *N*-terminal fragment ion series observed in the C5Q derivative's spectrum provide significantly more analytic information than can be obtained from the nonderivatized peptide's spectrum.

TMPP–Ac derivatization is carried out in a single 15-min step (Scheme 14.3) on peptide amounts down to low picomole levels. Although it is necessary to synthesize the reagents for TMPP–Ac derivatizations, they can be stored for months. As a rule, there is no need to purify TMPP–Ac derivatives by HPLC when the reaction mixture is to be analyzed by FAB or MALDI; purification is required, however, for analyses by ESI. TMPP–Ac derivatives have been analyzed successfully be FAB [77,78,98], ESI [72,99,100], and MALDI [77,78,80,81,100], methods. TMPP–Ac derivatization has only a small effect on the signal intensities of most peptide ions, but for certain

Figure 14.1. FAB high-energy CID fragment ion spectra produced on a JEOL JMSHX110HF/HX110HF tandem mass spectrometer: (a) 100 pmol of the tryptic peptide LHQLAFDTYQEFEEAYIPK (T4) of recombinant methionyl human growth hormone (hGH) displaying segments of several different series of fragment ions, none of which cover a substantial portion of the peptide's sequence; (b) 40 pmol of the dimethyloctylamine (DMOA) derivative of the T4 peptide of hGH displaying two nearly complete series of ions (a_n and d_n) but fewer peaks with greater signal-to-noise ratio. (*Note*: Some identified fragment ion signals are not labeled, and asterisks are not used to label the DMAA fragment ion peaks as recommended by Roth et al. [52].) The unlabeled high-mass ions correspond to losses from amino acid sidechains. The peak labeled d_{8a} is a doublet resulting from cleavage of the Thr sidechain, and the peak labeled d_7 is accompanied by a peak 2 u higher than that corresponding to a_7-CO_2. [Reprinted with permission from J. T. Stults, J. Lai, S. McCune, and R. Wetzel, *Anal. Chem.*, **65**, 1703–1706 (1993). Copyright 1993, American Chemical Society.]

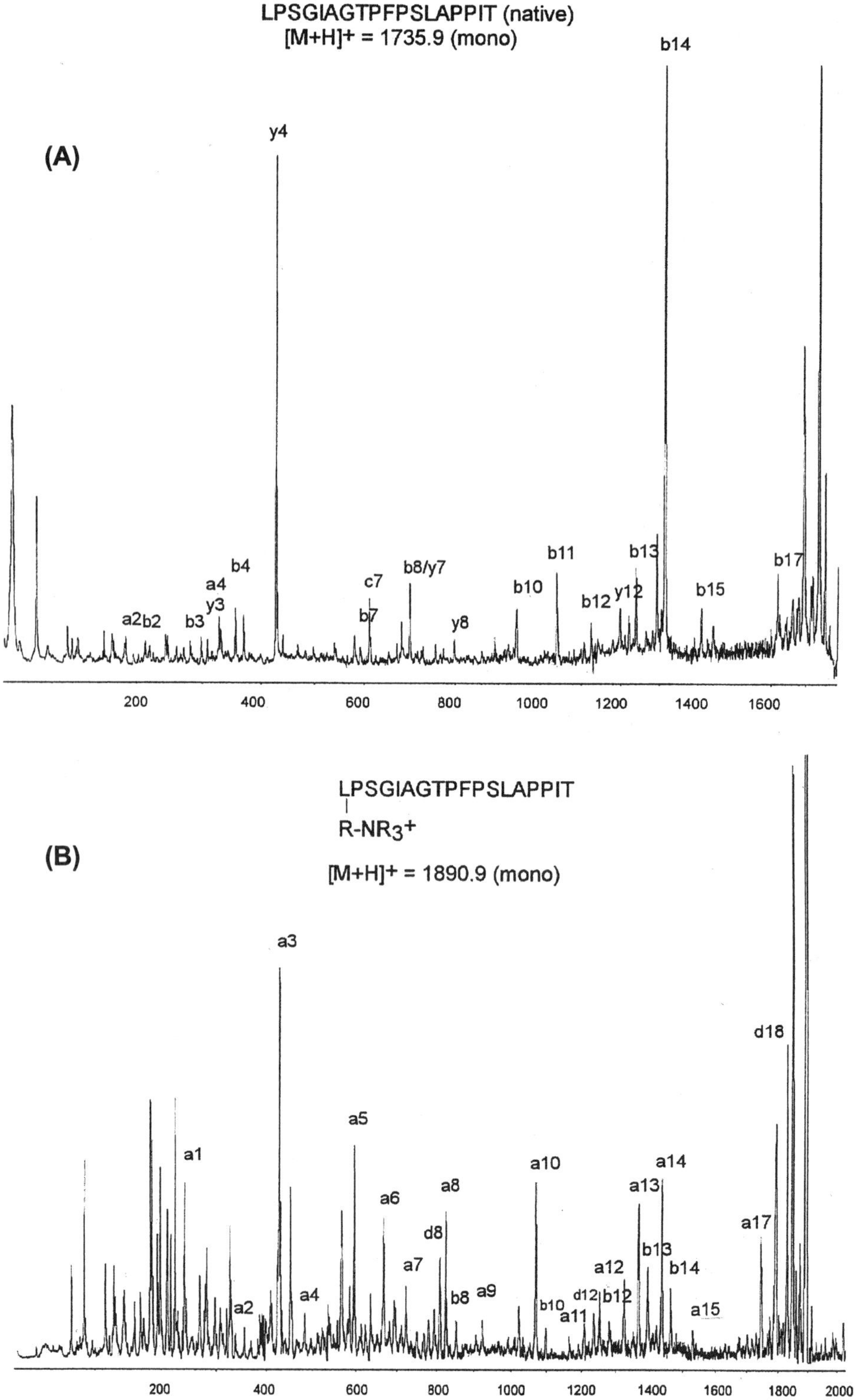
LPSGIAGTPFPSLAPPIT (native)
[M+H]+ = 1735.9 (mono)
(A)
y4
b14
a2
b2
b3
a4
y3
b4
c7
b7
b8/y7
y8
b10
b11
b12
y12
b13
b15
b17
200
400
600
800
1000
1200
1400
1600
LPSGIAGTPFPSLAPPIT
R-NR3+
(B)
[M+H]+ = 1890.9 (mono)
a3
d18
a5
a1
a10
a14
a13
a6
a8
a17
d8
b13
a7
a12
b14
a2
a4
b8
a9
b10
d12
b12
a11
a15
200
400
600
800
1000
1200
1400
1600
1800
2000

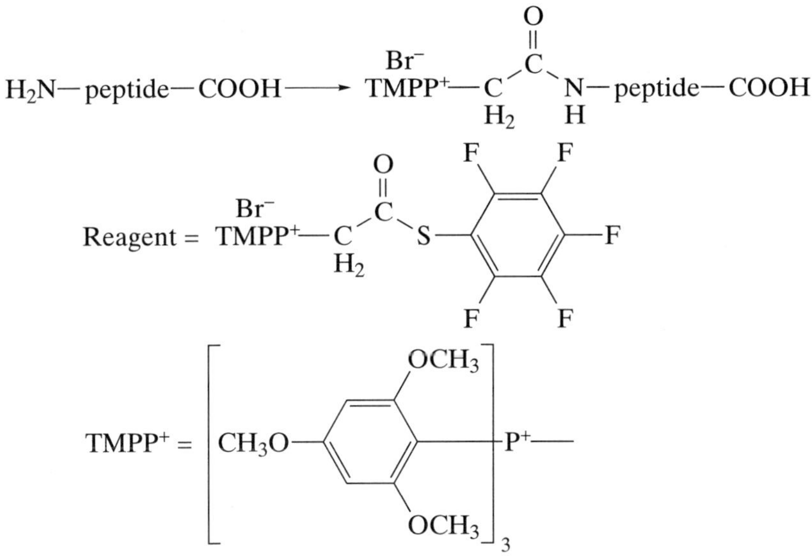

Scheme 14.3

hydrophobic peptides, the technique can significantly enhance signal strengths. Regardless of the mode of ionization or the level of energy used for activation, TMPP–Ac-derivatized peptides dissociate into numerous $*a_n$ fragments. Dissociation into other types of fragments, however, depends on the modes of ionization and energy activation. Specifically, FAB high-energy CID of TMPP–Ac derivatives produces $*d_n$ ions (comparable in number and abundance to the $*a_n$ ions) plus a few $*b_n$ and $*c_n$ ions; ESI low-energy CID generates $*b_n$ ions (comparable in number and abundance to the $*a_n$ ions) and a few, scattered, weak $*c_n$ and $*d_n$ ions; and MALDI PSD produces only a few, weak $*b_n$, $*c_n$, and $*d_n$ ions whose number and abundance are dependent on

Figure 14.2. MALDI PSD spectra of (a) the nonderivatized and (b) the C5Q-derivatized peptide LPSGIAGTPFPSLAPPIT. (*Note*: Some identified fragment ion signals are not labeled, and asterisks are not used to label the TMPP–Ac fragment-ion peaks as recommended by Roth et al. [52].) The spectrum of the nonderivatized peptide indicates formation of a mixture of *N*-terminal and *C*-terminal ions, whose probabilities of formation vary widely and whose identities do not account for much of the peptide's sequence. By contrast, the spectrum of the C5Q-derivatized peptide implies production of essentially only *N*-terminal fragment ions, whose probabilities of formation are relatively uniform and whose identities account for the peptide's entire sequence. [Reprinted with permission from B. Spengler, F. Luetzenkirchen, S. Metzger, P. Chaurand, R. Kaufmann, W. Jeffery, M. Bartlet-Jones, and D. J. C. Pappin, *Int. J. Mass Spectrom. Ion Process.* **169/170**, 127–140 (1997). Copyright 1997, Elsevier Science.]

laser power. As a rule, peptides containing basic amino acid residues are multiply protonated following ESI. Figure 14.3 shows the ESI fragment ion spectra of a doubly charged tryptic peptide and its TMPP–Ac derivative [99]. Because the precursor ions are doubly charged, it is not clear whether they fragment by charge remote, charge-directed, or a combination of both processes. Nevertheless, it is clear from Fig. 14.3 that TMPP–Ac derivatization can substantially increase the analytical content of a peptide's low-energy CID mass spectrum.

In summary, procedures for affixing positively charged derivatives to the *N* terminus of a peptide have been under investigation in conjunction with desorption mass spectrometry for many years. The three techniques discussed in the preceding section (two for quaternary ammonium groups and one for a phosphonium group) have been shown to improve fragmentation efficiency along the backbones of a variety of selected peptides. Despite this promising sign, none of these three methods has yet proved to be practical for sequencing peptides recovered from biological samples [52]. In MALDI, peptides generally become singly protonated and do not undergo efficient PSD along their backbones. This makes it difficult to extract de novo sequencing information from MALDI-PSD spectra. The potential for using the enhanced, simplified fragmentation patterns obtained with *N*-terminal fixed-charge derivatives as a remedy for this shortcoming has been demonstrated in a few cases, such as that shown in Fig. 14.2, but in general practice, this approach usually leads to poor-quality mass spectra by MALDI-PSD [101]. In ESI, a given peptide from a tryptic digest will typically acquire from one to four ionizing protons to create a set of ions with a fixed mass but a distribution of charge states. Nearly all electrospray mass spectrometers use low-energy activation in conjunction with tandem mass spectrometric analyses; therefore, these instruments form charge-remote fragment ions inefficiently from singly charged precursors. Positively charged derivatives have been shown to remedy this problem in a few special instances, but, as is the case for MALDI, results are for the most part either inconsistent or of poor quality. Low-energy CID of the doubly protonated native peptides by and large yield useful fragment ions but interpretation of the resulting spectra are often complicated by the occurrence of peaks corresponding to doubly protonated fragments. The analytic content of these spectra can be improved in some instances by derivatizing the peptide's *N* terminus with a positive charge, but the confounding presence of peaks due to double protonation, as seen, for example, in Fig. 14.3, usually persists.

14.4. NEGATIVELY CHARGED *N*-TERMINAL DERIVATIVES

Gaskell and co-workers discovered that the oxidation of cysteine to cysteic acid in peptides containing a *C*-terminal arginine strongly increases the yield

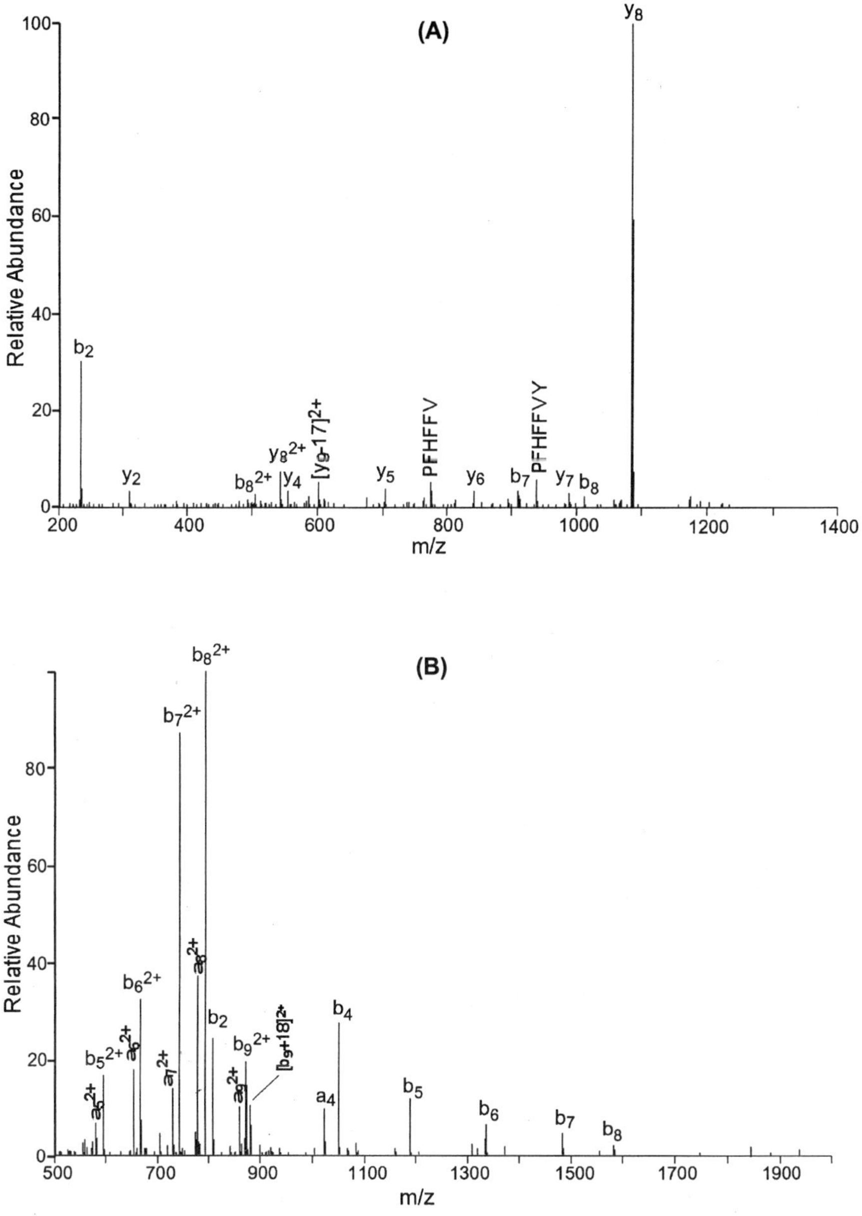

Figure 14.3. ESI low-energy CID spectra of (a) doubly charged PHPFHFFVYK (*m/z* 660), and (b) its doubly charged TMPP–Ac derivative (*m/z* 946). (*Note*: Some identified fragment ion signals are not labeled, and asterisks are not used to label the TMPP–Ac fragment ion peaks as recommended by Roth et al. [52].) The peptide's three basic amino acid residues (His_2, His_5, and Lys_{10}) provide four potential sites of protonation before and after charge derivatization (the three basic amino acids plus either the *N*-terminal primary amine before or the TMPP–Ac after). The spectrum of the protonated peptide displays both singly and doubly protonated b_n and y_n ions along with some intense internal fragment ions. The spectrum of the TMPP–Ac modified peptide contains a complete series of singly charged $*b_n$ ions, a half complete series of doubly charged $*a_n$ and $*b_n$ ions, and no internal fragment ions whatsoever. [Reprinted with permission from M. Adamczyk, J. C. Gebler, and J. Wu, *Rapid Commun. Mass Spectrom.* **13**, 1413–1422 (1999). Copyright 1999, John Wiley & Sons, Ltd.]

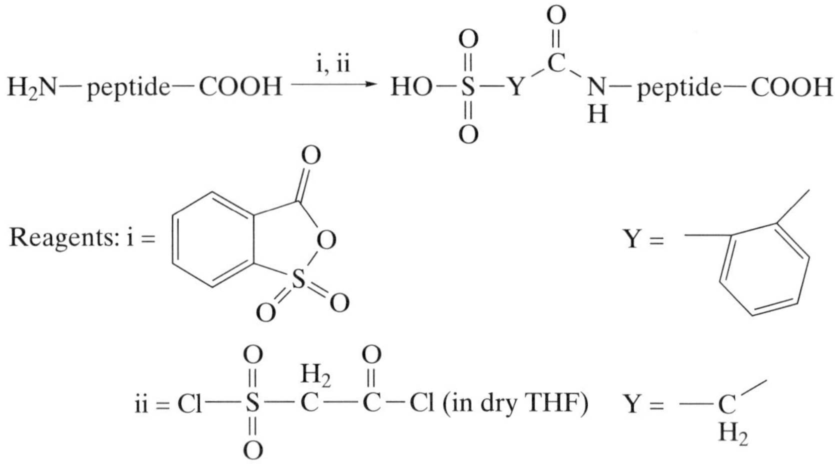

Scheme 14.4

of y_n fragment ions following low-energy CID of singly protonated precursors formed by ESI [65,102]. This finding as well as questions concerning the internal energy of precursor ions formed by MALDI under delayed extraction conditions plus compelling arguments for the mobile proton model of peptide fragmentation [63,66] motivated Keough and co-workers to investigate addition of a strongly acidic group, rather than a strongly basic group, to a peptide's *N* terminus as a means for promoting and directing fragmentation of peptides in MALDI-PSD mass spectrometry [101]. They succeeded in working out two, fast, one-step, derivatization chemistries for producing peptides with a *N*-terminal sulfonic acid (SA) that facilitates low-energy, charge-site initiated cleavage of backbone amide linkages but suppresses the detection of *N*-terminal fragment ions.

SA derivatization can be performed with commercial reagents either by reacting 2-sulfobenzoic acid cyclic anhydride (Scheme 14.4, reagent i) with the peptide in aqueous solution to introduce an aromatic sulfonic acid group directly to the *N* terminus or by combining chlorosulfonylacetyl chloride (Scheme 14.4, reagent ii) with the peptide in dry THF followed by addition of water to convert the chlorosulfonyl group into the desired sulfonic acid [101]. Either procedure takes about $1\frac{1}{2}$ h to complete, and either can be carried out on a quantity of isolated peptide as small as ~25 fmol or on a complete tryptic digest of as little as 500 fmol of a protein [103]. Chen et al. have found that chlorosulfonylacetyl chloride is easier to handle and has a higher coupling efficiency [104]. For a peptide with a *C*-terminal lysine, the reactions depicted in Scheme 14.4 can result in derivatization of the ε-amine on the lysine's sidechain as well as the α-amine at the peptide's *N* terminus. When it becomes necessary to protect against this side reaction with the sulfonation reagent,

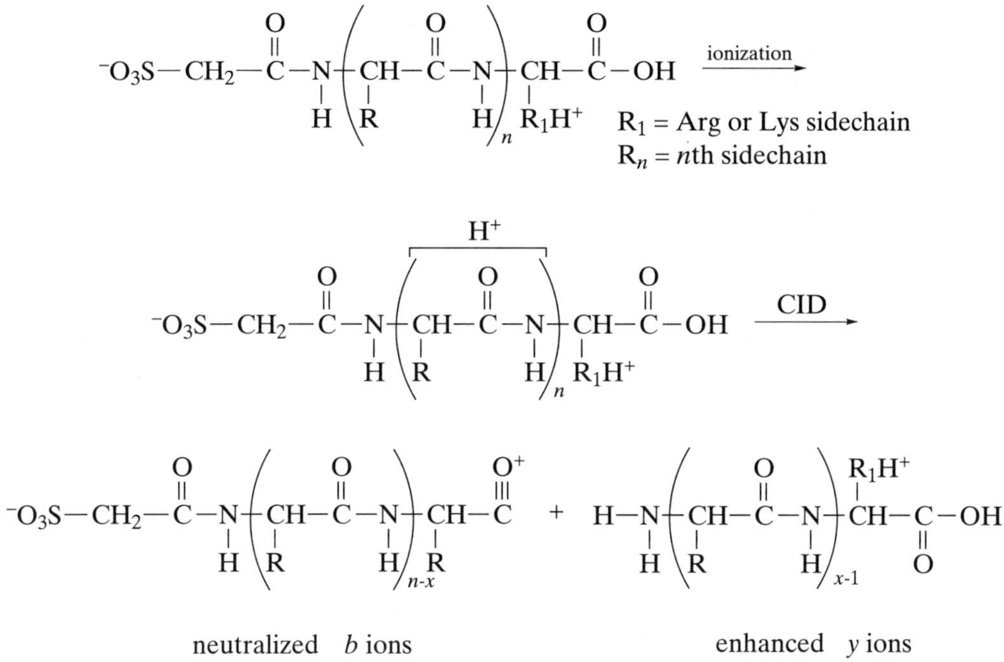

Scheme 14.5

the ε-amino group of a Lys-terminated tryptic peptide can be selectively guanidinated in a single step with *O*-methylisourea hydrogen sulfate to convert the Lys residue into a more basic homoarginine residue [105].

The addition of a sulfonic acid group to the *N* terminus of a peptide that has a strong basic group at its *C* terminus produces a species predominantly protonated at the *C* terminus and deprotonated at the *N* terminus (Scheme 14.5). To generate a singly charged ion for analysis by mass spectrometry, one additional proton must be added to this peptide during ionization. Because a proton is already attached to the most basic site in the molecule, the ionizing proton is presumably free to attach randomly to any of the uniformly, weakly basic, amide nitrogens along the peptide's backbone. These protonated amide bonds fragment extensively when subjected to PSD or low-energy CID conditions because the bond orders of protonated amides are substantially lower than those of nonprotonated amides [66]. Detection of *N*-terminal fragments is blocked because the negative charge on the sulfonate group neutralizes the ionizing proton that is transferred to the fragment's *C* terminus during cleavage of the amide bond; conversely, detection of *C*-terminal fragments is enhanced because decomposition of the amide linkage leaves them with a neutral *N* terminus and a protonated lysine or arginine on their *C* terminus (Scheme 14.5) [101]. Keough and co-workers investigated the ionizing proton's role in promoting cleavage of backbone amide bonds by comparing the fragmentation behavior of $[M + H]^+$, $[M + Na]^+$, and $[M - H]^-$ molecular ions and by comparing the fragmentation pattern of a tryptic peptide containing one arginine residue with another containing two arginine residues. By demonstrating respectively that no PSD y_n ions are formed in the absence of an ionizing proton or in the presence of an ionizing proton immobilized on a basic arginine residue, these two sets of experiments give impetus to the idea of a mobile ionizing proton.

Under typical MALDI conditions, singly protonated SA derivatives, each of which possesses a proton that is more or less free to randomly ionize backbone amide groups [66], fragment almost exclusively into series of y_n ions that can extend uninterrupted for sequences of 20 or more amino acid residues. As seen in Fig. 14.4, the PSD fragment ion spectrum of a SA derivative can exhibit dramatic increases in fragment ion signals that readily reveals the peptide's sequence. The relative abundance of the y_n ions decreases toward lower mass, and fragmentation efficiency decreases considerably at amide bonds on the *C*-terminal side of Pro. Essentially, these same mass spectral characteristics are exhibited by SA derivatives of peptides that contain a *C*-terminal lysine. Keough and co-workers observed that derivatives made with chlorosulfonylacetyl chloride are apparently more stable than those formed from 2-sulfobenzoic acid cyclic anhydride and that a protonated chlorosulfonylacetyl chloride derivative channels more of its internal energy into the production of structurally significant PSD-y_n ions than does a 2-sulfobenzoic acid cyclic anhydride derivative [101].

Using model peptides, Keough et al. have extensively demonstrated that SA derivatization can be used effectively in conjunction with MALDI-PSD to determine sites of phosphorylation, identify variant proteins, and analyze peptides with masses up to 3.6 kDa [101,103,105,106]. Of particular note is the fact that they regularly use MALDI-PSD of SA-derivatized tryptic peptides to identify proteins from biological samples [105,106]. The spectrum shown in Fig. 14.5 is typical for SA-derivatized peptides of a protein that has been isolated by 2D gel electrophoresis. When the sequence tag obtained from this particular mass spectrum was searched against the full library and all fragment ion types (a_n, b_n, $b_n + H_2O$, $b_n - NH_3$, $b_n - H_2O$, internal, and y_n) were allowed, 239 candidate proteins were returned. When the database search was repeated with the constraint that only y_n ions be allowed as fragments, only 6 candidate proteins (all mitochondrial aspartate aminotransferases) were returned. This search constraint is possible because *N*-terminal and internal cleavage ions are almost nonexistent in the PSD mass spectra of SA derivatives. Adding the MALDI PSD sequence tag from a second derivative at *m/z* 1916 to the search was sufficient to unambiguously establish the protein's identity.

Numerous comparative ESI analyses of singly and doubly protonated ions of native and SA-derivatized peptides have been performed on both ion trap and tandem quadrupole instruments [104,107,108]. Native peptides that

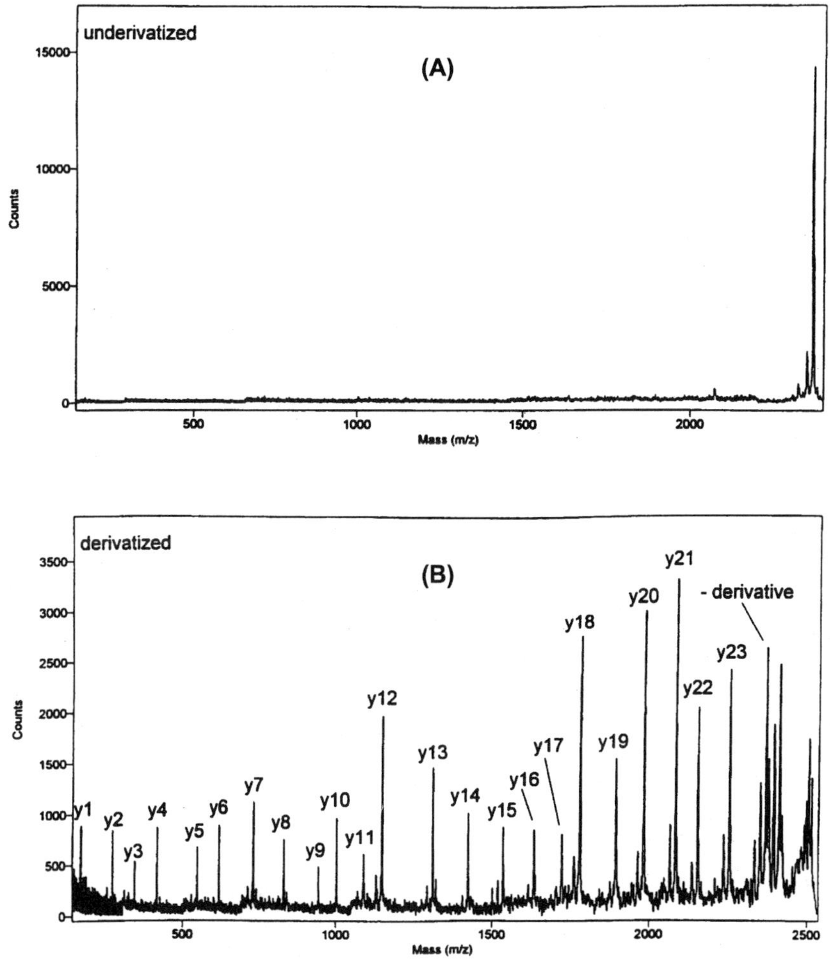

Figure 14.4. MALDI PSD mass spectra of the tryptic peptide NTATSLGSTNLYGS-GLVNAEAATR (a) before and (b) after SA derivatization with chlorosulfonylacetyl chloride (Scheme 14.4, reagent ii). The 24-amino acid peptide was isolated by reversed-phase HPLC from the tryptic digest of an enzyme obtained from a commercial laundry product. The native peptide's spectrum provides little structurally diagnostic information, but the derivative's spectrum, which was obtained under very similar experimental conditions, shows a complete series of y_n ions that enables rapid verification of this segment of the enzyme's amino acid sequence. [Reprinted with permission from T. Keough, R. S. Youngquist, and M. P. Lacey, *Proc. Natl. Acad. Sci.* (USA), **96**, 7131–7136 (1999). Copyright 1999, National Academy of Sciences of the USA.]

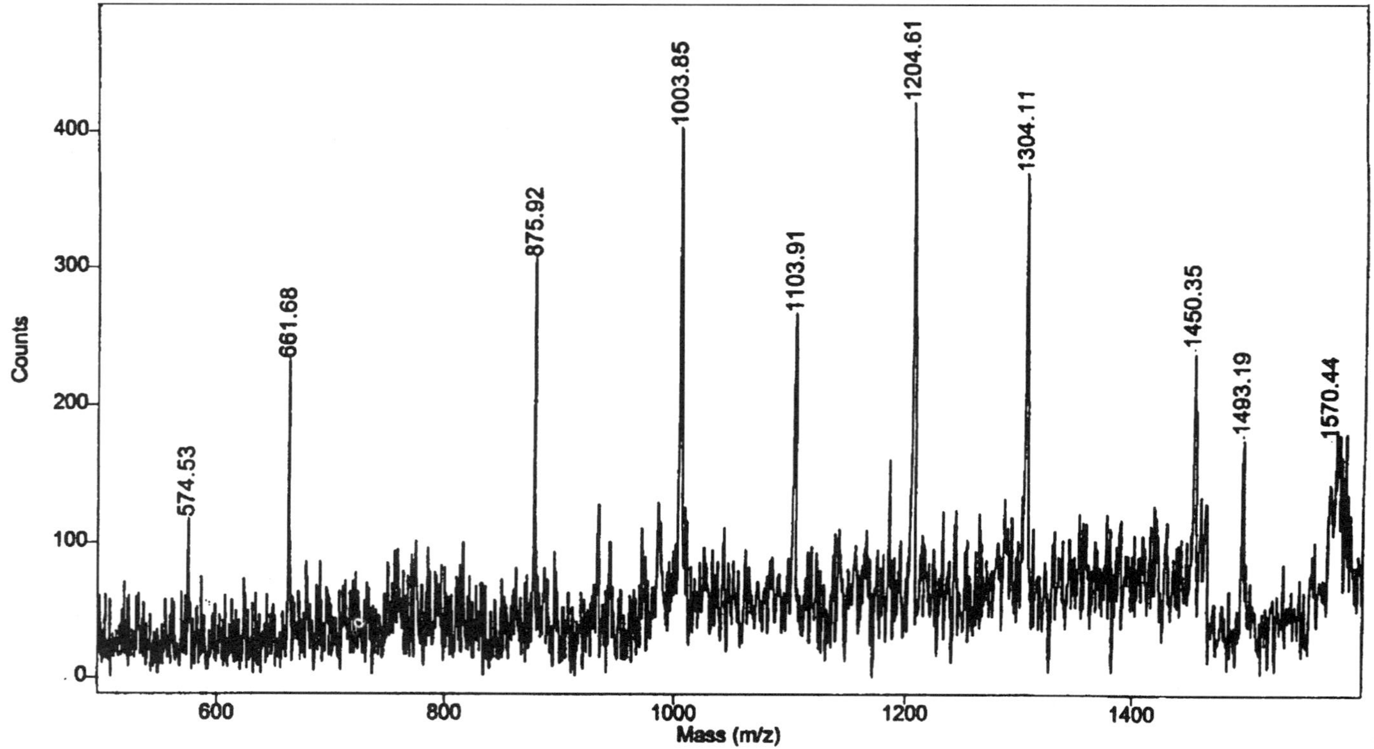

Figure 14.5. MALDI PSD fragment ion spectrum of a SA-derivatized, tryptic peptide from an unknown protein. The protein was isolated by 2D gel electrophoresis; the spot was subjected to in-gel tryptic digestion; and then the entire mixture of tryptic peptide was reacted with chlorosulfonylacetyl chloride (Scheme 14.4, reagent ii). A MALDI mass map of the derivatized digest was recorded, and PSD mass spectrometry was performed on the derivatized peptide appearing nominally in this map at *m/z* 1572. [Reprinted with permission from T. Keough, R. S. Youngquist, and M. P. Lacey, *Proc. Natl. Acad. Sci.* (USA), **96**, 7131–7136 (1999). Copyright 1999, National Academy of Sciences of the USA.]

become singly protonated by ESI generally fail to yield useful fragment ion spectra under the low-energy CID conditions typical of quadrupole instruments. By contrast, SA derivatives of peptides that possess a basic residue at their *C* terminus fragment into predominantly y_n ions when they become singly protonated by ESI—in fact, they fragment very much in the same manner as they do under MALDI PSD conditions. Doubly protonated native peptides generally yield useful, albeit complex, fragment ion spectra, in tandem quadrupole instruments. Doubly protonated SA derivatives behave basically the same way as their nonderivatized counterparts, but their low-energy CID spectra tend to exhibit more y_n ions and, as a consequence of this simplification, to be slightly easier to interpret [104]. The patterns of dissociation observed in the ESI fragmentation spectra of both singly and doubly protonated SA derivatives are for the most part consistent with the mobile proton model.

Figure 14.6 shows the low-energy CID spectra of an electrosprayed peptide that was found in a tryptic digest of the well-characterized DNA repair protein *E. coli* uracil glycosylase (Ung) [109]. The molecular mass of this particular peptide did not correspond to that of any of Ung's predicted proteolytic sequences. The native peptide's singly protonated precursor (*m/z* 1379) was particularly refractory to low-energy CID on an ESI ion trap mass spectrometer (Fig. 14.6a). The strong, nonsequence-specific $[M + H - NH_3]^+$ and $[M + H - 44]^+$ fragment ion peaks are especially evident in the spectrum, but it is nearly impossible to pick sequence-diagnostic peaks out of the spectrum's dense background of weak ion signals. In fact, diagnostically useful peaks are so faint that, even after the peptide's sequence became known, only three peaks in the spectrum could be unambiguously assigned to a sequence tag in the peptide. Performing sulfonic acid derivatization on an aliquot of the crude tryptic digest, selecting the unknown peptide's singly protonated precursor (*m/z* 1501), and subjecting it to low-energy CID produced a contiguous series of strong *y*-type fragments that unambiguously identified the peptide as Ung's tryptic peptide T2 minus its first four *N*-terminal amino acids (Fig. 14.6b). The added specificity that results from being able to restrict a database search solely to y_n ions, can in some instances enable a protein to be unambiguously identified from a single, contiguous series of y_n fragments that account for about two thirds of the sequence of only one of the protein's tryptic peptides [104,107]. This finding is similar to that stated in a preceding paragraph in reference to MALDI-PSD spectra like that shown in Fig. 14.5.

The presence of a proline, an aromatic amino acid, or a basic amino acid somewhere between the termini of a peptide that has been derivatized with SA profoundly effects the dissociation of the peptide's singly protonated precursor. Increased basicity of the backbone amide on the *N*-terminal side of a proline enhances fragmentation on its *N*-terminal side while retarding fragmentation on its *C*-terminal side [108]. Such peptides can be entirely sequenced by interrogating both their singly and doubly protonated

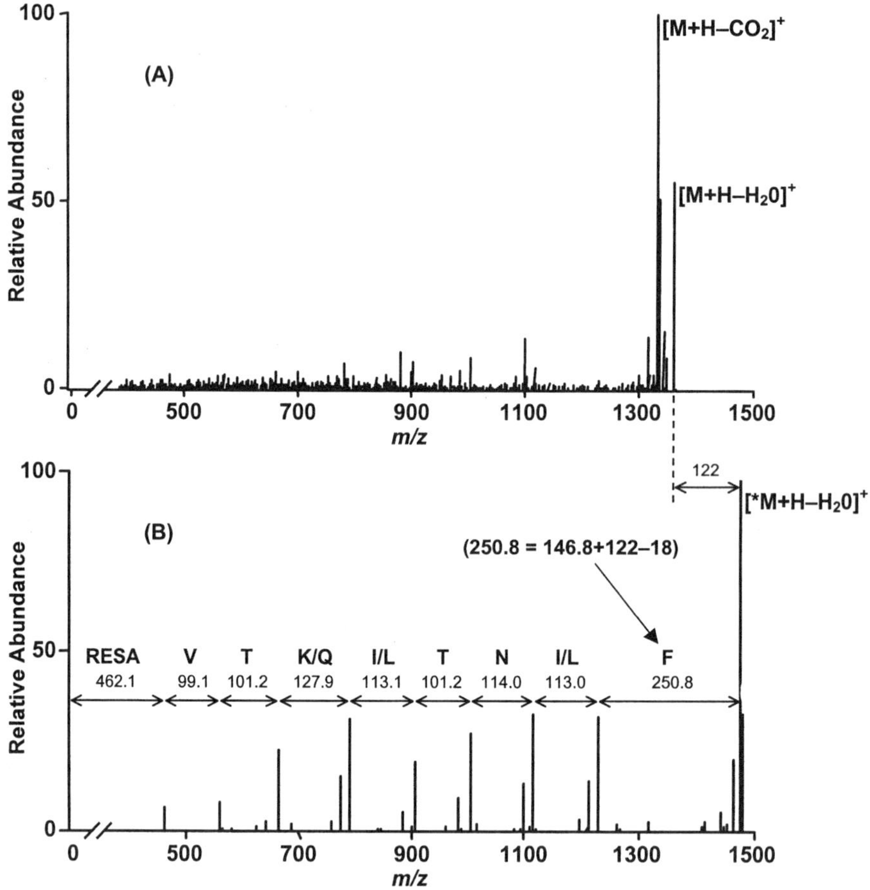

Figure 14.6. Positive, ion trap (Finnigan LC-Q) tandem mass spectra of the singly protonated precursor of an unknown peptide (a) before and (b) after SA derivatization. The peptide was found in a tryptic digest of *E. coli* uracil DNA glycosylase (Ung), which is an enzyme that specifically recognizes uracil in DNA and initiates base excision repair by hydrolyzing the glycosylic bond linking the uracil base to the deoxyribose sugar [109], but its observed molecular weight does not correspond to that of any of Ung's predicted tryptic peptides. The ion trap's collisional energy parameter was set at 30% of the highest available collision energy. Essentially no diagnostically useful peaks are observed in the spectrum of the native peptide, whereas a pronounced series of peaks composed exclusively of y_n ions is evident in the spectrum of the SA-derivatized peptide. The latter spectrum also displays a series of less prominent peaks that result from water losses off the y_n ions. The ion trap's low-mass cutoff [21] prevents signals corresponding to fragment ions smaller than y_4 from being observed. The SA-derivatized peptide's spectrum reveals the sequence F[I/L]NT[I/L][K/Q]TV, which exactly matches the middle eight residues of the tryptic peptide QQPYFLNTLQTVASER (T2) out of Ung's amino acid sequence.

precursor ions—the singly protonated precursor yielding abundant y_n ions *N*-terminal to the proline and the doubly protonated precursor yielding abundant y_n ions *C*-terminal to the proline. A basic amino acid, such as histidine, or an aromatic amino acid, such as tryptophan, located in the middle of a peptide can induce its singly protonated native form to dissociate predominantly into b_n fragments on the *C*-terminal side of the perturbating residue, thus revealing roughly half of the peptide's sequence. Attaching a sulfonic acid moiety to the peptide's *N* terminus can almost completely shift its fragmentation into y_n ions from the *N*-terminal side of the basic or aromatic residue and, thus, expose the other half of the peptide's sequence [104,107]. Sulfonic acid derivatives of arginine-containing peptides show increased in-source fragmentation, which could be useful for verifying a peptide's sequence or obtaining a sequence tag from it on single quadrupole instrument [108].

A sulfonic acid derivatization performed on the crude tryptic digest of a protein can produce both monoderivatized and bisderivatized forms of peptides that contain *C*-terminal lysines. This is potentially a problem because the tandem mass spectra of bisderivatized peptides are almost always too complex to interpret. To date, Chen et al. have found that they are able to easily find the singly derivatized peptides in mixtures of derivatization products [104]. Figure 14.7 exhibits the ion trap tandem spectra of a small, electrosprayed tryptic peptide whose *C*-terminal amino acid is lysine. The native peptide's singly protonated precursor (*m/z* 616) decomposes into nearly complete series of both b_n and y_n fragment ions (Fig. 14.7a). Although the native peptide's entire sequence can be deduced from its fragment ion spectrum, the latter is unduly complex for an amino acid sequence this short. By attaching a sulfonic acid functionality to the peptide's *N* terminus, the singly protonated precursor is induced to dissociate almost entirely into y_n ions, and in turn, the fragment spectrum is strikingly simplified (Fig. 14.7b). The peptide's entire sequence can be determined with much greater ease from the spectrum of its SA derivative than from that of its native form. In situations where no monoderivatized peptides can be found in a reaction mixture, one can still resort to selective guanidination of the ε-amino groups of the lysine-terminated tryptic peptides as described by Keough et al. [105].

In summary, adding a sulfonic acid moiety to the *N* terminal of a tryptic peptide leads to low-energy CID spectra that feature predominantly y_n fragment ions. The exceptional simplicity of these spectra and the added specificity gained from the highly directed fragmentation into y_n ions facilitates identification of peptide sequences and, when combined with on-line searching, can be a powerful aid to identifying proteins. The application of SA derivatization is not limited to tryptic peptides, but can be extended to a variety of others whose *C*-termini end in amino acids other than arginine or lysine [104]. In contrast to positively charged *N*-terminal derivatives, which have yet to be proven practicable, negatively charged *N*-terminal SA derivatives are beginning to find practical use in analyses of proteins isolated from biological samples [104–106].

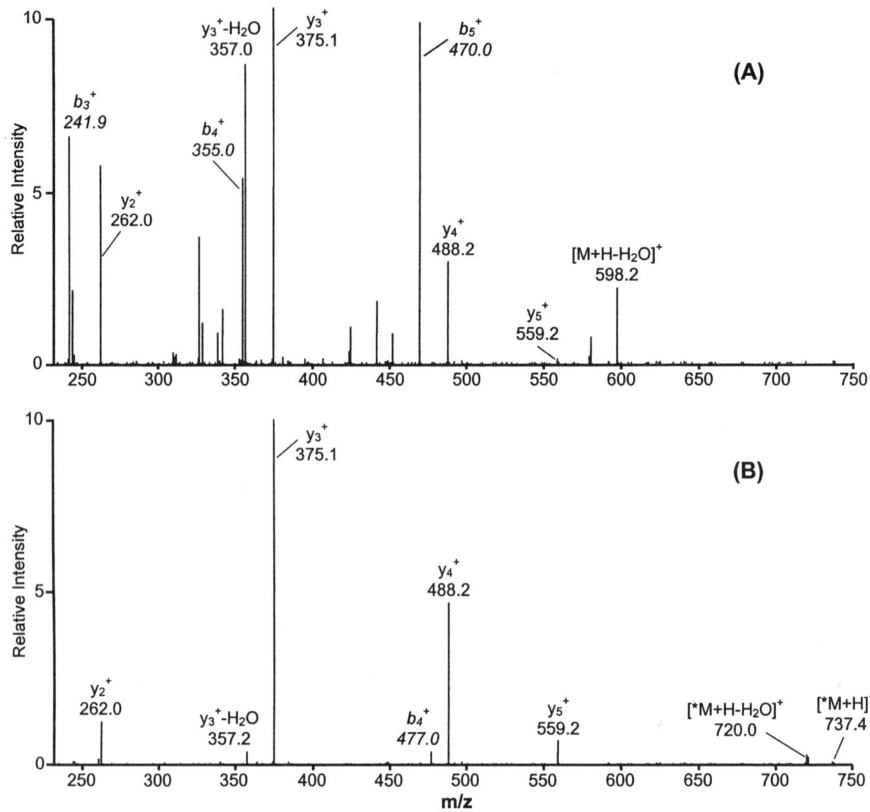

Figure 14.7. Positive, ion trap (Finnigan LC-Q) tandem mass spectra of the singly protonated precursor of the Lys-terminated, tryptic peptide GAIIDK (T13) from Ung (a) before and (b) after SA derivatization of the entire proteolytic digest. The ion trap's collisional energy parameter was set at 30% of the highest available collision energy. Detection of the smallest b_n and y_n fragments in either spectrum is prevented by the ion trap's low-mass cutoff [21]. With respect to sequence, the native peptide's spectrum is diagnostically complete but complex. By contrast, the SA-derivatized peptide's spectrum is equally complete but much simpler to interpret. It is readily evident from the tandem mass spectrum of the derivatized peptide that the SA moiety is attached to the *N*-terminal amine rather than to the *C*-terminal lysine's ε-amine because none of the y_n fragments have a 122 u mass increment due to the *N*-terminal SA group.

REFERENCES

1. C. K. Mathews and K. E. van Holde, *Biochemistry*, 2 ed., Benjamin/Cummings, Menlo Park, CA, 1996.
2. K. Biemann, F. Gapp, and J. Siebl, *J. Am. Chem. Soc.* **81**, 2274–2275 (1959).
3. D. J. Surman and J. C. Vickerman, *J. Chem. Soc., Chem. Commun.* 324–325 (1981).

4. M. Barber, R. S. Bordoli, R. D. Sedgwick, and A. N. Tyler, *J. Chem. Soc., Chem. Commun.* 325–327 (1981).
5. D. J. Surman and J. C. Vickerman, *J. Chem. Research* (*S*) 170–171 (1981).
6. M. Barber, R. S. Bordoli, R. D. Sedgwick, and A. N. Tyler, *Nature* **293**, 270–275 (1981).
7. K. Biemann and H. A. Scoble, *Science* **237**, 992–998 (1987).
8. R. S. Johnson, M. A. Stephen, K. Biemann, J. T. Stults, and J. T. Watson, *Anal. Chem.* **59**, 2621–2625 (1987).
9. R. S. Johnson, S. A. Martin, and K. Biemann, *Int. J. Mass Spectrom. Ion Process.* **86**, 137–154 (1988).
10. I. A. Papayannopoulos, *Mass Spectrom. Rev.* **14**, 49–73 (1995).
11. M. Yamashita and J. B. Fenn, *Iyo Masu Kenkyukai Koenshu* **9**, 203–206 (1984).
12. M. Yamashita and J. B. Fenn, *J. Phys. Chem.* **88**, 4451–4459 (1984).
13. M. Yamashita and J. B. Fenn, *J. Phys. Chem.* **88**, 4671–4675 (1984).
14. M. Karas, D. Bachmann, U. Bahr, and F. Hillenkamp, *Int. J. Mass Spectrom. Ion Process.* **78**, 53–68 (1987).
15. K. Tanaka, Y. Ido, S. Akita, Y. Yoshida, and T. Yoshida, *Iyo Masu Kenkyukai Koenshu* **12**, 219–222 (1987).
16. K. Tanaka, H. Waki, Y. Ido, S. Akita, Y. Yoshida, and T. Yoshida, *Rapid Commun. Mass Spectrom.* **2**, 151–153 (1988).
17. M. Karas and F. Hillenkamp, *Anal. Chem.* **60**, 2299–2301 (1988).
18. M. Karas, U. Bahr, and F. Hillenkamp, *Int. J. Mass Spectrom. Ion Process.* **92**, 231–242 (1989).
19. D. F. Hunt, J. R. Yates III, J. Shabanowitz, S. Winston, and C. R. Hauer, *Proc. Natl. Acad. Sci.* (USA) **83**, 6233–6237 (1986).
20. K. Biemann and I. A. Papayannopoulos, *Acc. Chem. Res.* **27**, 370–378 (1994).
21. K. R. Jonscher and J. R. Yates III, *Anal. Biochem.* **244**, 1–15 (1997).
22. O. A. Mirgorodskaya, A. A. Shevchenko, I. V. Chernushevich, A. F. Dodonov, and A. Miroshnikov, *Anal. Chem.* **66**, 99–107 (1994).
23. A. N. Verentchikov, W. Ens, and K. G. Standing, *Anal. Chem.* **66**, 126–133 (1994).
24. H. R. Morris, T. Paxton, A. Dell, J. Langhorne, M. Berg, R. S. Bordoli, J. Hoyes, and R. H. Bateman, *Rapid Commun. Mass Spectrom.* **10**, 889–896 (1996).
25. A. Shevchenko, I. Chernuchevich, W. Ens, K. G. Standing, B. Thomson, M. Wilm, and M. Mann, *Rapid Commun. Mass Spectrom.* **11**, 1015–1024 (1997).
26. R. J. Cotter, *Anal. Chem.* **71**, 445A–451A (1999).
27. I. V. Chernushevich, W. Ens, and K. G. Standing, *Anal. Chem.* **71**, 452A–461A (1999).
28. B. Spengler, D. Kirsch, and R. Kaufmann, *Rapid Commun. Mass Spectrom.* **5**, 198–202 (1991).
29. B. Spengler, D. Kirsch, and R. Kaufmann, *J. Phys. Chem.* **96**, 9678–9684 (1992).
30. B. Spengler, D. Kirsch, R. Kaufmann, and E. Jaeger, *Rapid Commun. Mass Spectrom.* **6**, 105–108 (1992).
31. B. Spengler and R. Kaufmann, *Analusis* **20**, 91–101 (1992).

32. R. Kaufmann, B. Spengler, and F. Lützenkirchen, *Rapid Commun. Mass Spectrom.* **7**, 902–910 (1993).
33. R. Kaufmann, D. Kirsch, and B. Spengler, *Int. J. Mass Spectrom. Ion Process.* **131**, 355–385 (1994).
34. C. K. G. Piyadasa, P. Håkansson, T. R. Ariyaratne, and D. F. Barofsky, *Rapid Commun. Mass Spectrom.* **12**, 1655–1664 (1998).
35. M. Vestal, P. Juhasz, W. Hines, and S. Martin, *Proc. 46th ASMS Conf. Mass Spectrometry and Allied Topics*, Orlando, FL, The American Society for Mass Spectrometry, 1998, p. 35.
36. D. L. Katz and D. F. Barofsky, *Proc. 47th ASMS Conf. Mass Spectrometry and Allied Topics*, Dallas, TX, 1999.
37. M. A. Baldwin, K. F. Medzihradszky, J. Campbell, X. C. Chen, P. Juhasz, M. L. Vestal, and A. L. Burlingame, *Proc. 47th ASMS Conf. Mass Spectrometry and Allied Topics*, Dallas, TX, The American Society for Mass Spectrometry, 1999.
38. K. F. Medzihradszky, J. M. Campbell, M. A. Baldwin, A. M. Falick, P. Juhasz, M. L. Vestal, and A. L. Burlingame, *Anal. Chem.* **72**, 552–558 (2000).
39. M. Vestal, J. Campbell, K. Hayden, and P. Juhasz, *Proc. 48th ASMS Conf. Mass Spectrometry and Allied Topics*, Long Beach, CA, The American Society for Mass Spectrometry, 2000.
40. B. P. Wolf, L. W. Sumner, J. Campbell, J. P. DeGnore, and P. Juhasz, *Proc. 48th ASMS Conf. Mass Spectrometry and Allied Topics*, Long Beach, CA, The American Society for Mass Spectrometry, 2000.
41. A. M. Falick, P. Juhasz, J. M. Campbell, and M. L. Vestal, *Proc. 48th ASMS Conf. Mass Spectrometry and Allied Topics*, Long Beach, CA, The American Society for Mass Spectrometry, 2000.
42. M. Vestal, P. Juhasz, W. Hines, and S. Martin, in *Mass Spectrometry in Biology & Medicine*, A. L. Burlingame, S. A. Carr, and M. A. Baldwin, eds., Humana Press: Totowa, NJ, 2000, pp. 1–16.
43. A. N. Krutchinsky, A. V. Loboda, V. L. Spicer, R. Dworschak, W. Ens, and K. G. Standing, *Rapid Commun. Mass Spectrom.* **12**, 508–518 (1998).
44. A. Loboda, A. Krutchinsky, V. Spicer, W. Ens, and K. Standing, *Proc 47th ASMS Conf. Mass Spectrometry and Allied Topics*, Dallas, TX, The American Society for Mass Spectrometry, 1999.
45. A. Shevchenko, A. Loboda, A. Shevchenko, W. Ens, and K. G. Standing, *Anal. Chem.* **72**, 2132–2141 (2000).
46. A. N. Krutchinsky, W. Zhang, and B. T. Chait, *J. Am. Soc. Mass Spectrom.* **11**, 493–504 (2000).
47. A. Verentchikov, K. Hayden, and M. Vestal, *Proc. 48th ASMS Conf. Mass Spectrometry and Allied Topics*, Long Beach, CA, The American Society for Mass Spectrometry, 2000.
48. A. M. Falick and D. A. Maltby, *Anal. Biochem.* **182**, 165–169 (1989).
49. W. V. Ligon, Jr., *Anal. Chem.* **58**, 485–487 (1986).
50. S. Naylor, A. F. Findeis, B. W. Gibson, and D. H. Williams, *J. Am. Chem. Soc.* **108**, 6359–6363 (1986).
51. K. L. Busch, S. E. Unger, A. Vincze, R. G. Cooks, and T. Keough, *J. Am. Chem. Soc.* **104**, 1507–1511 (1982).

52. K. D. W. Roth, Z.-H. Huang, N. Sadagopan, and J. T. Watson, *Mass Spectrom. Rev.* **17**, 255–274 (1998).
53. D. Renner and G. Spiteller, *Biomed. Environ. Mass Spectrom.* **13**, 405–410 (1986).
54. P. Roepstorff and J. Fohlman, *Biomed. Mass Spectrom.* **11**, 601 (1984).
55. K. Biemann, *Meth. Enzymol.* **193**, 886–887 (1990).
56. K. B. Tomer, F. W. Crow, and M. L. Gross, *J. Am. Chem. Soc.* **105**, 5487–5488 (1983).
57. N. J. Jensen, K. B. Tomer, and M. L. Gross, *J. Am. Chem. Soc.* **107**, 1863–1868 (1985).
58. M. L. Gross, *Int. J. Mass Spectrom. Ion Process.* **118/119**, 137–165 (1992).
59. D. A. Kidwell, M. M. Ross, and R. J. Colton, *Springer Ser. Chem. Phys.* (*Second. Ion Mass Spectrom., SIMS 4*) **36**, 412–414 (1984).
60. B. D. Bennett and R. A. Day, *Proc. 35th ASMS Conf. Mass Spectrometry and Allied Topics*, Denver, CO, The American Society for Mass Spectrometry, 1987, pp. 568–569.
61. K. Hirayama, S. Akashi, R. Yuji, U. Niitsu, and Y. Fujimoto, *Ort. Mass Spectrom.* **28**, 1516–1524 (1993).
62. D. S. Wagner, A. Salad, D. A. Gage, J. Leykam, J. Fetter, R. Hollingsworth, and J. T. Watson, *Biol. Mass Spectrom.* **20**, 419–425 (1991).
63. A. L. McCormack, Á. Somogyi, A. R. Dongré, and V. H. Wysocki, *Anal. Chem.* **65**, 2859–2872 (1993).
64. Á. Somogyi, V. H. Wysocki, and I. Mayer, *J. Am. Soc. Mass Spectrom.* **5**, 704–717 (1994).
65. K. A. Cox, S. J. Gaskell, M. Morris, and A. Whiting, *J. Am. Soc. Mass Spectrom.* **7**, 522–531 (1996).
66. A. R. Dongré, J. L. Jones, Á. Somogyi, and V. H. Wysocki, *J. Am. Chem. Soc.* **118**, 8365–8374 (1996).
67. J. C. Rouse, W. Yu, and S. A. Martin, *J. Am. Soc. Mass Spectrom.* **6**, 822–835 (1995).
68. R. S. Brown and J. J. Lennon, *Anal. Chem.* **67**, 1998–2003 (1995).
69. R. Kaufmann, P. Chaurand, D. Kirsch, and B. Spengler, *Rapid Commun. Mass Spectrom.* **10**, 1199–1208 (1996).
70. J. T. Stults, *Proc. 40th ASMS Conf. Mass Spectrometry and Allied Topics*, Washington, DC, The American Society for Mass Spectrometry, 1992, pp. 1815–1816.
71. N. E. Sherman, N. A. Yates, J. Shabanowitz, D. F. Hunt, W. Jeffery, M. Bartlet-Jones, and D. J. C. Pappin, *Proc. 43rd ASMS Conf. Mass Spectrometry and Allied Topics*, Atlanta, GA, The American Society for Mass Spectrometry, 1995, p. 626.
72. N. Sadagopan, J. Wu, Y. Yang, Z. H. Huang, and J. T. Watson, *Proc. 45th ASMS Conf. Mass Spectrometry and Allied Topics*, Palm Springs, CA, The American Society for Mass Spectrometry, 1997, p. 1404.
73. M. S. Cárdenas, E. van der Heeft, and A. P. J. M. de Jong, *Rapid Commun. Mass Spectrom.* **11**, 1271–1278 (1997).
74. M. Bartlet-Jones, W. A. Jeffery, H. F. Hansen, and D. J. C. Pappin, *Rapid Commun. Mass Spectrom.* **8**, 737–742 (1994).
75. P.-C. Liao and J. Allison *J. Mass Spectrom.* **30**, 511-512 (1995).
76. W. Hines, J. Peltier, F. Hsieh, and S. A. Martin, *Proc. 43th ASMS Conf. Mass Spectrometry and Allied Topics*, Atlanta, GA, The American Society for Mass Spectrometry, 1995, p. 387.

77. Z.-H. Huang, J. Wu, K. D. W. Roth, Y. Yang, D. A. Gage, and J. T. Watson, *Anal. Chem.* **69**, 137–144 (1997).
78. P.-C. Liao, Z.-H. Huang, and J. Allison, *J. Am. Soc. Mass Spectrom.* **8**, 501–509 (1997).
79. T. J. P. Naven, W. A. Jeffery, M. Bartlet-Jones, D. Rahman, and D. J. C. Pappin, *Proc. 45th ASMS Conf. Mass Spectrometry and Allied Topics*, Palm Springs, CA, The American Society for Mass Spectrometry, 1997, p. 318.
80. Z.-H. Huang, J. Wu, D. A. Gage, and J. T. Watson, *Proc. 45th ASMS Conf. Mass Spectrometry and Allied Topics*, Palm Springs, CA, The American Society for Mass Spectrometry, 1997, p. 1322.
81. J. R. Strahler, Y. Smelyanskiy, G. Lavine, and J. Allison, *Int. J. Mass Spectrom. Ion Process.* **169/170**, 111–126 (1997).
82. B. Spengler, F. Luetzenkirchen, S. Metzger, P. Chaurand, R. Kaufmann, W. Jeffery, M. Bartlet-Jones, and D. J. C. Pappin, *Int. J. Mass Spectrom. Ion Process.* **169/170**, 127–140 (1997).
83. K. Biemann, *Biomed. Environ. Mass Spectrom.* **16**, 99–111 (1988).
84. J. E. Vath and K. Biemann, *Int. J. Mass Spectrom. Ion Process.* **100**, 287–299 (1990).
85. S. A. McLuckey, *J. Am. Soc. Mass Spectrom.* **3**, 599–614 (1992).
86. O. Burlet, R. S. Orkiszewski, K. D. Ballard, and S. J. Gaskell, *Rapid Commun. Mass Spectrom.* **6**, 658–662 (1992).
87. K. M. Downard and K. Biemann, *J. Am. Soc. Mass Spectrom.* **5**, 966–975 (1994).
88. J. L. Jones, A. R. Dongré, Á. Somogyi, and V. H. Wysocki, *J. Am. Chem. Soc.* **116**, 8368–8369 (1994).
89. D. R. Mueller, M. Eckersley, and W. J. Richter, *Org. Mass Spectrom.* **23**, 217–222 (1988).
90. A. J. Alexander, P. Thibault, and R. K. Boyd, *Rapid Commun. Mass Spectrom.* **3**, 30–34 (1989).
91. P. T. Kenny, K. Nomoto, and R. Orlando, *Rapid Commun. Mass Spectrom.* **6**, 95–97 (1992).
92. X.-J. Tang, P. Thibault, and R. K. Boyd, *Anal. Chem.* **65**, 2824–2834 (1993).
93. V. H. Wysocki, J. L. Jones, A. R. Dongré, Á. Somogyi, and A. L. McCormack, in *Biological Mass Spectrometry: Present and Future*, J Wiley, New York, 1994, pp. 249–254.
94. R. S. Johnson, D. Krylov, and K. A. Walsh, *J. Mass Spectrom.* **30**, 386–387 (1995).
95. J. T. Stults, J. Lai, S. McCune, and R. Wetzel, *Anal. Chem.* **65**, 1703–1706 (1993).
96. J. Zaia and K. Biemann, *J. Am. Soc. Mass Spectrom.* **6**, 428–436 (1995).
97. B. Spengler, F. Luetzenkirchen, S. Metzger, R. Kaufmann, W. Jeffery, and D. J. C. Pappin, *Proc. 44th ASMS Conf. Mass Spectrometry and Allied Topics*, Portland, OR, The American Society for Mass Spectrometry, 1996, p. 352.
98. T. Lin and G. L. Glish, *Proc. 45th ASMS Conf. Mass Spectrometry and Allied Topics*, Palm Springs, CA, The American Society for Mass Spectrometry, 1997, p. 588.
99. M. Adamczyk, J. C. Gebler, and J. Wu, *Rapid Commun. Mass Spectrom.* **13**, 1413–1422 (1999).
100. N. Sadagopan and J. T. Watson, *J. Am. Soc. Mass Spectrom.* **12**, 399–409 (2001).

101. T. Keough, R. S. Youngquist, and M. P. Lacey, *Proc. Natl. Acad. Sci.* (USA) **96**, 7131–7136 (1999).

102. O. Burlet, C.-Y. Yang, and S. J. Gaskell, *J. Am. Soc. Mass Spectrom.* **3**, 337–344 (1992).

103. T. Keough, M. P. Lacey, A. M. Fieno, R. A. Grant, Y. Sun, M. D. Bauer, and K. B. Begley, *Proc. 48th ASMS Conf. Mass Spectrometry and Allied Topics*, Long Beach, CA, The American Society for Mass Spectrometry, 2000.

104. T.-F. Chen, P. R. Gafken, S. Kim, D. Griffin, E. Barofsky, B. Arbogast, and D. F. Barofsky, *Anal. Chem.* (in press).

105. T. Keough, M. P. Lacey, and R. S. Youngquist, *Rapid Commun. Mass Spectrom.* **14**, 2348–2356 (2000).

106. T. Keough, M. P. Lacey, A. M. Fieno, R. A. Grant, Y. Sun, M. D. Bauer, and K. B. Begley, *Electrophoresis* **21**, 2252–2265 (2000).

107. T.-F. Chen, P. R. Gafken, S. Kim, and D. F. Barofsky, *Proc. 48th ASMS Conf. Mass Spectrometry and Allied Topics*, Long Beach, CA, The American Society for Mass Spectrometry, 2000.

108. M. D. Bauer, Y. Sun, T. Keough, and M. P. Lacey, *Rapid Commun. Mass Spectrom.* **14**, 924–929 (2000).

109. D. W. Mosbaugh and S. W. Bennett, *Prog. Nucleic Acid Res. Mol. Biol.* **48**, 315–370 (1994).

CHAPTER 15

MALDI-TOF/TOF Technology for Peptide Sequencing and Protein Identification

PETER JUHASZ, JENNIFER M. CAMPBELL, and MARVIN L. VESTAL

Applied Biosystems, Framingham, Massachusetts, USA

CONTENTS

15.1. INTRODUCTION

Tandem mass spectrometric techniques, or MS/MS for short, have been used since the late 1980s to generate complete or partial structural information on peptides. The first applications were published at that time describing the various processes and pathways of peptide fragmentation and procedures to determine the primary sequence of peptides from the fragment data [1–3]. That time the favored method of ionization was fast-atom bombardment (FAB), and the most important selling point of the mass spectrometric approach was the ability to sequence modified and *N*-terminally blocked peptides not amenable for Edman degradation. Since then a few developments

Mass Spectrometry and Hyphenated Techniques in Neuropeptide Research, Edited by Jerzy Silberring and Rolf Ekman.
ISBN 0-471-35493-7

have dramatically increased the importance of mass spectrometry in the sequence analysis of peptides and proteins.

First, new ionization methods have been introduced, allowing the mass analysis of biological molecules at orders of magnitude better sensitivity. The significance of electrospray ionization [4] and matrix-assisted laser desorption/ionization (MALDI) [5] was recognized at about the same time. Almost immediately biochemists have gained access to three or more orders of magnitude higher sensitivity and now mass spectrometry clearly outperformed the classical chemical methods of peptide analysis. By today, using special sample introduction or sample preparation techniques for ESI and MALDI such as nanospray [6] or thin film preparation [7], the sensitivity limits of these ionization techniques are well in the realm of subfemtomole sample amounts.

The other, less spectacular and somewhat slower development was the ability to automate and speed up the operation of mass spectrometers. On today's commercial instrumentation hundreds and even thousands of mass spectra and MS/MS spectra can be recorded every day in an automated fashion [8,9]. The capability of generating data and characterizing peptides at such a rate has rendered the emerging field of large-scale protein characterization, or proteomics, one of the most exciting developments in the current life sciences. Biochemists and mass spectroscopists have realized that complete characterization of peptide sequences was not necessary as nature was using only a tiny subset of all combinatorially possible amino acid sequences. This subset is organized into genomic and protein sequence databases today [10]. As the coverage of the genomes of different organisms is becoming more and more complete, the better the potential of high-throughput mass spectrometry can be utilized [11–13].

Two types of MS experiments are used to identify a protein in a biological sample. In one case, where either the entire genome of an organism is known or the protein sequence of interest can be assumed to be in the database, masses of enzymatic degradation products—most frequently derived from trypsin digests—can be used to identify the source protein [14–17]. This approach fails if the database entries do not provide sufficiently extensive (DNA or amino acid) sequence coverage for the proteins, which is typical for the EST databases, or the sample contains peptides from many proteins (because no protein fractionation had been done previously) and individual peptide masses may belong to more than one protein in the sample. For such samples MS/MS is ideal an ideal tool to identify proteins [18,19]. Another important reason for the easy acceptance of mass spectrometry for these applications is the fact that the complete interpretation of MS/MS spectra is not necessary for the successful identification of proteins. The relatively complex interpretation procedure, perhaps the only perceived disadvantage of the MS/MS approach, can be abandoned.

Historically most of the MS/MS work has been done on instrumentation featuring electrospray ionization and ion fragmentation through multiple, low-energy collisions. The reason is quite likely the coincidence of the availability of scanning type instruments—triple quadrupoles and ion traps—and the need to connect to liquid chromatography requiring continuous ionization (demand for online LC/MS). On the other hand MALDI was the most successful coupled with time-of-flight mass analyzers (TOF), which are ideal instruments due to the pulsed nature of ion formation. Until the 1990s, the MS/MS capabilities of MALDI TOF were restricted to the utilization of metastable decomposition. This technique termed as postsource decay or (PSD) by its inventors [20] was hardly competitive with the ESI based methods because of its passive control of peptide fragmentation, and relatively modest data quality. As a significant step in combining MALDI ionization mode with high-performance MS/MS for proteomic applications, a quadrupole–TOF hybrid instrument have been introduced [21].

In this chapter we describe the efforts a new type of MS/MS technology that combines the attributes of mass spectrometric techniques to yield an optimum set of features required for high-speed protein characterization. It is based on the MALDI ionization method on a time-of-flight platform. The selection of the precursor ion is made in a time-of-flight analyzer as well. The methodology is referred to as MALDI-TOF/TOF. Its key characteristics are (1) MALDI-TOF capability in MS mode, (2) high-energy collision-induced dissociation (CID), and (3) time-of-flight mass analysis in MS/MS mode for high speed and high sensitivity. The most significant advantage of MALDI-TOF analysis is the possibility of generating peptide mass fingerprints of complex peptide mixtures very fast. High-energy CID will enable users to have access to more detailed structural information on peptide sequences: identifying modified amino acid residues, differentiating amino acid motifs that are indistinguishable in low-energy CID [22]. These features may come at the price of requiring more complex interpretation software for database searching and/or de novo sequence interpretation.

This chapter is divided into two sections: in the first the technical background of MALDI TOF/TOF technology will be described: details of the instrumentation and performance characteristics. In the second part the focus will be on the applications. First the fragmentation characteristics will be discussed, than a few applications of TOF/TOF for the purpose of protein identification and de novo sequencing will be presented.

15.2. MALDI-TOF/TOF INSTRUMENTATION

15.2.1. The Time-of-Flight/Time-of-Flight (TOF/TOF) Analyzer

Earlier versions of the PerSeptive BioSystems TOF/TOF tandem mass spectrometer have been described previously [23,24]. In this report the latest configuration of the instrument will be presented. Some elements of the hardware have not changed and more detailed description of these features can be found in the earlier references. The schematic of the instrument is depicted in

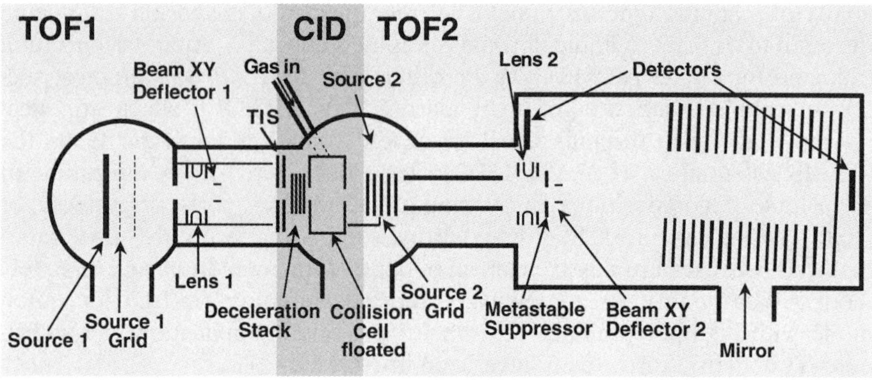

Figure 15.1. Schematic of the MALDI-TOF-TOF tandem mass spectrometer. TIS is the abbreviation for the timed ion selector.

Fig. 15.1. To illustrate the operation of the hardware, the timing diagram of a typical MS/MS experiment is provided in Fig. 15.2. The instrument can be conceived as consisting of three parts for the MS/MS operation. The first of these is a short linear time-of-flight analyzer, designated by TOF1 in Fig. 15.1. It serves the purpose of generating the collection of potential precursor ions and guiding them into the collision region marked as CID in Fig. 15.1. Here a precursor ion of a single *m/z* is isolated from the stream of ions produced in the source of TOF1. The isolated precursor ions undergo collisions with a neutral gas in the collision chamber. The third part of the instrument, TOF2, is a reflector type TOF analyzer intended to provide high-resolution mass spectra of the fragment ions. Appropriately timed extraction pulses applied on the "second ion source" accelerate the fragment ions and the unfragmented precursor ions into the second TOF analyzer to measure their mass-to-charge ratio. Each extraction pulse renders a complete MS-MS spectrum in the detector.

The ion source of TOF1 is identical to the source of a commercial *Voyager* STR MALDI time-of-flight mass spectrometer (PerSeptive Biosystems). In the MS/MS experiment the electric potentials applied in the two-stage ion source are typically 8.0 kV on the sample plate and 7.0 kV on the "extraction grid." As a modification of the commercial instrument, instead of a grid, a flat electrode was used with small holes cut in the center for letting the ion beam through and off the center for the laser beam and visual observation. The drift length of TOF1 is approximately 25 cm.

The extraction delay (τ_1 in Fig. 15.2)—in conjunction with the voltage settings in the ion source—is chosen such that the velocity focus for the precursor ions positioned approximately in the center of the timed ion selector device (TIS), which is used to isolate the desired precursor ions for MS/MS. Precise tuning of the extraction parameters is important because the breadth

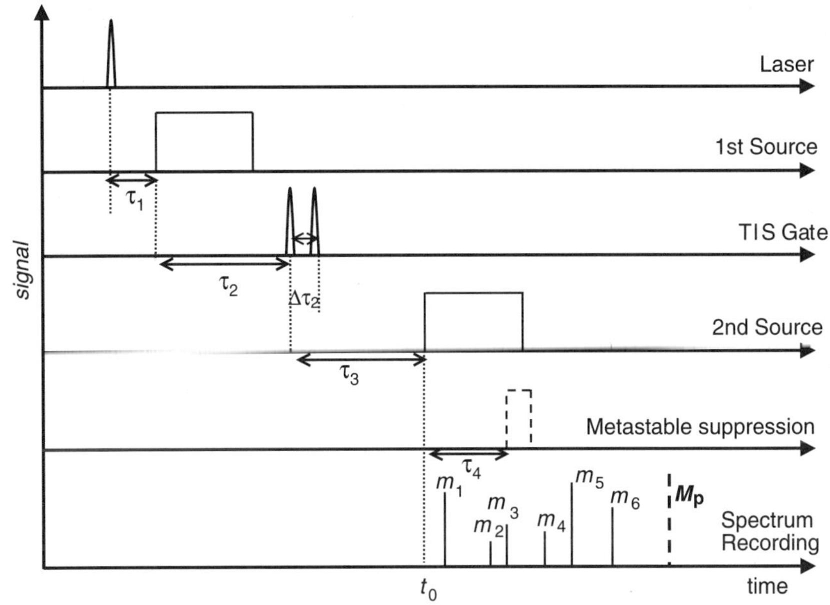

Figure 15.2. Timing diagram for an MS/MS experiment on the TOF-TOF platform. The laser pulse triggers the extraction pulse from the first source, with a delay τ_1. The delay τ_2 for the precursor ion selection is calculated from the voltages applied on the first source and from the mass of the precursor. With a time delay of $\tau_1 + \tau_2 + \tau_3$ after the laser shot, the extraction pulse is applied to the second ion source. This marks the starting time for the spectrum recording. Optionally, a metastable suppressor can be used to deflect metastable ions from the precursor generated in the drift tube of the second TOF analyzer.

of velocity focus of such a short linear TOF analyzer is quite tight [25]. The TIS consists of two deflection gates, separated by a distance of 1 cm. The first gate acts as a high-pass filter and is held at a deflecting potential (~1 kV) until the calculated arrival time of the precursor ion to the entrance of the TIS (τ_2 in Fig. 15.2). Consequently, ions with a mass lower than that of the precursor will be deflected and will not enter the collision cell. The second gate acts as a lowpass filter and is held at ground until the calculated time at which the precursor leaves the TIS ($\tau_2 + \Delta\tau_2$). At this time, the second gate is rapidly switched to a deflecting potential, preventing ions with a higher mass than the precursor from entering the collision cell. Hence the TIS region will be field-free for only the brief period of time during which the precursor ions are transmitted to the collision cell. The TIS provides a separation resolution ($m/\Delta m$) of about 200 at mass 1000.

The collision cell is electronically isolated from the flight tube and maintained at a potential chosen to provide the desired collision energy for the source operating conditions. This configuration allows the source acceleration energies to be altered independently of collision energies. Theoretically, collision energies can be varied from 0 eV almost up to the acceleration energy corresponding to the voltage in the first source. However, because of the compromised performance of the second analyzer (TOF2), this range is limited to 0–3 keV. Most frequently the collision cell is maintained at 7 kV, defining the collision energy as 1 keV in the lab frame. This means that precursor ions are decelerated substantially before entering the collision cell. To ensure efficient transmission through the collision region, the deceleration stack is configured as an inverted cathode lens. The pressure in the collision cell is maintained at a level where the majority of ion fragmentation is achieved through single collision events [26]. The gas inlet manifold facilitates easy switching between different collision gases. In the experiments reported below, argon was used exclusively. These collision conditions have proved to produce strong sequence ion coverage, as well as typical, "high energy" fragmentation patterns such as abundant low-mass fragments and internal fragment ions.

The 7-cm stretch of flight path between the collision cell and the second acceleration region is maintained at the potential of the collision cell, creating an effective field-free region between the collision cell and the second acceleration region. Ions (both the fragments and the precursor) in this region travel with a velocity distribution, which is, in the first approximation, independent of their mass-to-charge ratio. The resulting mass dependent energy distribution of fragment ions is determined by the collision energy and the mass-to-charge ratio of the precursor ion. For instance, assuming 1 keV collision energy for a precursor ion of *m/z* 1000, a fragment of *m/z* 100 will enter the second source region with kinetic energy of 100 eV. This mass dependent energy distribution will have to be efficiently compensated by an optimized design of the second TOF analyzer.

The acceleration region of the second source (source 2 in Fig. 15.1) also consists of two stages. The first plate (the one closest to the CID cell) is typically pulsed to 14 kV, and a second "grid" electrode is maintained at the potential of the collision cell (7 kV in most cases). Timing of the extraction pulses for the second source is calculated as the arrival time of the precursor ion to the center of the source region ($\tau_1 + \tau_2 + \tau_3$ in Fig. 15.2). The time of this pulse also defines the start time for recording MS/MS spectra in the analyzer TOF2. Optimizing the performance of the second TOF analyzer is carried out by the careful tuning of the extraction pulse delay and the reflector voltage. The overall ion transmission of the instrument can be optimized by two sets of *X–Y* deflection plates and two einzel lenses positioned next to both source 1 and source 2.

A more recent modification of the TOF/TOF platform was the addition of a Bradbury–Nielsen (BN) gate near the entrance to the ion mirror. This BN gate is labeled as "metastable suppressor" in Fig. 15.1, and located approximately 40 cm from the second source. The BN gate was added in response to the observation of spectral features and high chemical noise levels that originated from the metastable dissociation of precursor ions in the field-free

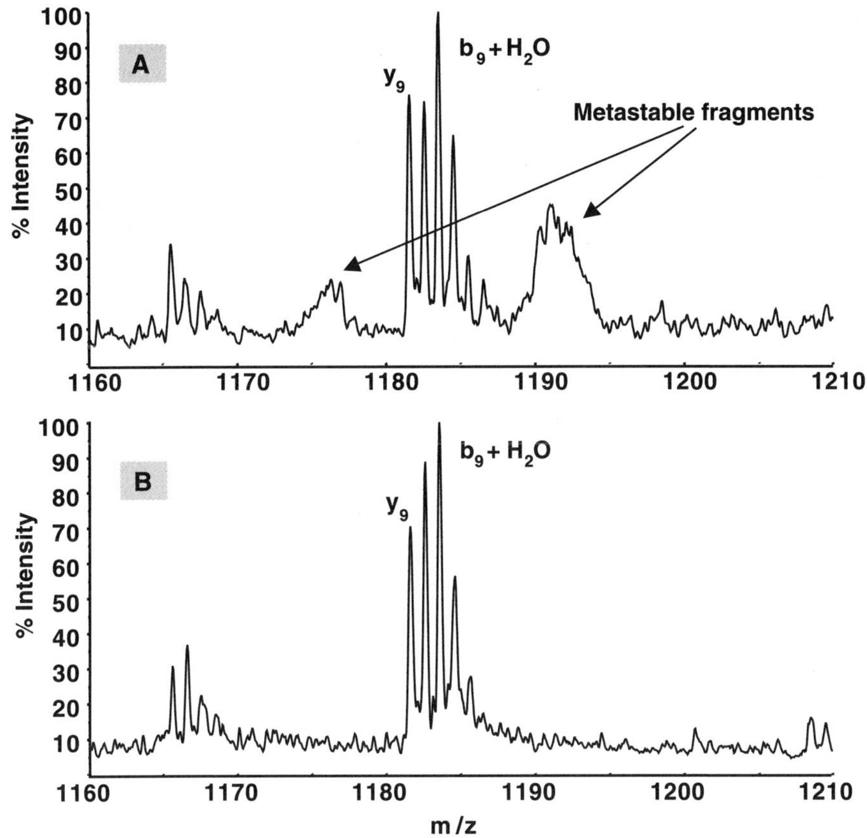

Figure 15.3. The effect of the metastable suppressor. Without the suppressor, see panel (a), the same y_9 and $b_9 + H_2O$ fragments formed as metastables, appear in the MS/MS spectrum as poorly resolved features potentially interfering with the well-resolved signal from fragments formed in the CID cell. The beneficial effect of the suppressor is shown in panel (b). The sample is angiotensin I [DRVYIHPFHL, MH^+ = 1296.685 Da (monoisotopic)].

region between the second source and the ion mirror. Fragment ions formed in this region have the same velocity as the precursor ion, but a kinetic energy proportional to the ratio of the mass of the fragment ion to the mass of the precursor ion. The BN gate is pulsed at the calculated arrival time of the precursor ions to the plane of the "metastable suppressor" (τ_4 in Fig. 15.2), preventing all ions traveling the velocity of the precursor (i.e., the precursor and metastable fragments) from entering the ion mirror. The efficiency of the metastable suppressor is demonstrated in Fig. 15.3. Panel (a) shows, in the 1160–1192-Da region, the MS/MS spectrum of angiotensin I (MH^+ = 1296.685 Da). The y_9 and $b_9 + H_2O$ fragments generated in the collision cell produce a well-resolved ion signal, whereas the same ion types as the result of metastable decay in the second TOF analyzer form poorly resolved features in the MS/MS spectrum. By activating the suppressor, the spectrum looks substantially cleaner in Fig. 15.3b. By careful tuning of the deflection plates in TOF2, it can be achieved that the precursor ion is not completely attenuated by the metastable suppressor. This practice allows one to detect the residual of the precursor ion signal and to utilize it for the mass calibration of the MS/MS spectrum.

15.2.2. Operating the TOF/TOF Instrument and Performance Characteristics

All MALDI spectra were recorded using an N_2 laser featuring a wavelength of 337 nm and operated at a 20-Hz repetition rate. As it is common in MALDI experiments many laser shots have been averaged to improve the signal-to-noise in the spectra. The typical spectrum represents an average of 1000 laser shots. In all the experiments α-cyano-4-hydroxycinnamic acid was used as matrix. This matrix material is prepared in 50% acetonitrile–water solution at 5–10 mg/mL concentration. The concentration and amount of sample depends on the biological application. It has been shown that 1 fmol of material deposited on the sample plate still produces MS/MS spectra of sufficiently good quality that database searching can identify the protein [26]. In most applications reported below the deposited sample amount was between 10 fmols and 1 pmol.

Variable electric potentials (on the extraction grids and on the ion mirror) and extraction delays offer a variety of ways to optimize MS/MS performance. In practice, the most important objective was to operate the instrument in a configuration where the performance was independent of the precursor ion. The performance characteristics reported below are reflective of this type of operating conditions even if careful tuning could give better performance for a particular precursor mass and fragment mass range. To demonstrate the instrument capabilities in terms of mass resolution and mass accuracy, Table 15.1 was compiled from the MS/MS spectrum of the peptide HQTVPQNTGGKNPDPWAK (MH^+ = 1974.979 Da, monoisotoic) derived from the tryptic digest of human transferrin. The Table lists the (almost complete) *b*- and *y*-ion series observed in this spectrum with the theoretical monoisotopic masses of the fragments, the signal-to-noise ratio, mass resolution, and the associated mass errors. Figure 15.4 provides blow-ups from four different regions of the MS/MS spectrum. As evident from Table 15.1 and Fig. 15.4, the mass resolution across the MS/MS spectrum varies between 2000 and 6000 at full width half maximum (FWHM). This ensures that across the entire mass region the isotopes are well resolved and the monoisotopic masses can be measured. The mass scale of MS/MS spectra was calibrated through an automated procedure using the precursor ion mass and one of the commonly observed low-mass peptide fragments as internal standards. Fragment masses in Table 15.1 have

TABLE 15.1. Signal-to-noise, Mass Resolution, and Mass Measurement Errors in the TofTof Spectrum of the Peptide HQTVPQNTGGKNPDPWAK (MH⁺ = 1974.979). See Figures 4 and 10B for Spectral Data and Fragment Annotation

Ion type	Observed mass (Da)	S/N	Mass resolution (FWHM)	Mass error (mDa)	Mass error of mass difference (mDa)
b_1	138.070	23	1300	3	3
y_1	147.112	17	~1000	−1	−1
y_2	218.135	56	1900	−15	−14
b_2	266.110	150	2200	−15	−18
b_3	367.134	118	3400	−39	−24
y_3	404.181	23	3300	−49	−34
b_4	466.216	78	2500	−25	14
y_4	501.257	103	2700	−26	23
b_5	563.263	9	3200	−31	−6
y_5	616.285	9	2560	−25	1
b_6	691.306	12	3100	−47	−16
y_6	713.330	44	3150	−32	−7
b_7	805.413	12	2140	17	64
y_7	827.388	24	2320	−17	15
b_8	906.581	9	~2000	138	119
y_8	955.623	7	~3000	123	140
b_9	963.516	12	2720	51	−87
y_9	1012.549	20	2880	27	−96
b_{10}	1020.544	9	~3000	58	7
y_{10}	1069.540	33	3500	−3	−30
b_{11}	1148.551	34	2850	−30	−88
y_{11}	1170.585	27	3100	−6	−3
b_{12}	1262.622	64	2520	−2	28
y_{12}	1284.632	27	4000	−2	4
b_{13}	1359.698	8	2550	−19	−17
b_{14}	1474.672	265	3600	−32	−13
y_{14}	1509.708	141	4000	−37	—
b_{15}	1571.606	4	3000	−151	119
y_{15}	1608.78	14	3900	−33	4
y_{16}	1709.835	30	5000	−26	7
b_{16}	1757.832	16	5600	−4	146
b_{17}	1828.804	19	4600	−31	−27
y_{17}	1837.774	19	6500	−146	−120
y_{18} = MH⁺	1974.979	93	6500	0	120

been determined by using the *m/z* 1974.979 and *m/z* 70.066 ions as internal references. The latter one is a characteristic fragment of proline residues (which also can originate from arginine). Mass errors are typically below 50 mDa across the spectrum provided the peaks have sufficiently high signal-to-noise ratio (S/N). As soon as the S/N declines below 20, the mass accuracy

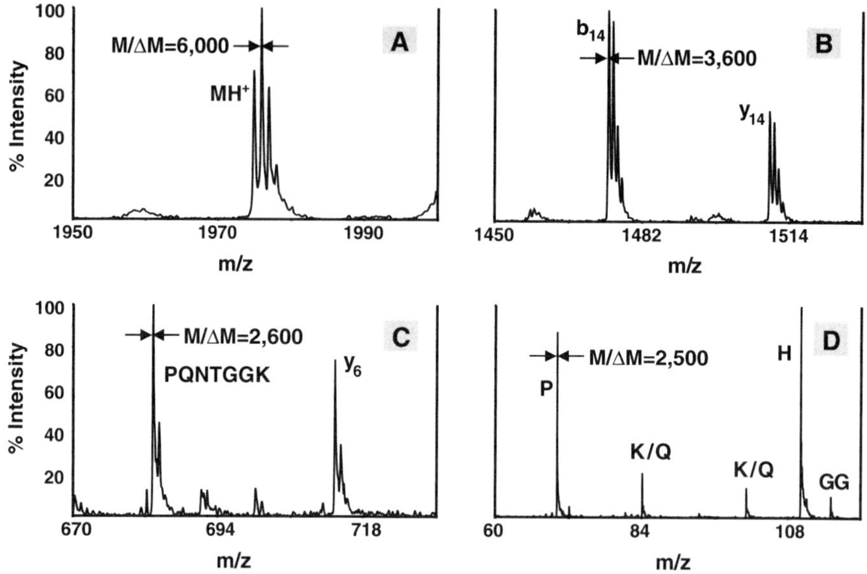

Figure 15.4. Typical mass resolution of the TOF-TOF tandem mass spectrometer in MS/MS mode. The sample is a tryptic peptide from human transferrin [HQTVPQNTGGKNPDPWAK, MH^+ = 1974.979 Da (monoisotopic)]. Four different zoom-ins are shown (a–d) to demonstrate distribution of mass resolution across the mass range.

quickly deteriorates because of poor ion statistics [27]. This dependence of the mass accuracy on the S/N is important to keep in mind when utilizing the MS/MS spectra for protein identification through database searching or for de novo sequence interpretation. Table 15.1 clearly indicates that mass resolution depends on the mass: the highest mass resolution is seen in the region near to the precursor mass, and it is gradually declining toward the lower masses. This behavior can be influenced to some extent by changing the tuning of the second TOF analyzer. However, changing the tuning in a way that compromises the precursor mass-independent performance of the instrument or sacrifices unit resolution in any region of the MS/MS spectra is of little practical value, if any.

15.3. APPLICATIONS OF THE MALDI-TOF/TOF TECHNOLOGY

15.3.1. Fragmentation Characteristics

Mass spectrometric techniques have been used for peptide sequencing for more than a decade (as of this writing), and the phenomenology of peptide fragmentation has been studied ever since [1,3,28–30]. The most widely used

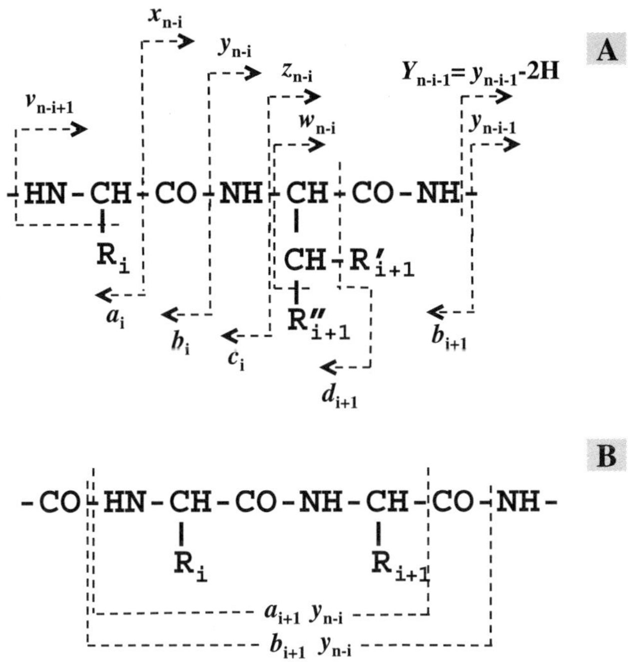

Figure 15.5. Fragmentation types observed in the CID of peptides. The annotation of fragment ions originates from Biemann [31]. Panel (a) presents fragment types resulting from a single cleavage of the backbone. In the case of *d*, *v*, and *w* ions this is accompanied by the loss of α- or β-carbon substituents. Fragments generated by the double cleavage of the amino acid backbone are referred to as *internal fragments*. Annotation for these is shown in panel (b).

naming convention for peptide fragmentation originates from Biemann [31], and this nomenclature is adopted here. A variety of chemical bonds can break during peptide fragmentation, and the most common fragmentation patterns are pictorially represented in Figs. 15.5a, b. In this chapter different fragment types will be grouped into four categories as described below:

1. *Single Cleavage of the Amino Acid Backbone.* These are the most commonly observed fragment types: *a*, *b*, *y* ions can be found in every kind of MS/MS spectra independent of the ionization mode or the collision energy. Other fragments such as the less frequently found *c*, *x*, and *z* fragments also belong here. However, comparison of TOF/TOF data and PSD spectra of the same compounds indicates that the formation of *x*, and to a lesser extent, *c* fragments are promoted by high-energy collisions. Fragments of the *z* type have identical composition with *y*-NH_3 fragments and, with the exception of high-energy CID data from four-sector instruments [32,33] and more recently from electron-capture dissociation experiments [34], no evidence is available

that this fragmentation occurs in TOF/TOF or low-energy CID experiments using electrospray ionization.

2. *Single Cleavage of the Amino Acid Backbone Accompanied by the Loss of Small Neutral Molecules*. Examples are water from S-, T-, D-, E-containing sequences or ammonia from R-, K-, Q-containing sequences. Quite frequently modified amino acids also produce neutral losses characteristic of the modification. For instance, the loss of phosphoric acid from phosphorylated serine, threonine, and tyrosine residues [35], or the loss of $SHOCH_3$ from oxidized methionine [36], are well documented in the literature. The common requirement for the formation of these fragment types is the presence of a critical amino acid residue in the fragment sequence that can account for the neutral loss. It is important to make sure that automated data interpretation procedures such as database search programs or sequencing algorithms are aware of the amino acid residues that can account for a particular neutral loss.

3. *High-Energy Fragments*. These are characterized by a single cleavage of the amino acid backbone accompanied by the loss of the α- or β-carbon substituents next to the cleavage site. Fragments of *d*,*v*,*w* types belong here. Whereas every standard amino acid residue, with the exception of glycine, can form a *v* ion by losing its α-substituent, the formation of *d* and *w* ions is restricted to residues having a β-substituent. Thus, *d* or *w* fragment formation is not possible with A, or G at the cleavage site. It has also been reported residues with aromatic β-substituents produce very weak or no *d*/*w* fragments [3]. There are two residues, isoleucine and threonine, with two different β-carbon substituents, giving rise to two different *d* and *w* fragments. Short of a standard notation, the loss of the lighter β-substituent—the CH_3 group in both cases—will be distinguished by using a prime sign (′) next to the fragment symbol. As pointed out in the literature, high-energy fragments form through a charge-remote mechanism and this requires the presence of a basic amino acid residue (preferably arginine) on the fragment sequence [3,22]. *Y* fragments are also categorized as high-energy fragments as evidenced by the comparison of TOF/TOF and PSD spectra where these fragments have not been observed (data not shown).

4. *Double Cleavage of the Amino Acid Backbone*. Double cleavage of the peptide backbone leads to internal fragments. The *C*-terminal side of these fragment ions may represent an *a*- or *b*-type cleavage, and the *N*-terminal side is a *y*-type cleavage in most cases (see Fig. 15.5b). Special occurrences of the $a_m y_n$ internal fragments consisting of a single amino acid residue are referred to as *immonium ions*. The majority of the immonium ions have a unique mass, so constraints on the amino acid composition of an unknown peptide can be readily determined, provided these ions are of sufficiently high abundance [37]. Internal fragments, similar to fragments from a single backbone cleavage, can exhibit losses of small neutrals if the composition of the fragment is compatible with such a loss.

All varieties and combinations of single and double cleavages would give rise to an enormous number of potential fragment ions from any given amino acid sequence. In reality, many of these combinatorial possibilities do not occur depending on the amino acid composition of the peptide, the ionization mode, and the collision energy. Relative abundances of other fragments, which appear in the mass spectra, can also exhibit a strong dependency on the same factors. Any application of MS/MS data, such as de novo peptide sequencing, protein identification, or just corroborating an ambiguous protein identity based on a peptide mass fingerprint, has to take these propensity rules into account in order to yield reliable answers.

Two fundamentally different fragmentation processes are employed in collision-induced dissociation (CID) of primary ions: (1) fragmentation through multiple, low-energy collisions in the collision gas where the timescale of fragmentation can extend from milliseconds to seconds; this process is referred to as low-energy collision activated dissociation (CAD) and (2) alternatively, the fragmentation process as a result of high-energy single-collision events where the timescale is well below a millisecond is referred to as *high-energy CID*. In addition to the collision energy, the mode of ionization also has a significant impact on the fragmentation pattern. Electrospray ionization (ESI) generates multiply charged ions from peptides, which undergo fragmentation more readily than do their singly charged counterparts. In MALDI singly charged ions are the most common and the fragmentation pattern of these ions exhibit substantially more loss of water and ammonia as observed in ESI. In the literature one can find detailed comparisons of these processes, and the reader is referred to these references (see Kaufmann et al. [38] and references cited therein). The focus below will be on the fragmentation types common in TOF/TOF fragment spectra of precursor ions generated by MALDI.

Collisions in the TOF/TOF instrument take place typically at 1-keV energy in the lab frame, so the fragment ions characteristic of high-energy CID are expected to occur: ions of the d, v, w, and Y type, and abundant internal fragments, and immonium ions. Also, as characteristic of the MALDI method of ionization, losses of water and ammonia (−17 and −18 Da, respectively) are anticipated to occur commonly [38]. As will be demonstrated, fragment types observed in TOF/TOF spectra bear strong similarity to those also observed in MALDI-PSD. The occurrence of high-energy CID fragments seems to be promoted by the presence of arginine residue(s) in the sequence [23]. This is a very important observation to consider when interpreting TOF/TOF spectra.

Similarities and differences between MALDI-PSD and MALDI-TOF/TOF spectra are illustrated in Figs. 15.6 and 15.7. In Fig. 15.6, fragmentation of a peptide is compared between PSD and TOF/TOF in the absence of arginine (PLYKKIIKKLLES, $MH^+ = 1573.010$ Da). The fragment types and even their relative abundances exhibit high similarity. (The suppressed b_{12} ion in the TOF/TOF spectrum is likely to be an artifact of the nonoptimized timing of the metastable suppressor). Also note the absence of high-energy fragments and neutral losses. The former effect is explained by the absence of arginine

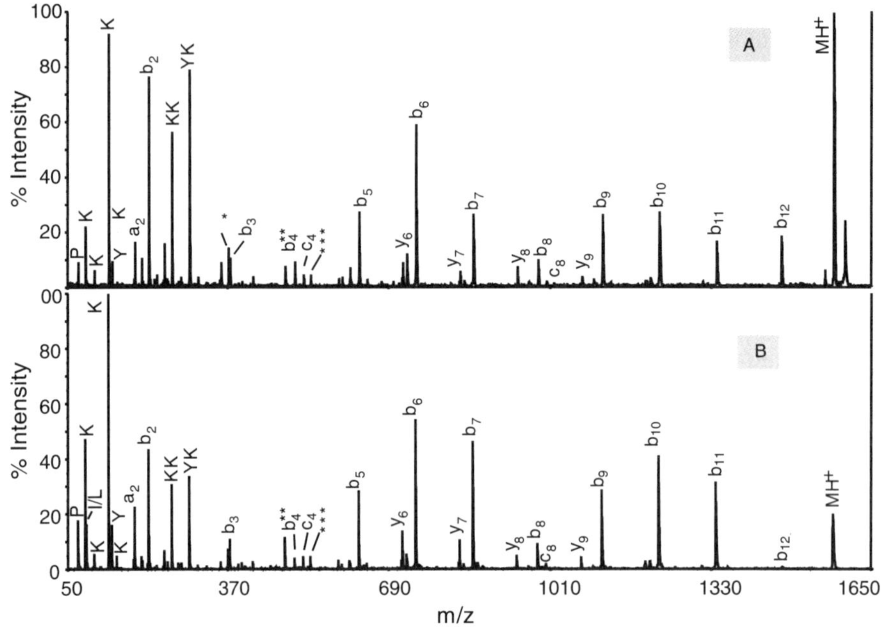

Figure 15.6. Comparison of post-source-decay (PSD) and TOF-TOF spectra of a peptide containing no arginine residues [PYLKKIIKKLLES, MH^+ = 1573.010 Da (monoisotopic)]. The PSD spectrum (a) and the TOF-TOF spectrum (b) are very similar in this case. Peaks labeled with asterisk(s) correspond to internal fragments with ambiguous sequence (LKK, IKK, KKI, KKL, etc.).

from the sequence. Abundant neutral losses are not expected either since the most of the fragment spectrum is dominated by *N*-terminal fragment ions and with the exception of K, no residues would be able to eliminate small neutrals. Ammonia loss from lysine residues is feasible, though, but the extent of this elimination reaction is substantially lower than it is from arginine-containing fragments.

Differences in the fragmentation patterns of PSD and TOF/TOF spectra of arginine-containing peptides may be very significant. The selected example in Fig. 15.7 has a strongly basic *N*-terminal part: HSRNSITLTNLT (MH^+ = 1356.724 Da). This accumulation of basic residues promotes the occurrence of many high-energy fragments in the TOF/TOF data, which are absent in the PSD spectrum. In the mass range shown in Fig. 15.7b, *d* ions are present from the cleavage of every peptide bond. In the case of d_6, d_7, and d_9 ions on the *C*-terminal sides of T, I, and T residues, both *d* and *d′* fragments occur as a result of the asymmetric β-carbon atom on these residues. This doublet of *d* ions differentiates I from L in this example. In the PSD spectrum (Fig. 15.7a) the *d* and *c* fragments are absent. (The d_9 ion may be an artifact in the PSD spectrum, as it does not exhibit the isotope pattern of the typical doublet as it

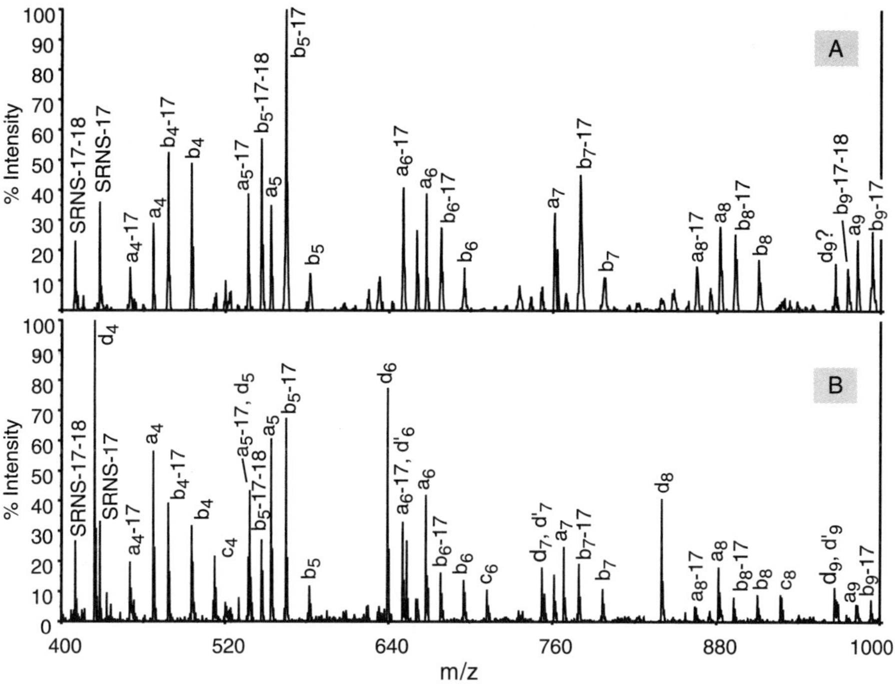

Figure 15.7. Comparison of PSD (a) and TOF-TOF (b) spectra of an arginine-containing peptide. The sequence is HSRNSILTLTNLT, $MH^+ = 1356.724$ Da, monoisotopic. Abundant high-energy fragments occur in the TOF-TOF spectrum, which are not observed in PSD.

occurs in the TOF/TOF spectrum). The case of *c* ions is particularly puzzling because in Fig. 15.6, in the absence of any high-energy fragments, the relative abundance of the c_4 and c_8 fragments looks similar in PSD and TOF/TOF. One can speculate that more than one mechanism may be involved in the formation of *c* ions.

The notion of various fragment ions being produced in a highly sequence and amino acid composition dependent manner requires the systematic organization and analysis of a large number of TOF/TOF spectra. Trends and conclusions drawn from such data sets can greatly enhance the accuracy of database searching and de novo sequencing algorithms. An initial compilation on some of the fragmentation trends in TOF/TOF spectra is provided in Table 15.2. The reader can refer to a number of previous reports on general fragmentation rules [22,39,40].

Internal fragment ions of the a_iy_{n-i} type have particular significance for establishing constraints on the amino acid composition of the fragmented peptides. The importance of these immonium ions in MS/MS spectra has been well explained in prior literature [37]. These ions carry valuable information on the

TABLE 15.2. Fragment Types Observed in TOF-TOF MS/MS Spectra. The Classification of Fragments is Described in Section 3.1

Class	Type	Typical abundance	Comments
Single backbone cleavage	*a*	strong	It appears together with the *b* ions. The abundance of *a* ions tends to be increased by the presence of basic residues on the fragment
	b	strong	In most MS/MS spectra these fragments are abundant—provided the N-terminus of the peptide is not modified
	$b + H_2O$	can be strong	It forms through the elimination of the C-terminal amino acid residue of the peptide. Usually very abundant if there is a basic residue at/near the N-terminus of the precursor
	c	weak	The occurrence is sequence specific [39]; In TOF-TOF it is also common on the N-terminal side of Q or between basic residues (H, K, R). c_1 is strong if the second residue in the sequence is Q
	x	weak	Rarely observed and comparison of PSD and TOF-TOF data suggests that *x* ions are produced in high-energy CID only. No obvious sequence specificity is obsered
	y	strong	Always abundant fragment ions provided they contain a basic amino acid residue (R, K, or H). Usually very strong ions arise from the cleavage of D-X and/or X-P bonds (X is an arbitrary but different amino acid)
	Y	weak	Frequently observed at the N-terminal side of P residues as a weak satellite of the corresponding *y* ion. Specific for high-energy CID
	z	not observed	No evidence found that *z* ions as backbone fragments form in the TOF-TOF instrument
Neutral loss	–17 (NH_3)	strong	Observed on *a*, *b*, *y* and internal fragments containing R, K, Q residues (the order of propensity being R > K > Q). Abundance of these ions is increased if the R, K, or Q is the terminal amino acid of the fragment ion

TABLE 15.2. *Continued*

Class	Type	Typical abundance	Comments
	–18 (H_2O)	weak, sometimes strong depending on the sequence	Observed on *a*, *b*, *y* and internal fragments containing S, T, D, E residues. Abundance of these ions is increased if the S, T, D, or E is the terminal amino acid of the fragment ion
	Other	some strong, others weak	Side chain loss of a few residues (e.g., R, F, V) from the MH^+ ions can be observed at low abundance. Other fragments such as –64 Da from oxidized M are abundant
High-energy fragments with side-chain loss	*d* *v* *w*	weak	All the "classic" high energy fragments seem to require the presence of R on the fragment ion. These fragments are typically weak and appear in a residue specific manner. For example, tryptic peptides with C-terminal R yield *w* ions at I/L residues very predictably.
Internal fragments and immonium ions	a_my_n	strong	Multi-residue ions appear along with their b_my_n counterparts. Single-residue ions—the immonium ions—are all of the a_my_n type with the exception of those formed from H, K, and P residues.
	b_my_n	strong	Multi-residue ions appear along with their a_my_n counterparts. Immonium ions of the b_my_n type are observed from H, K, P residues only.

amino acid composition of a peptide, which is readily extracted from the MS/MS spectrum—provided they can be detected. In TOF/TOF spectra immonium ions are usually among the most abundant fragments as demonstrated by Figs. 15.4, 15.6, and 15.8–15.10. As will be shown below, imposing the constraints evidenced by the immonium ions present in a TOF/TOF spectrum offers an invaluable help for spectral interpretation. The constant presence of immonium ions in the TOF/TOF spectra can also be utilized as for automatic internal mass calibration. Once the mass of the precursor ion—as obtained from an externally or internally calibrated MS spectrum—is precisely known, it can serve as a reference mass for the MS/MS spectrum. Another reference mass could be selected from the low-mass region of the spectrum from

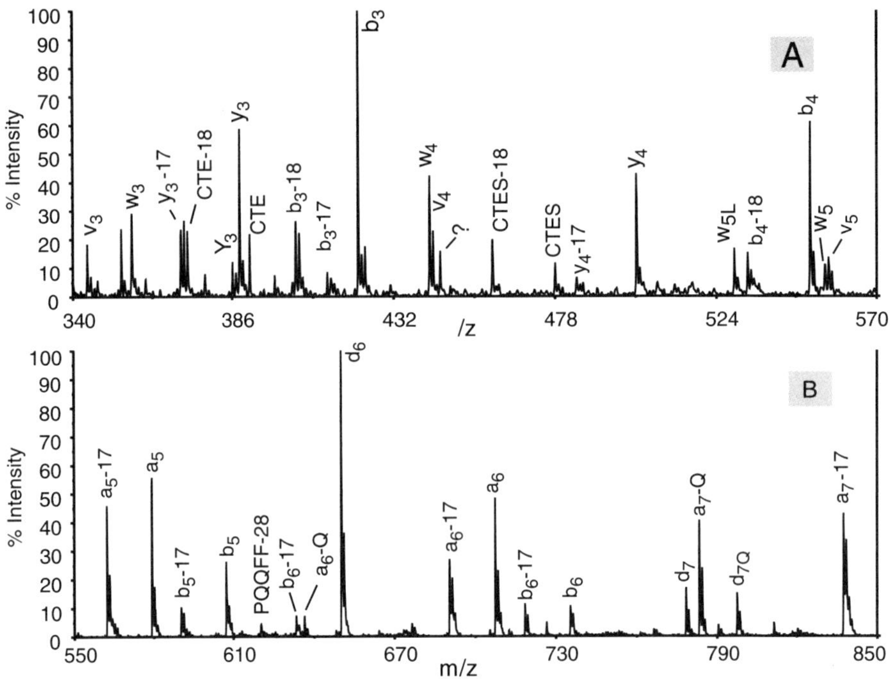

Figure 15.8. Two examples of TOF-TOF spectra exhibiting more complex fragmentation pathways. In panel (a) peptide CCTESLVNR [MH^+ = 1138.498 Da (monoisotopic)]; in panel (b) substance P [RPKPQQFFGLM-NH_2, MH^+ = 1347.736 Da (monoisotopic)] is shown. The complex high-energy fragments are interpreted in the text.

among the numerous unique masses found here. For instance, peptides containing any one of the I/L, K, R, H, P, F, Y, W, M, V residues would produce unique masses below 159 Da which can be utilized for this purpose. Accurate monoisotopic masses of immonium ions and related fragments are listed in Table 15.3.

In many cases the fragment types compiled in Table 15.2 do not account for every peak in an MS/MS spectrum. More complex fragmentation mechanisms have to be invoked to explain the origin of the unidentified peaks. One of the fragment types disregarded in most papers and interpretation software corresponds to multiple neutral losses. Figure 15.7 provides evidence on such fragmentation both in PSD and TOF/TOF spectra: if the fragment sequence contains multiple residues compatible with water/ammonia loss multiple neutral losses occur, indeed. The "SRNS" motif in this sequence is the source of consecutive loss of ammonia (from R) and water (from S). Interestingly, neither the b_5 18 fragment nor the SRNS-18 internal fragments are detected in Fig. 15.7, suggesting that the ammonia loss from the arginine residue is much faster than the water loss from the serine residues. Another "non-standard"

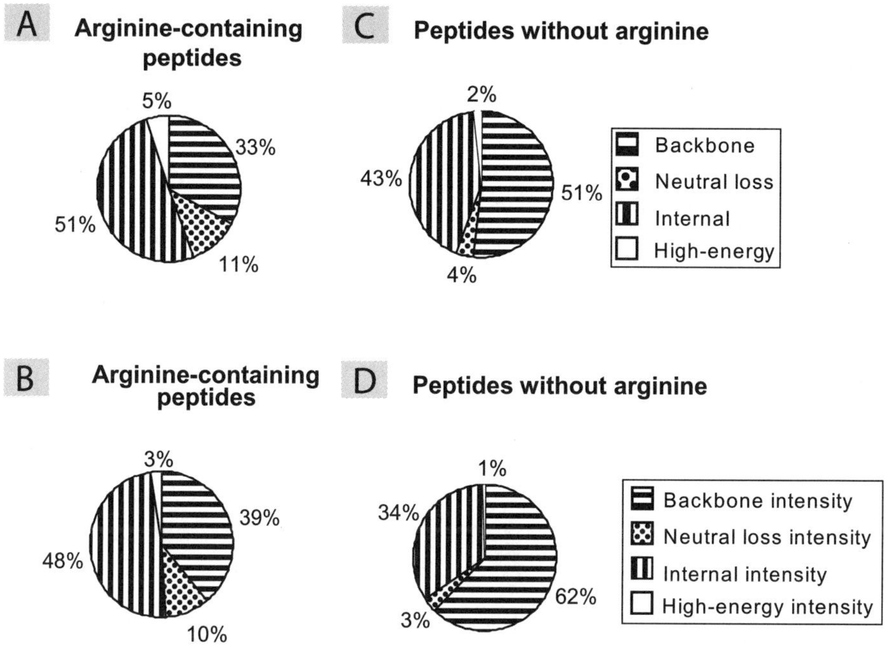

Figure 15.9. Distribution of various fragment types in the TOF-TOF spectra of arginine-containing peptides (a,b) and in peptides without arginine (c,d). The fragment categories are identical to those in Section 15.3.1 and in Table 15.2. Ten peptides were analyzed to create the charts. Charts (a) and (c) were generated from the number of fragment ions; charts (b) and (d) were calculated from the relative intensities of the fragments.

fragmentation path found in TOF/TOF spectra can be interpreted as the loss of the β-carbon substituent—as in *d* and *w* fragments—from one amino acid "downstream" from the backbone cleavage site. These fragment ions could be represented as d_{iaa}, where *aa* denotes the one-letter code of amino acid residue of which side chain is lost [3]. Two such examples are presented in Fig. 15.8. In Fig. 15.8a, the MS/MS spectrum of a trypic peptide CCTESLVNR is shown (MH^+ = 1138.498 Da with carbamidomethylated cysteines). The peak at *m/z* 442.22 is assigned as w_4 formed at the *N*-terminal side of L. There is a peak at *m/z* 529.26, however, which cannot be interpreted in terms of the fragment types compiled in Table 15.2. As the mass difference between these two peaks is exactly the mass of the S residue (87.032 Da) and the next residue toward the *N* terminus is a serine, it is postulated that the *m/z* 529 ion corresponds to w_{5L}. Similar assignments can be made in the MS/MS spectrum of substance P (RPKPQQFFGLM-amide, MH^+ = 1347.736 Da). This example has been studied in detail with four-sector instrumentation [3]. A quite significant peak at *m/z* 797.51 can be explained as d_{7Q}. Also, a couple of peaks in the spectrum

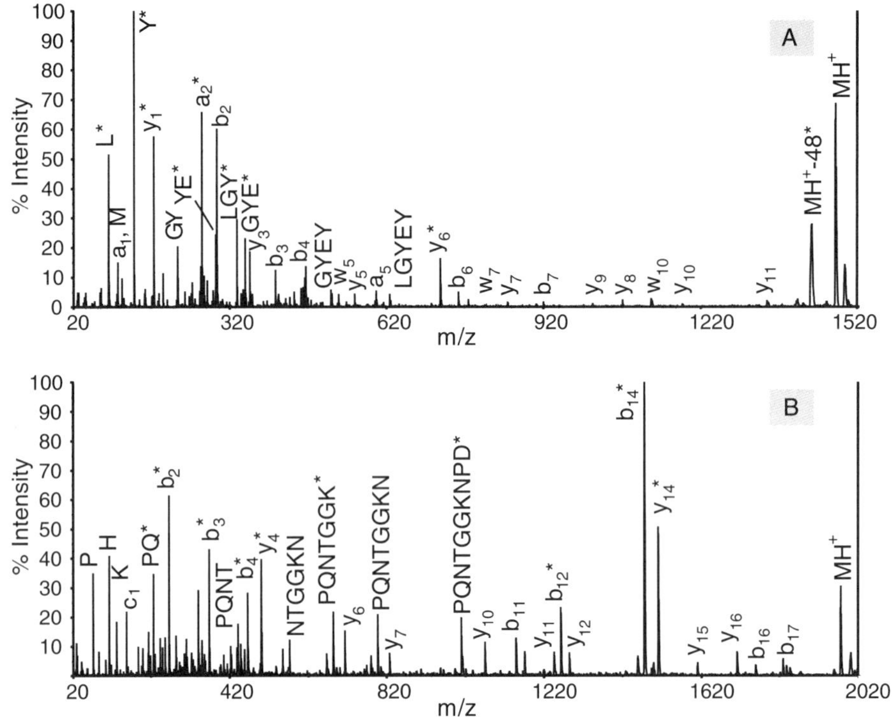

Figure 15.10. TOF-TOF spectra of two peptides from the tryptic digest of human transferrin. One of the peptides terminated with arginine (a), MYLGYEYVTAIR (MH^+ = 1478.735 Da, monoisotopic). The other one had a lysine at the *C* terminus: HQTVPQNTGGKNPDPWAK (MH^+ = 1974.979 Da, monoisotopic). The spectra were used to test the performance of publicly available database searching programs. The 10 most abundant peaks, marked by asterisk, were used from both spectra. Note that the measure of abundance was the total peak area of the isotope clusters (not the peak heights).

have been assigned as fragments formed from the respective *a* ions by subsequent losses of the α-carbon substituents. These fragments are denoted by a_i–*aa*. In particular, glutamine seems to exhibit this fragmentation path; both *a* ions in the substance P spectrum formed at the Q residues are accompanied by a sidechain loss of glutamine.

15.3.2. Database Searching and Protein Identification with TOF/TOF data

The objective of protein identification and database searching with MS/MS spectra is to find and establish the origin of a peptide in the protein, genomic or expressed sequence tag (EST) databases that can account for the MS/MS peak list. In comparison with the de novo interpretation of a spectrum where

TABLE 15.3. Monoisotopic Masses of Immonium and Other, Mono-Amino Acid Related Ions

Amino acid residue	Residue mass (Da)	Immonium ion mass (Da)	Related ion mass(es) (Da)	Abundance*
Ala (A)	71.037	44.050	—	weak or absent
Arg (R)	156.101	129.114	59.048, 70.066, 87.092, 95.061, 100.087, 112.087, 115.086,	112 and 129 strong; others may or may not be observed
Asp (D)	115.027	88.040	—	usually weak, occasionally strong
Asn (N)	114.043	87.056	—	weak
Cys (C)	103.009	76.022	—	weak
Gly (G)	57.021	30.034	—	absent
Gln (Q)	128.059	101.071	84.045	101 strong, 84 weak
Glu (E)	129.043	102.056	—	usually weak, occasionally strong
His (H)	137.059	110.072	138.067, 166.062	110 strong, 138 and 166 weak
Ile/Leu (I/L)	113.084	86.097	—	usually strong, occasionally weak
Lys (K)	128.095	101.108	84.081, 129.103	84, 129 strong, 101 weak
Met (M)	131.040	104.053	—	usually weak, occasionally strong
Phe (F)	147.068	120.081	91.054	120 strong, 91 weak or absent
Pro (P)	97.053	70.066	98.061, 126.060	70 strong, 98 and 126 weak
Ser (S)	87.032	60.045	—	weak or absent
Thr (T)	101.048	74.061	—	weak or absent
Trp (W)	186.079	159.092	130.066, 170.061	130 and 159 strong, 170 weak
Tyr (Y)	163.063	136.076	91.054, 107.049	136 strong, 91 and 107 weak or absent
Val (V)	99.068	72.081	—	weak

* The Abundance column in the table represents the "typical" behavior of immonium ions. Note that the intensity of these ions also depends on the positions of the residues in the peptide: immonium ions of the N-terminal residue tend to be more abundant than from an internal position. Immonium ions from the C-terminal position are usually suppressed.

the objective is to deduce the complete amino acid sequence, protein identification is a greatly simplified problem. To indicate the order-of-magnitude difference between these two tasks, one can take a 10-residue peptide as an example. De novo interpretation would have to validate 10^{20} amino acid sequences against the experimental data. However, the list of 10-residue amino acid motifs occurring in nature is many orders of magnitude smaller—probably less than 10^{10} individual sequences.

If MS/MS data in every possible scenario were of sufficiently high quality, so that at least one fragment ion were generated from every peptide bond, and if the mass accuracy were high enough, the majority of the spectra could be interpreted de novo by one of the previously published sequencing algorithms [32,33,39,41–44]. In reality, MS/MS data are sparse—not every peptide bond gives rise to an unambiguous fragment—and the reliability of mass assignments also varies from one instrument type to another. The dramatic simplification of the interpretation problem by matching only peptide sequences recorded in protein or genomic databases makes it possible to interpret "imperfect" MS/MS data, that is, to find the protein accommodating the peptide sequence that produced the experimental data. The success of protein identification by these database searching procedures critically hinges on the sophistication of these algorithms: how well they predict the generation of enzymatic hydrolysis products (peptides), or the fragmentation of the peptides in the mass spectrometer, and how well they incorporate the accuracy of mass measurement into their procedures.

Database searching engines publicly available through the World Wide Web have been used since the late 1990s to match MS/MS data to entries in the protein, genomic, and EST databases. In what follows a few of these engines will be analyzed, such as

- MASCOT from http://*www.matrixscience.com* [45]
- MS-TAG/Protein Prospector from http://*prospector.ucsf.edu* [46]
- PepFrag from http://*prowl.rockefeller.edu*
- PeptideSearch from http://*www.mann.emblheidelberg.de/Services/PeptideSearch/PeptideSearchIntro.html*

It has to be noted, however, that some of the most significant efforts to develop database searching programs have resulted in commercial products not available through the Web [18]. These programs will not be considered for further analysis.

The supported MS/MS features of these search engines are listed in Table 15.4. Comparing these features with the list of fragment types in Table 15.2, it can be concluded that none of these search programs support the full repertoire of fragments that potentially occur in TOF/TOF spectra. This is not necessarily a serious problem because even having some fragment peaks unmatched still could permit the successful identification of proteins. It is more

TABLE 15.4. Commonly Used Publicly Available Search Engines for Identifying Proteins Based on MS/MS Data

Search engine	Features
MS-Tag	— Instrument specific modes (i.e., MALDI TOF, ESI Trap, *etc.*) — MALDI TOF mode supports single and double backbone fragments including immonium ions — Variety of neutral losses (−17, −18, −64, *etc.*) — Version 3.4 supports *d*, *v*, and *w* ions through the TOF-TOF instrument option
PepFrag	— single backbone fragments: *a*, *b*, *c*, *x*, *y*, *z* — loss of ammonia — can be combined with amino acid information, but immonium ion masses are not supported
Mascot	— single backbone fragments: *a*, *b*, *y* — loss of ammonia — can be combined with amino acid content and sequence tag information — immonium ions indicated in results but not used for matching or scoring
PeptideSearch	— requires pre-defined sequence tags

important to know—at least in a gross approximation—how big a fraction of fragment ions would be supported by a given database search engine and what is the critical percentage of unmatched peaks that renders protein identification uncertain. For this purpose we have analyzed the most abundant fragment ions from the TOF/TOF spectra of numerous peptides resulting from different enzymatic hydrolysates. The procedure classified the 30-odd most abundant fragment ions from the individual spectra into the four categories according the scheme in Section 15.3.1. Comparing the fragment distributions from different peptides originating from tryptic or Endo Glu-C digests or from nonspecific hydrolysis, no significant difference has been detected. The only important difference seemed to be associated with the presence or absence of arginine in the peptide sequences.

Four charts are shown in Fig. 15.9 presenting the distributions of the four fragment categories for peptide sequences with or without arginine. These charts have been created based on the number of fragment ions and based on the intensity of fragment ions as well. Immonium ions were counted along with the internal fragments. This may have skewed the results slightly, overrepresenting the contribution of immonium ions, as the a_1 ion—resulting from a single backbone cleavage—is identical to the immonium ion of the same amino acid residue. As the most significant conclusion from these charts, one has to realize the importance of internal fragments/immonium ions in the TOF/TOF mass spectra. In the case of arginine-containing peptides, these fragments, based on either the number or the intensity of peaks, are equally or slightly more abundant than fragments from single backbone cleavage. The fragment distribution is shifted toward the single backbone cleavage fragments in the case of non-arginine-containing peptides but still the presence of internal fragments/immonium ions is significant. As seen from Table 15.4, with the exception of Protein Prospector, the support for internal fragments/immonium ions is missing. This is expected to cause more serious problems with arginine containing peptides, which, apparently, tend to produce more internal fragments. Neutral losses also seem to be more abundant for arginine-containing peptides. This trend could be explained by the higher the propensity of ammonia loss from arginine residues than similar losses from K or Q, or water loss from S, T, D, and E residues. All search engines support the ammonia loss, but only MS-TAG has the capability to interpret fragment ions resulting from water loss.

In consistency with the summary of Table 15.2, arginine-containing peptides exhibit more high-energy fragments. The chart based on the intensity of fragment ions shows a smaller contribution from the high-energy fragments than does that based on the number of fragment ions (3% in Fig. 15.9b vs. 5% in Fig. 15.9a). This suggests that even if high-energy fragments are present in the spectrum, these tend to be of low relative abundance. High-energy fragments are all but absent from peptides not containing arginine. Consequently, lacking support of *d*,*v*,*w* ion types on behalf of the database searching programs is very unlikely to "make or break" successful protein identification. MS-TAG/Prospector in its latest revision (Version 3.4) offers a TOF-TOF instrument option that comes with the support for high-energy fragments. In our experience, this option is not critical to match the correct protein to the MS/MS data.

Two MS/MS spectra have been selected to evaluate the efficiency of protein identification with the search engines listed in Table 15.4; both peptides were derived from the tryptic digest of human transferrin. TOF/TOF data are shown in Fig. 15.10. One of the peptides had an R at the *C* terminus, with the sequence MYLGYEYVTAIR ($MH^+ = 1478.735$ Da), the other was terminated with K: HQTVPQNTGGKNPDPWAK ($MH^+ = 1974.979$ Da). The ten most abundant fragment ions were selected and submitted along with the precursor mass for database searching. These peaks are designated in Fig. 15.10 by an asterisk attached to the peak label. (Submitting only 10 peaks was intended to simulate a scenario where weaker MS/MS spectra would not have resulted in more than ten good quality fragment peaks.) Although it was obvious that the internal fragments and immonium ions would not be matched successfully by the programs with the exception of MS-TAG, the question was rather, whether the remaining backbone fragments were sufficient for the unambiguous identification of the source protein. Queries for MS-TAG, Mascot, and PepFrag were submitted using 100-mDa mass tolerance for both the precursor and the fragment ion masses. Wherever possible, the NCBI non-redundant database was searched for eukaryotic entries. This option was not available on MS-TAG and PeptideSearch; here the entire database was searched instead.

The program PeptideSearch requires a user-provided sequence tag for the search [19]. The tag is a partially interpreted stretch of ion series in the spectrum complemented by its start and end mass values. As obvious from Fig. 15.10a, such a tag cannot be derived as none of the mass differences between these peaks could be mapped to amino acid residue masses. Therefore the first example in Fig. 15.10a could not be processed with PeptideSearch. For the second sequence in Fig. 15.10b, three *b* ions (b_2, b_3, and b_4) define a di(amino acid) motif "TV" with a start mass of 266.11 Da and with an end mass of 466.216 Da. These features were submitted for database searching with a 100-mDa mass tolerance.

The results are compiled in Table 15.5. In the table the top scoring proteins are shown down to the score obtained for the correct answer (i.e., human transferrin). Scoring is different for the individual programs. For PepFrag and MS-TAG the score is equal to the number of matched peaks/total number of peaks to match. The Mascot program determines a probability score, which is the measure of the likelihood to a match a given protein with randomly selected fragment masses [45]. As expected, all the search programs fared well with the example in Figure 15.10b, where only three of the 10 masses submitted belonged to internal fragments. Each search program identified human transferrin as the top-scoring candidate.

The results from the spectrum in Fig. 15.10a are more interesting. There are only four single backbone fragments, and the rest correspond to internal and immonium ions. Also there is a strong MH^+-48 fragment, quite likely arising from a partial sidechain loss of the *N*-terminal methionine ($SH\text{-}CH_3$ leaving group), which will not be matched up by any of the search engines. Still, apart from PeptideSearch, where we could not derive a sequence tag from the 10 most abundant peaks, all search programs found transferrin as the top-scoring candidate. With MS-TAG nine fragments out of 10—the MH^+-48 fragment was the exception—were successfully matched and transferrin was the only top-scoring protein. The confidence of such a match is excellent. Mascot and PepFrag could match the four backbone fragments only, and neither program could identify human transferrin unambiguously. Because of the way Mascot scores the matches (higher intensities matches bear higher weights), human transferrin is one of the two top-scoring proteins. However, the probability score is very low, suggesting that none of the two top candidates represent a good match. With the uniform weighting of individual matches in the case of PepFrag, there are five different sequences in the top-scoring position. One can conclude that the lack of support for internal fragments and immonium ion seriously compromises the accuracy of database searching. As a gross estimate it seems that more than 50% of the submitted fragments, or at least five fragments should be matched for unambiguous protein identification, whichever number is higher.

Lacking support for abundant fragment types is just on possible reason for a failure to identify the correct protein. Even if the MS/MS data quality is good, numerous other problems can spoil protein identification:

TABLE 15.5. Database Searching Results from the Ten Most Abundant Fragment Peaks from Figures 10A and 10B. Scoring for MS-TAG and PepFrag is Based on the Number of Matched Peptides vs. the Total Number of Fragment Masses Submitted. Mascot is Using a Probability Based Scoring Scheme. The Number in Parentheses Stands for the Score Necessary to Find the Right Match with High Confidence

Search engine	Sample spectrum	Protein matches	Peptide sequence	Score
MS-TAG	Fig. 10A	transferrin precursor—human	MYLGYEYVTAIR	9/10
	Fig. 10B	transferrin precursor—human	HQTVPQNTGGKNPDPWAK	10/10
PepFrag	Fig. 10A	S-adenosylmethionine synthetase—S. cerevisiae	GYDDSAKGFDYKTCNVL	4/10
		hypothetical protein F52E4.5—C. elegans	YVAQPNHNWGHR	4/10
		hypothetical protein H38K22.1—C. elegans	LPEEFELTGSSNR	4/10
		hypothetical protein ZK822.1—C. elegans	IPTPSSRHGNEKR	4/10
		transferrin precursor—human	MYLGYEYVTAIR	4/10
	Fig. 10B	transferrin precursor—human	HQTVPQNTGGKNPDPWAK	7/10
Mascot	Fig. 10A	probable eukaryotic translation initiation factor	YVTTSSTIWRHK	9 (27)
		transferrin precursor—human	MYLGYEYVTAIR	9 (27)
	Fig. 10B	transferrin precursor—human	HQTVPQNTGGKNPDPWAK	34 (35)
PeptideSearch	Fig. 10A	—	—	—
	Fig. 10B	transferrin precursor—human	HQTVPQNTGGKNPDPWAK	n/a

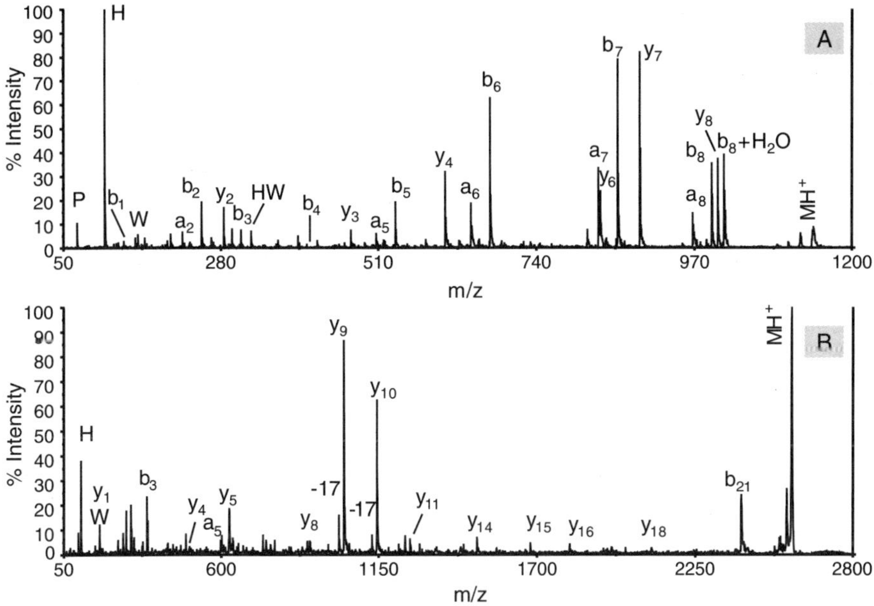

Figure 15.11. These two MS/MS spectra were generated from the tryptic digest of bovine carbonic anhydrase. The two peptides have four potential de-amidation sites. The corresponding positions are indicated in italics. The sequence of the peptide in panel (a) is H*N*GP*E*HWHK ($MH^+ = 1141.529$ Da, monoisotopic). It is unambiguously shown that the first site (N) is not deamidated but the second residue (Q → E) is. The other peptide is LVQFHFHWGSS*DD*QGSEHTVDR ($MH^+ = 2584.161$ Da, monoisotopic). The spectrum in panel (b) proves that both residues (in italics) are converted from N to D.

1. *The protein sequence derived from the database is incorrect.* This problem can frequently be encountered when using genomic or EST databases for search. Due to either errors or variations/mutations in the DNA sequence or due to post-translational changes in the amino acid sequence, no theoretically predicted peptides can be matched up to the MS/MS spectra. Proteins of highly variable sequence, such as immunoglobulins, represent an example for the first problem. As far as the second problem is concerned, perhaps de-amidation of the N and Q residues converting them into D and E (associated with a 1-Da change in the molecular weight of the peptide), are the most frequently encountered examples. Since de-amidation is a posttranslational event and the majority of database entries is derived from DNA sequences, this can pose a very serious problem. Even as commonly used protein standards as bovine carbonic anhydrase or α-lactalbumin contain ambiguous or erroneous sequences in the widely used non-redundant and/or SWISS-PROT databases. Figure 15.11 presents the MS/MS data on two tryptic peptides ($MH^+ = 1141.529$ Da and $MH^+ = 2584.161$ Da) from bovine carbonic anhydrase

containing four ambiguous amino acid residues as it is found in SWISS-PROT (accession number P00921). MS/MS data establish the de-amidation state of these four positions (amino acids 10-N, 13-Q, 100-D, and 101-D) with no ambiguity. A general solution to overcome problems associated with errors in the database entries could be to partially interpret a sufficiently large part of the MS/MS spectrum. With a sufficiently long stretch, the amino acid sequence, a BLAST-type search, can identify the protein where the peptide originated. In our hands this strategy has worked well for identifying immunoglobulin derived peptides. Unfortunately, there is an extra amount of work to be done with this approach, and the sequence coverage in the fragment spectrum may not be high enough in every case.

2. *There are other posttranslational modifications or chemically modified amino acids.* The number of different amino acid modifications imparted by the cell or the sample preparation procedures is enormous [47]. A systematic review of these is beyond the scope of this chapter; however, some of the most common modifications can be easily identified from the TOF/TOF mass spectra. For example

Oxidized methionine is readily identified from a very abundant MH^+-64 peak. Also a 64-Da loss from the most abundant backbone fragments is observed [36].

Phosphorylated S, T, and Y residues produce very abundant 98-Da neutral loss from the precursor ion [35].

Carbamylation of the N terminus can occur in urea containing solutions. This modification causes a strong neutral loss of 43 Da from the precursor.

Other modifications are more difficult to detect, such as the formation of *N*-terminal pyroglutamic acid either due to signal sequence removal in the cell or as a byproduct of enzymatic cleavage next to a glutamine residue, or proteins with an amidated *C* terminus. Although the former of these modifications is expected to produce an immonium ion at *m/z* 84, this ion is frequently masked by the presence of Q or K in the sequence. A potential b_1 ion (*m/z* 112) from pyroglutamic would be indistinguishable from the *m/z* 112 ion of arginine (see Table 15.3). Since most database search programs have support for the common *N*- and *C*-terminal modifications, failed search attempts should always be resubmitted for another search with these modifications enabled.

15.3.3. De Novo Peptide Sequencing with TOF/TOF

A common objective of de novo interpretation strategies is the identification of fragment peaks in the MS/MS spectra belonging to the same ion series. Once this is accomplished, finding the right sequence is reduced to the interpretation of mass differences in the ion series in terms of amino acid residue masses. The success of such strategies will depend on how wide a variety of fragmentation types occur in the mass spectra and how precise the mass measurement to assign amino acid residues to observe mass differences is. A good selection of papers is available for interested readers [32,33,39–45]. In this section the emphasis will be placed on how the fragmentation trends and rules can be applied to MS/MS data from unknown sequences rather than on the algorithmic aspects and automated computation of sequence candidates.

De novo interpretation of TOF/TOF MS/MS spectra has to deal with the variety of possible fragment types listed in Table 15.2. As was stated before, the mass resolution of the analyzer allows for detecting the monoisotopic peaks in most cases, and the mass measurement accuracy of the monoisotopic peaks is typically around ±50 mDa. In addition to the mass accuracy of individual peaks, it is equally important to know the accuracy of mass difference measurements: confidence in mass difference measurements can help to resolve potentially ambiguous amino acid assignments such as those corresponding to lysine vs. glutamine (128.095 vs. 128.059 Da) or to unambiguously identify peak pairs as belonging to the same ion series versus a coincidental mass difference arising from peaks from different ion series. The rightmost column in Table 15.1 lists the mass errors of the mass differences calculated between successive *b* and *y* ions. With these mass errors the unambiguous differentiation between D versus N, Q/K versus E should always be possible even if the peaks have very low signal-to-noise ratios. Resolving ambiguities of smaller mass differences may be more difficult. However, it is important to realize that the same mass differences can occur on more multiple ion series (e.g., *a*, *b* and *b*-17 series) and using mass difference measurements on di- and tri(amino acid) motifs the statistical significance of the problematic assignments can be further improved. The magnitude of mass errors in Table 15.1 would suggest the routine differentiation lysine and glutamine is questionable so averaging multiple mass difference measurements will be required. These points will be illustrated later.

Manual interpretation of MS/MS spectra utilizes a few simple relationships between different ions series. For a peptide constituted on *n* amino acid residues, these conditions are:

$$b_i - b_{i-1} = a_i - a_{i-1} = y_{n-i+1} - y_{n-i} = \cdots = i\text{th amino acid residue}$$
$$b_i + y_{n-i} = MH^+ + 1.008$$
$$c_i + y_{n-i} - 17 = MH^+ + 1.008$$
$$b_i - a_i = 27.995$$
$$c_i - b_i = 17.026$$
$$MH^+ - (b_{n-1} + H_2O) = C\text{-terminal amino acid residue}$$

Masses of the immonium and related ions are compiled in Table 15.3, and Table 15.6 lists the mass differences of various high-energy fragment ions from

TABLE 15.6. Monoisotopic Mass Differences of High-Energy Fragment Ions from the Respective a and y Ions (in Da)

Amino acid residue	d_i ions	d_i' ions	v_i ions	w_i ions	w_i' ions
Ala (A)	—	—	$y_i - 16.031$	—	—
Arg (R)	$a_i - 85.064$	—	$y_i - 101.095$	$y_i - 102.090$	—
Asp (D)	$a_i - 43.990$	—	$y_i - 60.022$	$y_i - 61.017$	—
Asn (N)	$a_i - 43.006$	—	$y_i - 59.037$	$y_i - 60.033$	—
Cys (C)	$a_i - 31.972$	—	$y_i - 48.004$	$y_i - 48.999$	—
Gly (G)	—	—	—	—	—
Gln (Q)	$a_i - 57.022$	—	$y_i - 73.053$	$y_i - 74.048$	—
Glu (E)	$a_i - 58.006$	—	$y_i - 74.037$	$y_i - 75.032$	—
His (H)	$a_i - 66.020$	—	$y_i - 82.050$	$y_i - 83.046$	—
Ile (I)	$a_i - 28.032$	$a_i - 14.016$	$y_i - 58.078$	$y_i - 45.058$	$y_i - 31.042$
Leu (L)	$a_i - 42.047$		$y_i - 58.078$	$y_i - 59.073$	—
Lys (K)	$a_i - 57.058$		$y_i - 73.089$	$y_i - 74.084$	—
Met (M)	$a_i - 60.003$		$y_i - 76.035$	$y_i - 77.030$	—
Phe (F)	$a_i - 76.031$		$y_i - 92.063$	$y_i - 93.057$	—
Pro (P)	$a_i - 26.016$		$y_i - 42.047$	$y_i - 43.042$	—
Ser (S)	$a_i - 15.995$		$y_i - 32.026$	$y_i - 33.021$	—
Thr (T)	$a_i - 15.995$	$a_i - 14.016$	$y_i - 46.042$	$y_i - 33.021$	$y_i - 31.042$
Trp (W)	$a_i - 115.032$		$y_i - 131.074$	$y_i - 132.069$	—
Tyr (Y)	$a_i - 92.026$		$y_i - 108.057$	$y_i - 109.052$	—
Val (V)	$a_i - 14.016$		$y_i - 44.063$	$y_i - 31.043$	—

the respective *a* and *y* ions. It is also useful to compile a table of di- or tripeptide mass differences and internal fragment masses up to about 250–300 Da. Many of these masses are unique and can be assigned from a TOF/TOF spectrum to a di- or tri(amino acid) motif with high confidence.

The concepts discussed above are illustrated in the example shown in Fig. 15.12 (see also Fig. 15.13). This MS/MS spectrum was obtained from an HPLC fraction from a very complex peptide mixture. With an accurate mass measurement in the MS spectrum, the mass of the precursor was found to be 1378.681 Da. Despite the good-quality MS/MS data, none of the database searching programs could identify the protein even if no enzyme specificity was required and the search was extended to all species. Searching EST databases or searching in the homology mode of MS-TAG was unsuccessful, too.

For the manual interpretation of the spectrum in Fig. 15.12, the following "protocol" was used:

- Establish the constraints on the amino acid composition from the immonium ion region (see the upper trace in Fig. 15.12).
- Identify *b*–*y* ion pairs.
- Look for accompanying *a* ions to identify *b* ions from the *b*–*y* pairs.

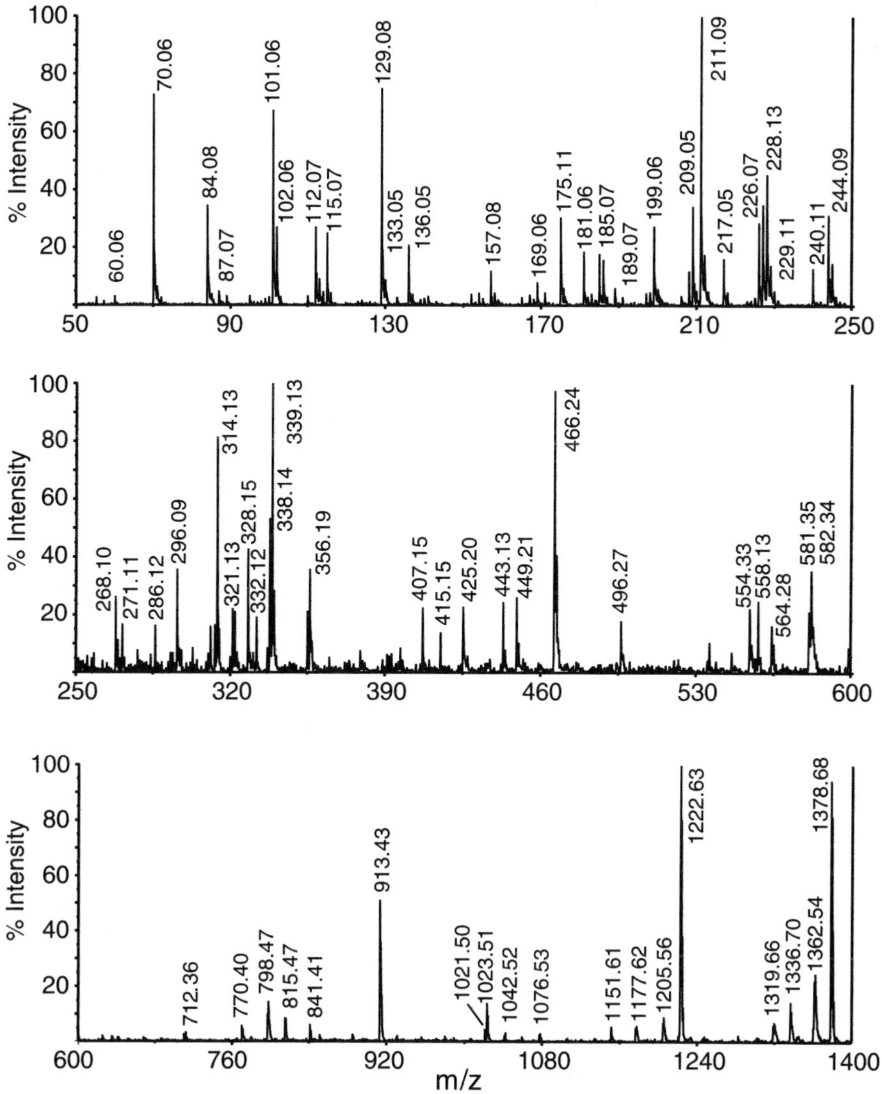

Figure 15.12. TOF-TOF spectrum of an unknown tryptic peptide. De novo interpretation of this spectrum is provided in Section 15.3.3. The spectrum was generated from an LC fraction spotted onto the MALDI plate and contained between 100 and 200 fmol of material.

- Try to extend *b*- or *y*-ion series in the absence of the complementing *y* or *b* ion.
- Resolve ambiguities by checking other ion types (internal fragments, high-energy fragments, etc.), applying fragmentation rules in Table 15.2.

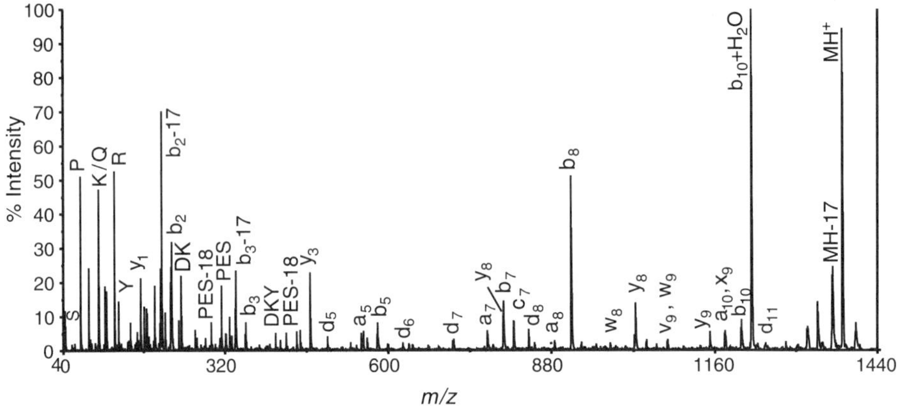

Figure 15.13. Annotated TOF-TOF spectrum of the unknown shown in Fig. 15.12. De novo interpretation of the spectrum concluded that the amino acid sequence was RAQPESEDKYR ($MH^+ = 1378.671$ Da).

In Fig. 15.12 the occurrence of the y_1 peak at *m/z* 175 suggests that the peptide was of tryptic origin with arginine at the *C* terminus. The arginine residue also accounts for the *m/z* 70, 87, 95, 112, 115, 129 peaks in the low-mass region (see Table 15.3). Strong (*m/z* 84, 101) peaks are the sign(s) of K and/or Q; strong 102 is indicative of E, 136 predicts Y, and the weak *m/z* 60 peak—although less convincingly—could be a sign of S in the sequence. On the basis of missing immonium ions the following residues can be declared as being absent: H (no 110), F (no 120), M (no 104), I/L (no 86), V (no 72), and probably no C (carbamidomethylated form of C gives an abundant 133 ion in most cases; Fig. 15.10b, *m/z* 133 looks too weak). Other amino acid residues of which presence cannot be concluded but cannot be ruled out, either, are P, N (the *m/z* 70 and 87 peaks can be masked by R), and A, G, D, T because immonium ions from these residues are weak or absent. The relative abundance of *m/z* 157 can be an important hint: weak 157 could form from the 175 y_1 ion of R as y_1-18, but this ion is usually weaker then the one at *m/z* 158 (y_1-17); the trend observed in Fig. 15.12 is just the opposite; the higher abundance of the *m/z* 157 peak can be an indicative of *N*-terminal arginine (b_1). If this is the right conclusion, then, because of a strongly basic residue at the *N* and *C* termini an abundant $b_{n-1} + H_2O$ ion would be expected [48]. In this case an ion arising from the elimination of the *C*-terminal arginine residue should be found at *m/z* *M*H-156.11 = 1222.57. Indeed, in the lowest trace of Fig. 15.12 the fragment ion with an observed *m/z* of 1222.61 is the most abundant fragment ion. So at this point it is concluded that both the *N* and *C* termini are R. When both termini are strongly basic it is expected that both *N*- and *C*-terminal fragments will be well represented in the MS/MS spectrum. Also, high-energy fragments are likely to appear due to the presence of *N*- and *C*-terminal arginine residues.

TABLE 15.7. Ion Pairs Found in the Spectrum in Figure 12 Combining to MH$^+$ + 1 (1379.69 Da). Asterisks Indicate the Presence of –28 Da Satellites Suggesting that These Peaks Correspond to *b* Ions

m/z (observed)	*m/z* (observed)	Sum
228.13	1151.61	1379.74
338.14	1041.56	1379.70
356.19*	1023.51	1379.70
449.21	930.42	1379.61
466.24	913.43*	1379.67
564.28	815.47	1379.75
581.35	798.47*	1379.82
582.33*	797.42	1379.75

As the second step in the interpretation procedure, complementing pairs of *b* and *y* ions will be looked for with a relation to the precursor mass as shown above. Masses of potential *b*–*y* peak pairs are listed in Table 15.7. It should be pointed out that combinations of complementary *y*-17 and *c* ions add up to *M*H$^+$ + 1 as well, and one also should not rule out coincidental matching of these peak pairs to MH$^+$ + 1. In the peak list of Table 15.7 it is not yet known which one of the pairs are *b* or *y* ions. The presence of –28-Da satellite, which would correspond to the related *a* ion, quite reliably identifies *b* ions. The masses for which the –28-Da satellite was found are labeled by an asterisk in Table 15.7. There remained two peak pairs that could not be identified at this point as either *b* or *y*: 449.21–930.42 Da, and 564.28–815.47 Da. A quick inspection reveals that these pairs are +/–17 Da satellites to the *b*–*y* pairs of 466.24–913.43 and 581.35–798.47 Da, respectively. Then these peaks should correspond to the appropriate *y*-17 and *c* ions.

Finding amino acid extensions to the already assigned *b* or *y* ions can yield further *y* or *b* assignments. This logic suggests that *m/z* 228.13 should represent the b_2 peak since its mass difference from b_1 (157.08) is 71.05 Da corresponding to alanine. The mass difference between *m/z* 228.13 and the postulated *b* ion of *m/z* 356.19 (see Table 15.7) is 128.06 Da, which corresponds to either Q (128.059) or K (128.095). Based on the smaller mass error, Q seems to be the better assignment, but further support will be looked for later to make this call statistically more significant. The combination of A and G also adds up to 128.059, so it has to be verified that no peaks at 228.13 + 57.02 (G) or 228.13+71.04 (A) are present. The peak list does not contain either *m/z* 285 or 299. However, bond cleavage is often weak at G or A residues, so further evidence was sought. The presence of K/Q could also be confirmed by the presence of specific high-energy fragments: d_3 ion or by a w_{n-2} ion. The mass of these ions can be looked up from Table 15.6. The mass of the d_3 ion should be around 271.1 Da. The mass of the w_{n-2} ion can be determined from the mass of the y_{n-2} ion, which is the complement ion of b_2, i.e., 1378.68 – 228.13 – 74.048 = 1076.50 Da. Both peaks, although of quite a low abundance, could be found in the spectrum, further corroborating the assignment of the third residue at the *N*-terminal side of the peptide as K/Q. Utilization of high-energy fragments can be very helpful in distinguishing sequence ambiguities such as N versus [G,G] or Q versus [A,G], in particular that the peptide bond cleavage of the *C*-terminal side of G is often weak or absent. (Square brackets are used to indicate the composition with unknown order of the residues.)

The next member of the *b*-ion series is very likely the ion of *m/z* 582.33. As seen from Table 15.7, it defines a mass difference 582.33 – 356.21 = 226.12 Da from the b_3 ion. The only amino acid compositions that can account for this mass difference are [E,P] or [I/L,I/L] (226.10 and 226.17 Da, respectively). The latter combinations can be discarded since I or L are incompatible with the immonium ions observed (see Fig. 15.12). The motif [E,P] seems more likely to be "PE" than "EP." Since fragmentations on the *N*-terminal side of acidic residues and on the *N*-terminal side of P are known to be preferential, a cleavage of an EP motif is expected to produce very abundant *b*–*y* pairing in this case [49]. Since such a pair or any of the corresponding *b* or *y* are not present in the spectrum, the hypothetical sequence is RA(K/Q)PE. From such a sequence—because of the same cleavage preferences—one would expect stronger b_3 (356.19 Da), weak or absent b_4 (453.257 Da) on the *C*-terminal side of P [39], and stronger b_5 (582.30 Da). These peaks were all found in the spectrum and follow a similar distribution. Further evidence can be found from the high-energy fragments again: *Y* ions are preferentially formed at the *N*-terminal side of proline, which is 2 Da below the corresponding *y* ion. Such a doublet is clearly detected at the y_{n-3} peak (*m/z* 1023.51 and 1021.47), suggesting that the fourth residue should be proline.

Still using Table 15.7, one can decipher the *C*-terminal part of the sequence, too. Peak 913 is postulated as a *b* ion, and, since 338 is not a *b* ion, its complement ion at *m/z* 1041 should be. The mass difference between 1041.56 and 913.43 Da is 128.13, suggesting an assignment to K or Q. Again, detecting the high-energy fragments from K/Q helped to rule out the presence of a [A,G] motif. Based on the mass error, K seems to be the more likely assignment in this case. Now, knowing that the *C*-terminal residue is arginine, the mass of the ion b_{n-1} should be 1378.68 – 18.01(H_2O) – 156.10(R) = 1204.57 Da. The mass difference between the b_{n-1} and 1041.56 is equal to 163.01 Da. This identifies tyrosine unambiguously. So the *C*-terminal part of the sequence is (Q/K)YR. The same conclusion can be arrived at from the complementing *y*-ion series. The complement of the 1041 *b* ion is a *y* ion of *m/z* 338.14, and this ion is 163.05 Da higher than the *m/z* 175.09 (y_1) species. Therefore, *m/z* 338.14 should be assigned as y_2. According to Table 15.7, y_3 would be at *m/z* 466.24, and the mass difference of 128.10 between y_3 and y_2 corresponds to Q/K. Since 128.13 Da was found from the *b*-ion series and 128.10 Da from the *y* series, averaging these numbers the assignment is made as K (128.095 Da) and not Q (128.059 Da).

Downward from the peak of *m/z* 913.43 (b_{n-3}) the next "*b* ion candidate" is that of *m/z* 798.47; the mass difference is 114.96 Da suggesting that aspartic acid could be the next residue in the sequence. This assumption is further corroborated by the fact that preferential fragmentation is frequently observed at the *C* termini of acidic, in particular, aspartic acid residues [49]. Figure 15.12 clearly indicates that the ion at *m/z* 913.43 formed preferentially, which is consistent with the interpretation that the *N*-terminal residue of this fragment is aspartic acid. Thus the *C*-terminal part of the sequence is now DKYR.

The only gap left uninterpreted in the sequence is defined by the mass difference between the *b* ions 798.47 and 582.33 Da (= 216.14 Da) or between the *y* ions 797.42 and 581.35 (= 216.07 Da). This mass difference is compatible with the amino acid combinations of [D,T] or [E,S] or [C,I/L]. However, cysteine residues in this sample have been carbamidomethylated so the last combination can be discarded. The [E,S] combination seems slightly more likely since there was a small peak at *m/z* 60 (immonium ion of S), and the immonium ion of T was not detected at *m/z* 74. However, further evidence is needed, as S and T usually produce too weak immonium ions to be of significant diagnostic value. To decide between [D,T] and [E,S], we tried to extend the *y* ion series from y_4 (*m/z* 582.33) onward. Three possible amino acid extensions were found): at *m/z* 668.31 (Δm = 86.96 Da), or at *m/z* 710.31 (Δm = 128.98 Da), or at *m/z* 712.36 (Δm = 131.03 Da). The first two would be assigned as S or E, and the third mass difference corresponds to methionine. Since no amino acid combination would yield a 216-Da mass difference with methionine in it, the last piece of the sequence must be the [E,S] motif in a as yet unknown order. The sequence reads as RA(Q/K)PE[E,S]DKYR.

The order of E and S residues in the middle of the sequence could not be established on the basis of any series of single backbone cleavages. Still internal fragments could help finish the interpretation process. Internal fragments are known to form efficiently at the *N*-terminal side of proline residues, so one could try to find an internal ion series starting with the proline at the fourth position in the sequence. The smallest internal fragment would be PE. Fortunately, both the a_5y_8 and the b_5y_8 forms of this fragment were found at *m/z* 199.07 and 227.07. Continuation of the internal fragment series could be also found: PES at *m/z* 286.12 (a_6y_8 form) and *m/z* 314.14 (b_6y_8 form). Although and extension corresponding to the PEE internal sequence could be found as well, these ions were of the same mass than the a_3 and b_3 ions. Another argument to rule out the presence of the partial sequence PEE was the failure to explain abundant fragment ions such as that at *m/z* 314. Therefore, the completed sequence looks like RA(Q/K)PESEDKYR.

The only remaining task is to convincingly resolve the ambiguity at the third position. The protonated molecular weight of the sequence RAQPESEDKYR is 1378.671 Da, and that of the sequence RAKPESEDKYR is 1378.708 Da. The experimentally observed value of MH^+ was 1378.681. Since with internal calibration, our mass accuracy is around 10 ppm (equivalent to ±0.014 Da) it is more likely that the RAQPESEDKYR sequence is correct. Independent mass measurements of the same mass difference were obtained from the a_3–a_2, b_3-17–b_2-17 peak pairs. The results were 128.04 and 128.05 Da, respectively, and both measurements support the assignment of Q to the third amino acid residue.

The overwhelming majority of peptides sequenced for proteomic projects originate from tryptic digests. This somewhat simplifies the challenge of de novo sequence interpretation. In most cases the y_1 ion can be found easily and further members of the *y*-ion series can be detected very likely, due to the basic character of the *C* terminus. However, the considerations presented in the little exercise above are equally well applicable to any peptide sequence with no enzyme specificity. The *C* terminus of these peptides may not be obvious from the low-mass fragments as the y_1 ions seldom show up, with the exception of R, K, and H. It is a useful trick in such a case to look for the $b_{n-1} + H_2O$ ion, which is present in the spectrum in most cases. The mass difference between this ion and the precursor ion is equal to the mass of the *C*-terminal amino acid residue. From that point on a procedure similar to the one presented above can be followed.

15.4. CONCLUSIONS AND FUTURE DIRECTIONS

The MALDI-TOF/TOF technique has been shown to produce high-quality fragment spectra from peptides. By virtue of its mass resolution and mass accuracy and producing a wide range of fragment ions, TOF/TOF can be an ideal platform for high-throughput protein characterization involving not only the identification of proteins but also determining complete amino acid sequences. The presence of characteristic high-energy fragment ions, such as internal fragments and fragments with sidechain losses, enables one to resolve ambiguities in the peptide sequence. Most frequently differentiation between leucine and isoleucine is mentioned as a benefit from high-energy CID. The capabilities in resolving ambiguities between diamino acid motifs having identical masses either by using immonium ion masses or by looking for *d*, *v*, or *w* ions were found to be even more useful. It is important to point out that even mass differences corresponding to monoamino acid residues can be ambiguous; for instance, N has the same mass as GG or Q has the same mass as GA or AG. High-energy CID is the best tool to resolve these structural details as bond cleavages between and next to G and A are frequently inefficient in low-energy collisions.

To fully realize the potential of TOF/TOF, two important areas should be developed further. As it has been shown, few database searching programs have sufficient support for the variety of fragment ions arising from high-energy CID. In particular, handling internal fragments and immonium ions is critical since these types account for about half of the fragment ion intensity in TOF/TOF spectra. However, the addition of new fragment types to the feature lists of these programs will not solve the problems automatically. An increasing number of coincidental matches will be concomitant with the extension of fragment types considered by database searching programs. This trend could be kept at bay by developing more sophisticated scoring methods. Since fragmentation behavior is determined by the chemical nature of the amino acid residues in the peptides, it should be possible to develop algorithms that can predict MS/MS spectra from any amino acid sequence. These predictions could then be utilized to eliminate random matches during database searching.

The other important area for development is the efficient combination of MALDI-TOF and TOF/TOF with chromatographic peptide separations. Despite the recognized strength of MALDI-TOF to deal with complex mixtures, it will be desirable to hyphenate MALDI-TOF/TOF to liquid chromatography (LC). The throughput of the MS/MS analysis will be the same but the sensitivity will be increased proportionally to the degree of separation. Online coupling of LC and MALDI is obviously more difficult than interfacing ESI and LC. However, offline coupling with MALDI can offer an unprecedented flexibility to run complex experiments. Samples deposited on MALDI sample plates can persist for several weeks, allowing the users to interleave data acquisition and data interpretation in a way that the maximum information can be gained. In the near future the speed of MALDI analyses is expected to exceed that of the chromatographic separation. In this case offline coupling will allow the use of one mass analyzer serving multiple separation devices without sacrificing sensitivity.

REFERENCES

1. D. Hunt, J. R. Yates, III., J. Shabanowitz, S. Winston, and C. R. Hauer, *Proc. Natl. Acad. Sci.* (USA) **83**, 6233–6237 (1986).
2. R. S. Johnson, S. A. Martin, J. T. Stults, J. T. Watson, and K. Biemann, *Anal. Chem.* **59**, 2621–2625 (1987).
3. R. S. Johnson, S. A. Martin, and K. Biemann, *Int. J. Mass Spectrom. Ion Process.* **86**, 137–154 (1988).
4. J. B. Fenn, M. Mann, C. K. Meng, S. F. Wong, and C. Whitehouse, *Science* **246**, 64–71 (1989).
5. M. Karas and F. Hillenkamp, *Anal. Chem.* **60**, 2299–2301 (1988).
6. M. Wilm and M. Mann, *Anal. Chem.* **66**, 1–8 (1996).
7. O. Vorm, P. Roepstorff, and M. Mann, *Anal. Chem.* **66**, 3281–3287 (1994).
8. O. N. Jensen, T. Houthave, A. Shevchenko, S. Cudmore, T. Ashford, and M. Mann, *Anal. Chem.* **67**, 1706–1714 (1997).
9. A. J. Link, J. K. Eng, D. D. Schieltz, E. Carmack, G. J. Mize, D. R. Morris, B. M. Garvik, and J. R. Yates, III., *Nat. Biotechnol.* **17**, 676–682 (1999).
10. A. Bairoch, in *Proteome Research: New Frontiers in Functional Genomics*, M. R. Wilkins, K. L. Williams, R. D. Appel, and D. F. Hochstrasser, eds., Springer, Berlin, 1997, pp. 93–132.

11. P. Berndt, U. Hobohm, and H. Langen, *Electrophoresis* **20**, 3521–3526 (1998).
12. A. Shevchenko, O. N. Jensen, A. V. Podtelejnikov, F. Sagliocco, M. Wilm, O. Vorm, and M. Mann, *Proc. Natl. Acad. Sci.* (USA) **93**, 14440–14444 (1996).
13. J. R. Yates, III., A. L. McCormack, and J. Eng, *Anal. Chem.* **68**, 534A–540A (1996).
14. W. K. Henzel, T. M. Billeci, J. T. Stults, S. C. Wong, C. Grimley, and C. Watanabe, *Proc. Natl. Acad. Sci.* (USA) **90**, 5011–5015 (1993).
15. M. Mann, P. Hojrup, and P. Roepstorff, *Biol. Mass. Spectrom.* 338–345 (1993).
16. D. D. J. C. Pappin, P. Hojrup, and A. J. Bleasby, *Curr. Biol.* **3**, 327–332 (1993).
17. J. R. Yates, III., S. Speicher, P. R. Griffin, and T. Hunkapiller, *Anal. Biochem.* **213**, 397–408 (1993).
18. J. K. Eng, A. L. McCormack, and J. R. Yates, III. *J. Am. Soc. Mass. Spectrom.* **5**, 976–989 (1994).
19. M. Mann and M. Wilm, *Anal. Chem.* **66**, 4390–4399 (1994).
20. B. B. Spengler, D. Kirsch, R. Kaufmann, and E. Jaeger, *Rapid Commun. Mass Spectrom.* **6**, 105–108 (1992).
21. A. Shevchenko, A. Laboda, A. Shevchenko, W. Ens, and K. Standing, *Anal. Chem.* **72**, 2132–2141 (2000).
22. K. Medzihradszky and A. L. Burlingame, *METHODS: A Companion to Meth. in Enzymolo.* **6**, 284–303 (1994).
23. K. Medzihradszky, J. M. Campbell, M. A. Baldwin, A. M. Falick, P. Juhasz, M. L. Vestal, and A. L. Burlingame, *Anal. Chem.* **72**, 552–558 (2000).
24. M. L. Vestal, P. Juhasz, W. Hines, and S. A. Martin, in *Mass Spectrometry in Biology and Medicine*, A. L. Burlingame, S. A. Carr, and M. A. Baldwin, eds., Humana Press, Totowa, NJ, 2000, pp. 1–16.
25. M. L. Vestal and P. Juhasz, *J. Am. Soc. Mass Spectrom.* **8**, (1998).
26. M. L. Vestal, J. Campbell, K. Hayden, and P. Juhasz, *Proc. 48th ASMS Conf. on Mass Spectrometry and Allied Topics*, June 11–15, 2000, Long Beach, CA, pp. 762–763.
27. K. F. Blom, *Anal. Chem.* **73**, 715–719 (2001).
28. P. Roepstorff and J. Fohlman, *Biomed. Mass Spectrom.* **11**, 601 (1984).
29. M. F. Bean, S. A. Carr, G. C. Thorne, M. H. Reilly, and S. J. Gaskell, *Anal. Chem.* **63**, 2824–2834 (1991).
30. B. Spengler, *J. Mass Spectrom.* **32**, 1019–1036 (1997).
31. K. Biemann, *Biomed. Environ. Mass Spectrom.* **16**, 99–111 (1988).
32. R. S. Johnson and K. Biemann, *Biomed. Environ. Mass Spectrom.* **18**, 945–987 (1989).
33. W. M. Hines, A. M. Falick, A. L. Burlingame, and B. W. Gibson, *J. Am. Soc. Mass Spectrom.* **3**, 326–336 (1992).
34. R. A. Zubarev, N. L. Kelleher, and F. W. McLafferty, *J. Am. Chem. Soc.* **120**, 3265–3266 (1998).
35. J. DeGnore and J. Quin, *J. Am. Soc. Mass Spectrom.* **9**, 1175–1188 (1998).
36. X. Jiang, J. B. Smith, and E. C. Abraham, *J. Mass Spectrom.* **31**, 1309–1310 (1996).
37. A. M. Falick, W. M. Hines, K. F. Medzihradszky, M. A. Baldwin, and B. W. Gibson, *J. Am. Soc. Mass Spectrom.* **4**, 882–893 (1993).

38. R. Kaufmann, D. Kirsch, and B. Spengler, *Int. J. Mass Spectrom. Ion Process.* **131**, 355–385 (1994).
39. R. E. Scarberry, Z. Zhang, and D. R. Knap, *J. Am. Soc. Mass Spectrom.* **6**, 947–961 (1995).
40. I. Papayannopoulos, *Mass. Spec. Rev.* **14**, 49–73 (1995).
41. Siegel and Bauman (1988) ••.
42. C. Bartels, *Biomed. Environ. Mass Spectrom.* **19**, 363–368 (1990).
43. J. A. Taylor and R. S. Johnson, *Rapid Commun. Mass Spectrom.* **11**, 1067–1075 (1997).
44. J. Fernandez-de-Cossio, Gonzalez, L. Betancourt, V. Besada, G. Padron, Y. Shimonishi, and T. Takao, *Rapid. Commun. Mass Spectrom.* **12**, 1867–1878 (1998).
45. D. N. Perkins, D. J. Pappin, D. M. Creasy, and J. S. Cottrell, *Electrophoresis* **20**, 3551–3567 (1999).
45. Zhang and McElvain (1999) ••.
46. K. R. Clauser, P. Baker, and A. L. Burlingame, *Proc. 44th ASMS Conf. Mass Spectrometry and Allied Topics*, Portland, OR, 1996, p. 365.
47. R. G. Krishna and F. Wold, in *Posttranslational Modifications. Proteins—Analysis and Design*, R. H. Angeletti, ed., Academic Press., San Diego, 1998, vol. 1, pp. 121–206.
48. G. C. Thorne, K. D. Ballard, and S. J. Gaskell, *J. Am. Soc. Mass Spectrom.* **1**, 249–257 (1990).
49. W. Yu, J. E. Vath, M. C. Huberty, and S. A. Martin, *Anal. Chem.* **65**, 3015–3023 (1993).

K. M. Downward and K. Biemann, *J. Am. Soc. Mass Spectrom.* **4**, 874–881 (1993).

CHAPTER 16

Quantitative Analysis of Neuropeptides by MALDI-TOF MS

JOHAN GOBOM and ECKHARD NORDHOFF

Max Planck Institute for Molecular Genetics, Berlin, Germany

CONTENTS

16.1. INTRODUCTION

While MALDI-TOF MS is an established technique, used extensively for molecular weight determination of biopolymers, its use has so far been limited largely to the generation of qualitative data. An increasing number of recent publications, however, demonstrate the technique's potential for quantitative

Mass Spectrometry and Hyphenated Techniques in Neuropeptide Research, Edited by Jerzy Silberring and Rolf Ekman.
ISBN 0-471-35493-7

analysis. Several aspects of MALDI-TOF MS make it attractive for quantitative analysis compared to established techniques. These are short analysis time, high detection sensitivity, and ease of automation.

In neuroscience, radioimmunoassay, Western blot, and immunohistochemical methods have, for several decades, been the predominant tools for studying the abundance and distribution of peptides and proteins of the nervous systems. These methods rely solely on the recognition specificity of an antibody or polyclonal antiserum for the detection and quantification of the antigen. The property measured by such antibody assays is not the quantity of a particular molecular species (the antigen). Instead, a property termed "antigen-like immunoreactivity" is recorded that, in addition to the targeted antigen, represents immeasurable contributions of all modified forms of the antigen as well as other structurally related molecules present in the sample, depending on their abundance and affinity for the antibody. It follows that the abundance of, for instance, a modified form of the antigen, which may be of functional importance, will evade detection, and variations in its concentration may be masked by other variations in the total antigen-like immunoreactivity. Using a mass spectrometry as a detector for immunoassays overcomes this lack of specificity. By separating the sample molecules and detecting their characteristic masses, their expected structure can be confirmed, and modifications can be detected. An example is as follows. Human hypothalamus extracts were immunoaffinity purified using an antiserum directed toward the neuropeptide neurotensin (NT). MALDI-TOF mass spectra of samples taken from two individuals are shown in Fig. 16.1. In addition to the NT 1–13 signal (m/z 1672.92 monoisotopic peak), several other signals were observed. Some of these had masses corresponding to various truncated forms of the peptide as indicated in the figure. Other signals (Fig. 16.1b) had m/z values that did not correspond to any part of the NT sequence, or that of its precursor, neuromedin N. These most likely represent other structurally related peptides, which cross-react with the used antiserum. The presence of such molecular heterogeneity and cross-reactivity is also bound to occur in radioimmunoassay experiments, but will remain undetected. The ability to combine the highly specific qualitative data generated by MALDI-TOF MS, with quantitative data on the analyte of interest would thus provide an attractive alternative to conventional assays.

Quantitative MALDI-TOF MS analysis is, however, associated with several problems. These include low shot-to-shot reproducibility and a nonlinear detector response. Sample molecules compete for incorporation into the matrix crystals during sample preparation and for charge during the MALDI process, resulting in unpredictable discrimination and suppression effects. Furthermore, the nonuniform nature of the matrix/analyte solid mixture results in sample spot-dependent signal intensities. Overcoming these problems requires the use of internal standards, and a proper MALDI sample preparation technique and data acquisition protocol.

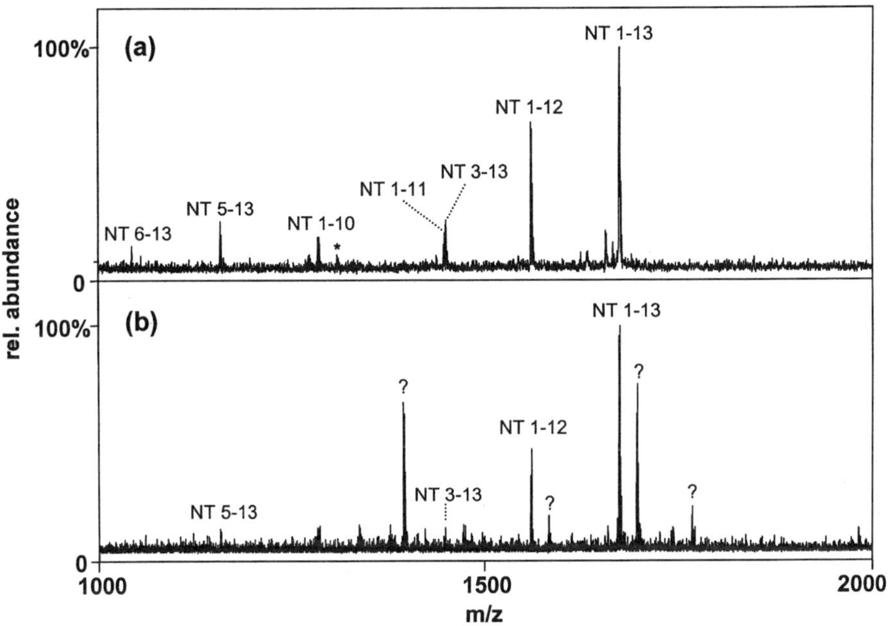

Figure 16.1. MALDI-TOF mass spectra of NT immunoprecipitates from hypothalamus samples from two individuals. The spectra indicate a varying abundance of partially degraded NT. Some signals in (b) have m/z values that do not match any cleavage product of NT or its precursor protein. These signals, marked with "?," most likely represent structurally related peptides, which cross-react with the antiserum used for immunopurification.

This chapter describes the general procedure and considerations for quantitative MALDI-TOF MS analysis of peptides. Examples are shown from a study on the quantification of NT immunoaffinity purified from human brain tissue extracts performed at the Institute of Clinical Neuroscience, Göteborg University [1]. The current status of this research field is reviewed, and some strategies for quantitative mass spectrometric analyses are discussed.

16.2. THE USE OF INTERNAL STANDARDS

The signal intensity of detected ions in MALDI-TOF MS depends on many parameters. These include the MALDI matrix and sample preparation used, the chemical properties of the analyte molecules, and instrument parameters. Furthermore, signal suppression of analyte ions may occur because of the presence of other sample components. Because several of these parameters are difficult to control, the use of internal standards is required. Thus, instead of using the absolute intensity of the analyte signal for quantification, the ratio of the intensity of the analyte signal to the standard signal is used. For quantitative analysis of molecules isolated from biological sources, internal standards may fill the additional purpose of accounting for analyte losses during sample pretreatment and purification.

16.2.1. Selection of Standard

When selecting the internal standard, it must be established that there is a linear relationship between measured signal intensity ratios (analyte ion to standard signals) and the analyte concentration, and over which range of concentration ratios this relationship is valid. Discrepancies occur if the analyte and standard differ in their cocrystallization with the MALDI matrix, their desorption/ionization efficiency, and their degree of metastable decomposition. Therefore, the standard molecules should be as structurally and chemically similar to the analyte molecules as possible. Peptide quantification by MALDI-TOF MS has been demonstrated using homologous compounds as internal standards. However, the established linear correlation may be altered by unpredictable factors, such as varying abundance of other sample molecules, and must therefore be carefully validated for each type of sample to be analyzed.

16.2.2. Stable Isotope-Incorporated Standards

In order to avoid the abovementioned problems, the ideal standard should be structurally and chemically identical to the analyte, but have a distinguishable mass difference. These criteria are met by using a stable isotope-labeled analog of the analyte molecule as the internal standard. Having the same number of protons and electrons, and the same structure as the analyte of interest, the standard differs only in the number of neutrons in the labeled atoms. For instance, if ^{2}D, ^{13}C, ^{15}N, or ^{18}O replace a number of ^{1}H, ^{12}C, ^{14}N, or ^{16}O atoms in a peptide, the resulting isotopic variants are not discriminated during the necessary sample purification and preparation steps, but are baseline-separated in state-of-the-art MALDI-TOF mass spectrometers.

For mass spectrometric analysis of peptides isolated from biological extracts, the detection sensitivity is most often limited not by the mass spectrometric instrumentation, but by analyte losses due to nonspecific surface adsorption (e.g., to sample vials and pipette tips) during the sample purification. The relative analyte loss increases with decreasing concentration. Inclusion of a stable isotope-incorporated internal standard increases the total concentration of analyte plus standard, thereby reducing such losses. Similarly, if the stable isotope-labeled standard is added to a crude cell or tissue extract, it serves to reduce the amount of analyte degraded by proteases present.

Stable isotope-labeled peptide standards can be obtained in several ways, such as by peptide synthesis using isotope labeled amino acid derivatives; by chemical derivatization of the peptide [2]; by hydrogen/deuterium exchange of acidic groups on the peptide [3]; or by proteolytic cleavage of the precursor protein in $H_2{}^{18}O$, whereby one ^{18}O atom is incorporated at the carboxyl-terminus of the generated peptides [4]. Another option is to express the peptide or its recombinant precursor protein in *Escherichia coli* or yeast host cells grown in ^{13}C, ^{15}N, or $^{13}C/^{15}N$ medium, followed by proteolytic cleavage to generate the desired product, as is common practice in protein structure elucidation using NMR.

16.3. MALDI-TOFMS SAMPLE PREPARATION

Choosing a suitable active matrix compound and sample preparation technique are important parameters for successful quantitative MALDI-TOF MS analysis. Whether the measured analyte to standard signal intensity ratio reflects the sample composition depends on the total number of generated and detected analyte and standard molecular ions. Many thousands are required to provide proper statistics based on which the quantification is performed. In MALDI-TOF MS, this can necessitate accumulation of many single-shot spectra. How many are required depends on the analyte-to-matrix ratio, the matrix compound used, the ion stability (life time) of the analyte ions, the transmission of the TOF analyzer used, the detector, as well as the applied laser pulse energy and irradiated sample spot size. Because of the many parameters involved, the minimum number of single-shot-spectra that should be accumulated cannot be predicted, and instead needs to be established experimentally (see text below).

The MALDI matrix 2,5-dihydroxybensoic acid (DHB) allows long series of single-shot spectra of similar signal intensity to be recorded from a single position on the sample, which suggests its use for quantitative analyses [5,6]. DHB sample preparations are, however, sensitive to sample impurities, which may impair the use of this matrix for the analysis of peptide extracts of biological samples. Furthermore, the crystalline morphology of DHB samples prepared according to the dried droplet sample preparation protocol [7] is highly heterogeneous, rendering spot-finding necessary, and thereby impairing automatic data acquisition. Other sample preparation protocols that yield more homogeneous samples, require that the matrix compound is little or not soluble in acidified water, and are therefore not applicable to DHB.

The MALDI matrix α-cyano-4-hydroxycinnamic acid (CHCA) is significantly more tolerant toward contaminants than DHB. Homogeneous MALDI samples can be prepared with this matrix by using the fast evaporation technique [8] or variants thereof [9], or the CHCA surface affinity sample preparation method. The latter includes peptide purification, concentration, and preparation for MALDI in a single step (see text below). All protocols start with the preparation of a thin homogeneous layer of CHCA crystals by depositing onto the sample support a small aliquot of CHCA dissolved in a volatile organic solvent. Rapid evaporation of the solvent leaves behind a dense carpet of small CHCA crystals, on top of which an aliquot of acidified aqueous peptide solution is placed and allowed to dry [8] or removed after a short incubation time during which the peptides bind to the surface of the CHCA crystals [10]. The fast evaporation technique has been successfully applied for peptide quantification. The main advantages of the named sample preparation techniques are good spot-to-spot and intersample signal reproducibility [11], facilitating automatic data acquisition. For samples derived from biological sources, which are often dilute and contain salts, it can be useful to combine this sample preparation technique with a micro-reversed-phase purification and enrichment step. Peptides bound to the column are eluted onto the pre-formed matrix layer in a 1–2-μL aliquot saturated CHCA solution, containing 45% acetonitrile [1].

16.4. MALDI-TOF MS ANALYSIS

16.4.1. Instrument Parameters

Homogeneous MALDI sample preparations and the use of a stable isotope-labeled internal standard does not necessarily yield reproducible and precise quantification of the analyte by MALDI-TOF MS, mainly for two reasons: (1) an insufficient number of molecular ions of the analyte and standard that reach the detector and (2) a nonlinear response of the detector. It is not known how many ions of a given type typically hit the detector in a single MALDI-TOF MS experiment. However, the observation that the relative signal intensities of an analyte's isotopic envelope in a single-shot spectrum may deviate significantly from the predicted isotopic distribution, suggests that the number may be rather small. By slightly defocusing the laser beam, analyte molecules are desorbed from a larger sample area [12]. If the laser beam attenuation is adjusted accordingly (higher pulse energy to keep the laser irradiance constant), this adjustment improves the quantification results. Possible reasons for this are an increase of the desorption/ionization yield per laser shot and an increase of the ion-impact area on the detector, which in turn improves the linearity of the detector response. The price for this modification is a decrease of the instrumental resolving power.

The dual microchannel plate detectors used for ion detection on most MALDI-TOF MS instruments give a rapid response and are sensitive, but display a large variation in the signal response on the impact of a given number of ions. Furthermore, the regeneration time for individual channels following the impact of ions is in the range of milliseconds, and thus the detector gives a significantly weaker response during this time period, following the impact of a large ion package. Keeping the laser fluency at a minimum and using a lower detector voltage than normal, improves the linearity of the detector response and minimizes the chemical and electronic noise in the recorded mass spectra. Unfortunately, these adjustments directly affect the detection sensitivity, and therefore should be tuned carefully. The linearity of the detector response can be further improved by slightly defocusing the ion beam so that it impacts on a larger area of the detector. The consequence of this modification is loss of resolving power. Using state-of-the-art instruments, however, which for peptides provide an instrumental resolving power exceeding 20,000 (FWHM), decline of the resolving power to 50% or less of the maximum possible can often be accepted, especially if the molecular mass of the analyte falls below 2500 Da. The overall performance of MALDI-TOF mass spectrometers is often subject to systematic drifts over time due to temperature changes, especially if these include electronic components. For improved reproducibility, the laser, ion source, and detector power supplies should be switched on minimum 30 min before the analysis.

16.4.2. Data Acquisition

As discussed above, poor ion statistics and a nonlinear detector response cause variations in the intensity of the recorded signals of the analyte and the internal standard. Figure 16.2a shows four MALDI-TOF mass spectra of the same sample, containing neurotensin (NT) and synthetic stable isotope-labeled NT, in which six carbon atoms in Leu^3 are replaced by ^{13}C ($[^{13}C_6]$ NT). Each spectrum is the sum of 10 single-shot spectra. The samples were prepared using the nano-reversed-phase sample purification and preparation technique described above, and CHCA as the MALDI matrix. As visible in the figure, the NT-to-$[^{13}C_6]$ NT signal intensity ratio varies by more than 20%. Averaging of a large number of single-shot spectra from a sample is thus necessary in order to establish a more precise signal intensity ratio. Figure 16.2b shows the cumulative signal intensity ratios as a function of the number of accumulated spectra, obtained from the same sample. Sums of 10 single-shot spectra were acquired in succession. The intensity ratio was calculated and plotted for each addition of a 10-shot spectrum. As indicated by the diagram, a variation of <2% was obtained after accumulation of 400 single-shot spectra.

Accumulation of the large series of spectra necessary for precise quantification is a task well suited for automation. This is possible with state-of-the-art MALDI-TOF mass spectrometers using acquisition software that continuously monitors the quality of the recorded spectra [13]. The software controls the MALDI probe movement and adjusts the laser intensity using real-time fuzzy-logic feedback control. After acquiring a small number of single-shot spectra, the control algorithm evaluates signal intensity and mass resolution of the base peak. It then regulates the laser attenuation to keep the ion signal intensity within the dynamic range of the detection system, while maintaining high mass resolution. When the signal intensity is decreasing on one spot on the sample, it moves the MALDI probe a small distance to irradiate a fresh spot. In this manner, a large number of spectra can be summed up from several positions on the sample.

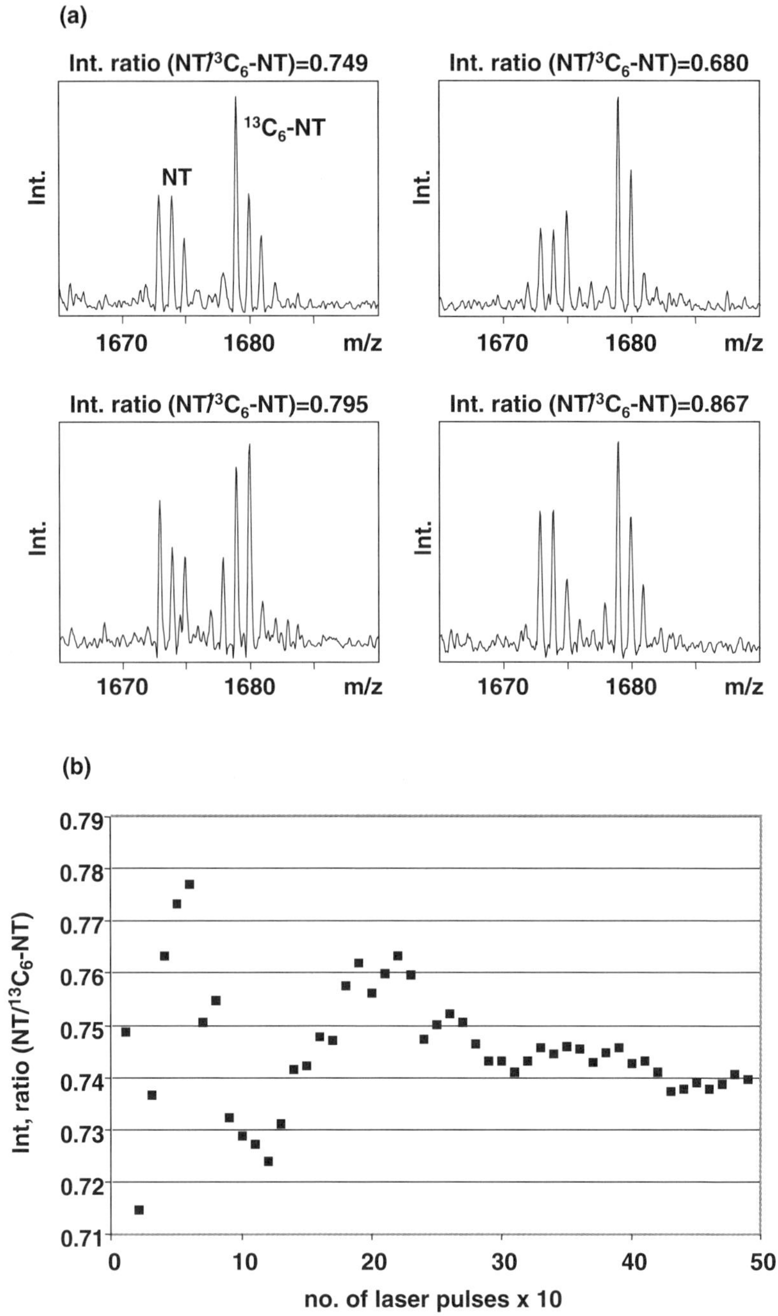

Figure 16.2. (a) Four 10-shot MALDI-TOF mass spectra acquired from a sample containing 20 fmol NT and 25 fmol [$^{13}C_6$] NT. The measured signal intensity ratios vary more than 20%. (b) Measured signal intensity ratios as a function of the number of accumulated 10-shot spectra. After accumulation of a total of 400 single-shot spectra, a variation of less than ± 2% was observed. [Reproduced with permission from *Anal. Chem.* **72**, 3320–3326 (2000). Copyright 2000 American Chemical Society.]

16.5. DATA ANALYSIS

16.5.1. Mass Spectra Processing

Typical operations performed on the acquired mass spectra include signal smoothing and baseline subtraction, both of which are implemented in most spectra processing software. Signal smoothing, can help to reduce errors due to rapidly varying electronic noise, whereas baseline subtraction reduces slowly varying (chemical) noise. To calculate the ratio of analyte to internal standard signal intensity, either the peak heights or the peak areas can be used. Considering the small number of data points covering a single monoisotopic peptide signal in a MALDI-TOF mass spectrum, calculating the integral of the peaks seldom improves the results. In most studies published to date, peak heights were used for peptide quantification instead. In the quantification of NT (see discussion above), the signal intensities were calculated as the average of the peak heights of the three first monoisotopic signals of NT and the internal standard [$^{13}C_6$] NT, respectively.

16.5.2. Construction and Use of the Standard Curve

The following example describes the procedure for constructing the standard curve for quantification of NT. A solution of NT was prepared and the concentration was determined by UV absorption at 280 nm. A solution of the internal standard (synthetic [$^{13}C_6$] NT) was prepared and its concentration adjusted so that the signal intensities of [$^{13}C_6$] NT and NT were roughly equal when an aliquot of the [$^{13}C_6$] NT solution mixed with an equal volume of the NT solution was analyzed by MALDI-TOF MS.

A series of solutions were prepared, all containing an equal volume of the [$^{13}C_6$] NT solution, and NT concentrations ranging between 0.1 and 20 fmol/μL. An aliquot of each solution was analyzed by MALDI-TOF MS as described above. The standard curve, shown in Fig. 16.3, was constructed by plotting the ratio of the NT to [$^{13}C_6$] NT signal intensities. Crude tissue extracts, in which the NT concentration was to be determined, were spiked with the same volume of the [$^{13}C_6$] NT solution or a fraction thereof. Following affinity purification and MALDI-TOF MS, the NT concentration in the crude extract could be obtained directly from the standard curve. Note that it is not necessary to determine precisely the concentration of the internal standard; only the concentration of the NT solution used for constructing the standard curve is needed.

The standard deviation of the mean gives an indication of the precision that can be achieved. The standard curve, however, is generated using pure analyte and internal standard. If the biological samples to be analyzed contain impurities that impair proper MALDI sample crystallization, thereby affecting the ion yield, the achievable precision may be lower. Therefore, both the NT and the [$^{13}C_6$] NT solution used for the standard curve were prepared in phosphate

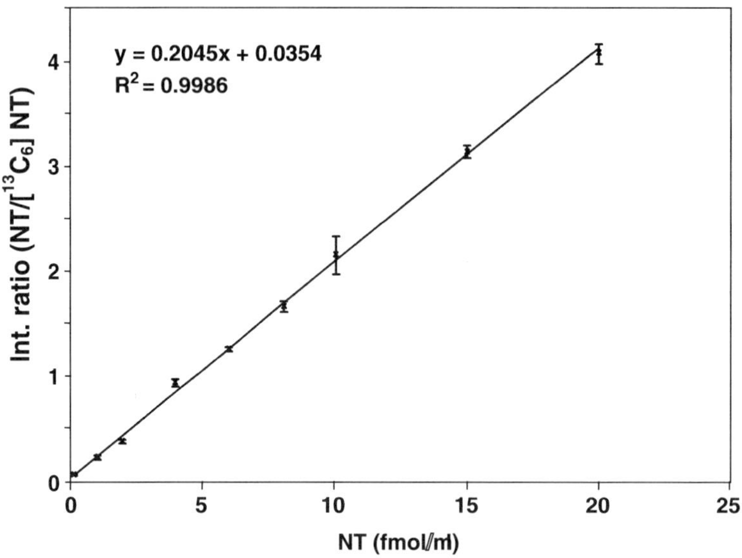

Figure 16.3. Standard curve for quantification of neurotensin. The relative standard deviation of the slope was 1.42%. The error bars indicate the 95% confidence interval (± 1.96 standard deviations around the mean) for each concentration ($n = 3$). [Reproduced with permission from *Anal. Chem.* **72**, 3320–3326 (2000). Copyright 2000 American Chemical Society.]

buffer and purified using the microscale reversed phase purification step described above. This was done to mimic as closely as possible the composition of the affinity purified samples to be quantified.

16.6. PERFORMANCE

16.6.1. Current Status

An increasing number of reports document the applicability of MALDI-TOF MS for quantitative analysis of peptides. A study by Jespersen et al. [5] investigated the use of different standards for quantification of peptides/proteins in the mass range of 1000–12,000 Da. When using peptides that were structurally analogous to the analyte as standards, a linear correlation for the standard curve was obtained if the peak height ratios of the analyte to standard were between 4:1 and 1:4. The relative standard deviation for the individual measurements varied significantly, depending on the internal standard and analyte. Improved linear correlation for the standard curve and lower standard deviations for the measurements were observed when a stable isotope incorporated analyte analog was used as the internal standard.

Several researchers have pointed out the importance of MALDI sample homogeneity. Nicola et al. [11] showed improved shot-to-shot reproducibility using the fast evaporation sample preparation technique [8]. The authors reported a relative standard deviation of the slope of ~2.2% for peak height ratios over two orders of magnitude. A detection sensitivity in the mid attomole range for the quantification of angiotensin II was achieved, using a homologous peptide as the internal standard. Axelsson et al. [14] and Hensel et al. [15] demonstrated that improved shot-to-shot reproducibility as well as sample spot-to spot reproducibility can be achieved by electrospray deposition of the analyte/matrix mixture. The latter group utilized this technique in a developed assay for the determination of desamino-[8-D-arginine] vasopressin in human plasma. Preston et al. [16] reported improved reproducibility of analyte ion yields by adding nitrocellulose to the MALDI matrix solution. The nitrocellulose is believed to provide crystal nucleation sites, thereby increasing the sample homogeneity. The authors used measured absolute signal intensities for quantification of a number of peptides, including bradykinin, substance P, angiotensin I, gramicidin S, and melittin. Other reported methods to improve the homogeneity of the matrix crystals include recrystallization of the MALDI sample in a volatile solvent [17] and the use of multicomponent matrices [6,18].

Gusev et al. [12] investigated the effect of instrument parameters on quantitative MALDI-TOF MS analyses. The study demonstrated improved shot-to-shot reproducibility by defocusing the laser beam and taking care to avoid detector saturation. Desiderio et al. reported the quantification of the mu (μ) opioid receptor agonist DAMGO in ovine plasma, using a deuturated analog as the internal standard [19]. In this study, verification of the identity of the detected peptide was achieved by MALDI post-source-decay (PSD) analysis. The authors reported improved detection sensitivity (50 fmol) by using the peak height ratios of some of the detected fragment ions for quantification.

16.6.2. Limitations and Further Improvements

As can be observed in the standard curve shown in Fig. 16.3, the error bars, indicating ± 1.96 standard deviations of the individual samples, are in some cases small compared to the deviation of the measurement from the best linear fit. This indicates that a large error in the quantification stems not from the reproducibility of the mass spectrometric determination of the signal intensity ratios, but is caused by imprecision in the liquid handling. Standard pipettes usually have an error of at least 10% for volumes <10 μL. Accuracy can be gained by handling larger volumes, by diluting the samples, but at the cost of detection sensitivity. The use of solenoid valve or piezo dispensing technologies allows precise deposition of nanoliter volumes and is expected to improve the precision of quantitative measurements [20–22].

It has been observed that the signal reproducibility is improved when the MALDI sample diameter falls below 250 μm, because a large portion of the sample is simultaneously irradiated by the desorption/ionization laser [23,24]. Such samples are well suited for automatic data acquisition because only one or a few sample positions have to be investigated to obtain a mass spectrum that is representative of the sample. The more recent development of prestructured MALDI sample supports [23] simplifies the preparation of microscale samples of well-defined spot size and location. In combination with the CHCA-affinity sample preparation technique [10], a large number of highly homogeneous samples can be prepared in a short time with little effort. These features were shown to improve signal reproducibility, render unnecessary variation of the desorption/ionization laser irradiance during data acquisition and thereby facilitate rapid, fully automated analyses.

As discussed above, to obtain a high precision in quantitative MALDI-TOF MS analyses, many single-shot spectra need to be accumulated from each sample, resulting in long measurement times. Although the duration of a single MALDI-TOF MS experiment is very short (50–100 μs), the data acquisition rate is limited to a few single-shot spectra per second due to the maximum laser pulse repetition frequency provided by the nitrogen lasers used in the commercial instruments (maximum 20 Hz) or transfer of the externally digitized raw data to the computer (e.g., maximum 10 Hz for the Bruker Reflex III used in our laboratory). The data acquisition rate, however, is expected to increase significantly (up to 100 Hz) in the near future because of the use of faster lasers and data transfer technology and protocols.

16.7. ANALYTIC STRATEGIES FOR QUANTITATIVE MALDI-TOF MS

As stated in the introduction, the high detection specificity obtainable by the combined use of affinity purification of selected molecules, and mass spectrometric separation and detection, presents a valuable research tool. The direct combination of affinity sample enrichment with MALDI-TOF MS was introduced by Hutchens and Yip [25]. Agarose beads with immobilized single-strand DNA were used to capture lactoferrin in untreated preterm infant urine. After incubation with the urine sample, the beads were washed, placed on the MALDI sample support, and analyzed. The selective detection of phosphopeptides and metal binding proteins has been shown, using metal chelate affinity chromatography medium [26]. Brockman and Orlando developed a technique termed *probe-immobilized affinity chromatography/mass spectrometry*, in which the MALDI sample plate was used as support for the affinity purification step [27]. Antibodies were immobilized directly onto the sample plate surface. After incubation with sample, and a washing step, the MALDI matrix was applied. Nelson et al. introduced the concept of using affinity purification and MALDI-TOF MS for analyte quantification [28]. The method, termed *mass spectrometric immunoassay* (MSIA), uses antibodies covalently linked to a solid support [similar to enzyme-linked immunosorbent assay (ELISA)]. Following incubation with the antigen-containing sample, and a washing step, the sample is eluted onto the MALDI sample support with matrix solution. The performance of the method was demonstrated by the detection and quantification of myotoxin *a* in human whole blood, using as the internal standard the analog H-myotoxin. Using multiple antibodies enabled the simultaneous detection of several antigens, as demonstrated with myotoxin *a* and Mojave toxin basic subunit.

One reported strategy for quantitative mass spectrometric analysis of complex protein mixtures is the use of isotope-coded affinity tags (ICATs) [29]. This technique uses a reagent, which consists of three elements: an affinity tag (biotin), a linker region that can incorporate stable isotopes, and a reactive group with specificity toward the thiol group of cysteines. The ICAT reagent exists in two forms: a heavy form containing eight deuterium atoms and a light form without deuterium. Two cell lysates, which are to be compared, are labeled with the reagent, one with the light form and one with the heavy form. Aliquots of the lysates are combined, followed by digestion of the contained proteins by a specific protease, for instance, trypsin. ICAT-labeled peptides are then isolated by avidin affinity chromatography and, after chromatographic separation, analyzed by mass spectrometry. Fragment ion analysis of detected peptides identifies the protein sequence that they were part of. Pairs of peptides tagged with the light and heavy ICAT reagents, respectively, serve as mutual internal standards for accurate relative quantification.

Mirgorodskaya et al. [4] presented a method for protein and peptide quantification based on the use of stable isotope incorporated internal standards generated by proteolytic cleavage of proteins in $H_2{}^{18}O$. A protein to be quantified is digested with a specific protease in $H_2{}^{16}O$, after which a known amount of its corresponding ^{18}O-labeled digest is added to the sample. Following MALDI-TOFMS analysis, each pair of detected $^{18}O/^{16}O$-containing peptides is used for protein quantification. Furthermore, this method can be used to determine the kinetic rate constants for the enzymatic cleavage reaction. For this application, the protein is subjected to controlled partial proteolysis; specifically, aliquots of the digestion reaction are removed at different time-points. The aliquots are spiked with the internal standard and analyzed. The peak height ratios of each peptide pair can then be used to calculate the rate of proteolysis at each specific cleavage site.

REFERENCES

1. J. Gobom, K.-O. Kraeuter, R. Persson, H. Steen, P. Roepstorff, and R. Ekman, *Anal. Chem.* **72**, 3320–3326 (2000).

2. D. M. Desiderio and M. Kai, *Biol. Mass Spectrom.* **10**, 471–479 (1983).
3. J. J. Kusmirez, R. Sumrada, and D. M. Desiderio, *Anal. Chem.* **62**, 2395–2400 (1990).
4. O. A. Mirgorodskaya, Y. P. Kozmin, M. I. Titov, R. Korner, C. P. Sonksen, and P. Roepstorff, *Rapid Commun. Mass Spectrom.* **14**, 1226–1232 (2000).
5. S. Jespersen, W. M. A. Niessen, U. R. Tjaden, and J. Vandergreef, *J. Mass Spectrom.* **30**, 357–364 (1995).
6. A. I. Gusev, W. R. Wilkinson, A. Proctor, and D. M. Hercules, *Fresenius J. Anal. Chem.* **354**, 455–463 (1996).
7. K. Strupat, M. Karas, and F. Hillenkamp, *Int. J. Mass Spectrom. Ion Process.* **111**, 89–102 (1991).
8. O. Vorm, P. Roepstorff, and M. Mann, *Anal. Chem.* **66**, 3281–3287 (1994).
9. M. Kussmann, E. Nordhoff, H. Rahbek-Nielsen, S. Haebel, M. Rossel-Larsen, L. Jakobsen, J. Gobom, E. Mirgorodskaya, A. Kroll-Kristensen, L. Palm, and P. Roepstorff, *J. Mass Spectrom.* **32**, 593–601 (1997).
10. J. Gobom, M. Schuerenberg, M. Mueller, D. Theiss, H. Lehrach, and E. Nordhoff, *Anal. Chem.* **73**, 434–438 (2001).
11. A. J. Nicola, A. I. Gusev, A. Proctor, E. K. Jackson, and D. M. Hercules, *Rapid Commun. Mass Spectrom.* **9**, 1164–1171 (1995).
12. A. I. Gusev, W. R. Wilkinson, A. Proctor, and D. M. Hercules, *Appl. Spectrosc.* **47**, 1091–1092 (1993).
13. O. Jensen, P. Mortensen, O. Vorm, and M. Mann, *Anal. Chem.* **69**, 1706–1714 (1997).
14. J. Axelsson, A. Hoberg, C. Waterson, P. Myatt, G. L. Shield, J. Varney, D. M. Haddleton, and P. J. Derrick, *Rapid Commun. Mass Spectrom.* **11**, 209–213 (1997).
15. R. R. Hensel, R. C. King, and K. G. Owens, *Rapid Commun. Mass Spectrom.* **11**, 1785–1793 (1997).
16. L. M. Preston, K. K. Murray, and D. H. Russell, *Biol. Mass Spectrom.* **22**, 544–550 (1993).
17. D. J. Harvey, *Rapid. Commun. Mass Spectrom.* **7**, 614–619 (1993).
18. A. I. Gusev, W. R. Wilkinson, A. Proctor, and D. M. Hercules, *Anal. Chem.* **67**, 1034–1041 (1995).
19. D. M. Desiderio, U. Wirth, J. L. Lovelace, G. Fridland, E. S. Umstot, T. M. D. Nguyen, P. W. Schiller, H. S. Szeto, and J. F. Clapp, *J. Mass Spectrom.* **35**, 725–733 (2000).
20. E. Nordhoff, H. Eickhoff, M. Horn, T. Przewieslik, V. Egelhofer, P. Giavalisco, D. Theiss, H. Lehrach, and J. Gobom, *Electrophoresis* (in press).
21. P. Önnerfjord, S. Ekström, J. Bergquist, J. Nilsson, T. Laurell, and G. Marko-Varga, *Rapid Commun. Mass Spectrom.* **13**, 315–322 (1999).
22. M. Mueller, H. Eickhoff, E. Nordhoff, H. Lehrach, and M. Kalkum, *48th ASMS Conf. Mass Spectrometry and Allied Topics*, Long Beach, CA, 2000.
23. M. Schuerenberg, C. Luebbert, H. Eickhoff, M. Kalkum, H. Lehrach, and E. Nordhoff, *Anal. Chem.* **72**, 3436–3442 (2000).
24. J. Gobom, E. Nordhoff, E. Mirgorodskaya, R. Ekman, and P. Roepstorff, *J. Mass Spectrom.* **34**, 105–116 (1999).
25. T. W. Hutchens and T. T. Yip, *Rapid Commun. Mass Spectrom.* **7**, 576–580 (1993).
26. D. I. Papac, J. Hoyes, and K. B. Tomer, *Anal. Chem.* **66**, 2609–2613 (1994).

27. A. H. Brockman and R. Orlando, *Anal. Chem.* **67**, 4581–4585 (1995).
28. R. W. Nelson, J. R. Krone, A. L. Bieber, and P. Williams, *Anal. Chem.* **67**, 1153–1158 (1995).
29. P. P. Gygi, B. Rist, S. A. Gerber, F. Tureck, M. H. Gelb, and R. Aebersold, *Nat. Biotechnol.* **17**, 994–999 (1999).

PART 4

SELECTED APPLICATIONS IN BIOSCIENCES

CHAPTER 17

Characterization of Peptides Bound to the Rat Class I MHC Molecule RT1-A1^c

R. C. JONES and S. J. GASKELL

Michael Barber Centre for Mass Spectrometry, Department of Chemistry, UMIST, Manchester, UK

J. STEVENS

Department of Molecular Biology, The Scripps Research Institute, La Jolla, California, USA

R. S. BORDOLI (deceased October 24, 2000)

Micromass (UK) Ltd., Manchester, UK

G. W. BUTCHER

Molecular Immunology Programme, The Babraham Institute, Cambridge, UK

E. JOLY

UPCM, Toulouse, France

CONTENTS

Mass Spectrometry and Hyphenated Techniques in Neuropeptide Research, Edited by Jerzy Silberring and Rolf Ekman.
ISBN 0-471-35493-7

17.1. INTRODUCTION

Antigenic peptides generated principally by the proteasome are transported into the endoplasmic reticulum (ER) via the heterodimeric transporter associated with antigen processing (TAP), where they associate with newly synthesized MHC class I molecules [1]. In the laboratory rat, the TAP2 subunit of the transporter exists as two main allotypes (TAP2-A and TAP2-B), and it appears that class I MHC alleles, encoded by the conventional class locus RT1-A, are genetically associated with only one form of the transporter. It has been shown that the two forms of the transporter display different peptide selectivities, such that TAP-A binds and transports peptides with many different amino acids present at the *C* terminus while TAP-B generally allows only hydrophobic and aromatic residues in that position [2]. Differences in sensitivity based on peptide length between rat TAP-A and TAP-B have also been reported [3,4], although the findings so far are not in agreement. Despite extensive studies on TAP peptide selectivities, few peptide-binding motifs for rat class Ia molecules (RT1-A) have been reported. Thus far, the binding motifs (anchor residues) of only two TAP-A-associated molecules, RT1-A^l [5] and RT1-A^a [6], and, more recently, two TAP-B-associated molecules, RT1-A^u and RT1-A1^c, have been determined [7]. Of these it is only RT1-A^a where both natural and synthetic peptides have been identified. RT1-A1^c was selected for study as a representative TAP-B-associated allele. Peptides identified in this study will allow confirmation of the adaptive significance of binding selectivities postulated for TAP2 subunits. This will also enable insight into the nature and source of peptides associated with this MHC molecule and confirmation of the binding motif proposed using a pool sequencing approach (Section 17.2). Once a natural peptide has been identified, its synthesis will allow the crystal structure of the heterotrimeric assembly of peptide, RT1-A1^c heavy chain, and β_2-microglobulin to be determined in order to visualize the conformation of peptide within the binding groove.

There is also additional interest in RT1-A1^c as a strong inhibitor of natural-killer (NK) cell function [8]. It has become increasingly evident that cytotoxic CD8$^+$ T-lymphocytes are not the only cell types that recognize class I MHC molecules. Natural-killer (NK) cells, another lymphocyte subset in the mammalian immune system, are now known to play important roles in both "innate" immunity and the regulation of the adaptive immune response. In particular, cells that fail to express MHC class I molecules are sensitive to NK lysis. NK cells are early entrants into cellular inflammatory sites where they can kill virally infected or neoplastic cells and release cytokines, notably substantial quantities of interferon-γ. There is increasing evidence that they can play an important role in the response to and rejection of transplants, particularly of bone marrow cell transplants [9], but also of solid organ allo- and xenografts [10]. Studies of the mechanisms by which NK cells are regulated have led to the discovery of families of NK cell surface receptors that either inhibit [killer inhibitory receptor (KIR)] or activate [killer activatory receptor (KAR)] NK cell cytotoxicity via recognition of MHC class I molecules. On human NK cells, the KIR and KAR receptors are members of the immunoglobulin superfamily, whereas on rodent NK cells the same functions are served by Ly-49 receptors which are homodimers of the C-type lectin family [11].

The peptide dependencies of class I MHC interactions with both types of receptors have been studied, and both were found to interact with the α_1/α_2 codomain of the class I molecule [12]. In humans, studies have suggested that KIRs can recognize HLA-C ligands in the presence and absence of bound peptides [13], although it is clear that different residues at positions 7 and 8 of peptides bound to HLA-B*2705 can influence the interaction between KIR and class I complexes [14]. For the mouse Ly-49A receptor, experiments have shown that that recognition is affected by the absence of peptides, but not by the presence of different peptides, thus suggesting that recognition is peptide-dependent but not peptide-specific [15]. This is not the case, however, for every mouse NK receptor since more recent studies with Ly-49l, another mouse NK cell receptor, have shown that it is able to discriminate peptides bound to H-2K^d [16]. Such complexity of interactions seen in both mouse and humans should also be expected in the rat. Although rat NK cells also recognize conventional (RT1-A-encoded) class I MHC molecules with inhibitory receptors [8], a unique feature, described only in the rat, is that some NK cells are specifically activated by certain polymorphic, non-classical (RT1-C/E/M-encoded) class I MHC molecules. Interest in the TAP-B associated molecule RT1-A1^c has increased since it was shown to inhibit rat NK cell activity via interaction with a Ly-49 lectin-like receptor, recognized by the antibody STOK2 [8]. However, any structural studies on the effect of peptides on RT1-A1^c/NK cell interactions have been hindered by the lack of known specific peptides able to bind to this molecule. In order to circumvent this problem, two approaches were pursued for the identification of binding peptides: (1) on the basis of the previously published binding motif of RT1-A1^c [7], peptides were designed, synthesized and tested for binding to RT1-A1^c when expressed in the TAP2-deficient mouse cell line, RMA-S; and (2) studies were performed in order to identify natural peptides presented by RT1-A1^c on rat splenocytes.

17.2. POOL SEQUENCING AND SYNTHETIC PEPTIDE DESIGN

Pool sequencing was used to determine the binding motif for RT1-A1^c [7] and these data were used to synthesize an "ideal" peptide based on the most abundant residue at each position (Fig. 17.1) [1,7]. At each cycle of Edman degradation, amino acids are listed as the highest percentage increase (compared to the previous cycle) first. An increase of >300% was regarded as indicative of an anchor residue during cycles 1–6; for later cycles an increase of 150% was considered similarly significant.

Dominant residues appeared at positions 2, 3, 4, and 9; there was an almost exclusive preference for proline at position 2. There was an observed preference for positive residues at position 3 with arginine and histidine dominating. Hydrophobic leucine at the *C* terminus is consistent with the peptide

	Position								
	1	2	3	4	5	6	7	8	9
Anchor	(N) (R)	P	R H Y	-	-	-	-	Y	L
Others	-	V I	N F Q K A S	K E M D Q T I F V	V Y L F H E	T Q A E	A I V M P	E K D	
"Ideal" peptide	N	P	R	K	V	T	A	Y	L

Figure 17.1. Results of pool sequencing of peptides bound to RT1-A1^{c} and sequence of the ideal RT1-A1^{c} peptide determined from the anchor and most abundant residues.

preference of the restrictive TAP-B transporter genetically linked to RT1-A1^{c} [2]. Assignment of the *N*-terminal residue is based on residues with the greatest yields, in this case asparagine and arginine.

The peptide NPRKVTAYL, hereafter described as the "ideal" peptide, was synthesized. The ability of this peptide to bind to and stabilise vacant RT1-A1^{c} molecules was then tested. Briefly, the gene for RT1-A1^{c} heavy chain was transfected into a mouse cell line called RMA-S. This is TAP2-deficient, and as a result peptides are not delivered across the ER membrane for the class I molecules to refold properly. If these cells are cultured overnight at 26°C rather than the usual 37°C, then "empty" class I molecules can be detected on the cell surface. If the cells are returned to 37°C, the empty molecules will fall apart: they may be stabilized, however, if suitable binding peptides are added to the culture medium. Fluorescence-activated cell sorting (FACS) analysis is then performed, using the RT1-A1^{c}-specific antibody, YR5/310 [17], to detect how much class I remains and therefore indicate the effectiveness of a peptide at stabilizing the molecule. Results are expressed as the percentage of stabilization compared to cells maintained at 26°C, with 100% corresponding to the level of fluorescence recorded with no peptide at 26°C (MHC molecules stable) and 0% corresponding to the fluorescence with no peptide at 37°C (MHC molecules unstable or absent). By varying peptide concentration in the medium, one can get a measure of what concentration is needed to achieve 50% of the level of cell surface expression compared to cells maintained at 26°C (C_{50}). As a negative control peptide (sequence with no propensity to bind to RT1-A1^{c}), we used a peptide designed to bind to the rat class I molecule

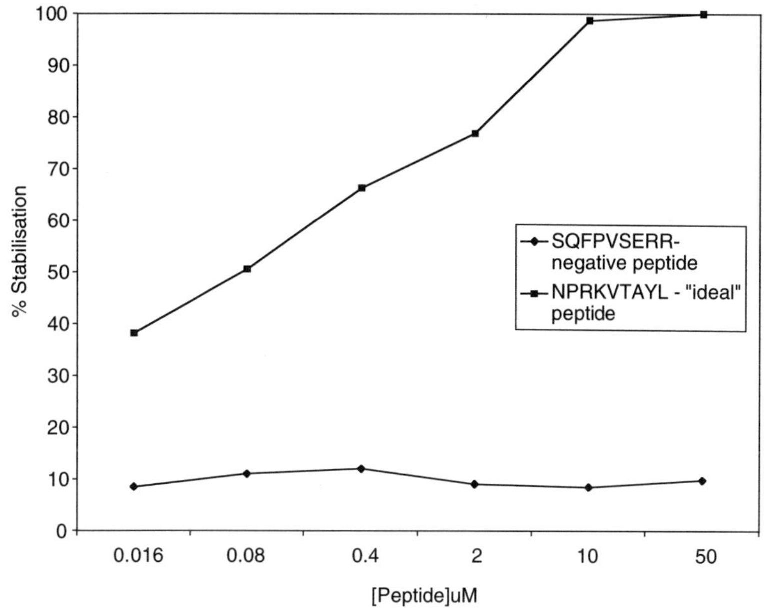

Figure 17.2. Binding of "ideal" peptide to RT1-A1[c] in stabilization assay; negative peptide known not to bind RT1-A1[c]; ideal peptide designed on basis of pool sequencing data.

RT1-A[a] [7]. The mean C_{50} was determined as 76nM (n = 3) for the ideal peptide (Fig. 17.2). Single amino acid substitution variants of this peptide were then synthesized, tested for stabilization, and compared to this ideal peptide. Results showed that (see Fig. 17.4 of Ref. 1), of the 31 peptide variants tested, most amino acid changes resulted in a reduction in binding affinity of the peptide compared to the "ideal" peptide. Only Phe or Asp substitutions at P1, Ser and His substitutions at P3, and a Thr substitution at P8 produced a statistically significant improvement in binding ($p < 0.05$) and, of these small improvements, only those variants at P3 of the peptide had been identified as secondary anchors in the motif (Fig. 17.1). Most peptides tested showed poorer binding ability, and statistically significant increases in C_{50} were observed with substitutions at P2 (Lys, Asp), P4 (Phe, Asp), P8 (Gln, Lys), and P9 (Tyr, Phe, Lys, Asp, Ile).

17.3. IDENTIFICATION OF NATURAL PEPTIDES BY MASS SPECTROMETRY

Spleen cells from PVG.R20 rats were used as the source of natural peptides. Following cell lysis, RT1-A1[c] immunoprecipitation and peptide elution, the

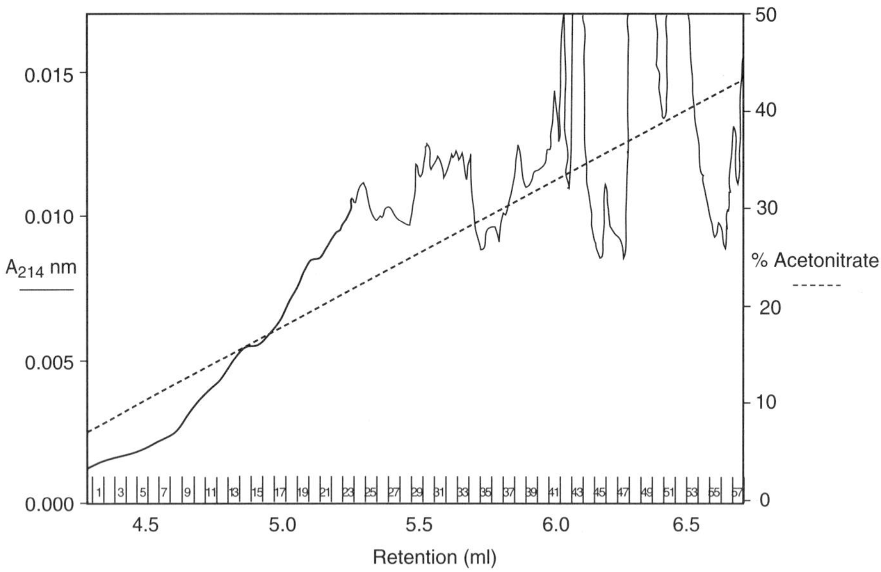

Figure 17.3. HPLC/UV chromatogram (214 nm) of total RT1-A1^c peptide extract; *x* axis indicates volume of elution solvent (mL).

total acid extract was fractionated by HPLC. The UV chromatogram is shown in Fig. 17.3. Peptides eluted between 6.75 and 40.5% acetonitrile were collected as 60 × 42-μL fractions; the fraction numbers are shown on the UV chromatogram. (In a separate experiment the large peaks covering fractions 48–55 were pooled and subjected to protein sequencing but yielded no sequence data, suggesting the bulk of this material was either *N*-terminally blocked or non-peptidic in nature.)

Three mass spectrometric strategies were implemented for the analysis of these samples; the use of each is described.

17.3.1. Nanospray Approach

Multiple fractions were screened by matrix-assisted laser desorption/ionization (MALDI) mass spectrometry (MS). The analysis of two fractions, 32 and 24, is described.

17.3.1.1. Fraction 32. MALDI/MS revealed 11 components with a signal-to-background ratio of greater than 2; many more components were present in the noise (Fig. 17.4). For the commitment of approximately 2% total fraction volume, this is an invaluable step. The analysis of the ion at *m*/*z* 1013.54 (component A) is described. Following lyophilization and reconstitution in electrospray (ES) solvent, 50% of the sample was loaded into a gold-coated

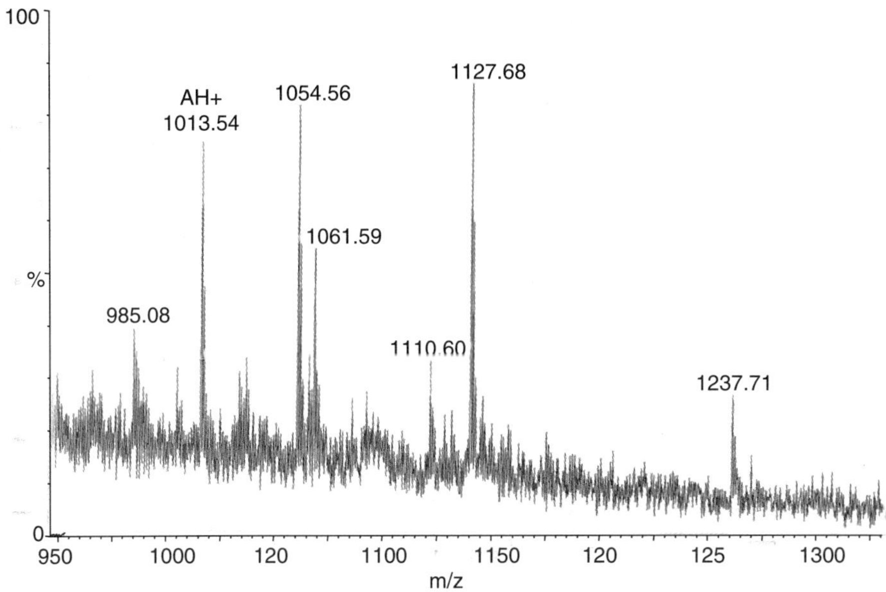

Figure 17.4. Fraction 32: MALDI-MS spectrum.

capillary and an ES/MS spectrum was acquired on a hybrid quadrupole/time-of-flight (Micromass Q-ToF) mass spectrometer (Fig. 17.5). Individual spectra were integrated over 2.5 s, and the ES mass spectrum represents the summation for a 1-min acquisition. Few multiply charged ions were observed given the number of components in the MALDI-MS spectrum; this may be attributed to either sample losses during workup or the presence of a species (such as detergent) that has preferential ionization. Preliminary knowledge of peptide masses (from the MALDI spectrum) nevertheless enabled interrogation of the ES/MS spectrum for corresponding multiply charged ions. A doubly charged ion at *m/z* 507.28 (corresponding to $[A + 2H]^{2+}$) was observed and selected for tandem MS.

Figure 17.6 shows products of *m/z* 507.28. The collision offset employed was 25 V, and spectra were accumulated for 2.2 min.

High-sensitivity tandem MS experiments were possible with low flow rate ES Q-ToF even though ions observed in the MS spectra were poorly observed within chemical noise. A prominent *b*-ion series was observed from b_8 to b_5, allowing assignment of the four *C*-terminal residues as [K/Q]A[L/I][L/I]. Selection of the precursor ion population (3 Th selection window) included an apparent nonpeptide component (separated from the peptide of interest by 1 Th); the observation of ions with *m/z* unassignable to the peptide sequence suggests that this component may have contributed to the product ion spectrum. Following *N*-acetylation of the remaining 50% of sample, an ES/MS spectrum was acquired (Fig. 17.7); data were summed over 1 min.

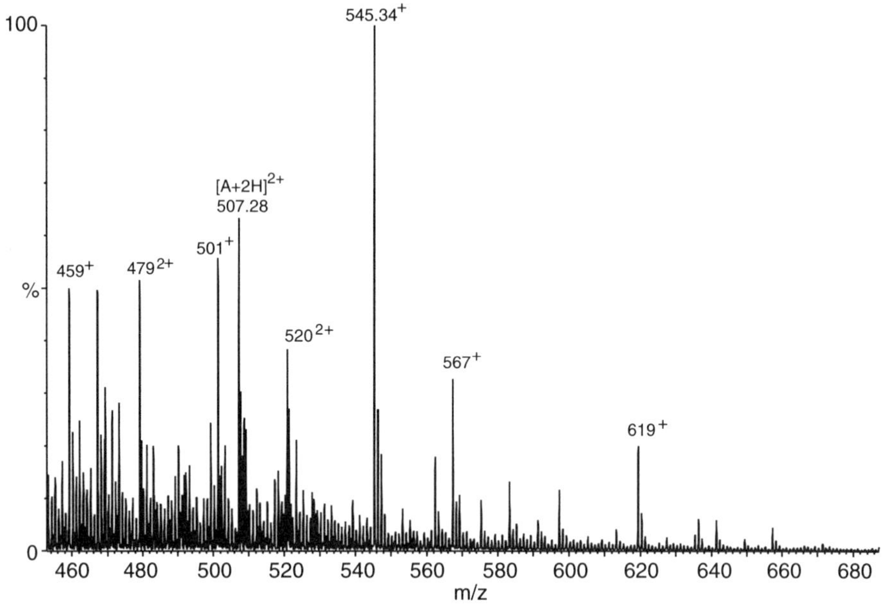

Figure 17.5. Fraction 32: ES/MS spectrum.

A doubly charged ion was observed at m/z 528.3 that corresponds to singly acetylated component A (Ac-A). The equivalent singly charged ion at m/z 1055.55 was also present; a reduction in basicity of the amine group is consistent with the observation of this ion only following derivatization. There were no ions corresponding to two sites of acetylation. Despite the very poor signal intensity, the m/z 528.3 ion could be selected for tandem MS. Figure 17.8 shows the corresponding product ion spectrum. Data were acquired for 6 min, and the collision offset was 25 V.

From the shift in product ion masses compared with underivatized peptide, the site of acetylation was shown to be the *N* terminus, and P6 was identified as glutamine by inference. There was no evidence for two sites of acetylation in the ES/MS spectrum. A dominant *b*-ion series was again observed, with the corresponding *a*-ion series from a_4–a_8. The shift in m/z by 42 Th allowed differentiation of *N*-terminal ions from *y* and internal ions. The observation of an internal ion series following cleavage *N*-terminal to proline was critical to full sequence elucidation. In both product ion spectra the presence of an *N*-terminal proximal arginine (P3) strongly influenced the degree of fragmentation at adjacent amide bonds; poor fragmentation was observed. This could be explained by the sequestration of an ionizing proton on the arginine sidechain, whereby it would then be electrostatically unfavorable for a second ionizing proton to reside at either adjacent amide oxygen (and promote fragmentation at that position). Losses of NH_3 were seen for b_2–b_8; this was a prominent loss

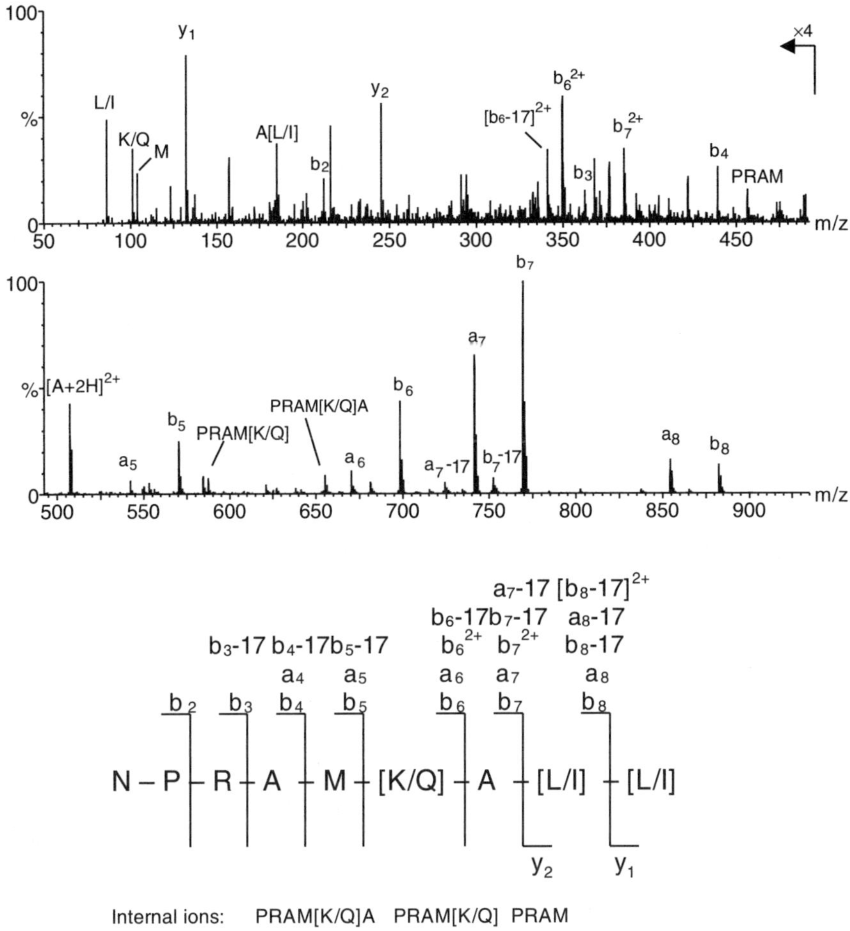

Figure 17.6. Fraction 32: products of *m/z* 507.28 ($[A + 2H]^{2+}$). The observed product ions are summarized; full sequence was based on combined data from products of $[A + 2H]^{2+}$ and of $[\text{Ac-}A + 2H]^{2+}$ following acetylation.

in b_3 and b_4 and is consistent with the presence of arginine at P3 as is the prevalence and stability of the *b*-ion series. The lack of a basic residue proximal to the *C* terminus would explain the lack of a *y*-ion series in both spectra; the observation of only y_2 and y_1 is consistent with the sequential decomposition of high member *y* ions. *N*-Terminal acetylation promoted observation of the b_1 ion, which, with the internal ion series, allowed assignment of asparagine at P1. Consideration of all product ion data enabled deduction of the primary sequence NPRAMQA[L/I][L/I].

All four isomers were synthesized, and their binding to RT1-A1[c] was assessed in the RMA-S stabilization assay; results revealed that all four

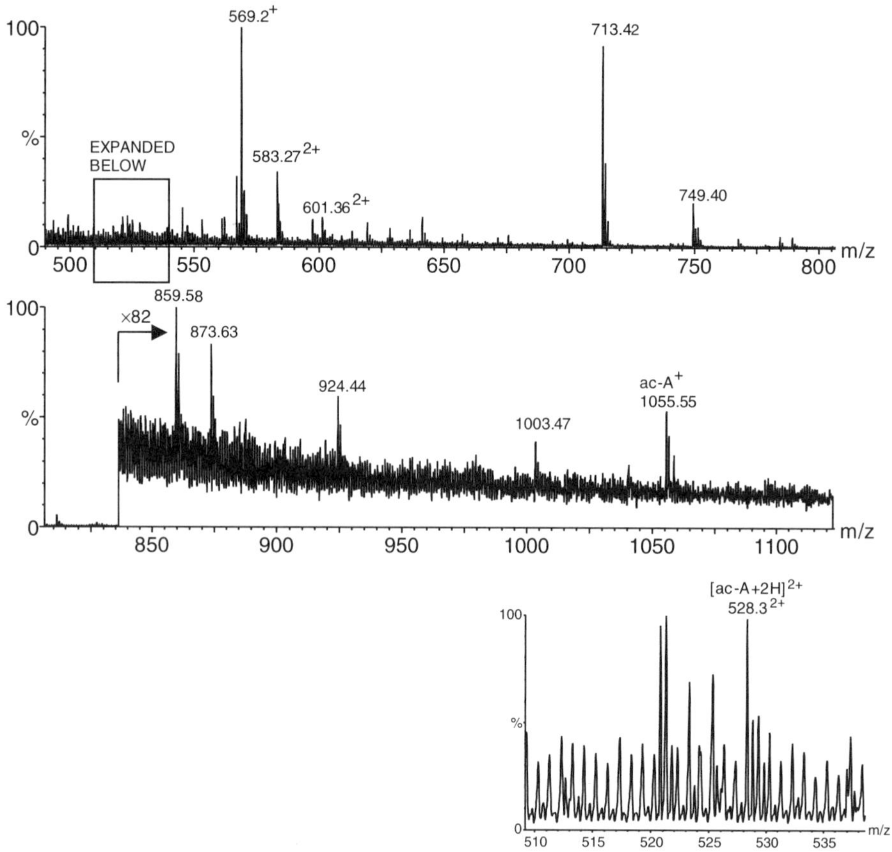

Figure 17.7. Fraction 32 following *N*-acetylation: ES/MS spectrum; expanded region shows $[\text{Ac-}A + 2\text{H}]^{2+}$.

peptides bound similarly, requiring a mean concentration of less than 20 *n*M ($n = 3$) to achieve 50% stabilization. This was a lower concentration required than the idealized peptide (Fig. 17.2), <20 *n*M compared with 76 *n*M. Because these results were all close to the lowest peptide concentration tested (16 *n*M), no distinction could be made between L/I preference at either position. Figure 17.9 shows the binding assay for the—LL isomer.

A search of the DNA (http://*www.ncbi.nlm.nih.gov/BLAST/*) and protein (MS-Edman: http://*prospector.ucsf.edu/*) databases with these four possible peptides identified NPRAMQALL as part of a rat protein associated with a tumor-suppressor protein (Accession No. D87950). Homologs were identified in both humans and *Xenopus*, indicating conservation of this protein across species. Table 17.1 summarizes the MS Edman search results.

The synthetic peptide NPRAMQALL was analyzed by HPLC/UV (Fig. 17.10) under the same conditions as the total RT1-A1^{C} peptide extract

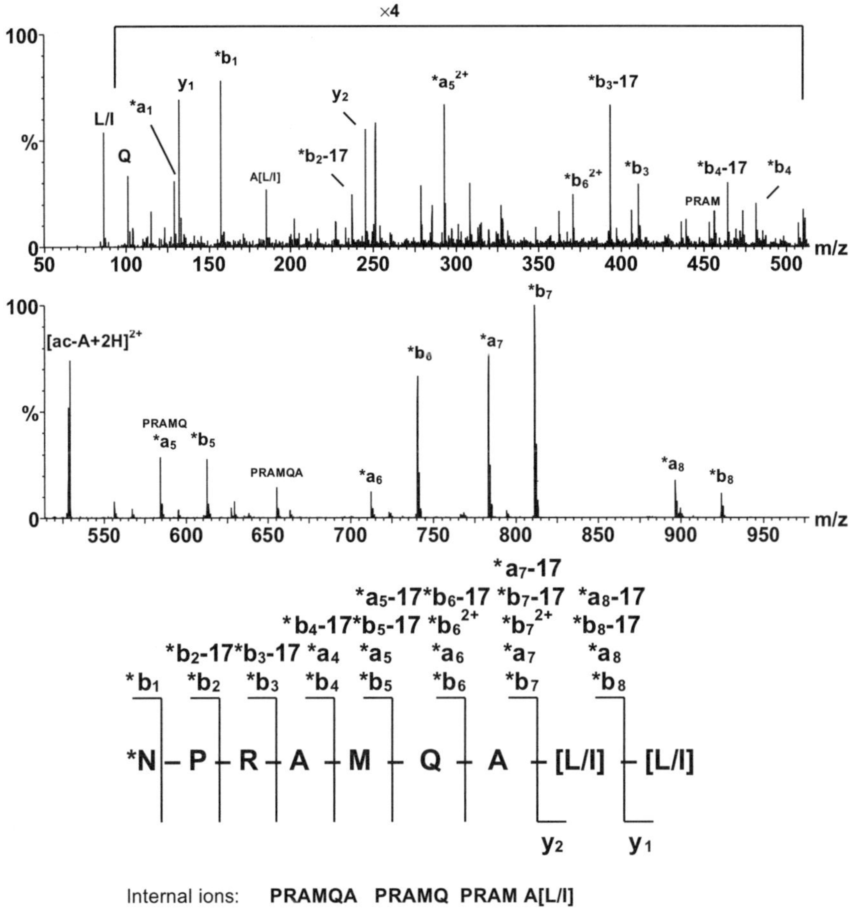

Figure 17.8. Fraction 32 following *N*-acetylation: products of *m*/*z* 528.30 ([Ac-*A* + 2H]$^{2+}$); observed product ions are summarised. The full sequence was based on combined data from products of [A + 2H]$^{2+}$ and [Ac-*A* + 2H]$^{2+}$.

(Fig. 17.3). A total of 0.1 µg was loaded on-column, and fraction 32 was collected and screened by MALDI-MS.

The spectrum (Fig. 17.11) revealed the expected *m*/*z* 1013.6, *M*H$^+$, with sodiated and potassiated adducts; the remainder of the fraction was analyzed using low-flow-rate Q-ToF MS/MS. Figure 17.12(1) shows products of *m*/*z* 507.30 [*M* + 2H]$^{2+}$ for confirmation of sequence and as a direct comparison to products of *m*/*z* 507.28 ([*A* + 2H]$^{2+}$) from the biological isolate [Fig. 17.12(2)].

The collision offset employed for the synthetic peptide (25 V) was the same as for [*A* + 2H]$^{2+}$. The product ion spectrum for the synthetic NPRAMQALL shows close similarity to products of [*A* + 2H]$^{2+}$. The observation of a singly charged component cotransmitted with the precursor ion population for the

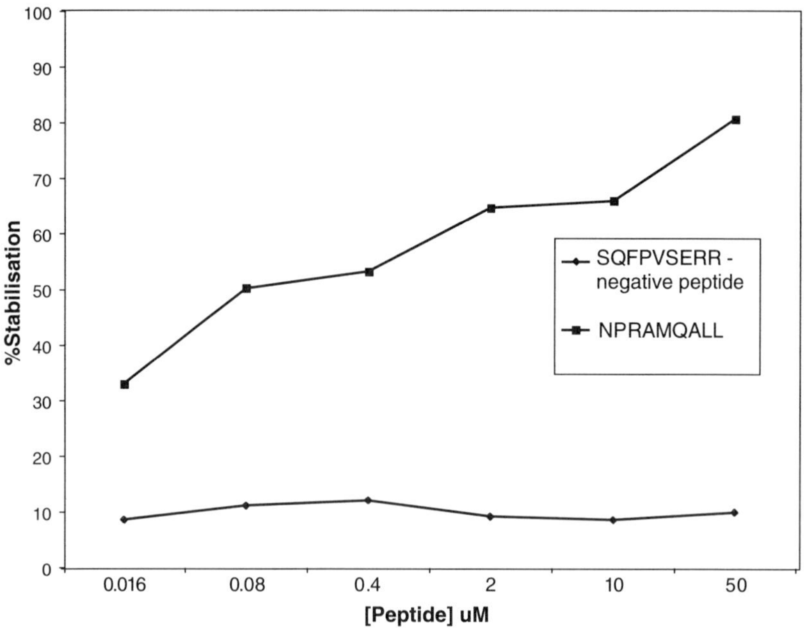

Figure 17.9. Binding of NPRAMQALL to RT1-A1^c in stabilization assay.

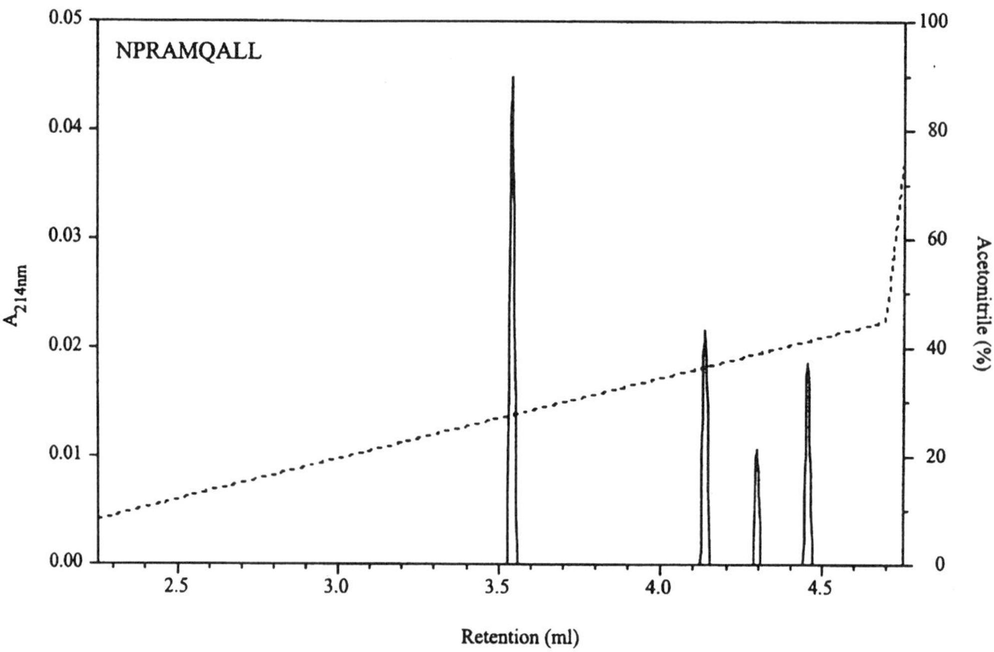

Figure 17.10. HPLC/UV chromatogram (214 nm) of 0.1 μg synthetic NPRAMQALL; the largest peak was isolated in fraction 32.

TABLE 17.1. MS-Edman Search Results (Search Mode: Sequence and Mass) for NPRAMQA[LI][LI] and MH$^+$ at *m/z* 1013.54

	Parent Ion	Matching Sequence	$M + H$ Matched	Delta Mass (Da)	Protein MW (Da)/pI	Species	Accession #	Protein Name
1	1013.54	(S)NPRAMQALL(Q)	1013.556	–0.016	62013.8/5.11	*Xenopus laevis*	5668703	(AB030502) XDRP1
2	1013.54	(S)NPRAMQALL(Q)	1013.556	–0.016	63184.0/5.14	*Homo sapiens*	7305607	(AF176069) ubiquilin
3	1013.54	(T)NPRAMQALL(Q)	1013.556	–0.016	47943.6/4.82	*Homo sapiens*	6606156	(AF113544) HRIHFB2157–like protein
4	1013.54	(S)NPRAMQALL(Q)	1013.556	–0.016	62072.0/4.87	*Rattus norvegicus*	7209586	(D87950) DA41

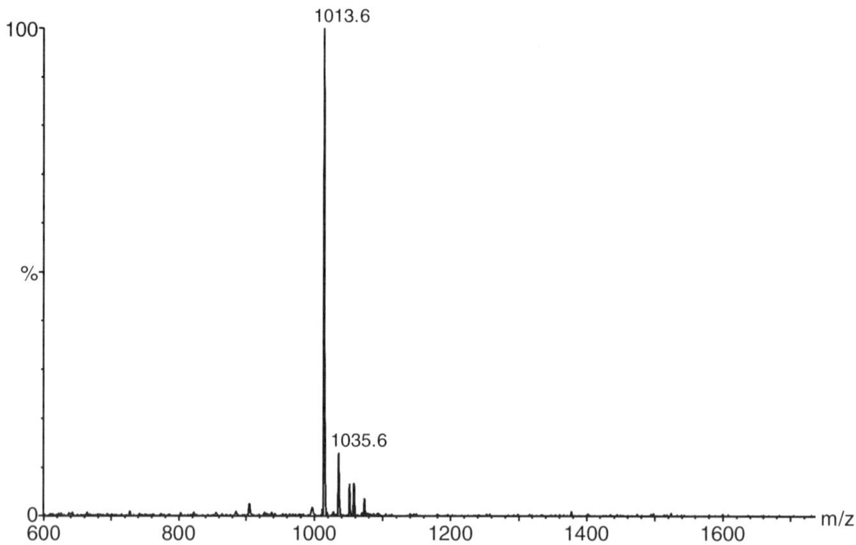

Figure 17.11. Fraction 32 from HPLC isolation of synthetic NPRAMQALL: MALDI/MS spectrum.

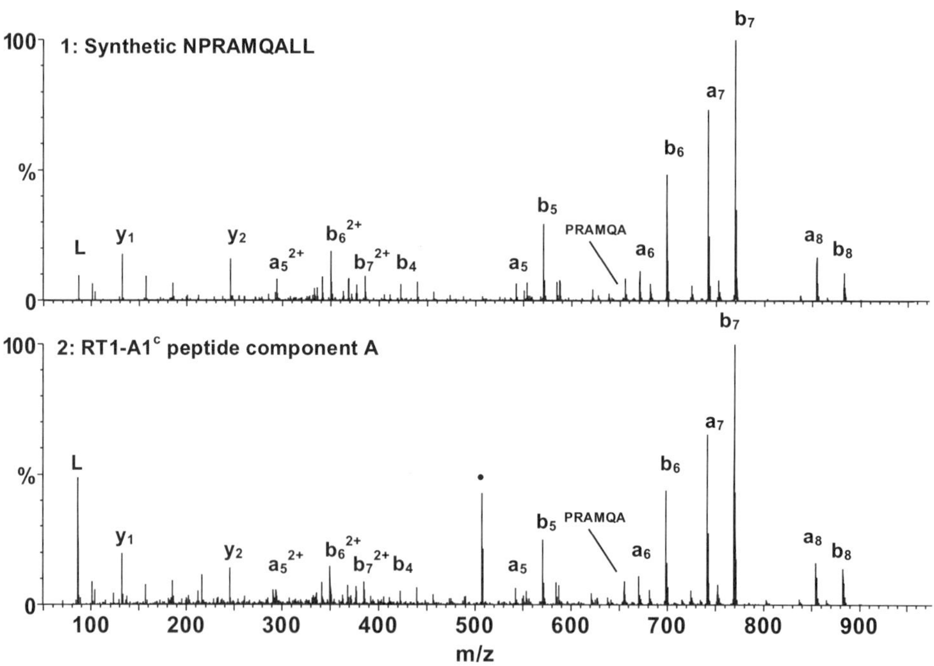

Figure 17.12. Fraction 32 products of m/z 507.3 for synthetic NPRAMQALL (1) and RT1-A1[c] component A (2; Fig. 6). • denotes coisolated singly charged component in precursor ion population.

RT1-A1^{c} sample represents the major difference between the two spectra. This is understandable with a biological isolate; components may derive from reagents used in workup or inherent species.

All the evidence suggests that the peptide NPRAMQA[L/I][L/I] was identified. Database matching to a rat tumor-suppressor protein allows proposal of the sequence NPRAMQALL. It is useful to summarize the contributory evidence:

1. Underivatized and *N*-acetylated tandem MS data.
2. Corollary with binding motif data. The observation of amino acid residues asparagine, proline, arginine, and leucine at positions 1, 2, 3 and 4, respectively, was consistent with the previously identified binding motif.
3. Binding assay data. Peptides showed excellent binding in a stabilization assay, better than the idealized peptide, NPRKVTAYL. This did not enable distinction between the four leucine/isoleucine isomers.
4. Database match to a known protein. This is the only evidence to suggest leucine/isoleucine discrimination.
5. Elution profile by HPLC of synthetic peptide. The synthetic peptide was observed in fraction 32 and confirmed by MALDI and ES tandem MS. The same peptide also carried over into fraction 33 for both RT1-A1^{c} sample and synthetic peptide (data not shown). This demonstrates a disadvantage to the methodology; the total available peptide is not isolated in a single fraction. This did not, however, influence peptide identification.
6. Product ion spectra of the synthetic peptide. The observation of the same product ions at the same relative abundances is conclusive evidence of the proposed primary sequence (leucine/isoleucine distinction aside).

17.3.1.2. Fraction 24. MALDI/MS revealed 14 components with a signal-to-background ratio of >2; many more components were present in the chemical noise (Fig. 17.13). The analysis of the ion m/z 1036.54 (component B) is described. Following lyophilisation and reconstitution in ES solvent, 50% of the sample was loaded into a gold-coated capillary and an ES/MS spectrum was acquired (Fig. 17.14). Individual spectra were integrated over 2.5 s, and the data represent the summation of a 4.9 min acquisition. Ten distinct doubly charged components were observable at low relative intensities; all corresponded to observed singly charged ions in the MALDI-MS spectrum. Three doubly charged ions that were not observed as singly charged species by MALDI were also present, this may relate to differing response factors by MALDI and ES. Again, it is preliminary detection of peptide masses by MALDI-MS that enabled interrogation of the ES/MS spectrum for corresponding multiply charged ions. A doubly charged ion at m/z 518.77 (corresponding to $[B + 2H]^{2+}$) was observed and selected for tandem MS.

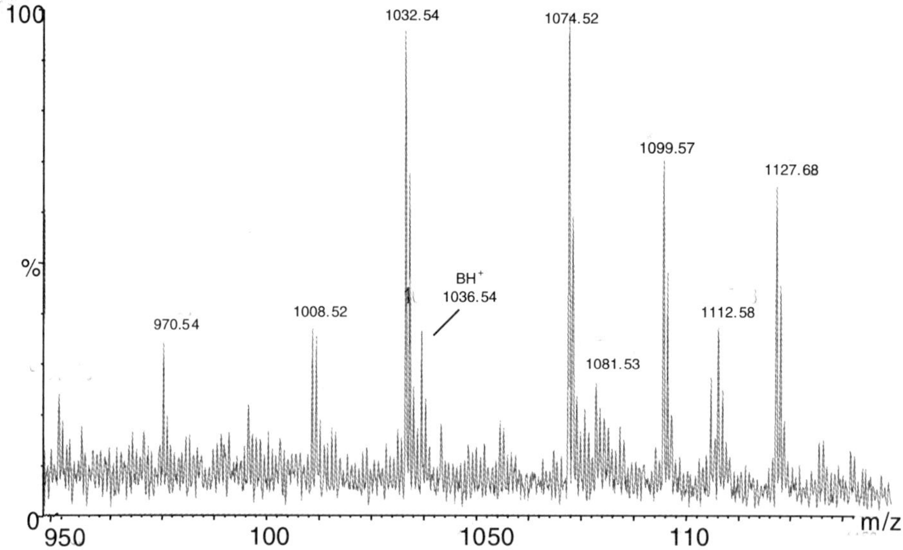

Figure 17.13. Fraction 24: MALDI/MS spectrum.

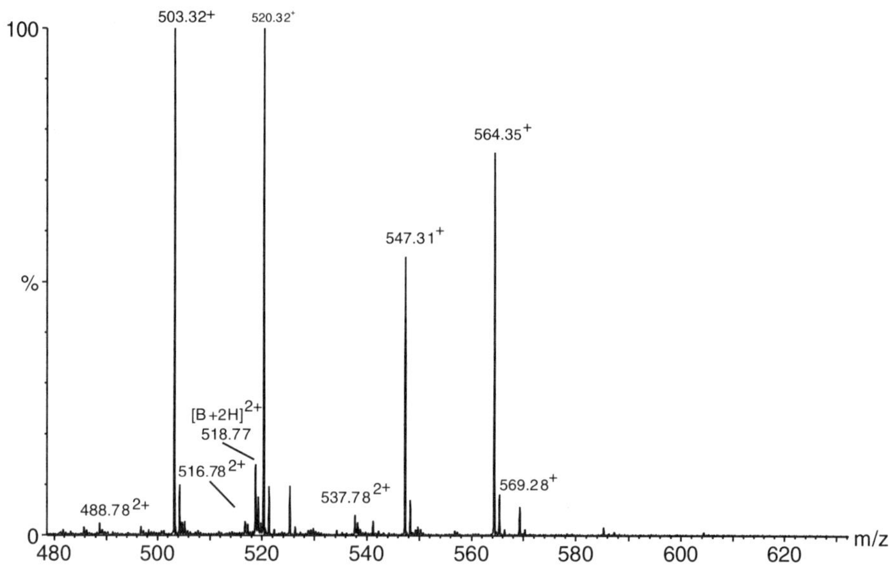

Figure 17.14. Fraction 24: ES/MS spectrum.

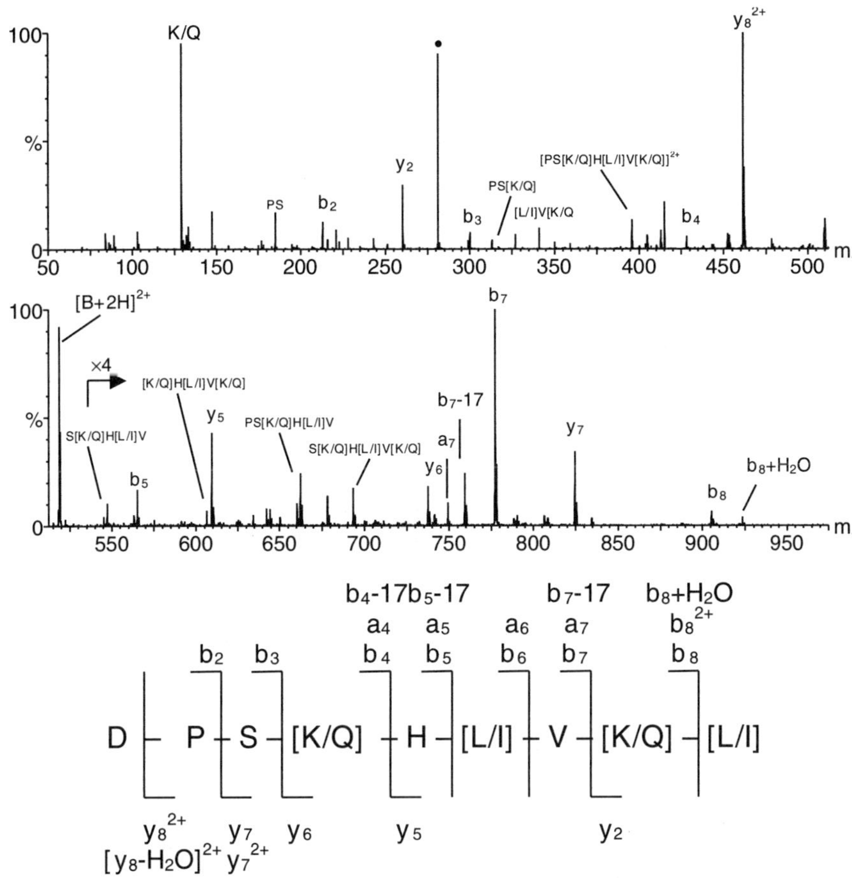

Internal ions: $[PS[K/Q]H[L/I]V[K/Q]]^{2+}$ PS[K/Q]H[L/I]V[K/Q] S[K/Q]H[L/I]V[K/Q] PS[K/Q]H[L/I]V [K/Q]H[L/I]V[K/Q] S[K/Q]H[L/I]V PS[K/Q]H[L/I] H[L/I]V[K/Q] S[K/Q]H[L/I] H[L/I]V [L/I]V[K/Q] PS[K/Q] H[L/I] S[K/Q] [L/I]V PS

Figure 17.15. Fraction 24: products of *m/z* 518.77 ($[B + 2H]^{2+}$); observed product ions are summarized. The full sequence is based on combined data from products of $[B + 2H]^{2+}$ and $[Ac\text{-}B + 2H]^{2+}$ following acetylation. (• denotes an unassignable product ion).

Figure 17.15 shows products of *m/z* 518.77. The collision offset employed was 21 V, and spectra were acquired for 2.4 min. A high-quality product ion spectrum was obtained. The fragments observed appear to be influenced by the presence of a basic histidine residue at P5. High members of both *b*- and *y*-ion series (b_5–b_8 and y_5–y_8) have high relative abundances compared to lower member series ions. The gas-phase basicity of histidine would allow sidechain sequestration of an ionizing proton and stabilization of product ions. The

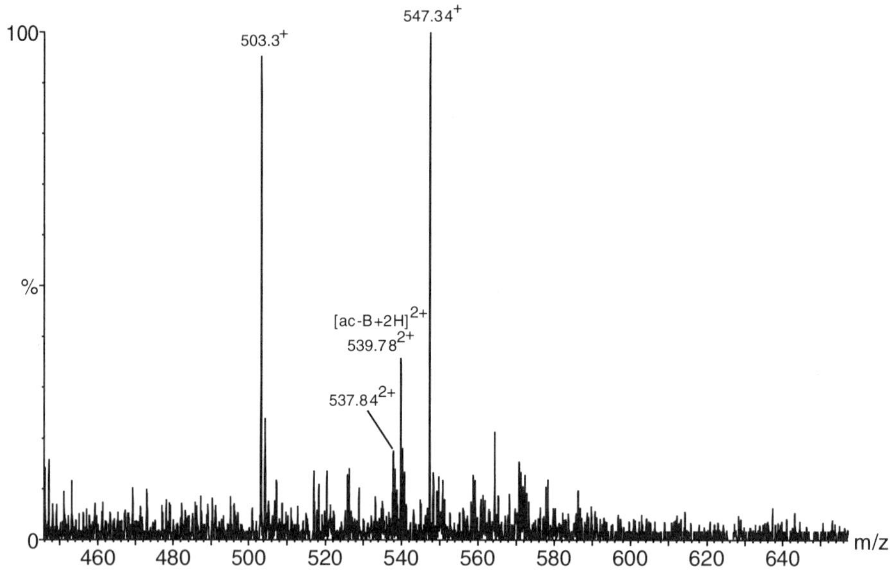

Figure 17.16. Fraction 24 following *N*-acetylation: ES/MS spectrum.

high prevalence of internal ion series is dominated by, but not exclusive to, cleavage *N*-terminal to proline. There were four amino acid ambiguities, two lysine/glutamine (P4 and P7) and two leucine/isoleucine (P6 and P8). Selection of the precursor ion population (3 Th selection window) included an apparent nonpeptide component (separated from the monoisotopic peptide of interest by 0.5 Th). The observation of ions with *m/z* unassignable to the peptide sequence suggests this component may have contributed to the product ion spectrum.

Following *N*-acetylation of the remaining 50% of sample, an ES/MS spectrum was acquired (Fig. 17.16); data were summed over 1.5 min. A doubly charged ion was observed at *m/z* 539.78, which corresponds to singly acetylated component *B*. There was no evidence for two or three sites of acetylation. Interestingly, the equivalent singly charged ion observed previously [fraction 32 (Fig. 17.7)] by ES on *N*-acetylation was not observed here. Although this would be sequence-dependent, the peptide NPRAMQALL had a more basic arginine residue and proline at P3, a more basic combination than the histidine and proline present here. The ion at *m/z* 539.78 was selected for tandem MS (Fig. 17.17); data were acquired for 2 min, and the collision offset was 21 V.

The shift in product ion masses when compared to the underivatized peptide showed the site of acetylation to be the *N* terminus; P4 and P8 were identified as glutamine by inference. There was no evidence for two or three sites of acetylation in the ES/MS spectrum. With no sidechain derivatization,

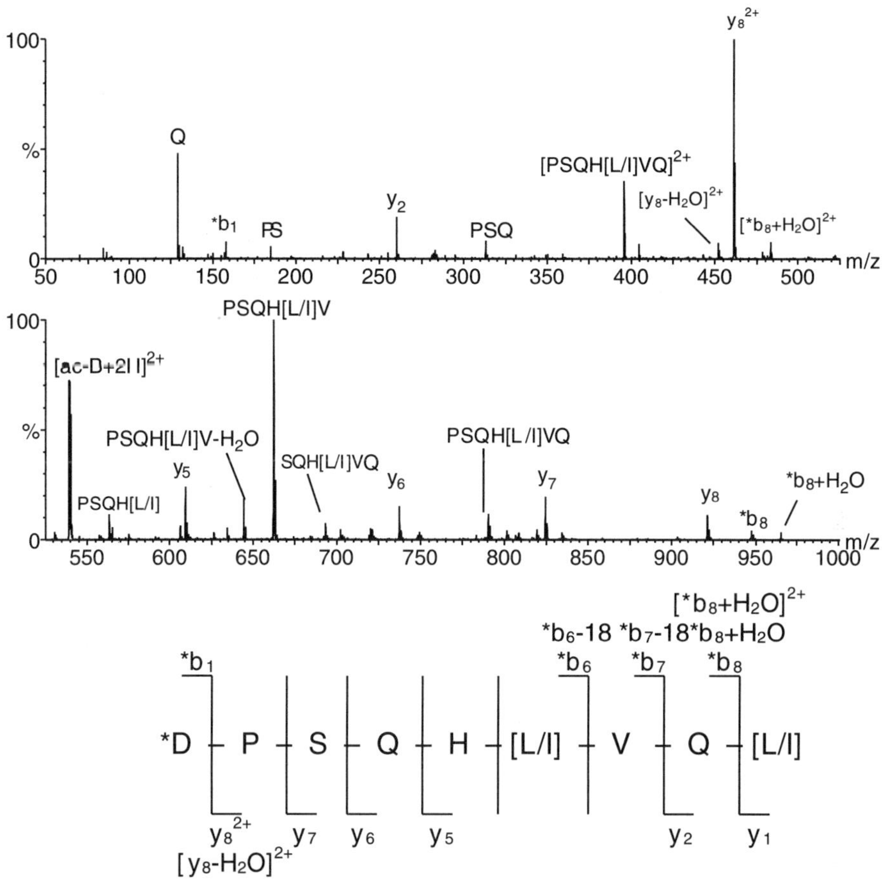

Internal ions: $[PSQH[L/I]VQ]^{2+}$ PSQ]H[L/I]VQ SQH[L/I]VQ PSQH[L/I]V QH[L/I]VQ SQH[L/I]V PSQH[L/I] H[L/I]VQ PSQH QH[L/I] H[L/I]V [L/I]VQ PSQ H[L/I] VQ SQ PS

Figure 17.17. Fraction 24: products of m/z 539.77 ($[\text{Ac-}B + 2\text{H}]^{2+}$); observed product ions are summarized. The full sequence is based on combined data from products of $[B + 2\text{H}]^{2+}$ and $[\text{Ac-}B + 2\text{H}]^{2+}$ following acetylation.

the shift in *N*-terminal-containing fragments (*N*-terminal derivatization) allowed distinction between *b*- and internal or *y*-ion series. Multiple internal ion series were prevalent. The observation of b_1, y_8^{2+}, and internal ions formed *N*-terminal to proline allowed confident identification of aspartic acid at P1. Histidine had strongly influenced the fragmentation pathways observed; only *b*- and *y*-ion series members containing histidine were stable: y_5–y_8 and b_6–b_8. Lower member product ions, with no basic residue to sequester the ionizing proton, were able to access further fragment ion pathways ultimately yielding stable b_1, y_2, and y_1.

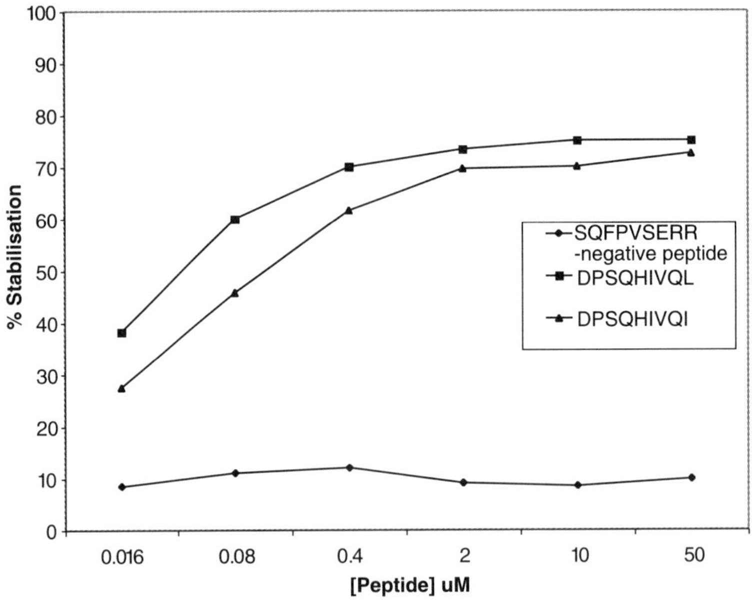

Figure 17.18. Binding of DPSQHIVQL and DPSQHIVQI to RT1-A1^c in stabilization assay.

In both derivatized and underivatized product ion spectra, a *C*-terminal rearrangement ion, $b_8 + H_2O$, was observed for doubly and singly charged $[\text{Ac-}B + 2H]^{2+}$, and only for singly charged $[B + 2H]^{2+}$. Leucine/isoleucine as a stable leaving group was previously unreported.

Consideration of all product ion data allowed deduction of the primary sequence DPSQH[L/I]VQ[[L/I]; leucine/isoleucine distinction was not possible by low-energy collision-induced dissociation as performed here.

All four isomers were synthesized, and their binding to RT1-A1^c was assessed. Results revealed that a preference for leucine was found at the *C* terminus; a mean concentration of 39 nM (*C*-terminal leucine) versus 100 nM (*C*-terminal isoleucine) was required to achieve 50% stabilization (Fig. 17.18). This is consistent with the known *C*-terminal anchor, leucine. The variation of leucine/isoleucine at P6 did not have a significant effect on binding (data not shown).

A search of the DNA and protein databases (with all four L/I possibilities) did not identify any proteins from which this peptide could be derived. The synthetic peptide DPSQHIVQL was used to compare the fragmentation pattern of the isolated component *B*. Figure 17.19 shows products of m/z 518.77 for the synthetic peptide (1) and the ion m/z 518.77 ($[B + 2H]^{2+}$) from the biological isolate (2); the collision offset was 21 V in both cases.

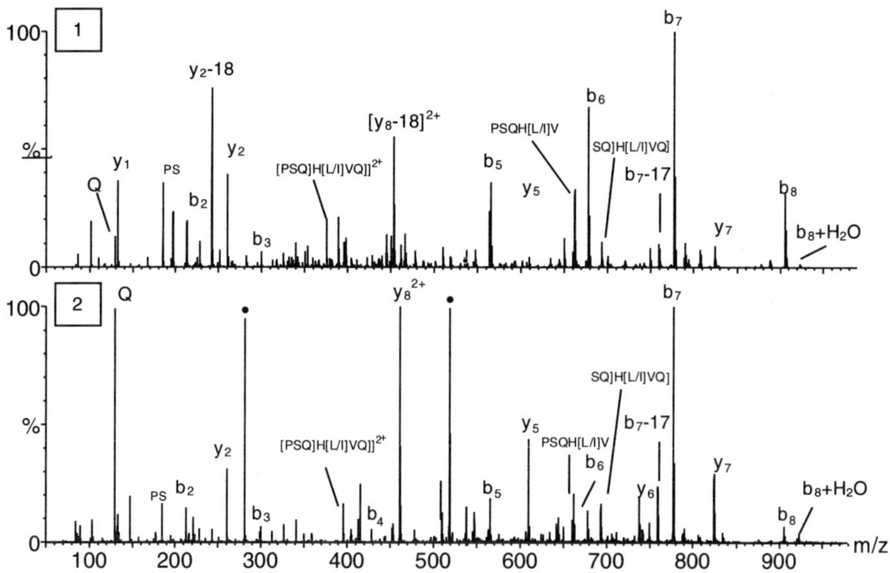

Figure 17.19. Fraction 24 products of *m/z* 518.77 for synthetic DPSQHIVQL (1) and RT1-A1^c component *B* (2). [• denotes coisolated singly charged component in precursor ion population and (possibly) related product ion].

The two product ion spectra show a high level of similarity, although there are some key differences. The component coisolated with *B* possibly yields two ions not observed for the synthetic peptide. The doubly protonated synthetic peptide also yielded higher relative abundance of $[y_8\text{-}18]^{2+}$ versus y_8^{2+} than did $[B + 2H]^{2+}$, though both ions are observed in both product ion spectra. The observation of y_2-18 is exclusive to the synthetic peptide.

Consideration of all evidence allowed confident assignment of the sequence DPSQH[L/I]VQ[L/I]. Binding assay data (and known anchor residue) suggest that the *C* terminus is leucine.

17.3.2. Single-Shot Strategy

Following concentration, a selection of fractions were analyzed using a single-shot approach. The analysis of fraction 37 is described. The total fraction was concentrated into 1.5-μL ES solvent in a gold-coated capillary, and a small fraction was transferred to the MALDI target using a low-flow-rate probe from the Quattro I mass spectrometer [18]; for comparison, a 1-μL portion was taken prior to concentration for "conventional" MALDI loading.

While MALDI/MS analyses revealed multiple components, loading from the ES capillary (Fig. 17.20a) resulted in an increase in the number of ion counts and components over conventional sample preparation (Fig. 17.20b). There were 15 components apparent in the conventional analysis with a signal-

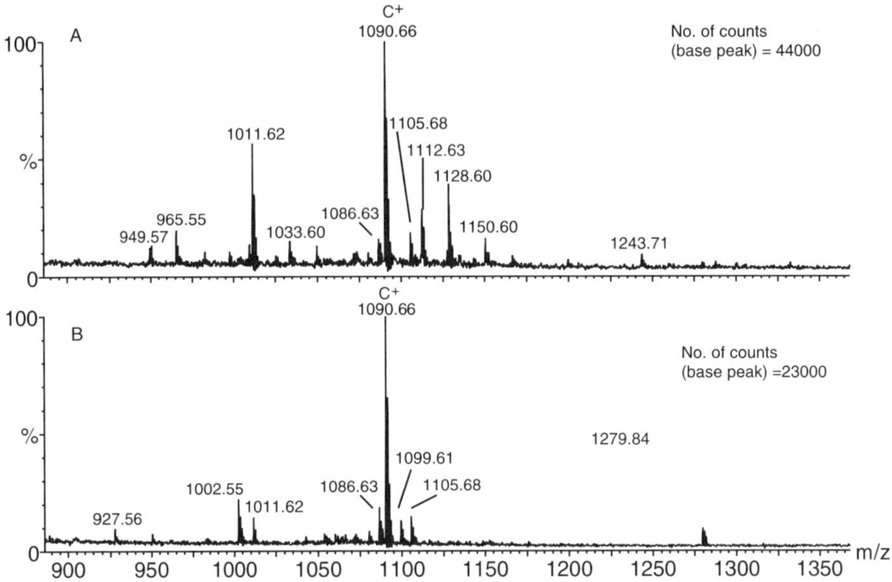

Figure 17.20. Fraction 37 MALDI/MS: conventional loading on the MALDI target (b) and following loading from a gold-coated capillary (a).

to-background ratio of ≥2, whereas with the single-shot strategy there were 22 such components; ions present in both spectra showed a 90% average increase in number of counts in the latter analysis. Four peptides were not observed following concentration, these may have been lost during the lyophilization process. The observation of peptides only when loaded from a gold-coated capillary was attributed to the concentration of sample and not the ES process. It is important to emphasize that transfer of a portion of the sample to the target occurs on initiation of ES and is apparently not associated with stable ES deposition.

Further analysis of the ion observed at m/z 1090.66 (component C) is described. The capillary containing the remaining sample was transferred to the Q-ToF instrument, and an ES spectrum was acquired (Fig. 17.21). Individual spectra were integrated over 2.5 s, and each spectrum (A and B) represents the summation for a 1 min acquisition. The upper and lower spectra (A and B) show the acquisition with the MCP detector set to 2800 V and 2600 V, respectively. Lowering the MCP voltage discriminates against lower charge states by raising the threshold of the pulse height required to register an ion event. The observation of doubly charged components is greatly enhanced and allows visualization of component C at m/z 545.7 $[C + 2H]^{2+}$: several doubly charged ions were observed that corresponded to detergent (confirmed following tandem MS; data not shown). A polyethylene glycol (PEG)-based detergent (Nonidet P-40) used in sample workup carried

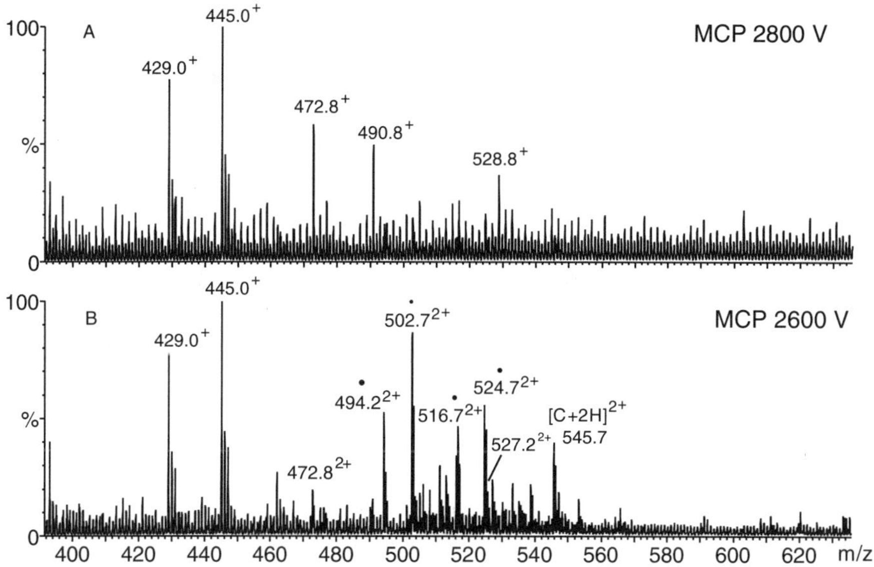

Figure 17.21. Fraction 37 ES/MS: MCP 2800 V (a) and 2600 V (b); • denotes a detergent ion.

through to the mass spectrometric analysis. The repeating unit of PEG, —CH_2OCH_2—, gives rise to ions separated by 44 Th or, in case of doubly charged species observed here, 22 Th. Such ions are normally problematic in ES ionization on account of the preference of the hydrophobic detergent species to reside on the droplet surface during ion formation, which prevents the formation and observation of peptide species contained within the same droplet. Here, the concentration of detergent was evidently low enough to allow formation of peptide ions by ES.

The ion *m/z* 545.7 was selected for tandem MS. Figure 17.22 shows the product ion spectrum obtained. The collision offset employed was 23 V, and spectra were accumulated for 1.4 min. A prominent *y*-ion series was observed from y_8 to y_4, allowing assignment of the five *N*-terminal residues as NPHD[L/I]. Detergent ions separated by 44 Th were present at low *m/z*; the selection of a detergent-containing component in the precursor ion population could be related to an ES cluster ion phenomenon (unpublished observations). The *C*-terminal residue was identified as leucine/isoleucine following the observation of b_8, $b_8 + H_2O$, and y_1 ions. P8 was also identified as leucine/isoleucine with the observation of b_7.

Following the further tandem MS analysis of peptidic components (data not shown), the remaining sample was derivatized (*N*-acetylation) in-tip and immediately returned to the Q-ToF instrument. An ES spectrum was acquired (Fig. 17.23). Data were summed over 50 s, and the MCP was set at 2600 V.

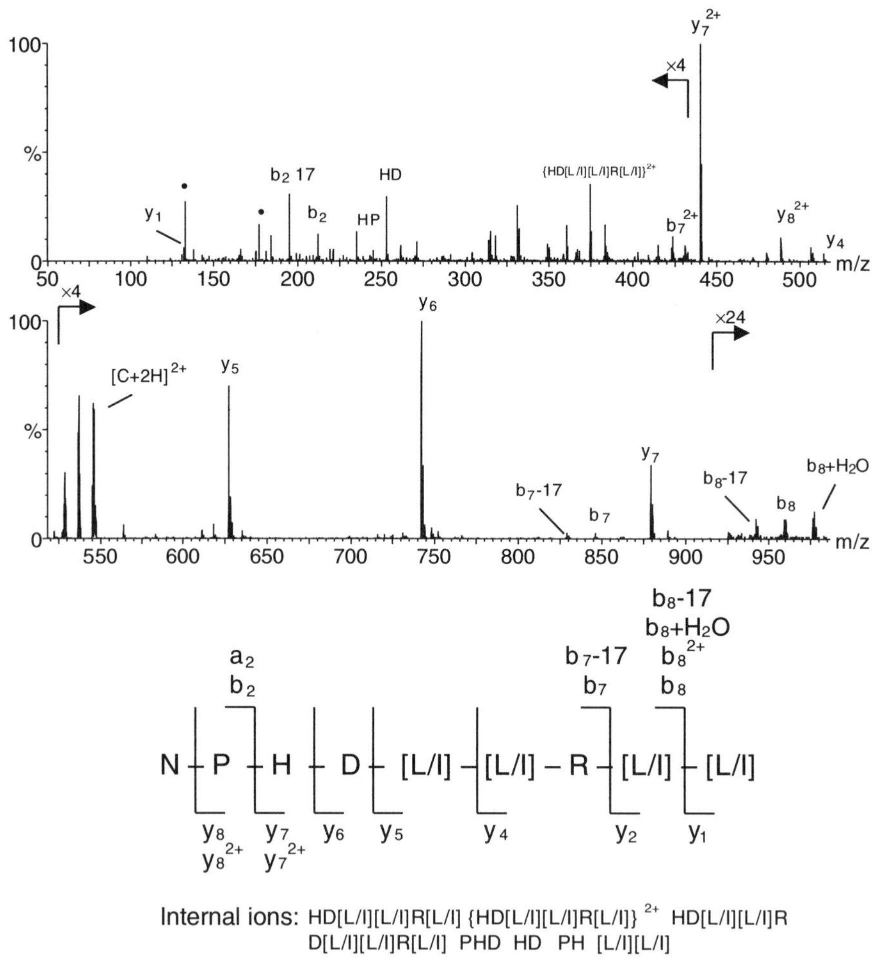

Figure 17.22. Fraction 37: products of m/z 545.7 ($[C + 2H]^{2+}$); observed product ions are indicated. The full sequence was based on combined data from $[C + 2H]^{2+}$ and [Ac-$C + 2H]^{2+}$; • denotes a product ion derived from a detergent ion.

Again, detergent ions were prevalent: note that the m/z values of these ions have not shifted following derivatization.

A doubly charged ion was observed at m/z 566.7 that corresponds to singly acetylated component C. Figure 17.24 shows the product ion spectrum: data were acquired for 3.4 min and the collision offset was 24 V.

Acetylation of the N terminus gave a shift in m/z by 42 Th of all N-terminal-containing fragments and confirmation of b- and a-ion formation. The y-ion series observed in the underivatized peptide was again prevalent, allowing identification of P1–P5, P8, and P9. The two residues at P6 and P7 could be identified as [L/I]R by the presence of b_5–b_8 and the observation of

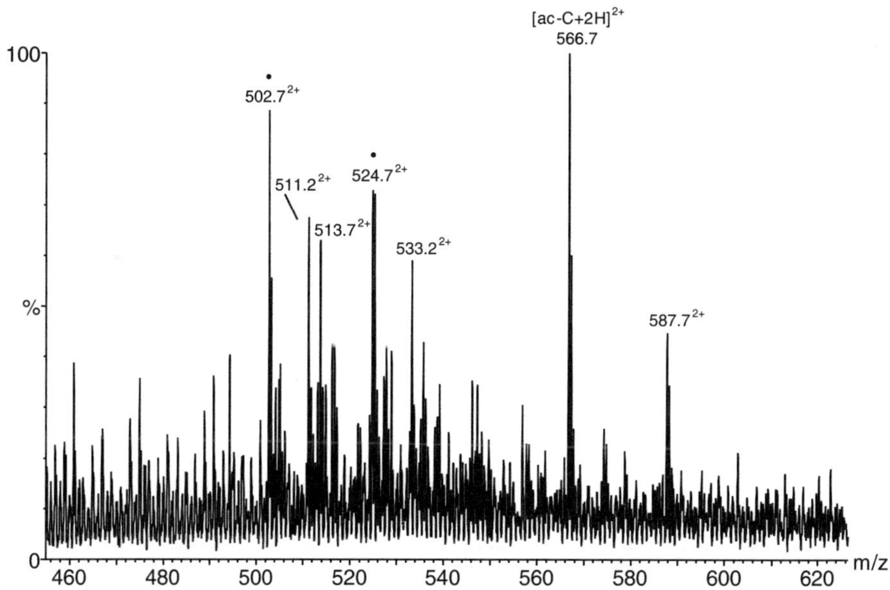

Figure 17.23. Fraction 37 following in-tip acetylation: ES/MS spectrum; • denotes a detergent ion.

a series of internal ions (generated, in part, by cleavage *N*-terminal to proline). The presence of arginine at P7 is consistent with several observations:

1. The stable and abundant *y*-ion series is rationalized by sequestration of an ionizing proton on the arginine sidechain.
2. Two generations of *C*-terminal rearrangement, both with leucine/isoleucine as the leaving group.
3. Poor fragmentation at the amide bonds adjacent to the arginine; no fragmentation was observed *N*-terminal to arginine for the underivatised peptide.

The presence of histidine at P3 allows stable formation of the b_3 ion. *N*-acetylation promoted observation of b_1 and confirmation of the assignment of asparagine at P1. Consideration of all product ion data allowed confident assignment of the sequence NPHD[L/I][L/I]R[L/I][L/I]. Identification of asparagine, proline, histidine and leucine/isoleucine (assuming leucine) at positions 1, 2, 3, and 9, respectively, conforms to the known RT1-A1^c binding motif. No further work characterizing this peptide was performed.

Multiple peptides were identified using a single-shot strategy. It was shown with MHC peptides of natural abundance that sequencing could be performed following concentration of the total sample amount into 1.5 μL of ES solvent in a gold-coated nanospray capillary. Total analysis involved sequential:

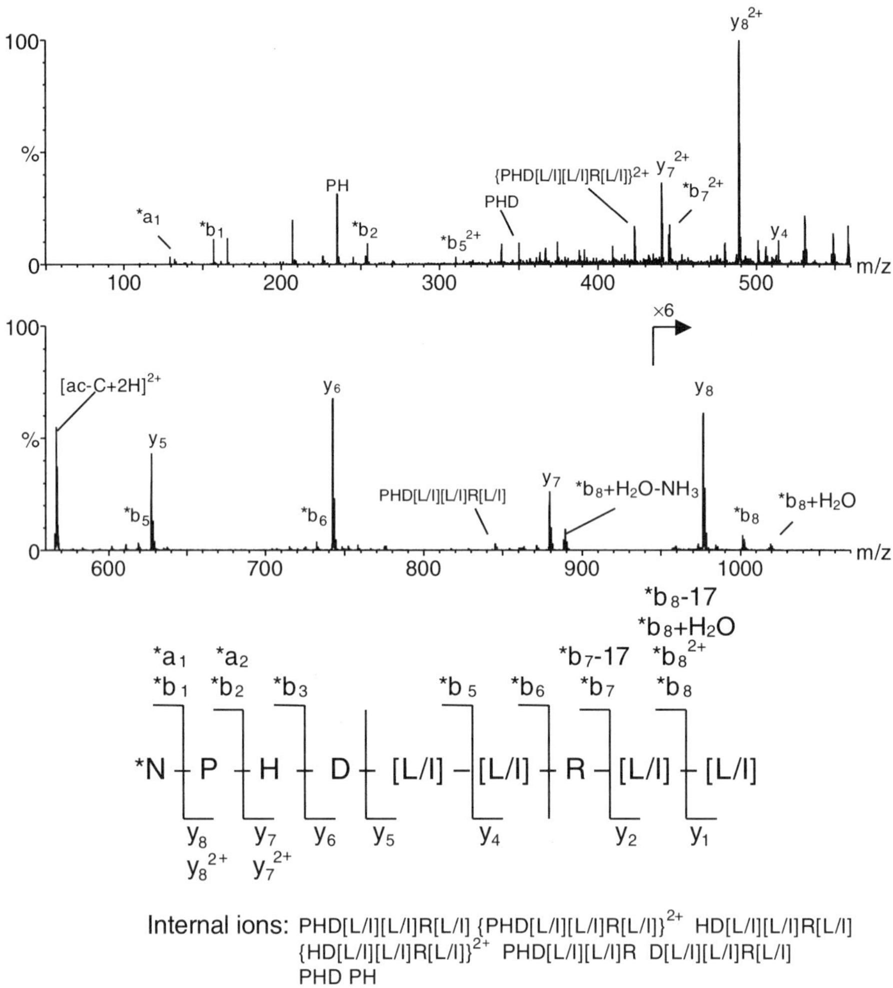

Figure 17.24. Fraction 37 following in-tip acetylation: products of m/z 566.7 ([Ac-C + 2H]$^{2+}$); observed product ions are summarized. The full sequence was based on the combined data from products of [C + 2H]$^{2+}$and [Ac-C + 2H]$^{2+}$.

1. Loading of a small percentage of sample onto a MALDI target using the Quattro I low-flow-rate ES probe
2. Transfer of the capillary to the Q-ToF and performing initial MS and tandem MS analyses
3. Derivatisation (N-acetylation) of the remaining sample within the capillary by addition of ~0.2 μL of acetic anhydride
4. Reanalysis (MS and tandem MS) by Q-ToF

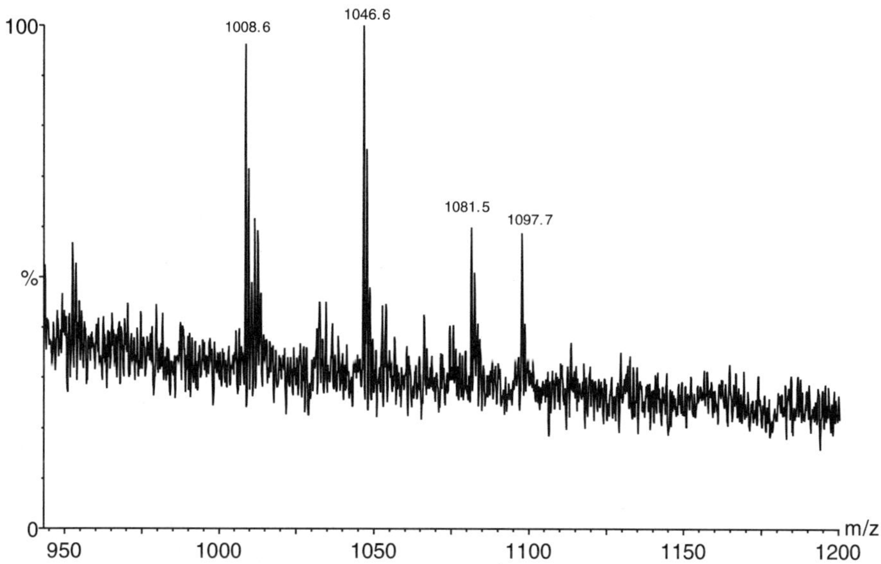

Figure 17.25. Fraction 26: MALDI-MS.

17.3.3. Liquid Chromatography/Mass Spectrometry (LCMS)

This methodology was investigated in the analysis of fraction number 26; screening by MALDI-MS (Fig. 17.25) revealed multiple components.

LCMS analyses were performed using the Q-ToF instrument with the electrospray interface modified for low-flow-rate LC. A 75-μm-i.d. × 375-μm-o.d. section of fused-silica capillary was heat-extruded manually using a microtorch to a 10 ± 5-μm tip. A slurry of POROS 10 R2 in methanol was loaded into a pressure bomb device and pumped into the capillary; packing was monitored under a microscope. A 5-cm section was cleaved with a ceramic scribe and connected to a low dead-volume stainless-steel union, to which the electrospray high voltage was subsequently applied. A Rheodyne injector fitted with a 13-μL PEEK sample loop was connected to the union with a 60-cm section of fused silica (25 μm i.d. × 375 μm o.d.). An isocratic pump (Kontron T-414) was used to deliver solvent.

The isolated fraction 26 was dried sufficiently to remove volatile acetonitrile, acidified and halved. Half of the total fraction was injected in 13 μL and washed for 5 min with 0.1% aqueous formic acid. The column was then eluted using a 13 μL "plug" of 1:1 methanol/water (0.1% formic acid). One MS/MS acquisition channel was used. The total ion chromatogram (TIC) for the MS survey channel is shown in Fig. 17.26. Acquisition was initiated during column washing in case any hydrophilic peptides eluted during this time; the total analysis time shown on the x axis (~28 min) is therefore misrepresentative of deadtime in the system.

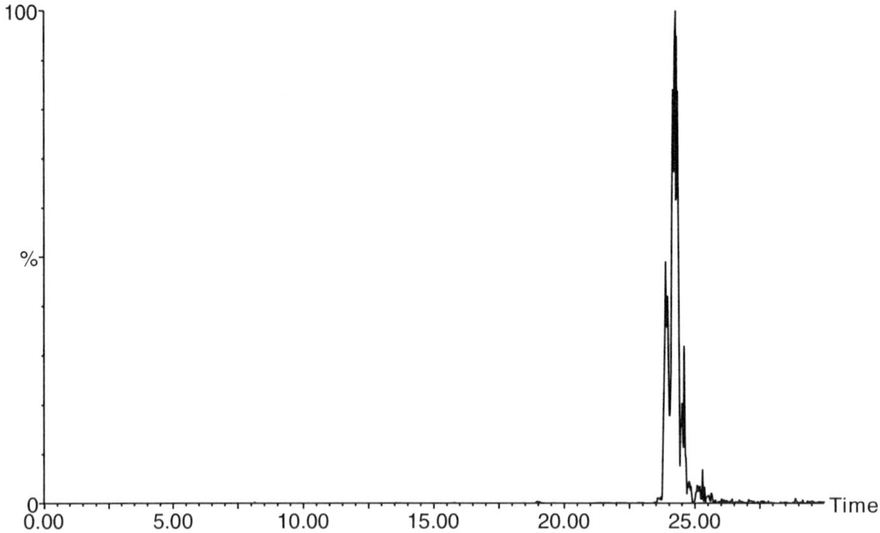

Figure 17.26. Fraction 26: LC-MS total ion chromatogram.

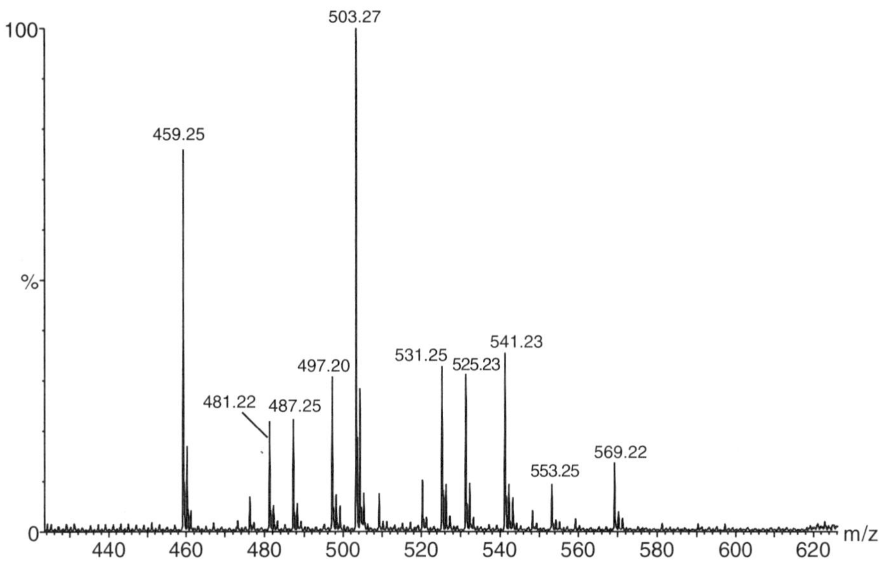

Figure 17.27. Fraction 26: LC-MS survey scan after 24.3 min.

All MS spectra represented by the TIC in Fig. 17.26 were dominated by the presence of detergent. A typical MS spectrum is shown in Fig. 17.27 (MS survey scan at 24.3 min); there is no evidence of multiply charged species. Highly abundant singly charged ions separated by 44 Th dominate the spectrum; subsequent product ion spectra following automated function switching reveal only nonpeptidic components.

The use of a C18 reversed-phase material concentrates detergent material present in the sample. From the initial MALDI experiment it is known that peptide components were present; the elution characteristics of the detergent are, under the conditions employed, similar to those of peptide material. This prevents the analysis of this sample using LC/MS methodology. More judicious choice of detergent for cell lysis is required in future studies. Commercially available detergent with differing elution characteristics to peptides (e.g., Zwittergent, from Sigma, St. Louis, MO) would overcome this problem.

17.4. SUMMARY

Nanospray and single-shot strategies were successfully applied to analysis of peptides eluted from the class I rat MHC molecule RT1-A1^{C}. By contrast, the presence of detergent in samples (from cell lysis) prevented the analysis using reversed-phase (C18) LC/MS and MS/MS. Table 17.3 summarizes the fractions sampled and the peptide sequences deduced. Of the 22 peptides analyzed, only 7 full sequences were successfully determined: NPH[L/I]GE[L/I]PR[L/I] (from fraction 21), DPSQH[L/I]VQ[L/I] (fraction 24), NVAQDYVR[L/I] (fraction 27), NPAMQA[L/I][L/I] (fraction 32), N[L/I]PPH[L/I][L/I]R[L/I] (fraction 35), NPHD[L/I][L/I]R[L/I][L/I] (fraction 38), and NP[RS/SR] WEVV[L/I] (fraction 38). Of these, three revealed hits in the databases, of which two can be attributed to known proteins:

1. Fraction 32: NPRAMQALL (discussed in Section 17.3.1)
2. Fraction 35: NLPPHIIRL; Table 17.2 summarizes the results for an MS Edman search of the NCBInr protein database. Although human proteins are displayed in this search, a rat homolog for this gene is also

TABLE 17.2. MS Edman Search Results (Search Mode: Sequence and Mass) for N[LI]PPH[LI][LI]R[LI] and MH^{+} 1072.64

	Parent Ion	Matching Sequence	$M + H$ Matched	Delta Mass (Da)	Protein MW (Da)/pI	Species	Accession #	Protein Name
1	1072.64	(E)NLPPHIIRL(V)	1072.663	−0.023	24185.9/8.78	*Homo sapiens*	7657046	(M91670) ubiquitin carrier protein
2	1072.64	(E)NLPPHIIRL(V)	1072.663	−0.023	26463.4/9.39	Unreadable	345829	> gi|345829| pir||B42856 ubiquitin carrier protein E2—human

TABLE 17.3. Summary of RT1-A1^c Peptides Sequenced or Partially Sequenced by Mass Spectrometry[a]

Fraction Number	m/z $[M + H]^+$	Deduced Sequence	Origin
21	1008.6	DPSQV—	Unknown
21	1081.6	—EHTA[L/I][L/I]	Unknown
21	1103.6	NPHQE—	Unknown
21*	1145.6	NPH[L/I]GE[L/I]PR[L/I]	Unknown
24	1032.6	—[L/I]HPV[L/I]	Unknown
24*	1036.5	DPSQH[L/I]VQ[L/I]	Unknown
24	1074.6	VPR—DPT[L/I]	Unknown
27	981.4	—[L/I]SNV[L/I]	Unknown
27	1022.4	—VP[L/I] [L/I]	Unknown
27*	1077.4	NVAQDYVR[L/I]	Unknown
30	808.4	PSP—	Unknown
30	907.4	VPSP—	Unknown
30	1000.6	VPS—F[L/I]	Unknown
30	1035.6	R[L/I]—[L/I] [L/I] [L/I]	Unknown
32*	1013.4	NPRAMQALL	Tumour-suppressor associated protein[1]
35	1045.6	NP—[L/I]F[L/I]	Unknown
35*	1072.6	NLPPHI IRL	Ubiquitin carrier protein[2]
36	1040.3	NPY—	Unknown
38	1002.6	QPAS—FD[L/I] or NP—FD[L/I]	Unknown
38*	1090.7	NPHD[L/I] [L/I]R[L/I] [L/I]	Unknown
38*	1099.6	NP[RS/SR]WEVV[L/I]	Unknown[3]
41	1056.6	D—NV[L/I] [L/I]	Unknown

[a] *Key:* * denotes full sequence determined; 1, 2, 3 are described in this section.

described in the EST database (Acc. No. H33742) showing conservation of this protein across species.

3. Fraction 38: NPSRWEVVL; a mouse EST (Acc. No. AU079261) exists which codes for this peptide. However its sequence shares no homology with any other database entry. Mass spectrometric sequencing was unable to resolve an [SR/RS] ambiguity at positions 3 and 4. The EST hit suggests the SR combination, although this is in disagreement with the previous identification of arginine as an anchor residue at P3.

17.5. CONCLUSIONS

Single substituted variants of the "ideal peptide" for RT1-A1^c (designed on the basis of pool sequencing data) were made and tested for binding to the RMAS-A1^c cell line. Of the 31 peptides synthesized, only 5 were significantly better binders, and the improvement in binding affinity was always less than twofold. The majority of peptides tested exhibited reduced binding affinity. Results revealed that changes at the P1 and P3 positions had little effect on peptide binding, whereas the P2 position appeared unable to accommodate charged residues, particularly aspartic acid (C_{50} = 1.32 μM). In addition, all except one of the residue changes in the P4, P5, P8, and P9 positions reduced binding affinity. Such an intolerance to single residue substitutions suggests that there are peptide sidechain–MHC interactions along the length of the peptide. This would be consistent with previous studies of peptide length preferences for this molecule which found a restricted preference for peptides of 9–12 amino acids in length [19]. The greatest effect was seen at the P9 position where both charged and aromatic substitutions resulted in a 10–100-fold reduction in peptide binding. Even an isoleucine substitution produced a threefold reduction in binding. This indicates a high level of selectivity for leucine by the F pocket in the binding groove and consequently the importance of the sidechain of the *C*-terminal residue for complex stability.

Studies were also performed in order to identify natural peptides presented by RT1-A1^c on rat splenocytes. Peptides bound to RT1-A1^c were isolated by immunoprecipitation, purified by acid elution and reversed-phase chromatography, and subjected to sequence analysis by mass spectrometry. A sample (0.5 μL) from each fraction was pooled and analyzed by MALDI/MS to determine the range of peptide length in the pool. Results (not shown) showed that the bulk of peptides were in the *m*/*z* range of 1000–1150 Th, which correspond to peptides of 9–10 amino acids in length. This observation is in agreement with previous length preference studies [19]. However, longer species were also visible in the spectrum above 1150 Th as discrete peaks (see Fig. 17.20b, *m*/*z* 1279.84, for example), suggesting that this class I molecule can also accommodate some longer peptides, although the proportion of these species was low compared to the total. Thus, this molecule does not bind longer peptides in significant amounts in vivo.

Of the peptide species analysed, only seven full sequences were successfully determined. There were two major reasons: (1) detergent carryover from the immunoprecipitation step prevented successful analysis of many fractions—here, preferential ionization of the detergent by electrospray resulted in low peptide ion yields; co-selection of detergent ions with peptides for low-energy collision induced dissociation also complicated tandem MS data; and (2) the presence of a midchain arginine residue appears to reduce the efficiency of cleavage of the peptide backbone at proximal sites. This reduction was observed with six of the seven fully sequenced peptides, the exception not containing arginine. The consequent reliance on low-intensity signals for determination of portions of each sequence prevented full structure determination using the lower-intensity product ion spectra recorded for other examples. Such peptides yielded either high member *y*- or *b*-ion series only. Amongst the seven full sequences, only three revealed hits in the databases, two of which can be attributed to known proteins. Both peptides originate from proteins with probable high turnover rates. Consequently, such proteins would be expected to provide peptides in sufficient amounts to be detected. As the peptides under analysis were of rat origin, it is not surprising that few positive identifications were possible. Sequence information from the rat is more limited than that from either mice or humans. It is fortunate that both the identified peptides were derived from sequences conserved across species. This appears not to be the case for the other five full sequences identified in this study, although they may be components of as yet unknown conserved proteins.

In summary, natural RT1-A1^c binding peptides were identified. One of these, NPRAMQALL, has already enabled the generation of crystals for the heterotrimeric RT1-A1^c heavy-chain/β_2m/peptide complex with excellent X-ray-diffraction characteristics (work in progress; Dr. J. Speir and Prof. I. A. Wilson, Scripps Research Institute, San Diego, USA). The peptides identified will allow future studies to probe the effect of peptide variants on possible natural-killer cell interactions using both biophysical and cell-based assays.

REFERENCES

1. J. Stevens, R. C. Jones, R. S. Bordoli, J. Trowsdale, S. J. Gaskell, G. W. Butcher, and E. Joly, *J. Biol. Chem.*, **275**, 29217–29224 (2000).
2. F. Momburg, J. Roelse, J. C. Howard, G. W. Butcher, G. J. Hämmerling, and J. J. Neefjes, *Nature* **367**, 648–651 (1994).
3. M. T. Heemels and H. L. Ploegh, *Immunity* **1**, 775–784 (1994).
4. J. O. Koopmann, M. Post, J. J. Neefjes, G. J. Hämmerling, and F. Momburg, *Eur. J. Immunol.* **26**, 1720–1728 (1996).
5. B. Reizis, H. Schild, S. Stefanovic, F. Mor, H. Rammensee, and I. R. Cohen, *Immunogenetics* **45**, 278–279 (1997).
6. S. J. Powis, L. L. Young, E. Joly, P. J. Barker, L. Richardson, R. P. Brandt, C. J. Melief, J. C. Howard, and G. W. Butcher, *Immunity* **4**, 159–165 (1996).
7. J. Stevens, K. H. Wiesmüller, P. J. Barker, P. Walden, G. W. Butcher, and E. Joly, *J. Biol. Chem.*, **273**, 2874–2884 (1998).
8. C. Naper, J. C. Ryan, M. C. Nakamura, D. Lambracht, B. Rolstad, and J. T. Vaage, *J. Immunol.* **160**, 219–224 (1998).
9. W. J. Murphy and D. L. Longo, *Immunol. Rev.* **157**, 167–176 (1997).
10. J. O. Manilay and M. Sykes, *Curr. Opin. Immunol.* **10**, 532–538 (1998).
11. L. L. Lanier, *Cell* **92**, 705–707 (1998).
12. F. M. Karlhofer, R. K. Ribaudo, and W. M. Yokoyama, *Nature* **358**, 66–70 (1992).
13. O. Mandelboim, H. T. Reyburn, M. Vales-Gomez, L. Pazmany, M. Colonna, G. Borsellino, and J. L. Strominger, *J. Exp. Med.* **184**, 913–922 (1996).
14. M. Peruzzi, K. C. Parker, E. O. Long, and M. S. Malnati, *J. Immunol.* **157**, 3350–3356 (1996).
15. I. Correa and D. H. Raulet, *Immunity* **2**, 61–71 (1995).

16. T. Hanke, H. Takizawa, C. W. McMahon, D. H. Busch, E. G. Pamer, J. D. Miller, J. D. Altman, Y. Liu, D. Cado, F. A. Lemonnier, P. J. Bjorkman, and D. H. Raulet, *Immunity* **11**, 67–77 (1999).
17. E. Joly, L. Leong, W. J. Coadwell, C. Clarkson, and G. W. Butcher, *J. Immunol.* **157**, 1551–1558 (1996).
18. R. C. Jones, M. Ferro, S. G. Summerfield, and S. J. Gaskell, *J. Mass Spectrom.* **33**, 1261–1264 (1998).
19. J. Stevens, K. H. Wiesmüller, P. Walden, and E. Joly, *Eur. J. Immunol.* **28**, 1272–1279 (1998).

CHAPTER 18

Analysis of Neuropeptides in Cerebrospinal Fluid and Neural and Neuroendocrine Tissues by MALDI-TOFMS

CAROL L. NILSSON

Institute of Medical Biochemistry, Göteborg University, Göteborg, Sweden

CONTENTS

18.1. INTRODUCTION

The analysis of neuropeptides from biological sources presents special challenges, because of the typically low concentration of analyte (picomoles per liter) and the presence of salts, lipids, and free amino acids in the millimolar range. Analysis of these molecules in neuroscience has traditionally relied on antibody-based detection methods, such as radioimmunoassay (RIA) and immunohistochemistry. While exquisitely sensitive, these assays are limited by the fact that antibodies can only recognize a single epitope of three to seven amino acid residues [1] in the molecule of interest. Antibodies can cross-react with molecules other than the desired species, if these share the same epitope.

Mass Spectrometry and Hyphenated Techniques in Neuropeptide Research, Edited by Jerzy Silberring and Rolf Ekman.
ISBN 0-471-35493-7

On the other hand, if the protein epitope is modified or blocked, for example by phosphorylation, sulfation, or glycosylation, it may escape recognition. Antibodies cannot give structural information outside the epitope, and thus cannot resolve truncated or elongated forms of proteins or peptides. High-performance liquid chromatography (HPLC) separation prior to RIA is sometimes performed to improve the specificity of RIA measurements. This combination is techniques can indicate sample heterogeneity, but cannot always be used to unambiguously determine the identity of a peptide (or mixture of peptides) in a chromatographic peak.

A typical example of the difficulties in accurately measuring neuropeptides is the measurement of cholecystokinin (CCK) in biological fluids [2] such as plasma. The human CCK precursor is 95 amino acids long, and may be differentially processed into as many as 10 different bioactive forms (five different CCK peptides, each of which may or may not be sulfated). Difficulties in measuring CCK accurately by RIA, presented by the heterogeneity of CCK peptides, are compounded by the presence of another neuropeptide, gastrin, which has a high degree of sequence homology with CCK and also can exist in a sulfated form. Gastrin also circulates in higher amounts in plasma than CCK peptides.

Mass spectrometry is a technique first described around 1900, in which gas-phase ions are separated and detected at their mass-to-charge ratio (m/z). Mass spectrometry techniques were long restricted for use in physics and chemistry, because ionization techniques for large biopolymers did not exist. However, when two new technologies, matrix-assisted laser desorption/ionization time-of-flight mass spectrometry (MALDI-TOFMS) [3,4] and electrospray ionization mass spectrometry (ESI MS) [5], emerged as sensitive techniques for the desorption and ionization of large molecules, mass spectrometry began to gain widespread use in the peptide research community.

An advantage of MALDI-TOFMS analysis in neuroscience is the technique's ability to detect molecules in complex mixtures, in the presence of large molar excesses of salts, buffers, and other contaminants. MALDI-TOFMS can be used to identify posttranslational modifications of peptides; cyclical neuropeptides can be resolved from their linear forms by virtue of the 2u mass difference. Common modifications of terminal amino acids observed in neuropeptides that make chemical sequencing difficult, such as pyroglutamic acid and *C*-terminal amidation, do not hinder sequencing by mass spectrometry.

18.2. POST-SOURCE DECAY ANALYSIS OF PEPTIDES

Sequence information from ions under 2000u can be obtained directly, through post-source-decay (PSD) analysis [6,7]. PSD refers to the unimolecular dissociation of an ion that occurs in the first field free region in the mass spectrometer (Fig. 18.1). The amino acid sequence of a linear peptide can be

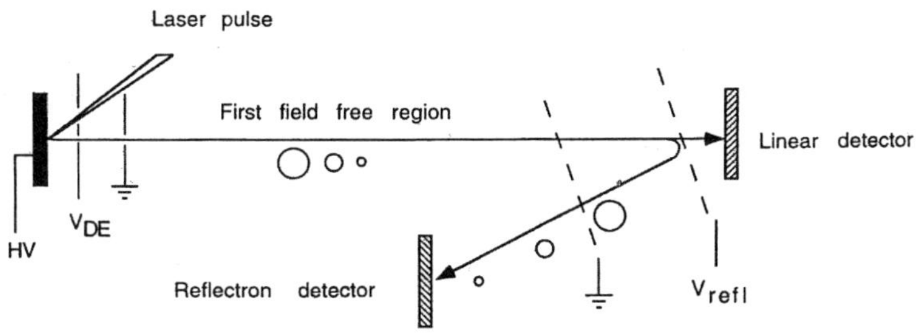

Figure 18.1. Schematic drawing of a MALDI-TOF mass spectrometer. Post-source decay refers to the unimolecular dissociation of an ion in the first field-free region.

determined through the series of product ions formed in this process. The PSD spectra obtained from the dissociation of cyclical peptides are also useful, although their analysis is more complex [8].

The theoretical and practical aspects of PSD analysis in peptide sequencing have been extensively studied [6,7,9–12]. Laser-desorbed ions undergo metastable decay in the first field-free region of the time-of-flight mass spectrometer, resulting from uni- and bimolecular collisions [11]. Ion fragments due to both the loss of small neutral molecules (water and ammonia) and cleavage of peptide bonds arise through metastable decay [11,13]. The fragments retain the same velocity as the parent ion, and thus reach the detector simultaneously in linear mode. In reflectron mode, fragments of different kinetic energies take trajectories of different lengths through the reflectron, and can be resolved on the basis of differing times of flight through the mass spectrometer. It is necessary to acquire a series of mass spectra at a constant acceleration voltage and a series of decreasing reflectron potentials, in order to resolve all fragment ions (with widely differing kinetic energies). A PSD spectrum is then constructed from these segments by the mass spectrometer's computer software, using user-supplied calibration parameters based on the PSD spectrum of a known test substance. Acquisition of a complete PSD spectrum takes longer than acquiring a simple mass spectrum, and consumes more sample material. A comprehensive nomenclature, as suggested by Roepstorff and modified by Biemann, is used to describe the product ions detected in PSD-MS [14,15]. This nomenclature is described in Fig. 18.2.

Because the interpretation of PSD spectra of unknown samples is time-consuming, PSD analysis is most often used to confirm peptide sequences. The loss of small neutral molecules such as water and ammonia may aid in the interpretation of PSD spectra. Also, peptides containing an oxidized methionine have been observed to undergo facile metastable decay during MALDI or ESI analysis, resulting in the diagnostic neutral loss of methanesulfenic acid [16–18] and the appearance of a peak at a m/z approximately 64 u lower than that of the intact peptide.

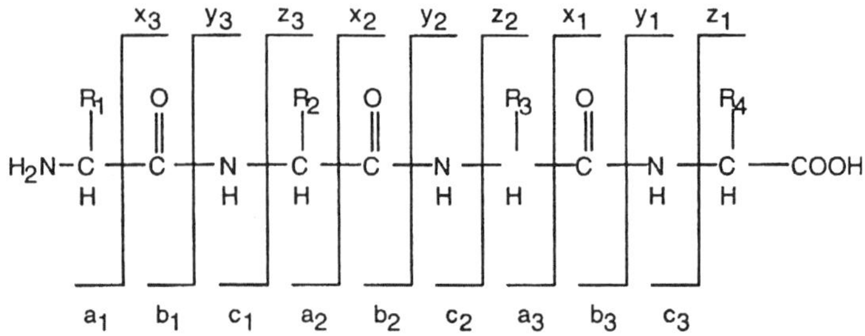

Figure 18.2. Peptide fragment nomenclature as proposed by Roepstorff and as modified by Biemann. Small letters denote dissociation sites, and subscript numbers refer to the number of amino acid residues in the ion. In the case of *a*-, *b*-, and *c*-type ions, the charge is retained on the *N* terminus, whereas in *x*-, *y*-, and *z*-type ions, the charge is retained on the *C* terminus of the peptide fragment.

The combination of peptide mapping and PSD analysis has been successfully used by several investigators to identify proteins in databases [19–24]. PSD analysis has also been used to sequence glycopeptides [25] and analyze disulfide-linked peptides [8,26]. PSD analysis can be performed in both positive and negative ion modes, yielding different sets of sequence ions [27]. The sequence ions obtained also depend on the laser power used [28] and the amino acid constituents of the peptide. Peptides containing Asp are easily cleaved, due the labile proton in this amino acid's sidechain [29]. Because cleavage of the sidechains of amino acids is rarely observed in PSD analysis (a low-energy process), the isobaric residues Leu and Ile cannot be distinguished from each other.

18.3. INFLUENCE OF SAMPLE PREPARATION ON MALDI-TOFMS ANALYSIS

One of the key features of MALDI-TOFMS is its ability to perform rapid, sensitive measurements of proteins and peptides in complex mixtures, in the presence of common biological contaminants such as salts, lipids, and amino acids. However, all of these contaminants can affect the sensitivity of measurement, by retarding desorption/ionization. Proper sample preparation can overcome some of these difficulties. The method used to prepare samples for MALDI-MS analysis will greatly determine which analytes are detected. This is especially prominent when complex biological samples are analyzed. Sample preparation techniques differ depending on the nature of the analyte and must be optimized empirically; no sample preparation technique has yet been found that is ideal for all samples [30]. In the original description of MALDI by Karas and Hillenkamp [3], nicotinic acid was used as a matrix. This compound has two major disadvantages: it is fairly volatile and sublimes in the ion source, and it has a strong tendency to form adducts with proteins, decreasing peak resolution. Several other matrix compounds have been discovered with improved characteristics compared to nicotinic acid.

The cinnamic acid derivatives 3,5-dimethoxy-4-hydroxycinnamic acid (sinapinic acid), 3-methoxy-4-hydroxycinnamic acid (ferulic acid), and α-cyano-4-hydroxycinnamic acid (CHCA) are particularly useful for the analysis of proteins and peptides [31,32]. CHCA is relatively selective for peptides, and is also useful for the analysis of biological mixtures [32]. It is a useful matrix for post-source-decay (PSD) analysis, because it induces strong metastable fragmentation in peptides [33]. Two other matrices, 2,5-dihydroxybenzoic acid and 2,4,6-trihydroxyacetophenonehave, have also been used successfully for the analysis of peptides. All MALDI matrices give significant background in the lower mass range (<700 u) of the mass spectrum. This is normally overcome by pulsed deflection, which can render studies of smaller peptides (<500 u) difficult by MALDI-MS.

The manner in which the sample and matrix are prepared together for MALDI analysis has been shown to profoundly affect the resulting signal intensity and mass resolution. The first sample preparation method used was the dried drop technique, in which a dilute (1–100 μM) analyte solution is mixed with the matrix solution, a small volume of the mixture is applied to the metal probe, and the droplet is allowed to dry under ambient conditions [3]. This technique is often sufficient for the analysis of pure compounds, but can be problematic for biological samples. Incorporation of analyte molecules into the matrix crystals is inhomogeneous using the dried drop method [34], which may explain why poor shot-to-shot reproducibility is obtained. Also, the sample surface is uneven, which can negatively affect mass accuracy. A great number of different sample preparation techniques have been devised in the past to overcome these problems. Improved mass accuracy has been reported using the thin-layer method, which produces a uniform carpet of microcrystals on the target [35,36]. Comatrix dopants (fucose and 5-methoxysalicylic acid) together with quick drying of the sample was found to improve signal reproducibility, ion current intensity, and resolution [37]. This method has been successfully used for the analysis of tryptic peptides from recombinant proteins [38].

The analysis of peptides containing thermolabile modifications such as phosphorylation or sulfation can be especially challenging using MALDI-MS. The matrix additive diammonium hydrogen citrate (DAHC) has been demonstrated to sequester metal cations (e.g., sodium) and enhance the desorption/ionization of phospho- and sulfapeptides in the presence of nonmodified peptides in MALDI-MS analysis [39–41]. An illustration of the effects of DAHC on the analysis of CCK sulfate in the presence of nonsulfated CCK is illustrated in Fig. 18.3. However, the most robust method of identifying phospho- and sulfapeptides is by analyzing the sample in both positive reflectron and negative linear mode, as demonstrated in a study of crustacean sulfakinins [42]. In the complete structural identification of chromogranin A from

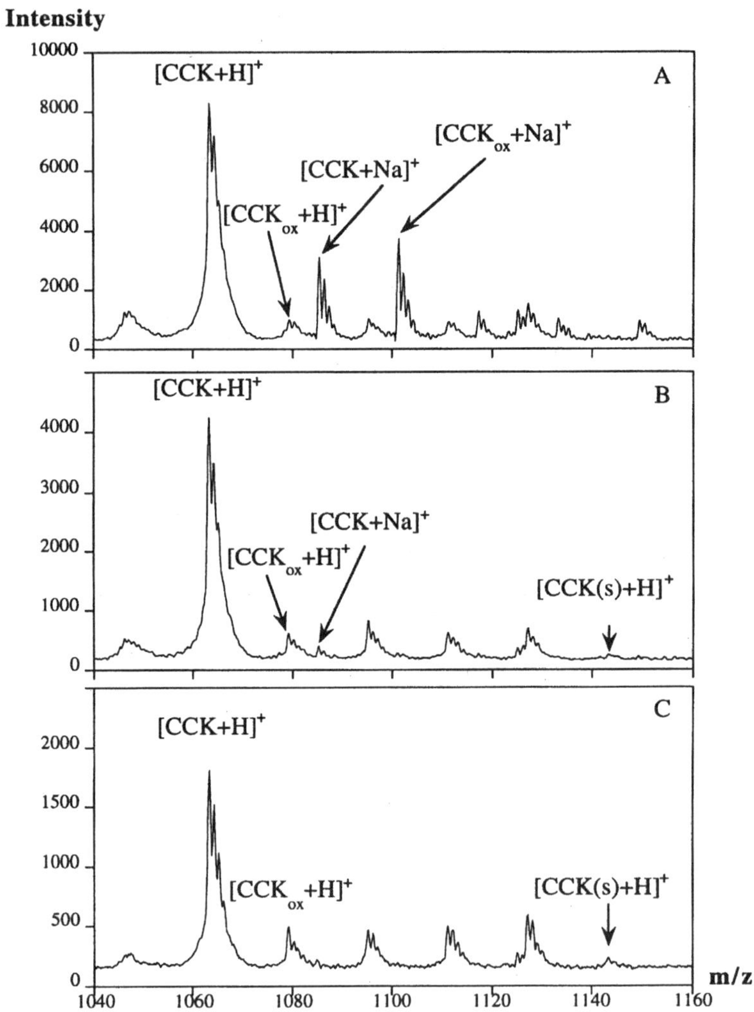

Figure 18.3. Effects of diammonium citrate on MALDI-MS analysis of sulfapeptides. As illustrated in parts (a–c), the desorption/ionization of sulfated cholecystokinin-8 (CCK8) is enhanced by the addition of 25 mM diammonium citrate, allowing for the observation of this peptide, in an equimolar mixture with unmodified CCK8, in positive linear mode. Overall peak intensities decrease with the addition of 10 and 25 mM diammonium citrate. The scavenging effect of citrate on sodium ions is also apparent. In (c), no sodium adducts can be observed. Equimolar (10 pmol/μL) amounts of sulfated and nonsulfated CCK8 in CHCA were analyzed. (a) Without matrix additive; (b) with addition of 10 mM diammonium citrate; (c) with addition of 25 mM diammonium citrate.

bovine adrenal medulla, peptide mapping, enzymatic dephosphorylation, and chemical deglycosylation were used to confirm the identity of peptide modification [43]. In summary, the choice of sample preparation depends largely on the sensitivity of analysis that is required, the composition of the sample, and the MALDI probe geometry. Further sample preparation strategies for neurochemical investigation using MALDI-MS are described in the following sections.

18.4. MALDI-MS ANALYSIS OF CEREBROSPINAL FLUID

Cerebrospinal fluid (CSF) plays a significant role in protecting the brain and maintaining ventricular pressure. CSF reflects metabolic processes in the brain and is used for diagnostic purposes, due to the free exchange of substances between the brain and CSF. The most abundant molecular species in CSF can be readily observed using MALDI-MS analysis without any pretreatment of the sample. The seed-layer method [44] has been shown to be particularly useful in the analysis of cerebrospinal fluid and other biological samples. In this method, a seed layer of matrix microcrystals is created by depositing a drop of dilute (1 g/L) solution of matrix dissolved in acetonitrile. When the seed layer has dried, the analyte solution is mixed with a second, saturated matrix solution containing acetonitrile and 0.1% TFA, and 0.5 μL is dropped onto the seed layer. A homogeneous layer of analyte/matrix crystals is formed rapidly on the surface of the probe. The seed-layer method gives homogeneous signal intensity, improved sensitivity, and improved reproducibility for the analysis of proteins and peptides in neat CSF [44–46], as well as samples that contain urea, ampholytes, and residual SDS [40,41,47].

Because neuropeptides typically are found in low concentrations, they may be impossible to detect by mass spectrometry in an untreated sample. Concentrating and desalting may be achieved either in a conventional chromatography system or a miniaturized setting [48,49]. However, hydrophilic peptides may be lost in the flowthrough of C18-based columns. The use of a graphite column for the high-resolution separation of small hydrophilic peptides has recently been described [50].

A study of the temporal processing of the neuropeptides somatostatin, neuropeptide Y and galanin by enzymes present in human cerebrospinal fluid (CSF) has been performed on samples to which synthetic peptides were added in order to make possible their detection by MALDI-TOFMS without any further sample treatment [45,46,51,52]. Differences in the processing of somatostatin (1–28) were observed through direct MALDI-MS analysis of the CSF of normal controls as compared to CSF from patients suffering from Alzheimer's disease or frontotemporal dementia (Fig. 18.4, Table 18.1). Recently, novel fragments of endogenous neuroendocrine peptides and other polypeptides in human CSF were identified using a separation strategy that

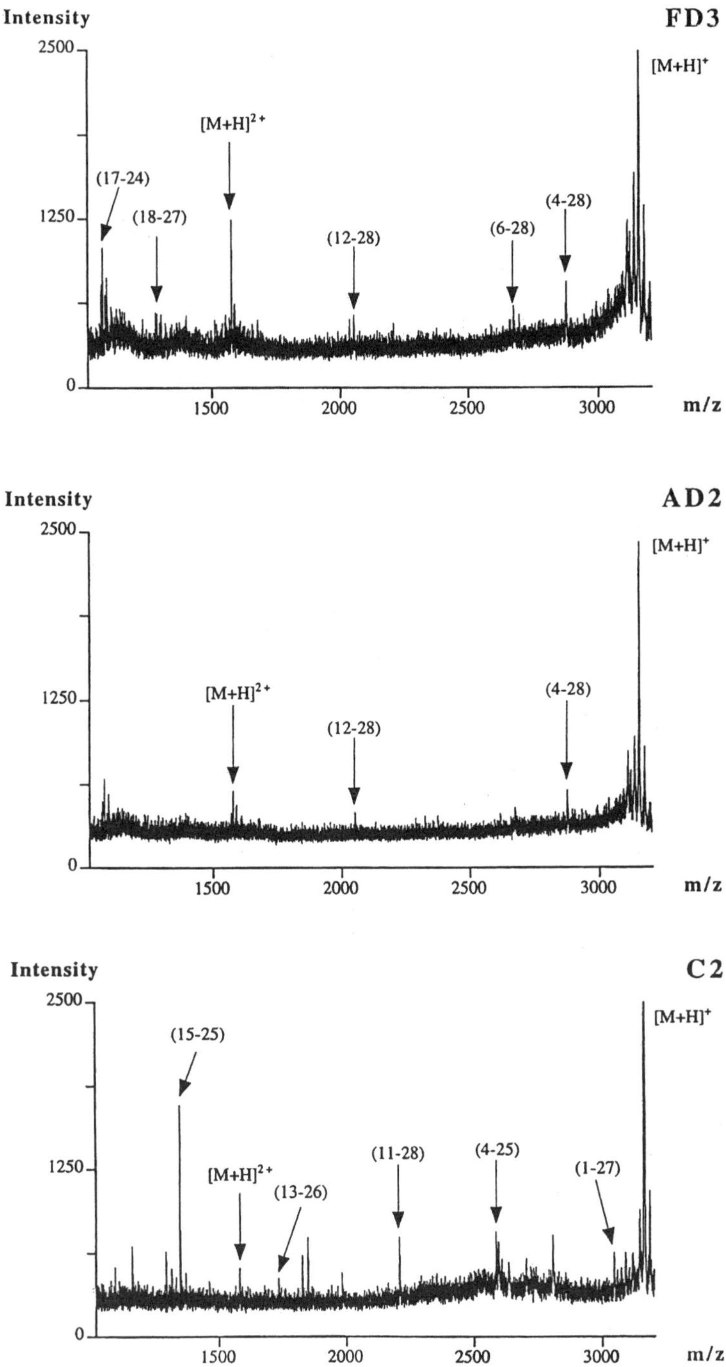

Figure 18.4. MALDI-TOF mass spectra of three CSF samples after 3 days of incubation with synthetic somatostatin (1–28). FD3: patient with frontotemporal dementia. AD2: patient with Alzheimer's disease. C2: normal control.

TABLE 18.1. Peptide Fragments Observed in CSF Incubated with 20 μm Somatostatin (1–28)

Fragment	m/z (calc.)	C1	C2	FD1	FD2	FD3	AD1	AD2	AD3
(1–13)	1400.5	•	•	—	—	—	•	•	•
(1–16)	1657.8	•	—	—	—	—	—	•	•
(1–27)	3046.5	—	•	—	—	—	—	—	—
(4–25)	2586.0	—	•	—	—	—	—	—	—
(15–25)	1347.6	—	•	—	—	—	—	—	—
(17–24)	1072.3	•	•	—	—	•	—	—	•
(18–27)	1305.5	—	—	—	—	•	—	—	—
(3–28)	2992.9	—	•	—	—	—	—	—	—
(4–28)	2879.0	•	•	•	•	•	•	•	•
(6–28)	2676.1	—	—	—	•	•	•	—	•
(9–28)	2375.7	—	—	—	—	•	—	—	—
(10–28)	2305.7	—	—	—	—	•	—	—	—
(11–28)	2208.6	—	•	—	—	•	—	•	•
(12–28)	2052.4	•	—	•	•	•	•	•	•
(16–28)	1566.1	—	—	•	—	—	•	•	—

included phase separation in chloroform/methanol/water and reversed-phase liquid chromatography [53].

Direct MALDI-TOFMS analysis of neuropeptides in neural tissues and cerebrospinal fluid, without prior sample cleanup, can be achieved under certain conditions. By simply introducing an on-probe rinsing step after the sample has dried, signal intensity can be improved dramatically.

18.5. MALDI-MS ANALYSIS OF NEURAL AND NEUROENDOCRINE TISSUES

The neurons of some marine animals contain large granulae rich in peptides. It has been shown that MALDI spectra can be obtained from single cells even in the presence of high salt concentrations [54]. In certain cases, PSD analysis can be performed on neuropeptides from a single such neuron, in order to confirm their identity. To date, the pituitary gland of the frog *Xenopus* [8], the cardiac ganglion [55] and penis nerve [56] of the snail *Lymnaea*, single neurons of the snail *Aplysia* [57], and single cells of the crayfish *Oronectes* [58] have been successfully analyzed using MALDI-MS. Analysis of the cells required optimization of matrix conditions (in some cases, even mixing different matrices) in order to obtain the best results.

Simple fractionation can improve the detection of analytes from a complex mixture. A micro–liquid chromatography procedure has been described in

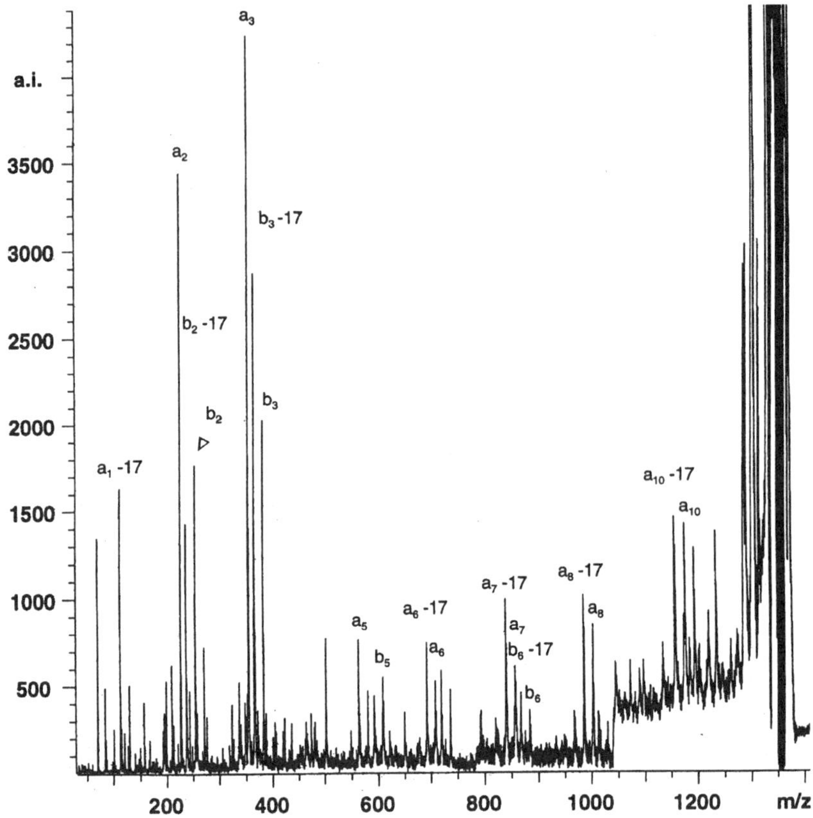

Figure 18.5. PSD spectrum of porcine brain substance P (*m/z* 1349). Intensity is given in arbitrary units.

which nanoliter deposition of fractions and matrix to a MALDI probe was employed [59]. The retention time and experimentally determined masses as well as PSD analysis were used to identify peptides from a single neuron. The use of PSD analysis can in some cases be replaced by the use of amino- or carboxypeptidases prior to MALDI-MS analysis. A novel calcitonin gene-related peptide from frog skin has been identified using carboxypeptidase Y digestion MALDI-MS [60]. Enzymatic ladder sequencing simplifies sequence determination, but all possible fragments may not be generated. Complete sequence coverage is especially critical in the characterization of novel peptides. A combination of ladder sequencing and PSD ("nested PSD") was used to identify neurotoxic peptides from single wasps [61].

Immunoaffinity sample cleanup procedures are useful for concentrating analytes from dilute biological mixtures prior to analysis by mass spectrometry. Immunoprecipitation of analytes followed by MALDI analysis was successful in the analysis of antigenic sites of monoclonal antibodies [62], neu-

TABLE 18.2. Fragment Ions from the Peptide at *m/z* 1349 Observed by PSD Analysis

	H-R[a]	P	K	P	Q	Q	F	F	G	L	*M*-NH_2
a-17	112	—	—	—	562	690	837	985	—	1155	—
a	—	226	354	—	579	707	854	1002	—	—	—
b-17	—	237	365	—	—	718	—	—	—	—	—
b	—	254	382	—	607	735	—	—	—	—	—

[a] Amino acid sequence.

rotensin and neurotensin fragments in peptide extracts from human hypothalamus [63], arginine vasopressin from human lymphocytes [64], and tachykinins from brain cortex [45,46]. In the last example, the identity of the substance P-like immunoreactive component was confirmed through PSD analysis (Fig. 18.5, Table 18.2, reproduced from J Chrom A).

18.6. SUMMARY

MALDI-TOFMS has become an important tool to the neuropeptide researcher, providing data that are complementary to that obtained immunologically based assays. It is possible to obtain accurate mass information for neuropeptides, even in the presence of common biological contaminants such as buffers, lipids, and free amino acids. The ability to perform sensitive and direct analyses of extremely complex samples, such as cells, is perhaps the largest advantage of MALDI over ESI mass spectrometry. The quality of data obtained in MALDI-MS analysis is dependent on the sample preparation technique employed. With a few notable exceptions, neuropeptides are typically encountered in biological fluids in a low concentration. Because mass spectrometry is frequently less sensitive than immunoassays, steps may need to be taken to desalt and concentrate the sample prior to analysis by MS. Through post-source-decay analysis, it is possible to obtain sequence information, and thus confirmation of peptide identity.

REFERENCES

1. I. Schechter, Mapping of the combining sites of antibodies specific for poly-L-alanine determinants, *Nature* **228**, 639–641 (1970).
2. J. F. Rehfeld, Accurate measurement of cholecystokinin in plasma, *Clin. Chem.* **44**, 991–1001 (1998).
3. M. Karas and F. Hillenkamp, Laser desorption ionization of proteins with molecular masses exceeding 10,000 Daltons, *Anal. Chem.* **60**, 2299–2301 (1988).

4. K. Tanaka, H. Waki, Y. Ido, S. Akita, Y. Yoshida, and T. Yoshida, Protein and polymer analyses up to m/z 100,000 by laser ionization time-of-flight mass spectrometry, *Rapid Commun. Mass Spectrom.* **2**, 151–153 (1988).
5. J. B. Fenn, M. Mann, C. K. Meng, S. F. Wong, and C. M. Whitehouse, Electrospray ionization for mass spectrometry of large biomolecules, *Science* **246**, 64–71 (1989).
6. R. Kaufmann, D. Kirsch, and B. Spengler, Sequencing of peptides in a time-of-flight mass spectrometer: Evaluation of post-source decay following matrix-assisted laser desorption ionisation (MALDI), *Int. J. Mass Spectrom. Ion Process.* **131**, 355–385 (1994).
7. R. Kaufmann, B. Spengler, and K. Lützenkirchen, Mass spectrometric sequencing of linear peptides by product-ion analysis in a reflectron time-of-flight mass spectrometer using matrix-assisted laser desorption ionization, *Rapid Commun. Mass Spectrom.* **7**, 902–910 (1993).
8. S. Jespersen, P. Chaurand, F. J. van Strien, B. Spengler, and J. van der Greef, Direct sequencing of neuropeptides in biological tissue by MALDI-PSD mass spectrometry, *Anal. Chem.* **71**, 660–666 (1999).
9. B. Spengler, D. Kirsch, and R. Kaufmann, Fundamental aspects of postsource decay in matrix-assisted laser desorption mass spectrometry. 1. Residual gas effects, *J. Phys. Chem.* **96**, 9678–9684 (1992).
10. B. Spengler, D. Kirsch, R. Kaufmann, and E. Jaeger, Peptide sequencing by matrix-assisted laser-desorption mass spectrometry, *Rapid Commun. Mass Spectrom.* **6**, 105–108 (1992).
11. B. Spengler, D. Kirsch, and R. Kaufmann, Metastable decay of peptides and proteins in matrix-assisted laser-desorption mass spectrometry, *Rapid Commun. Mass Spectrom.* **5**, 198–202 (1991).
12. B. Spengler, Post-source decay analysis in matrix-assisted laser desorption/ionization mass spectrometry of biomolecules, *J. Mass Spectrom.* **32**, 1019–1036 (1997).
13. J. A. Hill, R. S. Annan, and K. Biemann, Matrix-assisted laser desorption ionization with a magnetic mass spectrometer, *Rapid Commun. Mass Spectrom.* **5**, 395–399 (1991).
14. P. Roepstorff and J. Fohlman, Proposal for a common nomenclature for sequence ions in mass spectra of peptides, *Biomed. Mass Spectrom.* **11**, 601 (1984).
15. K. Biemann, Contributions of mass spectrometry to peptide and protein structure, *Biomed. Environ. Mass Spectrom.* **16**, 99–111 (1988).
16. K. C. Parker, J. I. Garrels, W. Hines, E. M. Butler, A. H. Z. McKee, D. Patterson, and S. Martin, Identification of yeast proteins from two-dimensional gels: Working out spot cross-contamination, *Electrophoresis* **19**, 1920–1932 (1998).
17. L. Betancourt, T. Takao, J. Gonzales, O. Reyes, V. Besada, G. Padron, and Y. Shimonishi, The metastable decomposition of a peptide containing oxidized methionine(s) in matrix-assisted laser desorption/ionization time-of-flight mass spectrometry, *Rapid Commun. Mass Spectrom.* **13**, 1075–1076 (1999).
18. X. Jiang, J. B. Smith, and E. C. Abraham, Identification of a MS-MS fragment diagnostic for methionine sulfoxide, *J. Mass Spectrom.* **31**, 1309–1310 (1996).
19. T. Larsson, J. Norbeck, H. Karlsson, K. A. Karlsson, and A. Blomberg, Identification of two-dimensional gel electrophoresis resolved yeast proteins by matrix-assisted laser desorption ionization mass spectrometry, *Electrophoresis* **18**, 418–423 (1997).

20. P. Dainese, W. Staudenmann, M. Quadroni, C. Korostensky, G. Gonnet, M. Kertesz, and P. James, Probing protein function using a combination of gene knockout and proteome analysis by mass spectrometry, *Electrophoresis* **18**, 432–442 (1997).

21. K. L. O'Connell and J. T. Stults, Identification of mouse liver proteins on two-dimensional electrophoresis gels by matrix-assisted laser desorption/ionization mass spectrometry of *in situ* enzymatic digests, *Electrophoresis* **18**, 349–359 (1997).

22. P. L. Courchesne, R. Luethy, and S. D. Patterson, Comparison of in-gel and on-membrane digestion methods at low to sub-pmol level for subsequent peptide and fragment-ion mass analysis using matrix-assisted laser-desorption/ionization mass spectrometry, *Electrophoresis* **18**, 369–381 (1997).

23. D. Fabris, M. M. Vestling, M. M. Cordero, V. M. Doroshenko, R. J. Cotter, and C. Fenselau, Sequencing electroblotted proteins by tandem mass spectrometry, *Rapid Commun. Mass Spectrom.* **9**, 1051–1055 (1995).

24. C. Nilsson, C. M. Murphy, and R. Ekman, Isolation and characterization of proteins from human lymphocyte nuclei using matrix-assisted laser desorption/ionization time-of-flight mass spectrometry and post-source decay analysis, *Rapid Commun. Mass Spectrom.* **11**, 610–612 (1997).

25. S. Goletz, M. Leuck, P. Franke, and U. Karsten, Structure analysis of acetylated and nonacetylated O-linked MUC1-glycopeptides by post-source decay matrix-assisted laser desorption/ionization mass spectrometry, *Rapid Commun. Mass Spectrom.* **11**, 1387–1398 (1997).

26. M. D. Jones, S. D. Patterson, and H. S. Lu, Determination of disulfide bonds in highly bridged disulfide-linked peptides by matrix-assisted laser desorption/ionization mass spectrometry with postsource decay, *Anal. Chem.* **70**, 136–143 (1998).

27. J. Jai-nhuknan and C. Cassady, Anion and cation post-source decay time-of-flight mass spectrometry of small peptides: Substance P, angiotensin II, and renin substrate, *Rapid Commun. Mass Spectrom.* **10**, 1678–1682 (1996).

28. Z. Szilágyi, J. E. Varney, P. J. Derrick, and K. Vékey, Dependence of matrix-assisted laser desorption/ionization post-source decay spectra on laser power, *Rapid Commun. Mass Spectrom.* **12**, 489–492 (1998).

29. W. Yu, J. E. Vath, M. C. Huberty, and S. A. Martin, Identification of the facile gas-phase cleavage of the Asp-Pro and Asp-Xxx peptide bonds in matrix-assisted laser desorption time-of-flight mass spectrometry, *Anal. Chem.* **65**, 3015–3023 (1993).

30. M. Kussmann, E. Nordhoff, H. Rahbek-Nielsen, S. Haebel, M. Rossel-Larsen, L. Jakobsen, J. Gobom, E. Mirgorodskaya, A. Kroll-Kristensen, L. Palm, and P. Roepstorff, Matrix-assisted laser desorption/ionization mass spectrometry sample preparation techniques designed for various peptide and protein analytes, *J. Mass Spectrom.* **32**, 593–601 (1997).

31. R. C. Beavis and B. T. Chait, Cinnamic acid derivatives as matrices for ultraviolet laser desorption mass spectrometry of proteins, *Rapid Commun. Mass Spectrom.* **3**, 432–435 (1989).

32. R. C. Beavis, T. Chaudhary, and B. T. Chait, α-Cyano-4-hydroxycinnamic acid as a matrix for matrix-assisted laser desorption mass spectrometry, *Org. Mass Spectrom.* **27**, 156–158 (1992).

33. M. Karas, U. Bahr, K. Strupat, F. Hillenkamp, T. Tsarbopoulos, and B. N. Pramanik, Matrix dependence of metastable fragmentation of glycoproteins in MALDI TOF mass spectrometry, *Anal. Chem.* **67**, 675–679 (1995).

34. S. J. Doktycz, P. J. Savickas, and D. A. Krueger, Matrix-sample interactions in ultraviolet laser-desorption of proteins, *Rapid Commun. Mass Spectrom.* **5**, 145–148 (1991).

35. O. Vorm and P. Roepstorff, Detector bias gating for improved detector response and calibration in matrix-assisted laser desorption/ionization time-of-flight mass spectrometry, *J. Mass Spectrom.* **31**, 351–356 (1996).

36. O. Vorm, P. Roepstorff, and M. Mann, Improved resolution and very high sensitivity in MALDI TOF of matrix surfaces made by fast evaporation, *Anal. Chem.* **66**, 3281–3287 (1994).

37. A. I. Gusev, W. R. Wilkinson, A. Proctor, and D. M. Hercules, Improvement of signal reproducibility and matrix/comatrix effects in MALDI analysis, *Anal. Chem.* **67**, 1034–1041 (1995).

38. T. M. Billeci and J. M. Stults, Tryptic mapping of recombinant proteins by matrix-assisted laser desorption/ionization mass spectrometry, *Anal. Chem.* **65**, 1709–1716 (1993).

39. J. M. Asara and J. Allison, Enhanced detection of phosphopeptides in matrix-assisted laser desorption/ionization mass spectrometry using ammonium salts, *J. Am. Soc. Mass Spectrom.* **10**, 35–44 (1999).

40. C. L. Nilsson, M. Puchades, A. Westman, K. Blennow, and P. Davidsson, Identification of proteins in a human pleural exudate using two-dimensional preparative liquid-phase electrophoresis and matrix-assisted laser desorption/ionization mass spectrometry, *Electrophoresis* **20**, 860–865 (1999).

41. C. L. Nilsson, A. Westman, T. Larsson, H. Karlsson, and K.-A. Karlsson, Effects of ammonium citrate as a matrix additive on MALDI-TOFMS analysis of sulfa- and phosphopeptides, *Proc. 47th ASMS Conf. Mass Spectrometry and Allied Topics*, Dallas, TX, June 13–17, 1999.

42. A. H. Johnsen, H. Duve, M. Davey, M. Hall, and A. Thorpe, Sulfakinin neuropeptides in a crustacean. Isolation, identification and tissue localization in the tiger prawn *Penaeus monodon*, *Eur. J. Biochem.* **267**, 1153–1160 (2000).

43. S. H. Bauer, X. Y. Zhang, W. Van Dongen, M. Claeys, and M. Przybylski, Chromogranin A from bovine adrenal medulla: Molecular characterization of glycosylations, phosphorylations, and sequence heterogeneities by mass spectrometry, *Anal. Biochem.* **274**, 69–80 (1999).

44. A. Westman, C. L. Nilsson, and R. Ekman, Matrix-assisted laser desorption/ionization time-of-flight mass spectrometry analysis of proteins in human cerebrospinal fluid, *Rapid Commun. Mass Spectrom.* **12**, 1092–1098 (1998).

45. C. L. Nilsson, E. Brodin, and R. Ekman, Substance P and related peptides in porcine cortex: whole tissue and nuclear localization, *J. Chromatogr. A* **800**, 21–27 (1998).

46. C. Nilsson, A. Westman, K. Blennow, and R. Ekman, Processing of neuropeptide Y and somatostatin in human cerebrospinal fluid as studied by radioimmunoassay and mass spectrometry, *Peptides* **19**, 1137–1146 (1998).

47. C. L. Nilsson, T. Larsson, E. Gustafsson, K.-A. Karlsson, and P. Davidsson, Identification of protein vaccine candidates from Helicobacter pylori using a preparative two-dimensional electrophoretic procedure and mass spectrometry, *Anal. Chem.* **72**, 2148–2153 (2000).

48. H. Zhang, P. E. Andrén, and R. M. Caprioli, Micro-preparation procedure for high-sensitivity matrix-assisted laser desorption ionization mass spectrometry, *J. Mass Spectrom.* **30**, 1767–1771 (1995).
49. J. Gobom, E. Nordhoff, E. Mirgorodskaya, R. Ekman, and P. Roepstorff, Sample purification and preparation technique based on nano-scale reversed-phase columns for the sensitive analysis of complex peptide mixtures by matrix-assisted laser desorption/ionization mass spectrometry, *J. Mass Spectrom.* **34**, 105–116 (1999).
50. E. T. Chin and D. I. Papac, The use of a graphitic carbon column for desalting hydrophilic peptides for MALDI-TOF mass spectrometric analysis, *Proc. 46th ASMS Confe. Mass Spectrometry and Allied Topics*, Orlando, FL, May 31–June 4, 1998, p. 173.
51. R. Ekman, P. Juhasz, M. Heilig, H. Ågren, and C. E. Costello, Novel neuropeptide Y processing in human cerebrospinal fluid from depressed patients, *Peptides* **17**, 1107–1111 (1996).
52. C. L. Nilsson, A. Westman, L. Minthon, K. Blennow, and R. Ekman, Processing of neuropeptide Y, galanin, and somatostatin in the cerebrospinal fluid of patients with Alzheimer's disease and frontotemporal dementia (in press).
53. M. Stark, O. Danielsson, W. J. Griffiths, H. Jörnvall, and J. Johansson, The peptide repertoire of human cerebrospinal fluid: Novel proteolytic fragments of neuroendocrine proteins (in press).
54. R. W. Garden, L. L. Moroz, T. P. Moroz, S. A. Shippy, and J. V. Sweedler, Excess salt removal with matrix rinsing: Direct peptide profiling of neurons from marine invertebrates using matrix-assisted laser desorption/ionization time-of-flight mass spectrometry, *J. Mass Spectrom.* **31**, 1126–1130 (1996).
55. C. R. Jiminez, K. W. Li, K. Dreisewerd, S. Spijker, R. Kingston, R. H. Bateman, A. L. Burlingame, A. B. Smit, J. van Minnen, and W. P. M. Geraerts, Direct mass spectrometric profiling and sequencing of single neurons reveals differential peptide patterns in a small neuronal network, *Biochemistry* **37**, 2070–2076 (1998).
56. K. Dreisewerd, R. Kingston, W. P. M. Geraerts, and K. W. Li, Direct mass spectrometric peptide profiling and sequencing of nervous tissues to identify peptides involved in male copulatory behavior in *Lymnaea stagnalis*, *J. Mass Spectrom.* **169/170**, 291–299 (1997).
57. L. Li, R. W. Garden, E. V. Romanova, and J. V. Sweedler, In situ sequencing of peptides from biological tissues and single cells using MALDI-PSD/CID analysis, *Anal. Chem.* **71**, 5451–5458 (1999).
58. V. Redeker, J.-Y. Toullec, J. Vinh, J. Rossier, and D. Soyez, Combination of peptide profiling by matrix-assisted laser desorption/ionization time-of-flight mass spectrometry and immunodetection on single glands or cells, *Anal. Chem.* **70**, 1805–1811 (1998).
59. S. Hsieh, K. Dreisewerd, R. C. van der Schors, C. R. Jiménez, J. Stahl-Zeng, F. Hillenkamp, J. W. Jorgenson, W. P. M. Geraerts, and K. W. Li, Separation and identification of peptides in single neurons by microcolumn liquid chromatographymatrix-assisted laser desorption/ionization time-of-flight mass spectrometry and postsource decay analysis, *Anal. Chem.* **70**, 1847–1852 (1998).

60. A. A. Seon, T. N. Pierre, V. Redeker, C. Lacombe, A. Delfour, P. Nicolas, and M. Amiche, Isolation, structure, synthesis, and activity of a new member of the calcitonin gene-related peptide family from frog skin and molecular cloning of its precursor, *J. Biol. Chem.* **275**, 5934–5940 (2000).

61. M. Hisada, K. Konno, Y. Itagaki, H. Naoki, and T. Nakajima, Advantages of using nested collision induced dissociation/post-source decay with matrix-assisted laser desorption/ionization time-of-flight mass spectrometry: Sequencing of novel peptides from wasp venom, *Rapid Commun. Mass Spectrom.* **14**, 1828–1834 (2000).

62. Y. Zhao and B. T. Chait, Protein epitope mapping by mass spectrometry, *Anal. Chem.* **66**, 3723–3726 (1994).

63. J. Gobom, K.-O. Kraeuter, R. Persson, H. Steen, P. Roepstorff, and R. Ekman, Detection and quantification of neurotensin in human brain tissue by matrix-assisted laser desorption/ionization time-of-flight mass spectrometry, *Anal. Chem.* **72**, 3320–3326 (2000).

64. R. Ekman, J. Gobom, R. Persson, P. Mecocci, and C. Nilsson, Arginine vasopressin in the cytoplasm and nuclear fraction of lymphocytes from healthy donors and patients with depression or schizophrenia, *Peptides* (in press).

CHAPTER 19

Neuropeptide Processing

PIOTR SUDER, AGNIESZKA SCIUBISZ, MAREK SMOLUCH, and JERZY SILBERRING

Neurobiochemistry Group, Faculty of Chemistry and Regional Laboratory, Jagiellonian University, Krakow, Poland

CONTENTS

19.1. INTRODUCTION

Posttranslational processing is an important factor during maturation, release and inactivation of bioactive molecules. Such events are crucial in apoptosis [1], cleavage of the amyloid precursor protein [2], and other important functions in the pathophysiology of the nervous system [3]. Neuropeptides are synthesized as inactive precursors—pre-propeptides. An example of the general scheme of a preprohormone composition is shown in Fig. 19.1, and more details on the neuropeptides and their functions can be found in Chapter 1 of this book.

The released fragments may have various functions, though they arrive from the same precursor [4]. For example, proopiomelanocortin, which is a product of a single gene, contains a copy of β-endorphin, which is involved, for instance, in pain mechanisms in the CNS, and also a copy of the adrenocorticotropic hormone (ACTH), which is further liberated from the pituitary and acts as a hormone, thus releasing steroids from the adrenal glands [5]. The first step in pro-precursor processing is the removal of signal peptide, which is

Mass Spectrometry and Hyphenated Techniques in Neuropeptide Research, Edited by Jerzy Silberring and Rolf Ekman.
ISBN 0-471-35493-7

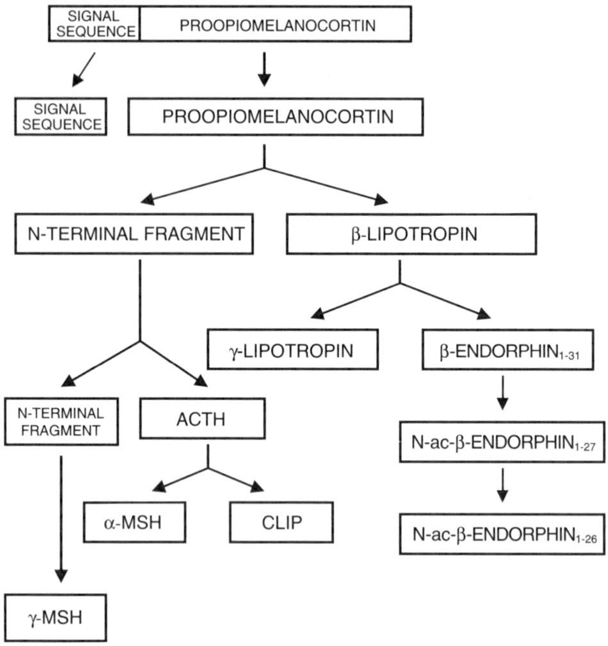

Figure 19.1. Preproopiomelanocortin precursor (ACTH—adrenocorticotropin; MSH—melanocyte stimulating hormone; CLIP—corticotropin-like intermediate lobe peptide).

located at the *N* terminus of the protein molecule. This signal peptide usually contains approximately 20–40 amino acid residues, and its role is to assure a proper folding and transport to the Golgi apparatus. The release of the bioactive peptides involves so-called prohormone convertases (PCs) that belong to the subtilisin/kexin family of proteins. Several excellent reviews are recommended for further reading [6–10]. The enzyme usually recognize pairs of basic amino acids; however, processing has been reported at single basic residues or in the vicinity of other, nonpolar amino acids. It is postulated that maturation of peptides in the CNS occurs along their axonal transport. Therefore, peptide precursors are often packed together with the appropriate set of enzyme in the secretory granules [11]. It has also been postulated that the pH inside granules changes during axonal transport, which may function as an activating/deactivating mechanism to trigger various enzyme responsible for each metabolic event.

A general pathway of the neuropeptide biosynthesis, metabolism, and degradation in body fluids or tissues of the central nervous system is presented in Fig. 19.2.

As can be deduced from Fig. 19.2, concentration of the particular bioactive fragment strongly depends on the proteolytic activities of two enzyme, the one

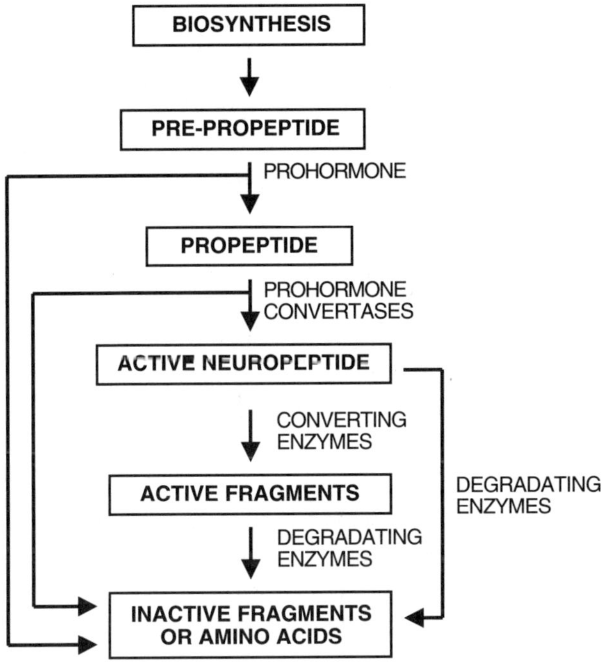

Figure 19.2. Processing of neuropeptide precursor.

that releases bioactive fragment from its elongated form, and the one that terminates or changes peptide action. Very often, a terminating enzyme is replaced by another peptidase, called *converting enzyme*. The role of such protein is to convert one peptide sequence possessing biological activity to another shorter one, of similar or different properties [12]. Therefore, changes in the concentration of each neuropeptide may, at least to some extent, depend on the enzymatic activities present in the nervous system. For example, elevated enzymatic activity of dynorphin B convertase was found in the spinal cord of rats subjected to chronic pain [13]. Interestingly, the increased activity was detected only in those cord segments where the pain signals were affected. This could be achieved by selective injection of the pain stimuli, such as Freund's adjuvant, into the appropriate paw [13]. Similar observation was described using the cortical cells in primary culture as a model, and subjected to chronic incubation with morphine [14]. Neuronal cells exposed to morphine caused a significant increase in the activity of dynorphin B convertase. Moreover, aminopeptidase activity truncating the *N*-terminal Tyr residue from dynorphins was also elevated as well as the enzyme cleaving orphanin FQ/nociceptin to shorter fragments [15]. This morphine-derived effect was even more pronounced (>300% increase in enzymatic activity) when the neuronal cells were subjected to the "withdrawal state" where the chronically added morphine was removed from the incubation medium. The effect was related to the opioid receptors because naloxone—an opioid antagonist—was able to reverse this process [15]. In these experiments, enzymatic activity was measured simultaneously with peptides concentration (by radioimmunoassay) to assure that the peptide levels followed changes in the enzyme activity, thus strongly suggesting the role of peptidases in the regulation of the peptides concentration. Therefore, it is advisable to measure both parameters, whenever possible, to draw proper conclusions from the results obtained. In most cases, determination of enzymatic activity is a better measure than quantitation of the enzyme (protein) concentration because the former reflects the real dynamic changes in the system.

Pharmacologically (see Fig. 19.2), it would be of an extreme interest to design specific tools, such as protease inhibitors or antisense oligodeoxynucleotides, to influence this system in hopes of designing novel drugs. There are many indications coming from the most recent literature that such strategy might be, or already is, very efficient for the treatment of hypertension (inhibitors of the angiotensin converting enzyme) [16], type II diabetes (inhibitors of dipeptidyl peptidase IV) [17,18], or HIV [19].

Enzymatic activity of a particular proteinase depends on many factors, as summarized in Fig. 19.3. Among them, pH changes, oxidation–reduction, endogenous inhibitors, and metal binding may play a significant role.

19.2. ENZYME INHIBITORS

Very little is known about the endogenous inhibitors of peptidases cleaving neuropeptides present in the central nervous system. The role of an inhibitor is to prevent unwanted cleavage, such as outside of the cell compartment, where enzymatic action should be terminated. Natural inhibitors may also serve as a template to design shorter, synthetic molecules serving as possible drug candidates. Therefore, the entire metabolic pathway cannot be fully revealed, without the knowledge on the factors affecting proteolytic events. Several inhibitors of serine proteinases were detected in the CNS, particularly in the cerebrospinal fluid [20,21]. The information, however, is not complete, and these proteins were usually detected by means of immunoelectrophoretic methods or by amino acid sequence of their tryptic fragments. The recently discovered protein, called neuroserpin, was found by the cloning techniques, and appears to be an axonally secreted serine proteinase inhibitor [22]. This brain-associated molecule inhibits tissue plasminogen activator and plasmin [23,24]. Morever, this molecule may be associated, among other proteins, with the early-onset dementia, reduction of cerebral infarct volume, and protection of neurons from ischemia-induced apoptosis [25–27]. The role of another endogenous protein, a Kunitz-type inhibitor, which is present in the full sequence of the amyloid precursor protein, is still obscure [28]. Our group has identified another inhibitor in the human cerebrospinal fluid, thus inhibiting dynorphin converting enzyme [6]. This peptidase was classified as a serine

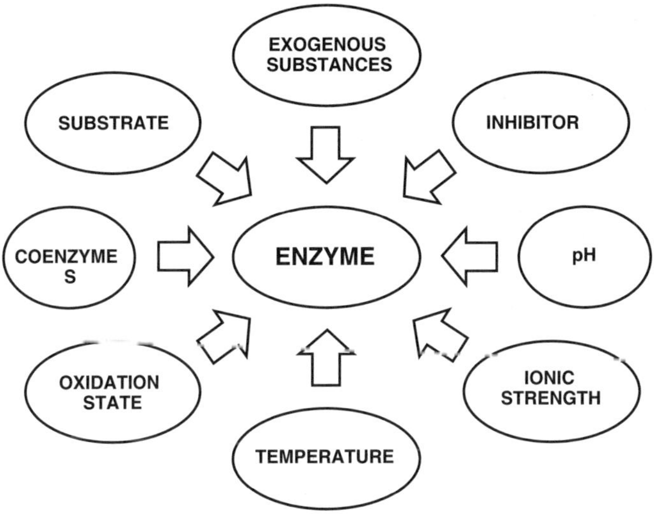

Figure 19.3. Factors that influence proteolytic activity.

enzyme and is also present in the human CSF. It is striking that both peptidase and its natural inhibitor are present in the same body fluid and the enzyme is still active toward dynorphins; however, its estimated K_m value is approximately 10 times higher than those of the other dynorphin convertases of a similar specificity [29,30].

Another group of endogenous inhibitors might be neuropeptides themselves or their fragments. These molecules can inhibit metabolism in either a competitive or noncompetitive manner. For example, substance P converting enzyme was inhibited by a structurally different calcitonin gene-related peptide (CGRP) [31]. This observation may have physiological significance in pain transmission [32,33]. Another metabolic pathway, involving conversion of dynorphins by a group of highly specific enzymes (dynorphin A and dynorphin B converting enzymes), was abolished by dynorphins [29,30]. Namely, dynorphin A convertase was inhibited by dynorphin B, and dynorphin B convertase was inhibited by dynorphin A. This observation might be explained based on the sequence similarities of both pepides. However, these peptides, acting as inhibitors, remained intact during the reaction [26,27,34,35]. The consequence of an existence of such a mechanism, such as in analgesia and drug abuse, might be a feedback loop, thus controlling the amount of the released enkephalins (delta ligands) from dynorphins (kappa ligands), and the amount of dynorphins itself [36,37].

Nevertheless, the growing interest in this rapidly expanding area will certainly bring more new exciting data on the complete regulation of neuropeptide metabolism in pathophysiology of neurodegenerative diseases and drug abuse.

19.3. DETERMINATION OF ENZYMATIC ACTIVITY

A vast number of methodologies concerning measurement of enzymatic activity was developed and described in the literature. The simplest techniques utilize chromogenic (less sensitive) or fluorogenic (more sensitive) substrates and UV/VIS (ultraviolet/visible spectrum) or fluorescence meters [38–41]. These approaches, although very useful, are not suitable in certain cases when the enzyme tested requires the entire sequence of a natural substrate for recognition, such as the dynorphin converting enzymes. In such case, HPLC in conjunction with an appropriate detector [42], internally quenched substrates [43], or radioactive substrates [29] may be used as an alternative [44]. The most versatile method that can be applied for such purposes is mass spectrometry linked, if possible, to a preseparation method such as capillary liquid chromatography or capillary-zone electrophoresis [45]. In any case, it should be stressed that the measurement of enzymatic activity strongly depends on the purpose of such analysis. During enzyme isolation, the detection method is used to follow purification scheme. The investigator will obtain an information on enzymatic activity at a given separation step. This value is related only to the total activity applied on the chromatographic column and to the efficiency of the chromatographic process. On the other hand, measurement of the changes in proteolytic activity in, for instance, pathophysiology of pain or drug addiction, should be performed in the environment closest to the natural [46,47]. This means that the process will be traced in the tissue homogenate or in the body fluid that were not handled (e.g., purified) previously. Such a procedure ensures that all factors potentially influencing enzymatic activity are still present in the test tube in the same or similar proportions as under the native conditions. No activators such as EDTA or 2-mercaptoethanol etc. should be added to the reaction mixture because the decreased or increased activity might be the result of the presence of such additives, thus, obscuring the final picture. Similarly, body fluids should be collected without addition of anticoagulants, protease inhibitors, or other agents.

An example of the analysis of enzymatic activity, cleaving nociceptin/orphanin FQ by the peptidase found in the rat spinal cord and in the rat cortical neurons, is presented in Fig. 19.4. The electrospray ionization mass spectrometry was applied in this case [48,49]. The main feature of this technique is a possibility of simultaneous determination of all fragments formed during proteolytic event. By a careful selection of the incubation time and other conditions (reagents concentrations, incubation volume etc.), it was also possible to reveal the primary and secondary cleavage sites. This approach is very rapid and convenient, but final verification of the released fragments should always be performed by the MS/MS experiments [50].

Recently, peptide libraries have been applied for the efficient determination of enzyme specificity. Such strategy, demands application of the very efficient preseparation methods such as capillary HPLC or capillary-zone

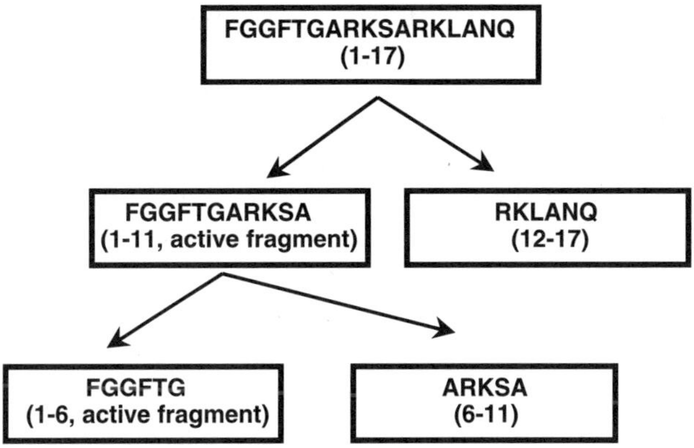

Figure 19.4. Truncation of nociceptin/orphanin FQ by the enzyme found in the rat's spinal cord.

electrophoresis, often linked *online* to the mass spectrometers [51–56]. These topics are described in detail in other chapters of this book.

19.4. CONCLUSION

In conclusion, neuropeptide processing plays an important role as this process may influence the peptide level(s). Therefore, increasing attention is paid to the full recognition of these phenomena. The rapid progress in genome and proteome analysis will soon accelerate an exciting challenge in the discovery of novel molecules involved in the metabolic pathways.

ACKNOWLEDGMENTS

This work was partially supported by the grant from the Polish Committee for Scientific Research (KBN 0584/P05/99/16), the S. Estreicher Foundation, and the A. Krzyżanowski Foundation.

REFERENCES

1. J. Zha, S. Weiler, K. J. Oh, M. C. Wei, and S. J. Korsmeyer, Posttranslational N-myristoylation of BID as a molecular switch for targeting mitochondria and apoptosis, *Science* **290**(5497), 1761–1765 (2000).
2. L. Pellegrini, B. J. Passer, M. Tabaton, J. K. Ganjei, and L. D'Adamio, Alternative, non-secretase processing of Alzheimer's beta-amyloid precursor protein during apoptosis by caspase-6 and -8, *J. Biol. Chem.* **274**(30), 21011–21016 (1999).

3. Y. P. Loh, M. J. Brownstein, and H. Gainer, Proteolysis in neuropeptide processing and other neural functions, *Annu. Rev. Neurosci.* **7**, 189–222 (1984).
4. L. Terenius and F. Nyberg, Neuropeptide-processing, -converting, and -inactivating enzymes in human cerebrospinal fluid, *Int. Rev. Neurobiol.* **30**, 101–121 (1988).
5. Y. P. Loh, H. A. Gritsch, and T. L. Chang, Pro-opiocortin processing in the pituitary: A model for neuropeptide biosynthesis, *Peptides* **3**(3), 397–404 (1982).
6. S. Persson, P. Le Greves, M. Thornwall, U. Eriksson, J. Silberring, and F. Nyberg, Neuropeptide converting and processing enzymes in the spinal cord and cerebrospinal fluid, *Prog. Brain Res.* **104**, 111–130 (1995).
7. N. G. Seidah, R. Day, M. Marcinkiewicz, S, Benjannet, and M. Chretien, Mammalian neural and endocrine pro-protein and pro-hormone convertases belonging to the subtilisin family of serine proteinases, *Enzyme* **45**(5–6), 271–284 (1991).
8. W. J. Van de Ven, A. J. Roebroek, and H. L. Van Duijnhoven, Structure and function of eukaryotic proprotein processing enzymes of the subtilisin family of serine proteases, *Crit. Rev. Oncol.* **4**(2), 115–136 (1993).
9. W. J. Van de Ven, J. W. M. Creemers, and A. J. Roebroek, Furin: The prototype mammalian subtilisin-like proprotein—processing Enzyme, *Enzyme* **45**, 257–270 (1991).
10. J. B. Denault and R. Leduc, Furin/PACE/SPC1: A convertase involved in exocytic and endocytic processing of precursor proteins, *FEBS Lett.* **379**(2), 113–116 (1996).
11. H. Gainer, J. T. Russell, and Y. P. Loh, The enzymology and intracellular organization of peptide precursor processing: The secretory vesicle hypothesis, *Neuroendocrinology* **40**(2), 171–184 (1985).
12. N. G. Seidah, Molecular strategies for identifying processing enzymes, *Meth. Neurosci.* **23** (1995).
13. J. Silberring, T. Sakurada, and F. Nyberg, Dynorphin converting enzyme in the rat spinal cord. Decresed activities during acute phase of adjuvant-induced arthritis, *Life Sci.* **50**, 839 (1992).
14. M. Vlaskovska, I. Nylander, M. Schramm, S. Hahne, L. Kasakov, J. Silberring, and L. Terenius, Opiate modulation of dynorphin conversion in primary cultures of rat cerebral cortex, *Brain Res.* **760**(1–2), 85–93 (1995).
15. M. Vlaskovska, L. Kasakov, P. Suder, J. Silberring, and L. Terenius, Biotransformation of nociceptin/orphanin FQ by enzyme activity from morphine-naive and morphine-treated cell cultures, *Brain Res.* **818**, 212–220 (1999).
16. N. D. Karnik, Y. K. Oza, S. P. Sane, R. Kaushik, A. D. Bhatt, K. P. Chawla, A. B. Vaidya, V. H. Yajnik, and R. C. Khokhani, An open clinical trial of benazepril—a new ACE inhibitor in mild-moderate hypertension, *J. Assoc. Physicians* **46**(3), 283–285 (1998).
17. T. E. Hughes, M. D. Mone, M. E. Russell, S. C. Weldon, and E. B. Villhauer, NVP-DPP728 (1-[[[2-[(5-cyanopyridin-2-yl)amino]ethyl]amino]acetyl]-2-cyano-(S)-pyrrolidine), a slow-binding inhibitor of dipeptidyl peptidase IV, *Biochemistry* **38**(36), 11597–11603 (1999).
18. R. P. Pauly, H. U. Demuth, F. Rosche, J. Schmidt, H. A. White, F. Lynn, C. H. McIntosh, and R. A. Pederson, Improved glucose tolerance in rats treated with the dipeptidyl peptidase IV (CD26) inhibitor Ile-thiazolidide, *Metabolism* **48**(3), 385–389 (1999).

19. B. A. Larder, K. Hertogs, S. Bloor, C. H. van den Eynde, W. DeCian, Y. Wang, W. W. Freimuth, and G. Tarpley, Tipranavir inhibits broadly protease inhibitor-resistant HIV-1 clinical samples, *AIDS* **14**(13), 1943–1948 (2000).
20. A. Rayford, J. S. Rao, and B. W. Festoff, Characterization of the serpin, alpha 1-antichymotrypsin, in normal human cerebrospinal fluid, *J. Neurochem.* **58**(1), 88–94 (1992).
21. W. Bode and R. Huber, Natural protein proteinase inhibitors and their interaction with proteinases, *Eur. J. Biochem.* **204**(2), 433–451 (1992).
22. T. Osterwalder, J. Contartese, E. T. Stoeckli, T. B. Kuhn, and P. Sonderegger, Neuroserpin, an axonally secreted serine protease inhibitor, *EMBO J.* **15**(12), 2944–2953 (1996).
23. G. A. Hastings, T. A. Coleman, C. C. Haudenschild, S. Stefansson, E. P. Smith, R. Barthlow, S. Cherry, M. Sandkvist, and D. A. Lawrence, Neuroserpin, a brain-associated inhibitor of tissue plasminogen activator is localized primarily in neurons. Implications for the regulation of motor learning and neuronal survival, *J. Biol. Chem.* **272**(52), 33062–33067 (1997).
24. T. Osterwalder, P. Cinelli, A. Baici, A. Pennella, S. R. Krueger, S. P. Schrimpf, M. Meins, and P. Sonderegger, The axonally secreted serine proteinase inhibitor, neuroserpin, inhibits plasminogen activators and plasmin but not thrombin, *J. Biol. Chem.* **273**(4), 2312–2321 (1998).
25. R. L. Davis, A. E. Shrimpton, P. D. Holohan, C. Bradshaw, D. Feiglin, G. H. Collins, P. Sonderegger, J. Kinter, L. M. Becker, F. Lacbawan, D. Krasnewich, M. Muenke, D. A. Lawrence, M. S. Yerby, C. M. Shaw, B. Gooptu, P. R. Elliott, J. T. Finch, R. W. Carrell, and D. A. Lomas, Familial dementia caused by polymerization of mutant neuroserpin, *Nature* **401**(6751), 376–379 (1999).
26. D. Vivien and A. Buisson, Serine protease inhibitors: Novel therapeutic targets for stroke? *J. Cerebral Blood Flow Metab.* **20**(5), 755–764 (2000).
27. M. Yepes, M. Sandkvist, M. K. Wong, T. A. Coleman, E. Smith, S. L. Cohan, and D. A. Lawrence, Neuroserpin reduces cerebral infarct volume and protects neurons from ischemia-induced apoptosis, *Blood* **96**(2), 569–576 (2000).
28. S. L. Heald, R. F. Tilton Jr., L. J. Hammond, A. Lee, R. M. Bayney, M. E. Kamarck, T. V. Ramabhadran, R. N. Dreyer, G. Davis, A. Unterbeck et al. Sequential NMR resonance assignment and structure determination of the Kunitz-type inhibitor domain of the Alzheimer's beta-amyloid precursor protein, *Biochemistry* **30**(43), 10467–10478 (1991).
29. J. Silberring, M. E. Castello, and F. Nyberg, Characterization of dynorphin A-converting enzyme in human spinal cord, *J. Biol. Chem.* **267**, 21324–21328 (1992).
30. J. Silberring and F. Nyberg, Isolation and characterization of the enzyme with high specificity towards dynorphin B, *J. Biol. Chem.* **264**, 11082–11086 (1989).
31. P. Le Greves, F. Nyberg, L. Terenius, and T. Hokfelt, Calcitonin gene-related peptide is a potent inhibitor of substance P degradation, *Eur. J. Pharmacol.* **115**(2–3), 309–311 (1985).
32. J. L. Henry, Relation of substance P to pain transmission: Neurophysiological evidence, *Ciba Found. Symp.* **91**, 206–224 (1982).
33. T. W. Smith, The mechanisms of pain and opioid-induced analgesia, *Mol. Aspects Med.* **7**(6), 509–545 (1984).

34. E. A. Young, J. M. Walker, R. Houghten, and H. Akil, The degradation of dynorphin A in brain tissue in vivo and in vitro, *Peptides* **8**(4), 701–707 (1987).
35. M. Vlaskovska, I. Nylander, L. Kasakov, J. Silberring, and L. Terenius, Opiate-sensitive dynorphin conversion in SH-SY5Y human neuroblastoma cells: Evidence for inhibition by dynorphin related peptides, *Analgesia* **1**, 821(1995).
36. K. Tan-No, A. Taira, T. Sakurada, M. Inoue, S. Sakurada, T. Tadano, T. Sato, C. Sakurada, I. Nylander, J. Silberring, L. Terenius, and K. Kisara, Inhibition of dynorphin-converting enzymes prolongs the antinociceptive effect of intrathecally administered dynorphin in the mouse formalin test, *Eur. J. Pharmacol.* **314**(1–2), 61–67 (1996).
37. I. Nylander, M. Vlaskovska, and L. Terenius, The effects of morphine treatment and morphine withdrawal on the dynorphin and enkephalin systems in Sprague-Dawley rats, *Psychopharmacology* (Berlin) **118**(4), 391–400 (1995).
38. B. Y. Karlan, A. S. Clark, and B. A. Littlefield, A highly sensitive chromogenic microtiter plate assay for plasminogen activators which quantitatively discriminates between the urokinase and tissue-type activators, *Biochem. Biophys. Res. Commun.* **142**(1), 147–154 (1987).
39. K. Kiuchi, T. Mutoh, and M. Naoi, A fluorometric microassay procedure for monitoring the enzyme activity of GM1-ganglioside beta-galactosidase by use of high-performance liquid chromatography, *Anal. Biochem.* **140**(1), 146–151 (1984).
40. D. Pintigny and J. Dachary-Prigent, Aprotinin can inhibit the proteolytic activity of thrombin. A fluorescence and an enzymatic study, *Eur. J. Biochem.* **207**(1), 89–95 (1992).
41. K. M. Carvalho, G. Boileau, A. C. Camargo, and L. Juliano, A highly selective assay for neutral endopeptidase based on the cleavage of a fluorogenic substrate related to Leu-enkephalin, *Anal. Biochem.* **237**(2), 167–173 (1996).
42. J. Silberring, P. Brostedt, M. Thörnwall, and F. Nyberg, An approach to study the proteinase specificity by FAB MS combined with photodiode-array UV detection, *J. Chromatogr.* **554**, 83–90 (1991).
43. M. D. Gomes, L. Juliano, E. S. Ferro, R. Matsueda, and A. C. M. Camargo, Dynorphin derived peptides reveal the presence of a critical cysteine for the activity of brain endo-oligopeptidase A, *Biochim. Biophys. Res. Commun*, **2**, 501–507 (1993).
44. L. Devi and A. Goldstein, Dynorphin converting enzyme with unusual specificity from rat brain, *Proc. Acad. Sci.* (SUA) **81**(6), 1892–1896 (1984).
45. S. Choi, Y. S. Lee, D. S. Na, and Y. S. Yoo, Determination of enzymatic activity and properties of secretory phospholipase A2 by capillary electrophoresis, *J. Chromatogr. A* **853**(1–2), 285–293 (1999).
46. F. Nyberg, C. Pernow, U. Moberg, and R. B. Eriksson, High-performance liquid chromatography and diode-array detection for the identification of peptides containing aromatic amino acids in studies of endorphin-degrading activity in human cerebrospinal fluid, *J. Chromatogr* **359**, 541–551 (1986).
47. R. J. Greenstein, L. R. Krakoff, and K. Felton, Activation of the renin system in acute pancreatitis, *Am. J. Med.* **82**(3), 401–404 (1987).
48. P. Suder, J. Kotlinska, M. T. Smoluch, M. Sallberg, and J. Silberring, Metabolic fate of nociceptin/orphanin FQ in the rat spinal cord and biological activity of its released fragment, *Peptides* **20**(2), 239–247 (1999).

49. P. Suder, J. Kotlinska, A. Legowska, M. Smoluch, G. Hohne, J. P. Chervet, K. Rolka, and J. Silberring, Determination of nociceptin-orphanin FQ metabolites by capillary LC-MS, *Brain Res. Brain Res. Protoc.* **6**(1–2), 40–46 (2000).
50. D. M. Desiderio, Mass spectrometric analysis of neuropeptidergic systems in the human pituitary and cerebrospinal fluid, *J. Chromatogr. B, Biomed. Sci. Appl.* **731**(1), 3–22 (1999).
51. J. Antal, G. Pal, B. Asboth, Z. Buzas, A. Patthy, and L. Graf, Specificity assay of serine proteinases by reverse-phase high-performance liquid chromatography analysis of competing oligopeptide substrate library, *Anal. Biochem.* **288**(2), 156–167 (2001).
52. K. Nishikawa, S. Sawasdikosol, A. D. Fruman, J. Lai, Z. Songyang, J. S. Burakoff, B. M. Yaffe, and C. L. Cantley, A peptide library approach identifies a specific inhibitor for the ZAP-70 protein tyrosine kinase, *Mol. Cell* **6**(4), 969–974 (2000).
53. P. D. Edwards, R. C. Mauger, K. M. Cottrell, F. X. Morris, K. K. Pine, M. A. Sylvester, C. W. Scott, and S. T. Furlong, Synthesis and enzymatic evaluation of a P1 arginine aminocoumarin substrate library for trypsin-like serine proteases, *Bioorg. Med. Chem. Lett.* **10**(20), 2291–2294 (2000).
54. M. Rinnova, M. Hradilek, C. Barinka, J. Weber, M. Soucek, J. Vondrasek, T. Klimkait, and J. Konvalinka, A picomolar inhibitor of resistant strains of human immunodeficiency virus protease identified by a combinatorial approach, *Arch. Biochem. Biophys.* **382**(1), 22–30 (2000).
55. F. Viti, F. Nilsson, S. Demartis, A. Huber, and D. Neri, Design and use of phage display libraries for the selection of antibodies and enzymes, *Meth. Enzymol.* **326**, 480–505 (2000).
56. G. Rosse, E. Kueng, M. G. Page, V. Schauer-Vukasinovic, T. Giller, H. W. Lahm, P. Hunziker, and D. Schlatter, Rapid identification of substrates for novel proteases using a combinatorial peptide library, *J. Comb. Chem.* **2**(5), 461–466 (2000).

CHAPTER 20

Measuring Neuropeptides with Single-Cell MALDI-MS

LINGJUN LI and JONATHAN V. SWEEDLER
Department of Chemistry and Beckman Institute, University of Illinois, Urbana, Illinois, USA

CONTENTS

20.1. INTRODUCTION

Chemical analysis at the single-cell level is of significant interest to elucidate complex interactions occurring in the biological microenvironments around particular cells. In biological systems, for example, many chemical events of interest are often restricted to the interior dimensions or exterior surfaces of a single cell, entities of micrometer-range dimensions. However, the small sample volume and minute quantity of material combined with the vast chemical complexity inherent in a cell present unique analytical challenges [1]. For cell-to-cell communication, even closely apposed, morphologically similar neurons often employ different messengers ranging from small molecule neurotransmitters to multiple peptides derived from one or multiple precursors. As a class of important chemical messengers, neuropeptides are perhaps by

Mass Spectrometry and Hyphenated Techniques in Neuropeptide Research, Edited by Jerzy Silberring and Rolf Ekman.
ISBN 0-471-35493-7

far the most diverse group of signaling modulators in the nervous system [2,3], and the functional consequence of its multiplicity and large diversity is under active investigation [4,5]. Clearly, chemical characterization of the full complement of neuropeptide modulators within individual neurons serves as the important first step toward peptide functional study. Conventional methods for cellular peptide identification involve extensive pooling and extraction of peptides from hundreds to millions of cells. This process is usually time-consuming and requires a large number of animals. As an example, 125,000 cockroaches were pooled and extracted to isolate and sequence a pentapeptide proctolin [6]. Another drawback of this pooling and averaging approach is the likelihood of losing chemical information from individual cells. Neurons in the central nervous system may exist in such a way that even adjacent cells use different sets of neurotransmitters; in some cases, only one specific cell contains a particular peptide in the entire animal, so that using the whole brain may result in no more of a rare peptide but a significant increase in the chemical complexity of the sample. Thus, highly sensitive analytic techniques that are capable of single-cell assays are needed. Alternatively, one can turn to a metabolic radiolabeling or immunocytochemistry approach to map neuropeptide synthesis and distribution [7]. While the sensitivity and selectivity of these approaches can be quite high, these methods suffer from several drawbacks, including the difficulty in unambiguous or simultaneous determination of multiple peptides and the need for the presence of a specific amino acid residue in the case of the radiolabeling approach or preselecting the compounds of interest as in immunotechniques.

Developed in 1988 [8,9], matrix-assisted laser desorption/ionization (MALDI) time-of-flight mass spectrometry (TOF MS) has provided an exciting alternative for rapid and accurate screening of peptides in complex biological samples. Among these applications, the most complex samples are perhaps the intact biological cells. MALDI-based single-cell peptide profiling features (1) compatibility with crude mixtures, (2) minimal sample cleanup, (3) no need for tagging or preselecting the peptide of interest, (4) simple instrumentation, (5) high speed and high sensitivity, (6) complementary information to immunochemical methods, and (7) compatibility with posttranslational modifications [10,11]. In contrast to the more commonly used tandem electrospray ionization (ESI) mass spectrometry, no tissue homogenization and separation is involved for single-cell MALDI assay. This, in turn, offers the unique capability of retaining the spatial resolution of peptide distribution within a cell and often provides higher sensitivity due to minimal dilution. Furthermore, the reduced sample handling inherent to the single-cell MALDI approach allows physical separation of the extracellular enzymes from the endogenous peptides, thus providing a more accurate profile of bioactive peptides present in a cell without degradation of the peptides of interest—a problem that often occurs in traditional peptide purification procedures [12]. Figure 20.1 illustrates the concept of "single-cell MALDI" where sample preparation has been limited to cell/tissue dissection and matrix application.

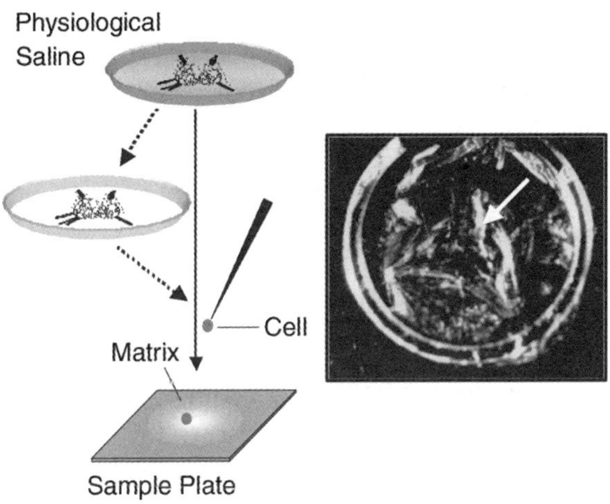

Figure 20.1. The strategy of cell isolation and sample preparation for single-cell MALDI assay. Briefly, following animal dissection, the tissue or ganglion of interest is pinned down in a dissection dish. Individual cells are then isolated either in the presence of physiological saline (freshwater specimens) or within a solution of MALDI matrix (marine specimens). Most cells are subsequently placed onto a MALDI sample plate containing a drop of matrix and allowed to air-dry. Shown on the right is a photomicrograph of a ~30-m diameter B8 neuron isolated from the marine mollusc, *Aplysia californica* (white arrow points to the position of the cell) crystallized with MALDI matrix on the sample probe.

This technique allows identification of neuropeptides on a cell-to-cell basis, providing great potential to explore the heterogeneity of cell signaling, as our understanding of cells to date is based primarily on biochemical measurements of large populations of cells. In addition to direct peptide profiling, which involves measuring molecular weight, several move recent enhancements in MALDI-MS technology, including in situ peptide sequencing as well as the direct spatial mapping of peptides in cells and tissues, are also highlighted.

20.2. SAMPLE PREPARATION

Although MALDI is relatively tolerant of impurities compared to other mass spectrometric techniques, the direct determination of peptides from single cells and biological tissues presents a significant challenge due to the presence of physiological concentrations of nonvolatile salts within the cells being analyzed. The problem is more severe with marine specimens wherein the cation ionic strength in the physiological fluids can be as high as 700 mM [13]. As the ability to obtain quality mass spectra is crucially dependent on crystallization of samples within a suitable matrix, different matrices and sampling protocols have been developed for various sample types [14,15]. Several commonly used matrices for peptide/protein profiling from cellular samples include 2,5-dihydroxybenzoic acid (DHB), α-cyano-4-hydroxycinnamic acid (CHCA), and 3,5-dimethoxy-4-hydroxycinnamic acid [sinapinic acid (SA)]. The sampling protocols reported encompass methods applicable to whole-cell lysates of bacteria [16–20] and human breast cancer cells [21] as well as techniques for blotting tissue onto organic membranes [22,23]. While several applications involve electrospraying the matrix over the tissue section [22,23], the most common sample preparation method used for single-cell analysis is the dried droplet method [11,13,24–29].

As illustrated in Fig. 20.1, following animal dissection and a mild protease treatment to remove excessive connective tissues, the tissue or cellular cluster of interest is pinned to a dissection dish for further cell manipulation and the cells are isolated in the presence of physiological saline or a solution of MALDI matrix. Most cells are subsequently transferred onto a MALDI sample plate containing a fresh drop (0.5–1.0 L) of matrix [13,25] via either micropipetts or fine tungsten needles. For single-cell MALDI studies, an aqueous solution of DHB (10 mg/mL) is the matrix used most commonly because it serves to aid in microdissections by stabilizing cell membranes, deactivating proteolytic enzymes, and reducing the effect of high salt concentrations in order to improve spectral quality [13]. A simple salt-removal procedure depicted in Fig. 20.1 has proved to work effectively with single neurons from marine species such as *Aplysia californica* [27,30]. In this approach, the DHB matrix is used as a cell rinsing reagent in addition to acting as the MALDI matrix. Both extracellular and intracellular salts can be removed in a stepwise fashion by spot-to-spot transfers (approximately 2–4 times) of isolated cell of interest to a fresh drop of matrix on the sample probe. An alternative simplified preparation method involves replacing the physiological saline with the DHB matrix solution during dissection, so that salt removal occurs simultaneously with neuronal isolation.

As a variant of this sample preparation method, Li et al. have demonstrated a novel on-plate microextraction procedure utilizing concentrated DHB matrix (50 mg/mL) dissolved in a mixture of acetone and water (4:1) to allow the detection of low-abundance gene expression products that are not detected using the regular DHB sampling protocol [29]. In this case, the cell is dissected as before and placed on the target, and the concentrated DHB is applied. This technique is analogous to the acidified acetone extraction procedure commonly used for peptide extraction and isolation from biological tissue samples for HPLC separation, except this was performed on plate, at the single-cell level. The use of mixed solvent not only provides more efficient peptide extraction but also creates differential evaporation to allow better incorporation of different peptides into matrix crystals, thus providing enhanced detection of several low-abundance species that are not detected otherwise.

While some of the initial single-cell MALDI studies were performed on individual molluscan neurons that are several hundred micrometers in size, many studies involve individual cells with diameters of 10–30m, including smaller mammalian cells. For example, direct analysis of mammalian cells using MALDI-MS has been demonstrated in which a ~90-fL volume single red blood cell was delivered to a MALDI sample probe utilizing a nanoliter-volume microspot sample delivery system consisting of a miniaturized sample capillary syringe assembly [31,32]. Subsequent MALDI mass spectra allowed identification of hemoglobin variants of sickle cell disease at the attomole level. Another interesting example involves MALDI-MS of mammalian cells cultured directly on a MALDI sample probe [33]. In this report, rat adrenal pheochromocytoma (PC-12) cells were assayed for peptides and proteins before and after stimulation with nerve-growth factor. Results of this study showed that the on-probe-grown cells yielded spectra of greater intensity compared with cells isolated and placed directly on the probe. While no identification of peptide or proteins was achieved with the mass spectra obtained from PC-12 cells, this on-target-grown cell technique has great potential for following biological events and facilitating cell fingerprinting using MALDI-MS.

As yet another elegant example of micromanipulation techniques for handling extremely small-volume samples for mass spectrometric analysis, Rubakhin et al. demonstrated MALDI-MS of single *Aplysia californica* atrial gland vesicles, thus pushing the sample size for MALDI analysis down to submicrometer and attoliter to femtoliter regimes [34]. In this work, a new sampling strategy was developed to allow handling 1–2-m-diameter 0.3–3-fL-volume vesicles with the use of picoliter-volume matrix solutions to obtain peptide profiles from these secretory granules. Within each vesicle, more than 10 peptides from at least four separate genes are detected. This study demonstrates that, with improved sampling methodology, dramatic enhancement in mass sensitivity (low-attomole range) is achievable even with commercial instruments, thus offering great potential for MS analysis of a wide range of small discrete samples, including small single cells, neuronal processes, synaptosomes, and individual organelles.

20.3. EXTRACTING CHEMICAL INFORMATION FROM SINGLE-CELL MALDI SPECTRA

With the appropriate sample preparation, a “library” of MALDI profiles from neurons, nerves, and glands is readily obtained. As illustrated in Fig. 20.2, an extraordinary glimpse into the peptide signal mediators present throughout the nervous system is achieved by constructing a peptide map in a cell- or tissue-specific manner. What information can we extract from such a mass spectrum? As MALDI is a soft ionization method, proteins and peptides usually ionize as intact, singly protonated ions so that the accurate molecular

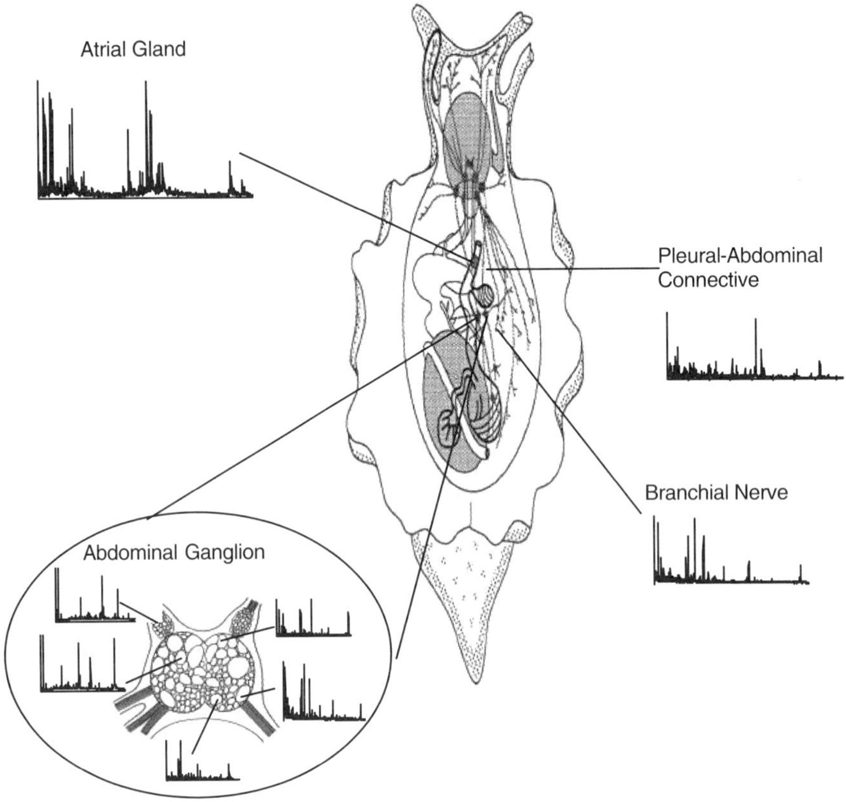

Figure 20.2. Diagram of *Aplysia californica* highlighting the CNS, with representative MALDI-MS spectra of cells, clusters, glands and nerves. (Adapted from Li et al. [11].)

weights of the peptides are easily acquired. Obtaining such molecular weight information is often the first step toward neuropeptide characterization. Neuropeptide genes are often regulated at successive steps, including post-transcriptional alternative mRNA splicing, as well as posttranslational processing of protein precursors and differential packaging of biologically active neuropeptides into secretory vesicles. This cascade of regulatory processes can give rise to differential, cell-specific neuropeptide expression in the nervous system and can also modify, quantitatively and qualitatively, the combinations of neuropeptides encoded by a common single gene, thus modulating the final neuronal physiological effect [35]. Therefore, the combination of experimentally derived molecular weight information in a mass spectrum with known cDNA often allows prediction and verification of prohormone processing and, in many cases, detection of unexpected novel processing products. A large number of single-cell MALDI studies have demonstrated the utility of this strategy [11,26,36–40]. As a few examples, Benjamin and co-workers reported that neuropeptide gene encoding FMRFamide-like peptides in the pulmonate mollusc *Lymnaea* is subjected to alternative splicing that generates cell-specific expression of distinct sets of peptides in the CNS via single-cell MALDI analysis [41,42]. Furthermore, MALDI profiling of individual bag cell neurons from *Aplysia californica* revealed numerous novel processing forms of egg-laying hormone and acidic peptide, resulting from unexpected cleavages at the Leu–Leu residues [27]. Results from such studies thus lead to new insights into unexpected expression and discovery of potential novel processing enzymes.

While some unexpected peptides may result from novel processing of prohormones, others may occur via posttranslational modifications. The accurate molecular weight information provided by MALDI assay, together with a residue-specific enzymatic treatment of isolated cells, often allows determination of the chemical nature and/or the site of modification. Figure 20.3 demonstrates an example of determining intermolecular disulfide bonding pattern of the APGWamide/CP1 gene product in *Aplysia* cerebral cells via on-target tryptic digestion [29]. By comparing the MALDI mass spectra before and after enzymatic treatment, a disulfide bond connecting pattern is readily determined.

In many other cases, single-cell MALDI assay detected unknown molecular masses that do not correspond to any known peptides, and no gene information is available. An example of the characterization of *Aplysia* insulin (AI) prohormone processing is highlighted here to demonstrate the strategy of novel peptide discovery and with the gene information, what types of experiments one can perform to understand the biochemistry of gene expression and peptide modification at the single-cell level. The F cluster of the cerebral ganglion of *Aplysia* is a group of neurons potentially responsible for initiating egg-laying behavior in the animal, and so has been studied with MS. Single-cell MALDI screening for peptides in the cerebral ganglion F-cluster neurons revealed a novel peptide unique to the top-layer cells. The same mass at 1714 Da was also observed in the anterior tentacular and upper labial nerves [43]. The peptide was then targeted for further isolation and characterization due to its axonal transport properties. Using LC purification and pooling the cells from multiple animals, the sequence of the peptides was obtained. Using the sequence of 1714-Da peptide with molecular biological approaches, the cDNA was characterized for the insulin gene. This encodes a 156 amino acid precursor and resulted in the first reported insulin gene in *Aplysia* [44].

The amino acid sequence of the precursor predicts several processing sites of the prohormone convertase/furin family that primarily cleaves at dibasic or RXXR sites. Furthermore, additional posttranslational processing such as amidation, pyroglutamate formation, disulfide bond formation, and additional proteolysis can further modify the peptides to produce the final bioactive products. When combined with the prohormone sequence, MALDI-MS confirms expected cleavages and determines unexpected processing steps. Figure 20.4A summarizes the complex processing of the AI prohormone as

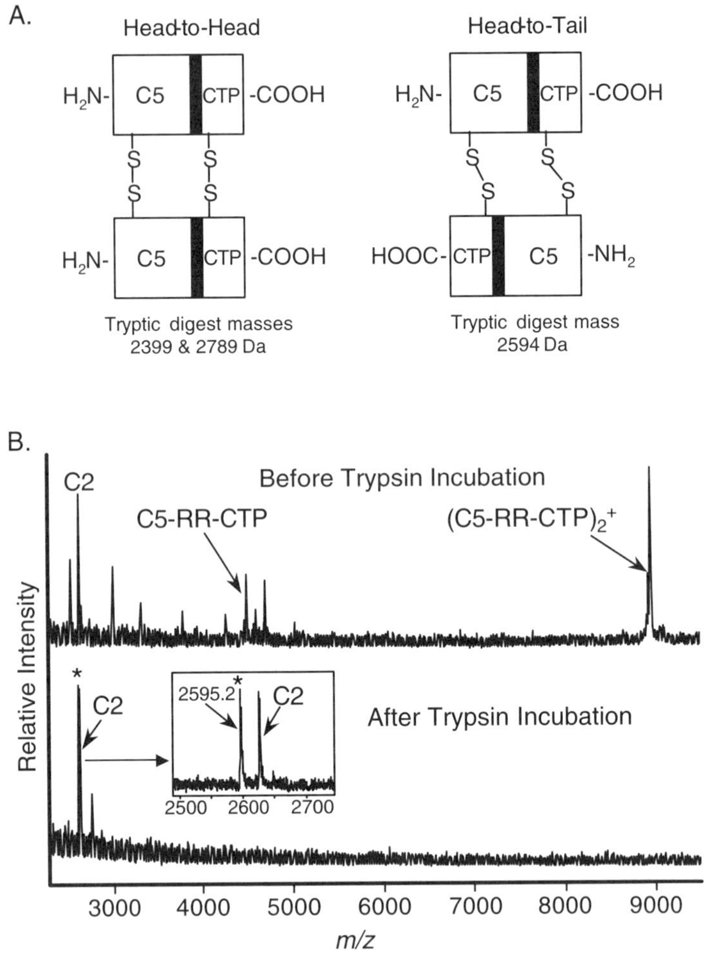

Figure 20.3. Direct characterization of intermolecular disulfide bonds from a cellular sample. (A) $[\text{C5-RR-CTP}]_2$ is a disulfide-bonded dimer encoded by the APGWamide/CP1 gene. Two possible disulfide bonding patterns are shown. If the disulfide bonds are connected via head-to-head orientation, tryptic digest peptide masses will be 2399 and 2789 Da; if head-to-tail, then a peak at 2595 Da will be generated on tryptic digest. [The black vertical bars indicate dibasic (–RR–) residues; S represents cysteine residues presumably involved in the formation of a disulfide bond.] (B) Acquisition of MALDI mass spectra of a single cerebral H-cluster homogenate before (upper trace) and after (lower trace) tryptic digest of the intact dimer reveals that this dimer has a head-to-tail orientation. (Adapted from Li et al. [29].)

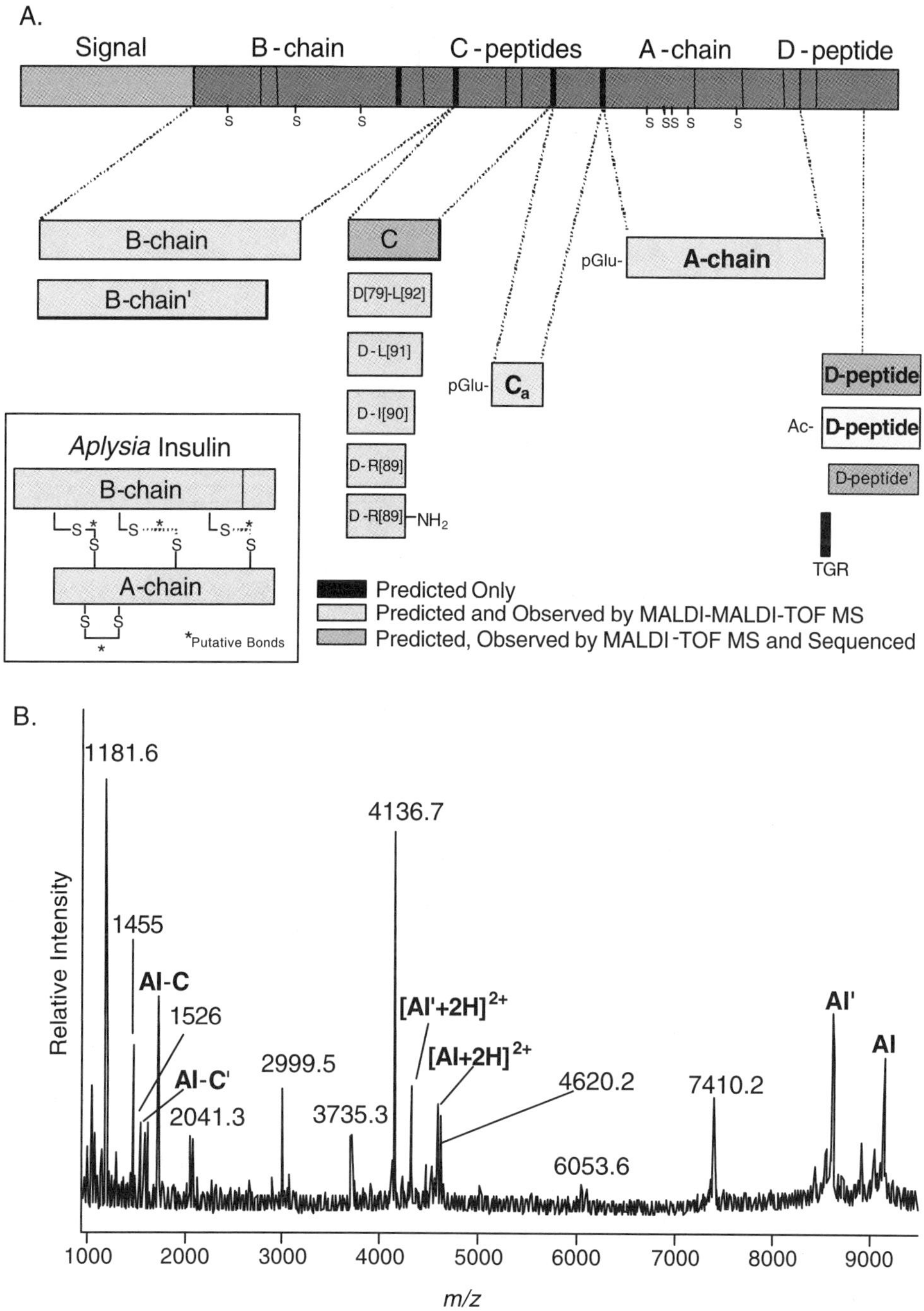
A.
Signal
B-chain
C-peptides
A-chain
D-peptide
B-chain
B-chain'
C
D[79]-L[92]
D-L[91]
D-I[90]
D-R[89]
D-R[89]-NH2
pGlu-
Ca
pGlu-
A-chain
D-peptide
Ac-
D-peptide
D-peptide'
TGR
Aplysia Insulin
B-chain
A-chain
*Putative Bonds
Predicted Only
Predicted and Observed by MALDI-MALDI-TOF MS
Predicted, Observed by MALDI-TOF MS and Sequenced
B.
Relative Intensity
1181.6
1455
AI-C
1526
AI-C'
2041.3
2999.5
3735.3
4136.7
[AI'+2H]2+
[AI+2H]2+
4620.2
6053.6
7410.2
AI'
AI
m/z

Figure 20.4.

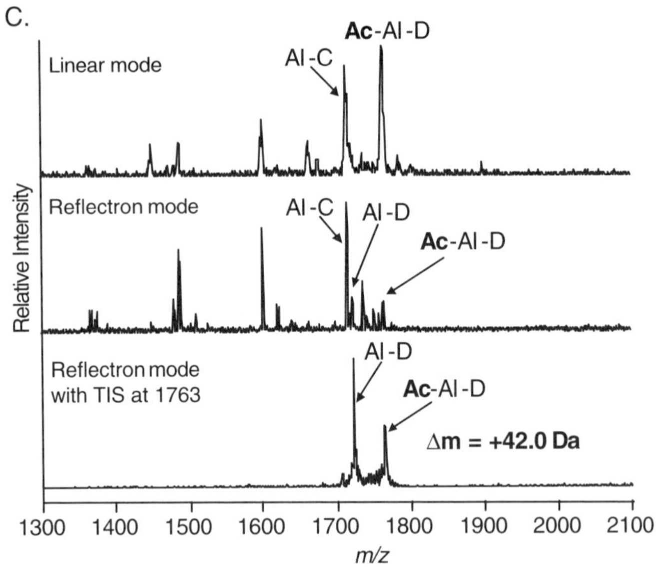

Figure 20.4. (A) Summary of insulin processing based on single-cell MALDI-MS and biochemical characterization. Inset shows *Aplysia* insulin (AI) with A and B chain connected with putative disulfide bonds. (B) Representative mass spectrum of a top-layer neuron from the F cluster in the cerebral ganglion. Peaks generally correspond to $[M + H]^+$, where M is the molecular weight of each peptide. AI, and its shortened form AI′, C_{β}, and its truncated form C_{β}', are labeled. (C) Confirmation of acetylation of the D peptide. In linear mode only the acetylated D peptide is observed. When the spectrum is acquired in reflectron mode, both native and acetylated D peptides are seen. The acetylation is further confirmed using the reflectron mode with a timed ion selector (TIS) set at the molecular weight of the acetylated peptide. (Adapted from Floyd et al. [44].)

revealed by single-cell MALDI assay of individual cerebral F-cluster neurons. The processing and modifications, including disulfide bonding between A and B chains, pGlu modification of the insulin C_{α} peptide, and insulin A chain, are observed with MALDI-MS analysis and have been confirmed with biochemical characterization. Figure 20.4B shows a representative spectrum from a single top-layer neuron isolated from *Aplysia* cerebral F cluster, with AI and insulin C peptide and their *C*-terminal truncated forms labeled. The AI appears to be unique among the insulin superfamily as it is the only insulin prohormone that contains a peptide after the A chain in the prohormone. As revealed in the AI prohormone schematic, the expression of the D peptide predicted by the DNA sequence is confirmed with MALDI-MS and biochemical sequencing. Interestingly, the D peptide appears to be acetylated as a peak exactly 42 Da higher was observed in both LC fractions and single cells. The confirmation of acetylation of the D peptide in single cells was achieved by altering MS configuration of the instrument. As detailed in Fig. 20.4C, single F cells were assayed with MALDI in both linear and reflectron modes. During the MALDI ionization process, sufficient energy is deposited into the peptide so that a fraction of the labile acetyl groups are removed in the flight tube. Thus, both versions of D peptides are detected in reflectron mode, while only acetylated D (Ac-D) peptide was observed in linear mode because the metastable fragment ions travel at the same speed as do their precursor ions in this instrument configuration. The observation was further substantiated by setting a mass selective ion gate at the acetylated mass in reflectron mode, wherein only the Ac-D peptide is selectively passed through the flight tube. The detection of both peptides (and no others) strongly suggests that the peptide is predominantly in the acetylated form and falls apart in the flight tube. Such characterization represents the first known acetylated neuropeptide in *Aplysia*, and also demonstrates the exciting capability of MALDI-MS for probing posttranslational modifications at the single-cell level [44].

While a wealth of chemical information can be extracted from single-cell MALDI assay with the available gene sequence, accurate molecular weight information alone is seldom sufficient to assign a peak to a particular peptide when prohormone sequence is not available. Consequently, there is substantial interest in developing the structural elucidation capabilities of MALDI-TOFMS. As one possible MALDI-based fragmentation technique, postsource-decay (PSD) analysis takes advantages of the metastable decomposition of the peptide precursor ions and mass analysis of the fragment ions using reflector TOF instruments to generate sequence information. This strategy has the advantage of simplicity and does not require additional enzymatic treatment or separation steps [45–47]. For example, direct PSD analysis from nerve tissues of *Lymnaea* has been reported [48,49]. In a move recent study, a microscale extraction was performed on a group of 100 cells from frog *Xenopus laevis* to allow PSD analysis of cell lysates without prior fractionation [50]. Furthermore, Benjamin and co-workers demonstrated the feasibility of using PSD to confirm a peptide sequence from a single neuron of *Lymnaea*, where partial sequence-specific fragment ions were obtained to allow the comparison with predicted fragmentation pattern [51]. Despite the numerous demonstrations of PSD analysis of tissue samples, one limitation that restricts single-cell PSD analysis is the DHB matrix employed. DHB is considered as "cold" Acold@ matrix for the PSD process in that it produces relatively stable ions, resulting in less fragmentation [52]. For single-cell analysis, however, we need to rinse the cells with DHB to allow peptide assay from high-salt-concentration marine specimens [13]. In contrast, CHCA is a "hotter" matrix that favors extensive fragmentation of peptides of interest [52]. In addition, CHCA also provides more homogeneous crystallization that is often advantageous for PSD analysis, as many fragment spectra are needed to obtain a composite PSD spectrum. However, it is relatively difficult to employ CHCA for single-cell dissection and isolation; the organic solvent required by the CHCA matrix causes severe cell lysis during manipulation

[27]. To overcome the problems associated with both matrices, Li et al. developed a new sampling protocol using the combination of both DHB and CHCA MALDI matrices. Briefly, the cellular cluster is placed in aqueous DHB matrix solution so that the salt removal occurs at the same time of cellular microdissection. The isolated single neurons are then transferred onto individual MALDI spots with CHCA matrix applied "on target." This two-matrix approach allows successful isolation of single neurons and also induces significant fragmentation for PSD analysis [53]. Figure 20.5 demonstrated the first complete de novo PSD peptide sequencing obtained from a single neuron prepared using this dual-matrix approach.

In addition to improving sample preparation and matrix selection to allow better PSD sequencing analysis, substantial effort is also directed toward modifications of MALDI instrumentation for higher-efficiency single-cell peptide sequencing. For example, a MALDI tandem double-focusing magnetic orthogonal acceleration time-of-flight mass spectrometer has been used to structurally characterize a tetrapeptide APGWamide from a single cerebral neuron [54], as well as a set of small cardioactive peptides from functionally related single neurons VD1 and RPD2 from *Lymnaea* [55]. The apparatus used in the above mentioned studies employing a MALDI source for high ionization efficiency, a trisector mass spectrometer as the first stage for high parent ion selectivity, a high-energy collision chamber for enhanced fragmentation efficiency, and finally a TOF mass spectrometer for highly sensitive detection of the fragment ions. These and even newer instruments such as MALDI–quadrupole TOF (MALDI-QqTOF) mass spectrometers will substantially improve sequencing capabilities of MALDI in the near future.

20.4. COMBINING SINGLE-CELL MALDI-MS WITH MICROSCALE SEPARATIONS

While single-cell MALDI features direct sampling of neurons without a separation, several advantages merit the incorporation of separation with single-cell MALDI assay in particular situations. Additional gains from coupling a microscale separation technique prior to MALDI assay include simplifying the cellular matrix such as removing the lipid interference, alleviating analyte suppression problem caused by vast chemical complexity, and eliminating the salt effects that often complicate direct cellular MALDI assay. The hyphenation of mass spectrometry with the separation approach thus generates an information-rich, two-dimensional analytic method. For example, while MS provides unparalleled structural information on low-level analytes, the separation produces approach additional elution time information, so as to enhance the confidence of peptide identification. Among various types of separation techniques, two microscale separation methods—capillary liquid chromatography (cLC) and capillary electrophoresis (CE)—have proved to be well suited for single-cell analysis. The typical inner diameters of the

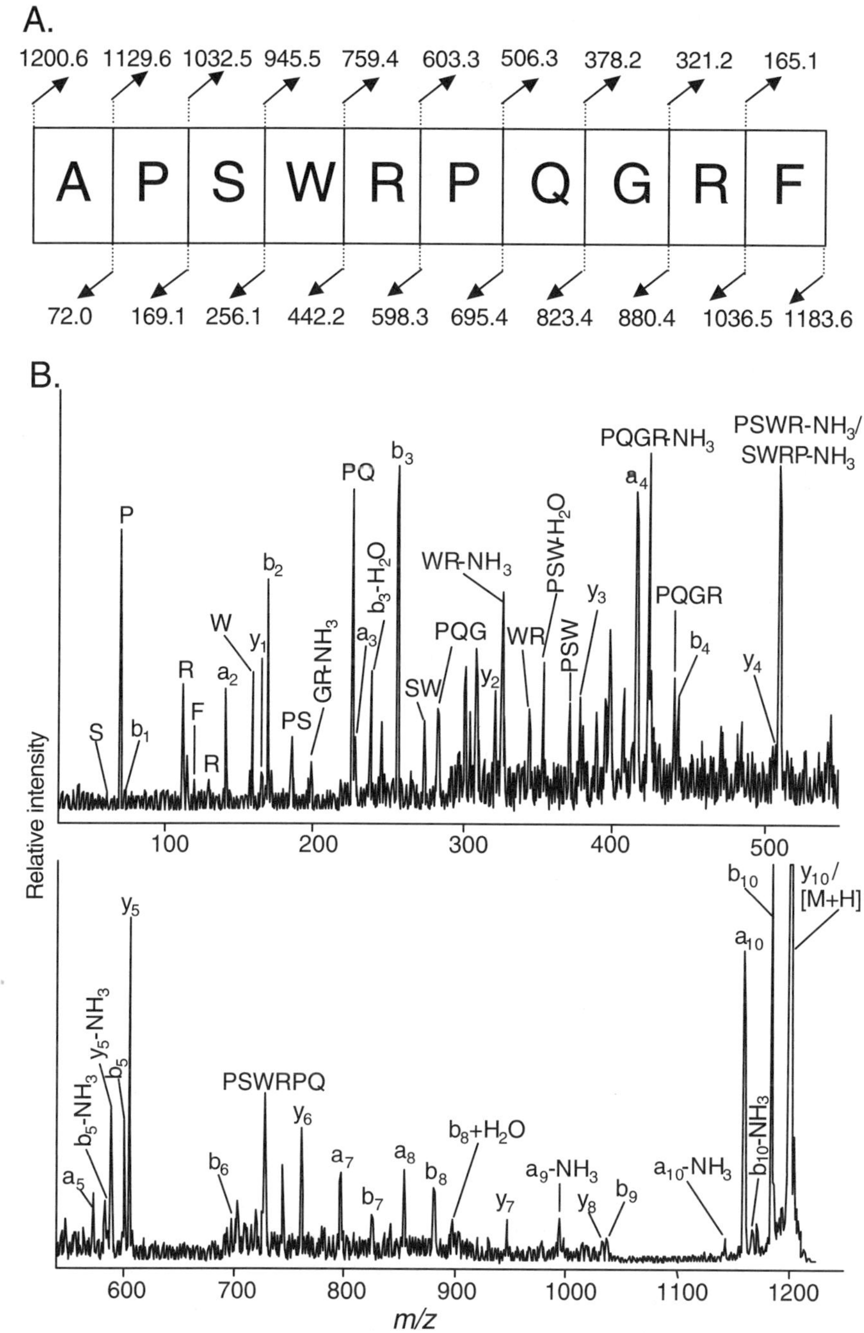

Figure 20.5. Direct sequencing of a peptide at *m/z* 1200.6 with MALDI-PSD of a single neuron from *Aplysia californica*. (A) Sequence analysis of precursor ion at 1200.6 u. The peptide contains 10 residues, with *N* terminal on the left, and *C* terminal on the right. Indicated by arrows are the observed *b*-type (bottom) and *y*-type (top) ion pairs. All the observed ions are shown in bold. Thus, as indicated, all the b-ions and all the *y* ions except y_9 (1129.6 u) are detected. (B) MALDI-PSD fragment ion spectrum of peptide precursor ion 1200.6 u. Mass range of 50–550 is shown in the upper trace, whereas mass range of 550–1250 is shown in the lower trace. *N*-Terminal ion series such as *a*/*b* ions and their loss of neutrals, *C*-terminal ion series such as *y* ions, and internal fragment ions as well as immonium ions are labeled in the spectrum. (Adapted from Li et al. [53].)

separation capillaries are on the scale of micrometers with similar dimensions of single cells, and injection volumes of picoliters to nanoliters correspond to the volume of individual cells typically on the order of femtoliter to low nanoliter. The small sample volume offered by these capillary-scale separation techniques also provides the advantages of minimal dilution and sample loss. The examples presented here therefore highlight the hyphenation of these two separation modes with MALDI-MS analysis of single cells.

One notable study by research groups led by Jorgenson, Hillenkamp, and Geraerts combined the resolving power of capillary LC with the structural information of MALDI-MS to characterize neuropeptides from a single identified neuron of the freshwater snail *Lymnaea stagnalis* [56]. In this study, the nanoliter volume LC eluent from a 50-m-i.d. × 30-cm LC capillary column packed with 5 μm octadecylsilane chemically modified silica particles mixed offline with nanoliter volume matrix solution delivered via a second capillary was deposited onto the plate every 60 s. Identification of peptides was based on LC elution order and the mass information. In addition, PSD was used to generate characteristic fragments to confirm peptide assignment. This appears to be the first example demonstrating that peptides contained in a single neuron can be fractionated by microcolumn LC and detected by MALDI-MS. Several benefits from microcolumn LC fractionation prior to MALDI assay include generation of "cleaner" spectra to facilitate PSD analysis for novel analytes or further confirmation of peptide species and the higher confidence for peptide assignment based on combined information from elution positions from LC columns and the measured molecular weight. This application also opens up the field for downscaling of peptide chemical characterization to the single-cell level.

Another approach that is attractive for single-cell assays involves coupling CE with MALDI-MS through postcapillary deposition techniques. Although a variety of coupling schemes have been demonstrated [57–62], the application of this methodology for single-cell assay was reported only in 2000 [63]. In this work, Page et al. described an offline CE-MALDI coupling scheme that integrates nanovial sampling system, CE separation and post-capillary deposition apparatus and MALDI-MS detection to achieve repeated assays on the CE effluent from single-neuron lysates. Briefly, a dissected neuron was placed in an individual CE nanovial (200 nL), followed by addition of mixed solvent of acetonitrile and water. After evaporation of the acetonitrile, most of the nanovial's contents, about 40 nL, was introduced into the capillary via hydrodynamic injection. Following CE separation, consecutive fractions were deposited onto a *xyz*-position-controlled MALDI sample plate at 30-s intervals for further analysis. The deposition of the CE effluent was achieved by joining the outlet of separation capillary with a second capillary delivering MALDI matrix via a syringe pump to allow a deposition of fractions along with the matrix. MALDI is essentially a nondestructive assay, in that most of the sample remained on the target after analysis. The authors have demonstrated several unique capabilities offered by offline coupling approach, such as performing multiple assays ranging from PSD sequencing analysis to on-target enzymatic digestion on the same sample spot over an extended period of time, thus maximizing the amount of information one can obtain from a single neuron.

While only several examples of MS coupled with separation for single-cell analysis are presented here, it is likely that small-scale hyphenation approaches will become more common in the future because of the diversity of information provided and the ability to directly assay complex neurologic tissues. The disadvantage of the separation–MALDI combination is that the femtoliter to picoliter volume cell with a high local concentration of peptide is diluted to the typically nanoliter or larger volume bands found in capillary separations. Thus, a peptide that may be barely detectable in single-cell MALDI may become undetectable with CE/MS because of this dilution. One must balance the advantages of the separation with the dilution disadvantage before deciding whether direct MALDI or capillary LC- or CE-MALDI is optimum for a particular application.

20.5. SEMIQUANTITATIVE ANALYSIS OF PEPTIDES IN CELLS AND TISSUES

Although MALDI-MS provides unprecedented chemical information from cells, its "Achilles heel," compared with other MS approaches, is quantitation. Several key limitations for quantitative detection with MALDI include poor shot-to-shot reproducibility, sample heterogeneity, signal suppression, and nonlinear detector response [64,65]. Several strategies involve the use of internal standards and sample preparation techniques that yield more homogenous samples [66,67]. However, these methods do not appear applicable to the detection of peptides at the single-cell level, and quantitation becomes even more problematic when working with cellular samples that contain literally thousands of components in concentrations ranging from millimolar to undetectable. The peak intensity of a peptide is dependent on the nature of the peptide, the matrix and sample preparation protocols used, as well as the presence or absence of contaminants such as salts, detergents, and denaturants. Nevertheless, it appears feasible to use MALDI-MS to study the regulation of a peptide in response to specific physiological conditions in a semiquantitative manner. For example, van Strien and co-workers showed that the peptide profiles of melanotrope cells from frogs adapted to different environmental light conditions are quantitatively distinct, with black-adapted *Xenopus* containing more (larger) POMC processing intermediates than cells of white-adapted *Xenopus* [68]. Additionally, Jiménez and co-workers developed a differential peptide display method to detect peptides that show semiquantitative changes in the neurointermediate lobe (NIL) of individual rats subjected to salt loading [69]. Garden et al. also demonstrated a spectral normalization approach that allowed semi-quantitative measurement from single bag cell neurons of

Aplysia by ratioing the peak intensities of all the peaks in the mass spectra with the peak intensity of a known peptide also found within the cell [27]. More recently, as a variant of the internal standard approach, Ekman and coworkers demonstrated quantification of neurotensin in human brain tissue using a stable-isotope labeling approach [64], yielding an error of less than 5%. As the labeled standard is essentially equivalent to the target molecule, the likelihood of discrimination effects between analyte and standard is minimized. With a low-femtomole detection limit, this technique shows promise for quantitative analysis at the single-cell level.

20.6. MALDI AS A COMPLEMENTARY TOOL TO IMMUNOCYTOCHEMISTRY

In addition to the chemical identity of the peptides present, the spatial distribution of the peptides within a neuronal network is important. Immunologic tools allow peptides to be identified on the basis of antibody recognition and not sequence, and immunocytochemistry can be combined with microscopy to provide high spatial details. The combination of MALDI-MS and immunologic tools, on the same sample, provides a valuable approach to more completely characterizing polypeptide expression and processing. Such a dual characterization approach has been performed in several examples ranging from the pheromone-biosynthesis-activating neuropeptide (PBAN) family in insect *Helicoverpa* [70] to *Aplysia mytilus* inhibitory peptide (MIP)-related peptides in the mollusc *Aplysia californica* [71]. One of the most impressive dual characterization examples is the study of the cellular expression of two hyperglycemic hormone (cHH) precursors in neuroseceretory cells from the crayfish *Orconectes limosus*, where immunocytochemistry identification and direct MALDI-MS peptide profiling was performed on the same single cHH-producing cell [72]. In some other cases, the combination of MALDI-MS and immunocytochemistry allows more complete and accurate peptide profiles to be obtained in a single neuron wherein some low-abundance peptide species may not be detected using one approach alone. As one example, immunocytochemistry showed the presence of APGWamide/CP1 immunoreactivity in *Aplysia californica* buccal neuron B1 and B2 cells. These two cells are known to synthesize a large amount of small cardioactive peptides (SCPs). Single-cell MALDI analysis of B1 and B2 cells always revealed the intense signals from SCPs without detection of other peptide species. The intriguing results from immunocytochemistry triggered the development of an improved MALDI-MS sample preparation method (as described previously) to allow the detection of trace-level APGWamide/CP1 gene products that were obscured originally because of the analyte suppression problem [29]. Such a study highlights the importance of using multiple complementary assays for complex samples such as biological cells.

While immunocytochemistry offers unique capability to map peptide distribution with spatial specificity, this technique is problematic in the study of individual peptides within complex peptide families wherein most peptides have similar structures and sequences. An area of increasing importance is the ability to acquire spatial information with the chemical-information-rich mass spectrometric technique. The MALDI imaging highlighted in Section 20.7 serves as an analogous technique to immunocytochemistry to map peptide species with high spatial resolution and chemical specificity in tissue samples.

20.7. MALDI-MS IMAGING

The imaging capability of MALDI-MS represents a significant advance in the field of mass spectrometry. In such imaging experiments, the mass spectra are acquired by rastering the laser across the sample in a stepwise manner to generate ion images of one or more *m/z* values, which can then be transformed and displayed as a spatial map of selected mass(es) of interest [22,73,74]. One of the key limiting factors for MALDI imaging of tissue samples is the need to apply a MALDI matrix without distorting the sample. Caprioli and co-workers described a procedure involving drying the tissue section followed by electrospray addition of matrix [22]. Using this protocol, mapping of insulin contained in an islet in a section of rat pancreas, hormone peptides in a small area of a section of rat pituitary, and a small protein bound to the membrane of human mucosa cells were demonstrated by creating MALDI molecular ion images from these tissue samples. Sweedler and co-workers have imaged individual neurons and even neuronal processes for peptides [75]. More recently the utility of the MALDI-based imaging technique as a diagnostic tool for examining the relationships between sample morphology, preparation strategy, and spectral quality was evaluated [74]. The ability to image lower-molecular-weight ions in a paramecium with secondary-ion mass spectrometry (SIMS) has also been demonstrated [76] and indicates the potential of SIMS to provide high spatial resolution MS images.

We expect MS microscopy to be an exciting new capability that will become a significant addition to the available microscopic chemical characterization methods. Future efforts to improve this methodology include obtaining higher-resolution images by decreased laser spot size, developing new matrix application methods to prevent analyte from spreading, as well as novel approaches for mass spectral acquisition and data display to accommodate the large amount of data generated in imaging experiments.

20.8. CONCLUSIONS AND PERSPECTIVES

In summary, MALDI mass spectrometry is a powerful tool for direct peptide profiling of single cells and complex biological tissues with high sensitivity and speed, which is important for the analysis of cell-specific expression and processing of neuropeptides. With the development of MALDI-based sequencing techniques, single-cell MALDI approaches have grown from simple mass profiling previously identified peptides to elegant experiments that enable the discovery and structural characterization of novel peptides. With improved sample handling and matrix application strategies, the single-cell MALDI technique should provide an expanded view of the diversity of peptide signal mediators at the subcellular level.

We expect this trend of rapid technological and methodological advances to continue and greatly increase the scope of MALDI for measuring peptides present in biological tissues, cells, and even subcellular organelles. As a general theme highlighted in this book, single-cell mass spectrometry and hyphenated capillary-scale separation techniques, in combination with multidisciplinary approaches involving single-cell molecular biology, immunocytochemistry, and conventional electrophysiology, will provide multidimensional information from complex ensembles of neurons. The development of such approaches will promise a better understanding of simultaneous interactions of multiple neurotransmitters and neuromodulators throughout the neuron and should greatly expand our knowledge of peptidergic communication at the cellular and subcellular levels.

ACKNOWLEDGMENTS

The authors gratefully acknowledge the support of the National Institutes of Health through Grant NS 31609 and the National Science Foundation with CHE98-77071.

REFERENCES

1. R. Dahlgren, J. S. Page, and J. V. Sweedler, Assaying neurotransmitters in and around single neurons with information-rich detectors, *Anal. Chim. Acta.* **400**, 13–26 (1999).
2. I. Kupfermann, Functional studies of cotransmission, *Physiol. Rev.* **71**, 683–732 (1991).
3. T. Hokfelt, C. Broberger, Z. Q. Xu, V. Sergeyev, R. Ubink, and M. Diez, Neuropeptides—an overview, *Neuropharmacology* **39**, 1337–1356 (2000).
4. D. R. N≅ssel, Peptidergic neurohormonal control systems in invertebrates, *Curr. Opin. Neurobiol.* **6**, 842–850 (1996).
5. V. Brezina and K. R. Weiss, Analyzing the functional consequences of transmitter complexity, *Trends Neurosci.* **20**, 538–543 (1997).
6. A. N. Starratt and B. E. Brown, Structure of the pentapeptide proctolin, a proposed neurotransmitter in insects, *Life Sci.* **17**, 1253–1256 (1975).

7. M. O'Shea and M. Schaffer, Neuropeptide function: the invertebrate contribution, *Annu. Rev. Neurosci.* **8**, 171–198 (1985).
8. M. Karas and F. Hillenkamp, Laser desorption ionization of proteins with molecular masses exceeding 10,000 daltons, *Anal. Chem.* **60**, 2299–2301 (1988).
9. F. Hillenkamp, M. Karas, R. C. Beavis, and B. T. Chait, Matrix-assisted laser desorption/ionization mass spectrometry of biopolymers, *Anal. Chem.* **63**, 1193A–1203A (1991).
10. R. Kaufmann, Matrix-assisted laser desorption ionization (MALDI) mass spectrometry—a novel analytical tool in molecular biology and biotechnology, *J. Biotechnol.* **41**, 155–175 (1995).
11. L. Li, R. W. Garden, and J. V. Sweedler, Single-cell MALDI: A new tool for direct peptide profiling, *Trends Biotechnol.* **18**, 151–160 (2000).
12. D. T. Chiu and R. N. Zare, Assaying for peptides in individual *Aplysia* neurons with mass spectrometry, *Proc. Natl. Acad. Sci.* (USA) **95**, 3338–3340 (1998).
13. R. W. Garden, L. L. Moroz, T. P. Moroz, S. A. Shippy, and J. V. Sweedler, Excess salt removal with matrix rinsing: direct peptide profiling of neurons from marine invertebrates using matrix-assisted laser desorption/ionization time-of-flight mass spectrometry, *J. Mass Spectrom.* **31**, 1126–1130 (1996).
14. J. J. Dalluge, Mass spectrometry for direct determination of proteins in cells: Applications in biotechnology and microbiology, *Fresenius J. Anal. Chem.* **366**, 701–711 (2000).
15. M. Kussmann and P. Roepstorff, Characterisation of the covalent structure of proteins from biological material by MALDI mass spectrometry—possibilities and limitations, *Spectroscopy* **14**, 1–27 (1998).
16. E. C. Lynn, M. C. Chung, W. C. Tsai, and C. C. Han, Identification of *Enterobacteriaceae* bacteria by direct matrix-assisted laser desorption/ionization mass spectrometric analysis of whole cells, *Rapid Commun. Mass Spectrom.* **13**, 2022–2027 (1999).
17. R. D. Holland, J. G. Wilkes, F. Rafii, J. B. Sutherland, C. C. Persons, K. J. Voorhees, and J. O. Lay, Jr., Rapid identification of intact whole bacteria based on spectral patterns using matrix-assisted laser desorption/ionization with time-of-flight mass spectrometry, *Rapid Commun. Mass Spectrom.* **10**, 1227–1232 (1996).
18. T. Krishnamurthy and P. L. Ross, Rapid identification of bacteria by direct matrix-assisted laser desorption/ionization mass spectrometric analysis of whole cells, *Rapid Commun. Mass Spectrom.* **10**, 1992–1996 (1996).
19. R. J. Arnold and J. P. Reilley, Fingerprint matching of *E. coli* strains with matrix-assisted laser desorption/ionization time-of-flight mass spectrometry of whole cells using a modified correlation approach, *Rapid Commun. Mass Spectrom.* **12**, 630–636 (1998).
20. A. M. Haag, S. N. Taylor, K. H. Johnston, and R. B. Cole, Rapid identification and speciation of Haemophilus bacteria by matrix-assisted laser desorption/ionization time-of-flight mass spectrometry, *J. Mass Spectrom.* **33**, 750–756 (1998).
21. B. E. Chong, D. M. Lubman, F. R. Miller, and A. J. Rosenspire, Rapid screening of protein profiles of human breast cancer cell lines using non-porous reversed-phase high performance liquid chromatography separation with matrix-assisted laser desorption/ionization time-of-flight mass spectral analysis, *Rapid Commun. Mass Spectrom.* **13**, 1808–1812 (1999).

22. R. M. Caprioli, T. B. Farmer, and J. Gile, Molecular imaging of biological samples: Localization of peptides and proteins using MALDI-TOF MS, *Anal. Chem.* **69**, 4751–4760 (1997).

23. P. Chaurand, F. Lützenkirchen, and B. Spengler, Direct profiling of proteins in biological tissue sections by MALDI mass spectrometry, *Anal. Chem.* **71**, 5263–5270 (1999).

24. P. A. van Veelen, C. R. Jiménez, K. W. Li, W. C. Wildering, W. P. M. Geraerts, U. R. Tjaden, and J. van der Greef, Direct peptide profiling of single neurons by matrix-assisted laser desorption-ionization mass spectrometry, *Org. Mass Spectrom.* **28**, 1542–1546 (1993).

25. C. R. Jiménez, P. A. van Veelen, K. W. Li, W. C. Wildering, W. P. M. Geraerts, U. R. Tjaden, and J. van der Greef, Neuropeptide expression and processing as revealed by direct matrix-assisted laser desorption ionization mass spectrometry of single neurons, *J. Neurochem.* **62**, 404–409 (1994).

26. K. W. Li, C. R. Jiménez, P. A. van Veelen, and W. P. M. Geraerts, Processing and targeting of a molluscan egg-laying peptide prohormone as revealed by mass spectrometric peptide fingerprinting and peptide sequencing. *Endocrinology* **134**, 1812–1819 (1994).

27. R. W. Garden, S. A. Shippy, L. Li, T. P. Moroz, and J. V. Sweedler, Proteolytic processing of the *Aplysia* egg-laying hormone prohormone, *Proc. Natl. Acad. Sci.* (USA) **95**, 3972–3977 (1998).

28. L. Li, T. P. Moroz, R. W. Garden, P. D. Floyd, K. R. Weiss, and J. V. Sweedler, Mass spectrometric survey of interganglionically transported peptides in *Aplysia*, *Peptides* **19**, 1425–1433 (1998).

29. L. Li, E. V. Romanova, S. S. Rubakhin, V. Alexeeva, K. R. Weiss, F. S. Vilim, and J. V. Sweedler, Peptide profiling of cells with multiple gene products: Combining immunochemistry and MALDI-mass spectrometry with on-plate microextraction, *Anal. Chem.* **72**, 3867–3874 (2000).

30. L. Li, R. W. Garden, P. D. Floyd, T. P. Moroz, J. M. Gleeson, J. V. Sweedler, L. Pasa-Tolic, and R. D. Smith, Egg-laying hormone peptides in the Aplysiidae family, *J. Exp. Biol.* **202**, 2961–2973 (1999).

31. L. Li, R. E. Golding, and R. M. Whittal, Analysis of single mammalian cell lysates by mass spectrometry, *J. Am. Chem. Soc.* **118**, 11662–11663 (1996).

32. R. M. Whittal, B. O. Keller, and L. Li, Nanoliter chemistry combined with mass spectrometry for peptide mapping of proteins from single mammalian cell lysates, *Anal. Chem.* **70**, 5344–5347 (1998).

33. J. Berquist, Cells on the target: Matrix-assisted laser-desorption/ionization time-of-flight mass-spectrometric analysis of mammalian cells grown on the target, *Chromatographia*(Suppl. I) **49**, S41–S48 (1999).

34. S. S. Rubakhin, R. W. Garden, R. R. Fuller, and J. V. Sweedler, Measuring the peptides in individual organelles with mass spectrometry, *Nature Biotechnol.* **18**, 172–175 (2000).

35. W. S. Sossin, M. J. Fisher, and R. H. Scheller, Cellular and molecular biology of neuropeptide processing and packaging, *Neuron* **2**, 1407–1417 (1989).

36. E. Kellett, S. J. Perry, N. Santama, B. M. Worster, P. R. Benjamin, and J. F. Burke, Myobodulin gene of *Lymnaea*: Structure, expression, and analysis of neuropeptides. *J. Neurosci.* **16**, 4949–4957 (1996).

37. M. J. Greenberg, K. E. Doble, W. Lesser, T. D. Lee, N. A. Pennell, C. G. Morgan, and D. A. Price, Characterization of myomodulin-related peptides from the pulmonate snail *Helix aspersa*, *Peptides* **18**, 1099–1106 (1997).

38. N. D. de With, K. W. Li, C. R. Jiménez, N. Vonk, K. Dreisewerd, F. Hillenkamp, M. Karas, and W. P. M. Geraerts, Intracellular degradation of C-peptides in molluscan neurons producing insulin-related hormones, *Peptides* **18**, 765–770 (1997).

39. C. R. Jiménez and A. L. Burlingame, Ultramicroanalysis of peptide profiles in biological samples using MALDI mass spectrometry, *Exp. Nephrol.* **6**, 421–428 (1998).

40. R. M. Vazquez-Martinez, M. M. Malagon, F. J. C. van Strien, S. Jesperson, J. van der Greef, E. W. Roubos, and F. Garcia-Navarro, Analysis by mass spectrometry of POMC-derived peptides in amphibian melanotrope subpopulations, *Life Sci.* **64**, 923–930 (1999).

41. N. Santama, K. W. Li, W. P. M. Geraerts, P. R. Benjamin, and J. F. Burke, Post-translational processing of the alternative neuropeptide precursor encoded by the FMRFamide in the pulmonate snail *Lymnaea stagnalis*, *Eur. J. Neurosci.* **8**, 968–977 (1996).

42. B. M. Worster, M. S. Yeoman, and P. R. Benjamin, Matrix-assisted laser desorption/ionization time of flight mass spectrometric analysis of the pattern of peptide expression in single neurons resulting from alternative mRNA splicing of FMRFamide gene, *Eur. J. Neurosci.* **10**, 3489–3507 (1998).

43. S. S. Rubakhin, L. Li, T. P. Moroz, and J. V. Sweedler, Characterization of the *Aplysia californica* cerebral ganglion F cluster, *J. Neurophysiol.* **81**, 1251–1260 (1999).

44. P. D. Floyd, L. Li, S. S. Rubakhin, J. V. Sweedler, C. C. Horn, I. Kupfermann, V. Y. Alexeeva, T. A. Ellis, N. C. Dembrow, K. R. Weiss, and F. S. Vilim, *Aplysia californica* insulin: Prohormone processing, distribution, and relation to metabolism, *J. Neurosci.* **19**, 7732–7741 (1999).

45. B. Spengler, D. Kirsch, R. Kaufmann, and E. Jaeger, Peptide sequencing by matrix-assisted laser-desorption mass spectrometry, *Rapid Commun. Mass Spectrom.* **6**, 105–108 (1992).

46. R. Kaufmann, B. Spengler, and F. Lützenkirchen, Mass spectrometric sequencing of linear peptides by product-ion analysis in a reflectron time-of-flight mass spectrometer using matrix-assisted laser desorption ionization, *Rapid Commun. Mass Spectrom.* **7**, 902–910 (1993).

47. B. Spengler, Post-source decay analysis in matrix-assisted laser desorption/ionization mass spectrometry of biomolecules, *J. Mass Spectrom.* **32**, 1019–1036 (1997).

48. K. Dreisewerd, R. Kingston, W. P. M. Geraerts, and K. W. Li, Direct mass spectrometric peptide profiling and sequencing of nervous tissues to identify peptides involved in male copulatory behavior in *Lymnaea stagnalis*, *Int. J. Mass Spectrom. Ion Process.* **169/170**, 291–299 (1997).

49. G. Critchley and B. Worster, Identification of peptides by matrix-assisted laser desorption ionization time of flight mass spectrometry (MALDI-TOF-MS) and direct analysis of the laterobuccal nerve from the pond snail *Lymnaea stagnalis*, in *Methods in Molecular Biology, Neuropeptide Protocols*, G. B. Irvine and C. H. Williams, eds., Humana Press, Totowa, NJ, 1997, pp. 141–152.

50. S. Jespersen, P. Chaurand, J. C. van Strien, B. Spengler, and J. van der Greef, Direct sequencing of neuropeptides in biological tissue by MALDI-PSD mass spectrometry, *Anal. Chem.* **71**, 660–666 (1999).

51. S. J. Perry, A. C. Dobbins, M. G. Schofield, M R. Piper, and P. R. Benjamin, Small cardioactive peptide gene: Structure, expression and mass spectrometric analysis reveals a complex pattern of co-transmitters in a snail feeding neuron, *Eur. J. Neurosci.* **11**, 655–662 (1999).
52. E. Stimson, O. Truong, W. J. Richter, M. D. Waterfield, and A. L. Burlingame, Enhancement of charge remote fragmentation in protonated peptides by high-energy CID MALDI-TOF MS using cold matrices, *Int. J. Mass Spectrom. Ion Processes* **169/170**, 231–240 (1997).
53. L. Li, R. W. Garden, E. V. Romanova, and J. V. Sweedler, In situ sequencing of peptides from biological tissues and single cells using MALDI-PSD/CID analysis, *Anal. Chem.* **71**, 5451–5458 (1999).
54. K. W. Li, R. Kingston, K. Dreisewerd, C. R. Jiménez, R. C. van der Schors, R. H. Bateman, and W. P. M. Geraerts, Structural eleucidation of a peptide from a single neuron by matrix-assisted laser desorption/ionization employing a tandem double-focusing magnetic-orthogonal acceleration time-of-flight mass spectrometer, *Anal. Chem.* **69**, 563–565 (1997).
55. C. R. Jiménez, K. W. Li, K. Dreisewerd, S. Spijker, R. Kingston, R. H. Bateman, A. L. Burlingame, A. B. Smit, J. van Minnen, and W. P. M. Geraerts, Direct mass spectrometric peptide profiling and sequencing of single neurons reveals differential peptide patterns in a small neuronal network, *Biochemistry* **37**, 2070–2076 (1998).
56. S. Hsieh, K. Dreisewerd, R. C. van der Schors, C. R. Jiménez, J. Stahl-Zeng, F. Hillenkamp, J. W. Jorgenson, W. P. M. Geraerts, and K. W. Li, Separation and identification of peptides in single neurons by microcolumn liquid chromatography-matrix-assisted laser desorption/ionization time-of-flight mass spectrometry and postsource decay analysis, *Anal. Chem.* **70**, 1847–1852 (1998).
57. K. M. Murray, Coupling matrix-assisted laser desorption/ionization to liquid separations, *Mass Spectrom. Rev.* **16**, 283–299 (1997).
58. R. D. Smith, J. H. Wahl, D. R. Goodlett, and S. A. Hofstadler, Capillary electrophoresis/mass spectrometry, *Anal. Chem.* **65**, 574A–584A (1993).
59. H. Zhang and R. M. Caprioli, Capillary electrophoresis combined with matrix-assisted laser desorption/ionization mass spectrometry; continuous sample deposition on a matrix-precoated membrane target, *J. Mass Spectrom.* **31**, 1039–1046 (1996).
60. H. Zhang, M. Stoeckli, P. E. Andren, and R. M. Caprioli, Combining solid-phase preconcentration, capillary electrophoresis and off-line matrix-assisted laser desorption/ionization mass spectrometry: intracerebral metabolic processing of peptide *E in vivo*, *J. Mass Spectrom.* **34**, 377–383 (1999).
61. J. Preisler, F. Foret, and B. L. Karger, On-line MALDI-TOF MS using a continuous vacuum deposition interface, *Anal. Chem.* **70**, 5278–5287 (1998).
62. J. Preisler, P. Hu, T. Rejtar, and B. L. Karger, Capillary electrophoresis—matrix-assisted laser desorption/ionization time-of-flight mass spectrometry using a vacuum deposition interface, *Anal. Chem.* **72**, 4785–4795 (2000).
63. J. S. Page, S. S. Rubakhin, and J. V. Sweedler, Direct cellular assays using off-line capillary electrophoresis with matrix-assisted laser desorption/ionization time-of-flight mass spectrometry, *Analyst* **125**, 555–562 (2000).
64. J. Gobom, K. O. Kraeuter, R. Persson, H. Steen, P. Roepstorff, and R. Ekman, Detection and quantification of neurotensin in human brain tissue by matrix-assisted

laser desorption/ionization time-of-flight mass spectrometry, *Anal. Chem.* **72**, 3320–3326 (2000).

65. D. M. Cannon, N. Winograd, and A. G. Ewing, Quantitative chemical analysis of single cells, *Annu. Rev. Biophys. Biomol. Struct.* **29**, 239–263 (2000).
66. A. J. Nicola, A. I. Gusev, A. Proctor, E. K. Jackson, and D. M. Hercules, Application of the fast-evaporation sample preparation method for improving quantification of angiotensin II by matrix-assisted laser desorption/ionization, *Rapid Commun. Mass Spectrom.* **9**, 1164–1171 (1995).
67. L. M. Preston, K. K. Murray, and D. H. Russell, Reproducibility and quantitation of matrix-assisted laser desorption ionization mass spectrometry: Effects of nitrocellulose on peptide ion yields, *Biol. Mass Spectrom.* **22**, 544–550 (1993).
68. F. J. C. van Strien, S. Jesperson, J. van der Greef, B. G. Jenks, and E. W. Roubos, Identification of POMC processing products in single melanotrope cells by matrix-assisted laser desorption/ionization mass spectrometry, *FEBS Lett.* **379**, 165–170 (1996).
69. C. R. Jiménez, K. W. Li, K. Dreisewerd, H. D. Mansvelder, A. B. Brussaard, B. B. Reinhold, R. C. Van der Schors, M. Karas, F. Hillenkamp, J. P. H. Burbach, C. E. Costello, and W. P. M. Geraerts, Pattern changes of pituitary peptides in rat after salt-loading as detected by means of direct, semiquantitative mass spectrometric profiling, *Proc. Natl. Acad. Sci.* (USA) **94**, 9481–9486 (1997).
70. P. W. K. Ma, R. W. Garden, J. T. Niermann, M. O'Connor, J. V. Sweedler, and W. L. Roelofs, Characterizing the Hez-PBAN gene products in neuronal clusters with immunocytochemistry and MALDI-MS, *J. Insect Physiol.* **46**, 221–230 (2000).
71. Y. Fujisawa, Y. Furukawa, S. Ohta, T. A. Ellis, N. C. Dembrow, L. Li, P. D. Floyd, J. V. Sweedler, H. Minakata, F. Nakamaru, F. Morishita, ••. Matsushima, K. R. Weiss, and F. S. Vilim, The *Aplysia mytilus* inhibitory peptide-related peptides: identification, cloning, processing, distribution, and action, *J. Neurosci.* **19**, 9618–9634 (1999).
72. V. Redeker, J.-Y. Toullec, J. Vinh, J. Rossier, and D. Soyez, Combination of peptide profiling by matrix-assisted laser desorption/ionization time-of-flight mass spectrometry and immunodetection on single glands or cells, *Anal. Chem.* **70**, 1805–1811 (1998).
73. M. Stoeckli, T. B. Farmer, and R. M. Caprioli, Automated mass spectrometry imaging with a matrix-assisted laser desorption ionization time-of-flight instrument, *J. Am. Soc. Mass Spectrom.* **10**, 67–71 (1999).
74. R. W. Garden and J. V. Sweedler, Heterogeneity within MALDI samples as revealed by mass spectrometric imaging, *Anal. Chem.* **72**, 30–36 (2000).
75. R. W. Garden, T. P. Moroz, J. M. Gleeson, P. D. Floyd, L. Li, S. S. Rubakhin, and J. V. Sweedler, Formation of N-pyroglutamyl peptides from H-Glu and N-Gln precursors in *Aplysia* neurons, *J. Neurochem.* **72**, 676–681 (1999).
76. T. L. Colliver, C. L. Brummel, M. L. Pacholski, F. D. Swanek, A. G. Ewing, and N. Winograd, Atomic and molecular imaging at the single-cell level with TOF-SIMS, *Anal. Chem.* **69**, 2225–2231 (1997).

CHAPTER 21

Dynamics of Cell Nucleus/ Lymphocyte Nuclei-Associated Peptides Studied by Mass Spectrometry and Future Peptidomic Aspects

JONAS BERGQUIST and ROLF EKMAN

Department of Analytical Chemistry, Institute of Chemistry, Uppsala University, Uppsala and Neurochemistry Section, Institute of Clinical Neuroscience, Göteborg University, Sahlgrenska University Hospital, Mölndal, Sweden.

CONTENTS

Mass Spectrometry and Hyphenated Techniques in Neuropeptide Research, Edited by Jerzy Silberring and Rolf Ekman.
ISBN 0-471-35493-7

TABLE 21.1. A Partial List of Some Methodologies for Study of Molecular Mechanism of CNS-Related Disorders–Diseases

Analysis of
Cerebrospinal fluid
Postmortem brain tissue
Animal models
Peripheral blood cells, lymphocytes

21.1. INTRODUCTION

Understanding the biology of the human brain is a challenging task. Up to the present, there has been no simple approach to study and follow the ongoing bioactivities in the brain from a neurochemical point of view over long periods of time. Although applications such as microdialysis, electrophysiology, and various imaging techniques are available for the trained specialist, they don't meet the necessary demands at a molecular level. For instance, we need novel strategies to follow the influence and outcome of different neurochemical perturbations over days, weeks, months, and even years as an effect of various psychosocial as well physiological stressors resulting in a psychiatric disease.

Table 21.1 lists some methodologies for the study of molecular mechanisms of psychiatric disorders. The use of autopsy material or even repeated cerebrospinal fluid (CSF) analysis do not meet these criteria in the respect of demonstrating different steps in the pathophysiological processes over a given time interval. There are now data supporting ongoing molecular interactions between the central nervous system (CNS) and the immune system under normal and diseased conditions reflected in protein or peptide alterations in circulating lymphocytes. The immune system mimicking different activities may act as a model of neurochemical activities in the brain [1], an upcoming research field.

The inside of a cell is a crowded and dynamic place, where proteins and peptides being created and discarded. Mass-spectrometry-based techniques have just entered into a new challenging field of cell biology, the biochemistry at the cell organelle level. This has been achieved through a combination of multidimensional separation methods and mass spectrometric techniques [2–6].

We want to suggest that the terminology given by us—*peptidomics* [7]—may be used as a name of this methodological approach when studying endogenous peptides. These approaches are important for mapping chemical modifications as well as structural changes. Furthermore, the identification of different molecular species interacting under normal conditions is important to get a better understanding of the complexity of the chemically modifications behind several common disorders as well as some psychiatric and stress-related disorders and diseases.

21.1.1. Neuropeptides

Although the role of neuropeptides in cellular communication and regulation is not well understood, neuropeptides represent the most diverse class of transmitter molecules [8]. They are known to serve as messenger molecules in both the nervous and immune systems through specific receptors common to both systems [1] (Fig. 21.1).

Several studies have demonstrated a nuclear localization of various neuropeptides as well as other molecular species. At present the mechanisms behind these data are unclear as to whether it is an unspecific or specific association with the nuclear membrane or internalization through a protein/receptor complex or an effect of both [9–14].

21.1.2. The Cell Nucleus

The cell nucleus is one of the best studied but least understood of the cellular organelle. Nuclear structures were first visualized microscopically in the late nineteenth century. Studies since 1990 have provided major insights into molecular mechanisms responsible for active transport of macromolecules in and out of the nucleus [15].

We now move from simple description of subnuclear structures to study how nuclear components interact with each other to carry out their critical steps and functions. Most of the nuclear peptides are synthesized in the cytoplasm and may either be actively transported into the nucleus through the nuclear pore as peptide protein/receptor complex or diffuse passively through nuclear pore complex or enter another rote for exocytosis [16,17]. Table 21.2 lists some peptides reported to be localized to the nucleus or undergo nuclear translocation [14,18].

21.1.3. The Molecular Dynamics of the Cell Nucleus

Neuropeptides exert their effects by either (1) binding to receptors on the cell membrane, with subsequent activation of post receptor second messenger systems, or (2) various mechanisms of transfer/internalization targeted to the cell cytoplasm and nucleus. Since the year 1999 considerable insight has been gained into how molecules are selected for nuclear import and export [19].

It has become more evident that growth factors (peptides and proteins), for instance, are involved in regulation of cell function and cell division, differentiation, and proliferation. The transfer of their messages to the nucleus occurs either indirectly by interaction with cell surface receptors via endosomes or

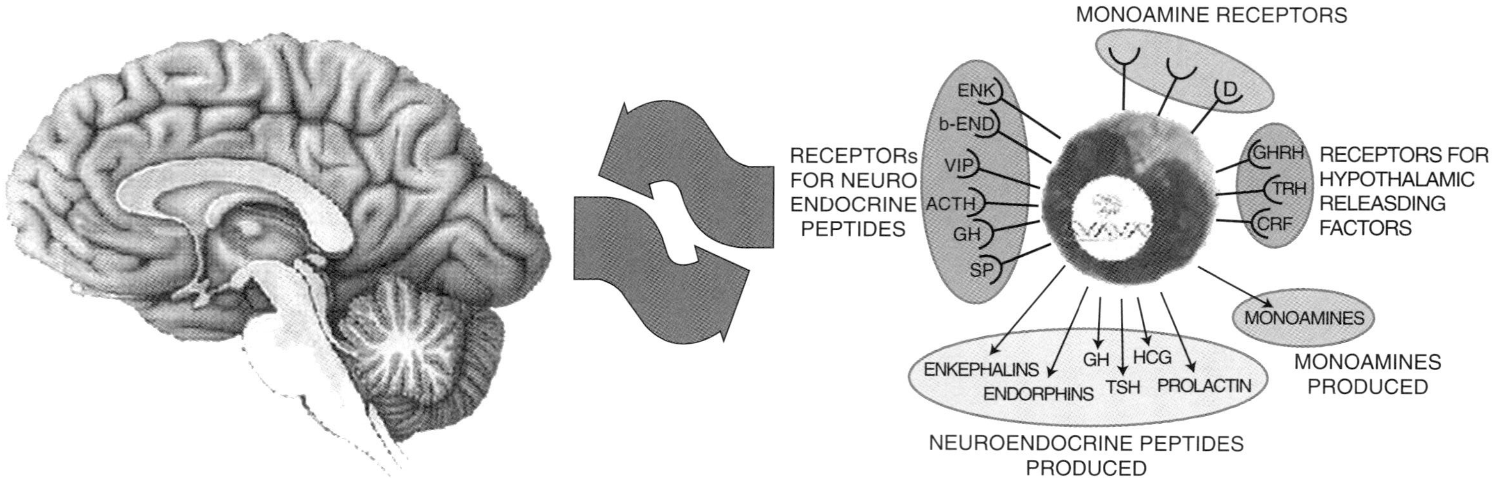

Figure 21.1. The ongoing crosstalk between the central nervous system and the immune system involves a number neuropeptides, hormones as well as catecholamines. Some of these transmitters not only act on membrane receptors but also have been demonstrated to enter the nuclear envelope and maybe directly interact with the transcription machinery. Another group of transmitter molecules not shown in this picture is the cytokines, which also play a major regulatory role in this communication.

TABLE 21.2. A Partial List of Neuropeptides with Nuclear Localization

Peptides	
AVP	Neurotensin
CRH	Somatostatin
Insulin	Substance P
Oxytocin	TRH
Neuropeptide Y	VIP

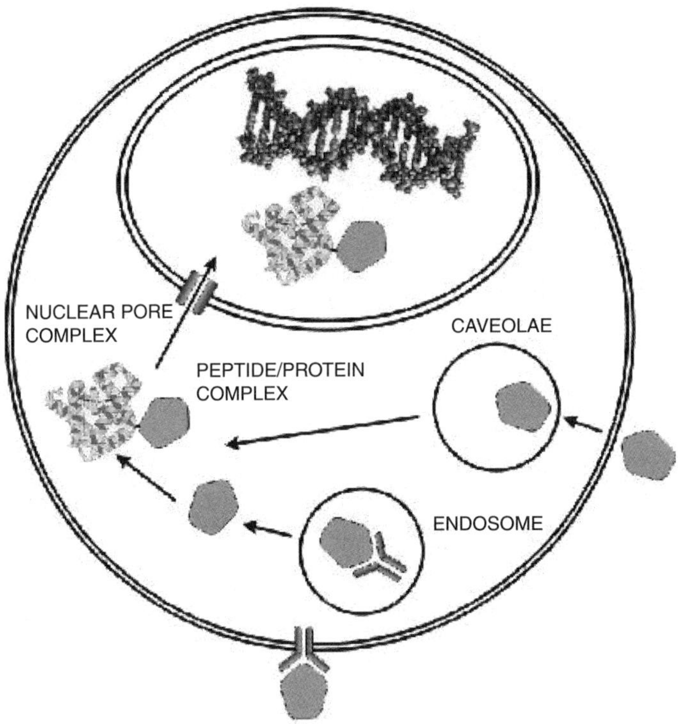

Figure 21.2. Mechanisms for targeting peptides through the cytoplasmatic compartment into the nucleus. Internalization may be via endosomes or via caveolae.

after being transported from the cell membrane via caveolae and through the cytoplasm by direct association to some carrier protein into the nucleus after passage of the nuclear pore complex (Fig. 21.2).

However, the mechanisms and control of actual pore passage are still the principal challenge in the field. Some of the imported peptides and proteins are randomly distributed throughout the nucleoplasm, whereas others are targeted to specific subcompartments or domains such as nucleoli, nuclear bodies, nuclear speckles, transcription and replication foci, and interchromatin granule clusters [20,21].

To understand the transcription regulation in health and its relation to disease is one of the real stimuli for the future and involves peptides as well as proteins transported from the cytoplasm into the nucleus. Understanding in molecular detail how different molecular species are coordinated and regulated in the nucleus is one major goal in understanding human disorders such as cancer, dementia, autoimmune disease, and particular forms of oxidative stress since reactive oxygen intermediates have been implicated in a variety of clinical conditions [22]. These conditions may be due to disturbances in the transcription process at various levels. It is also known that disturbances in transcription may cause different diseases and may well be the underlying factor in the pathophysiology of HIV, cancer, autoimmune diseases, and some of the most common mental disorders.

Several of the high-molecular-mass proteins have been characterized and identified, but there is limited information about structure and function of low-molecular-mass molecules, such as peptides, in the nucleus.

21.2. INTERACTION BETWEEN THE CENTRAL NERVOUS SYSTEM (BRAIN) AND THE IMMUNE SYSTEM

The ongoing accumulation of data of a molecular interaction between CNS and the immune system are reflected in monoamine and peptide/protein alterations as structural modifications as well as in concentration of the circulating cells of the immune system [9,14,23–25].

Our hypothesis is that human lymphocyte can be used as model for the study of molecular changes, and chemical modifications in stress-related disorders as well as psychiatric diseases, reflecting the same molecular perturbations in the cells of the brain as in the lymphocytes. Of course, the outcome of each molecular modification, depending on cell type, will give rise to different endpoint effects, although on the basis of the same modification but in a different compartment.

Detailed mechanisms underlying communication between the nervous and immune systems are presently not fully understood. There are many examples of direct communication between cells in the CNS and the immune system, involving cytokines, peptides, and monoamines as potential signaling molecules because of their regulatory functions in nervous tissue as well as in the immune system. Lymphocytes not only represent an easily obtained population of cells but also, on the basis of the now accepted communication between the nervous and immune systems, may serve as carriers of chemically modified molecules as markers of ongoing metabolic disturbances in the central nervous system [26]. The use of lymphocytes has been suggested by several other groups [27–30].

21.2.1. Our Strategy: Peptidomics of the Lymphocyte

The main goal of this strategy is to combine advanced analytic tools such as high-resolution mass spectrometry and microseparation techniques to study the cellular chemistry—represented by the proteome and peptidome (as we from now on would like to name the exploration of the peptides expressed by the genome, in analogy with the proteome terminology)—as well as smaller endogenous substances, such as amino acids, steroids, lipids, and monoamines. The technology is applied in the investigation of relevant biological questions in life science with focus on clinical neuroscience where a large number of proteins and peptides still are to be discovered and characterized. However, the technique may well be applied to any cell or tissue, either eukaryotic or prokaryotic. Since an emerging area within biomedical research is focusing on finding rapid and accurate methods for elucidating the etiology of and diagnosing diseases. The development and movement into proteomics and functional genomics has been very pertinent since the human genome project (HUGO) has almost reached its goal (Celera now claims to have finished 96% of the human genome). Now when the order of all 3.1 billion As, Ts, Gs, and Cs that make up the human DNA has approached completion, it is time to figure how to use that information. From knowing the code of a gene to the knowledge of where that specific protein or peptide is produced, what its function is, and how it may be modified during a diseased state, is a very large step. One such approach is to compare the proteome—the entire protein complement expressed by a genome—of healthy and diseased samples with two-dimensional polyacrylamide gel electrophoresis (2D-PAGE) the preferred method for generation of cellular protein maps, due to the high resolving power of the technique and mass spectrometry.

It is well known that disturbances in transcription may cause different diseases and may well be the underlying factor in the pathophysiology of some of the most common mental disorders. The protein patterns derived from the electrophoretic analysis of human samples may be used for clinical diagnoses. For example, Creutzfeldt–Jacobs disease and bovine spongiform encephalitis, as well as neonatal sepsis, has been diagnosed using 2D-PAGE. Even though 2D-PAGE at present is unchallenged for separation of complex protein mixtures, it does not permit mapping of all proteins in a cell. With the most high-resolving gels available, detection of roughly 10,000 proteins have been reported out of the predicted 80–120,000 proteins expressed by mammalian cells. Furthermore, the 2D-PAGE technology does neither allow the study of smaller proteins and peptides nor smaller organic molecules such as amino acids, steroids, lipids, and monoamines.

New methods have been developed using the combination of high-resolution mass spectrometry and separation techniques such as capillary electrophoresis and micro- or nanoscale liquid chromatography. This combination may well complement 2D-PAGE and provide a system for peptidome analysis with extremely high peak capacity and resolution and still keep the analytes in solution. The rapidly developing field of microfabricated/chip-based fluidic systems will also be useful in the future handling of ultrasmall sample volumes and high-throughput screening approaches.

We have used human peripheral blood lymphocytes to study nuclear low-molecular-mass proteins and peptides. The lymphocytes are not only easily obtained; they also contain a large nuclear "easy" to isolate. This model system was used because of our interest in the peptidergic interaction between the immune and nervous systems under normal and pathologic conditions as in patients with psychiatric diseases [14].

Peptides were isolated from whole cell and nucleus extracts using reversed-phase high-performance liquid chromatography (RP-HPLC). The HPLC fractions were further analyzed in order to characterize the peptide content, in some experiments by a combination with affinity-based purification methods [31] followed by matrix-assisted laser desorption/ionization time-of-flight (MALDI-TOF) mass spectrometry (Fig. 21.3). Selected fractions, giving clean

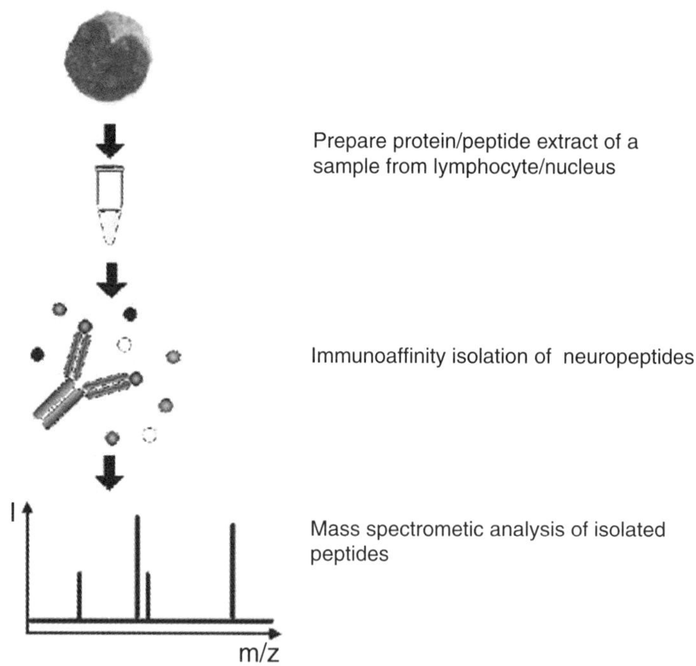

Figure 21.3. The peptidomic approach to study functional genomics may be to extract and analyze all endogenous peptides from a tissue, a body fluid, or a cell population, but may also involve as shown in this picture, the search for a specific peptide using affinity isolation and mass spectrometry.

and high-intensity signals, were also analyzed by electrospray ionization time-of-flight (ESI-TOF) and electrospray ionization Fourier transform ion cyclotron resonance (ESI-FTICR) mass spectrometry [13]. Microsequence analysis and database searches were used for identification of selected peptides.

21.3. ISOLATION OF HUMAN PERIPHERAL LYMPHOCYTE NUCLEI AND PROTEIN/PEPTIDE PURIFICATION

21.3.1. Isolation of Nuclei

Several criteria have been used to establish that the peptide content of the human lymphocyte nuclei are intrinsic components of the nuclei themselves and are not due to any preparation artifact. The purity of nuclear preparations was judged by light microscopy with nuclear staining. This was further confirmed by electron microscopy, and nuclei proved to be intact with both inner and outer nuclear membrane present and with negligible endoplasmic reticulum contamination. The nuclear preparations used demonstrated low contamination levels for marker enzymes for various cell fractions, as well as other obvious cytoplasmic contamination [13].

We have also studied normal human lymphocyte nuclear proteins from as little as 10-mL blood. When comparing the protein pattern from whole lymphocytes with isolated nuclei, we see a big difference. Nuclei associated proteins are highly enriched in the latter preparation, as expected. The results from this study strongly indicate that our nuclear preparation method works [6]. From 22 randomly chosen 2D gel electrophoresis spots from several hundreds, we identified 15 proteins recently (at the time of writing) described with nuclear localization and function; five proteins were identified for the first time with nuclear association, and only two proteins could not be identified by database search of mass spectrometric data.

21.3.2. Peptide Purification

Over the years, an impressive number of biologically active, short peptides have been purified from tissues, cells, and biological fluids [32]. Analytic and micropreparative techniques have been optimized since the mid-1990s, making analysis of low- to subpicomole amounts of peptides possible. As the peptide chemistry community is just approaching femtomole- and even attomole-level isolation and sequencing, the problem of minute losses during preparation, storage, and transfer has either not been fully recognized or has been attributed to unrelated factors.

Peptides were extracted from the nuclear pellets by acetic acid at 4°C and centrifuged, and the supernatant was dried and reconstituted and further purified using HPLC [11–14]. In some of the experiment the HPLC fractions were incubated with a specific antiserum and aliquots of immunoprecipitates were purified and concentrated on a nanoscale reversed-phase column and directly eluted onto the MALDI probe [31].

21.4. MALDI-MS OF WHOLE LYMPHOCYTE AND CELL NUCLEI PEPTIDE MATERIAL FROM HEALTHY DONORS AND PATIENTS WITH DEPRESSION

Several neurohypophyseal peptides have been studied in several immune tissues based on findings obtained with immunocytochemistry, radioimmunoassay, and Western blotting. While sensitive, these types of assays are limited by the fact that antibodies recognize only an epitope on the molecule of interest. Thus, there is an omnipresent risk of detecting antibodies cross-reacting with other molecules containing this epitope. Likewise, truncated or otherwise modified forms of the peptide may escape recognition.

MALDI-TOFMS has been utilized to provided molecular mass and sequence information. All mass spectra were acquired on a reflex MALDI-TOF mass spectrometer (Bruker–Franzen Analytik, Bremen, Germany). Positive ions were analyzed in reflector mode using delayed ion extraction. Mass spectra were analyzed using the Bruker software on a Sun Sparcstation, using external or internal calibration.

21.4.1. Chemokines and Defensins

The HPLC profile of the nuclear extract reveals a complex mixture of peptides, and the main peptide/protein found in HPLC fractions by MALDI-TOF analysis are listed in Table 21.3 [13]. The average molecular weights of proteins/peptides found in the HPLC fractions in MALDI-TOF analysis are shown. Only major peaks, including the monoisotopic molecular weights for the components in fraction 8, are listed. Of special interest is the finding of human defensins associated with the lymphocyte nuclear fraction as well as a small CXC chemokine, CTAP-III, from the same material. Interestingly, chemokines were originally identified through molecular biological techniques as putative factors, but have been studied mainly as secretory peptides. This class of molecules is known to stimulate DNA synthesis, mitosis, and formation of bioactive lipids in white blood cells [11].

From what is known today, it is most likely that defensins exist in the nucleus as a complex with other molecules of the nucleoplasm, and may be related to or involved in regulation of the transcription machinery, which, in turn, might influence the parts of the gene regulation procedure or DNA turnover in some stress-related disorders (unpublished observations).

TABLE 21.3. Molecular Masses of Proteins/Peptides Demonstrated in HPLC Fractions of Acidic Extract of Human Lymphocyte Nuclei

HPLC Fraction (number)	Molecular Weight (Da)
8	3440.0 ± 0.5; 3368.7 ± 0.5; 3483.9 ± 0.5 (HNP-1, HNP-2, HNP-3)
11	9301; 9115
13	9100; 9285 (CTAP III); 10,075; 10,259
14	9101; 9287
16	3713
18	10,099
19	9239
20	7437
23	10,612
24	10,536; 11,483
25	10,535; 11,728
26	10,842
27	4075; 4043; 8962
28	4566
35	13,785 (histone 2B)
36	13,899; 14,007

Source: Modified from Blomqvist et al. [13].

21.4.2. Depressed Patients

A variety of immunologic alterations have been observed in patients with depression in combination with dysfunction of the hypothalamus–pituitary adrenal (HPA) axis and its connection to the immune system [33,34]. We have chosen to study the nonapeptide arginine vasopressin (AVP) because of its involvement in the HPA axis and implication for the pathophysiology of depression. AVP has been studied and found to be involved in gene regulation of neurotrophic factors such as brain-derived neurotrophic factor BDNF or other neurotrophins [35].

We have investigated whether cytoplasmatic or nuclear extracts of human peripheral blood lymphocytes contain AVP in samples from healthy controls and patients diagnosed as depressed. AVP was detected in both cytoplasmic and nuclear fractions of human lymphocytes and HPLC, and MALDI-TOFMS (Fig. 21.4) confirmed the identity.

The finding of AVP in human lymphocytes demonstrates the existence of yet another neurohormone common to the immune and neuroendocrine systems. This is the first time that AVP has been demonstrated in the lymphocyte nucleus, and this may indicate an important role in modulating regulation of several disease states.

The mechanisms by which nuclear accumulation of peptides occur are unclear and are crucial issues of the cellular and molecular events sometimes

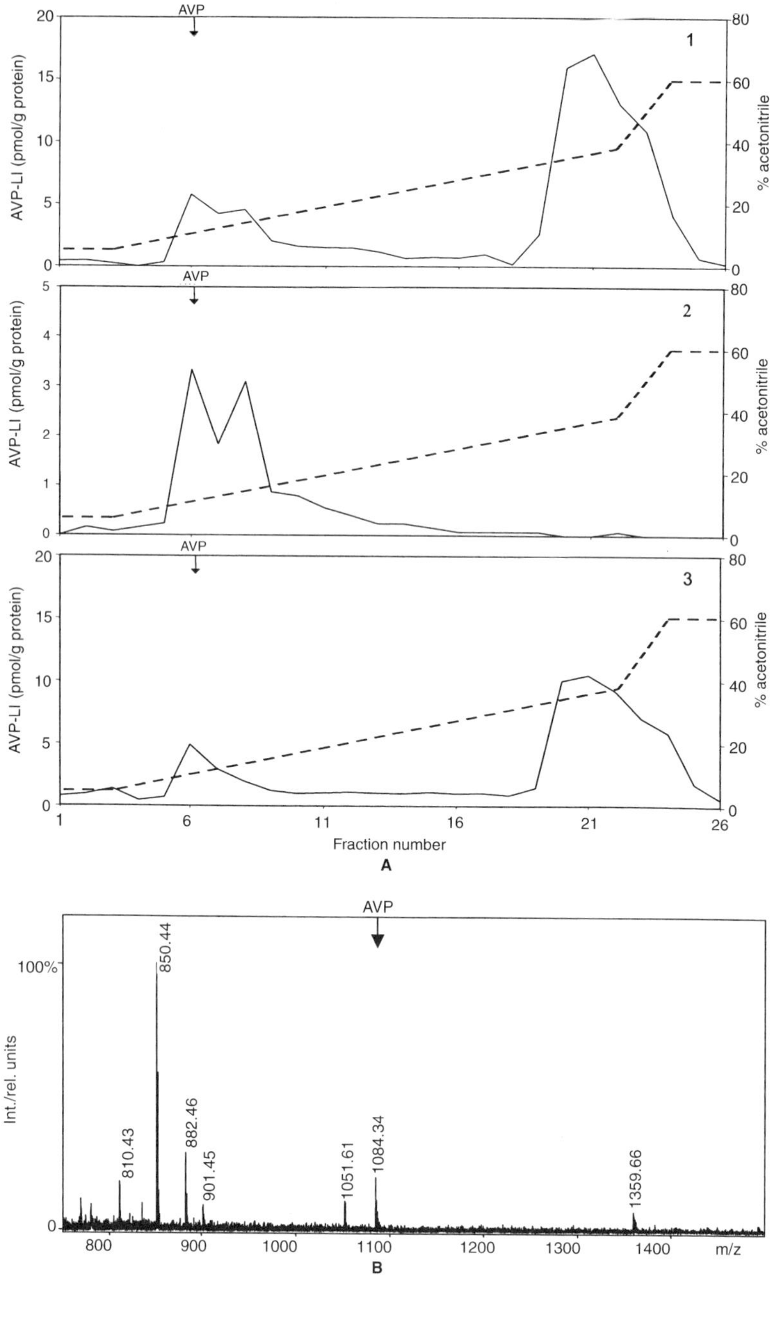

Figure 21.4. (A) Modified from Ref. 14, is the HPLC characterization of AVP immunoreactive material from extracted humane circulating lymphocytes. The arrow indicates the elution volume of synthetic AVP and the acetonitrile gradient is indicated by the dotted line (1—controls, 2—nuclei from control, 3—depressed patients). (B) A MALDI-TOFMS spectrum from HPLC fraction 6 after affinity pulldown.

preceding transformation of cell function into a pathophysiological state. It has been demonstrated that AVP regulate the transcription of mRNA for several nerve growth factors recently demonstrated to be expressed in lymphocytes [26].

The role of nuclear AVP and fragments of AVP are still to be resolved and may be involved in the regulation of specific nuclear molecules or interact with transcription.

21.5. PERSPECTIVES AND FUTURE DIRECTIONS

21.5.1. Nuclear Peptide Function

Considerable insight has been gained into how some molecules are selected for nuclear import and export [19,36]. The mechanisms of the actual nuclear pore passage are still the principal challenge in the field, and much effort has been put into identifying all the components of the nuclear pore complex [37]. It has also become clear that the nucleus may be much more dynamic than we ever thought since the nuclear envelope is in direct contact with many other compartments of the cell. The most recent data indicate that some of the transport proteins also bind to some transcription factors, suggesting not only an import/export function but also direct participation in the transcription. Understanding the roles of nuclear proteins and peptides will require determination of their structure and characterization of stoichiometry and binding specificities.

21.6. CONCLUSIONS

- The finding of AVP within lymphocytes is important because interactions between the brain and the immune system, although not fully understood, are known to exist.
- Mass spectrometry should play a significant role in future studies of the structures of peptides in neuroscience. These applications include peptide dynamics, peptide–protein interaction in the regulation of the transcription machinery, as well as the elucidation of mechanisms associated with normal and impaired neuron or immune function.
- An improved understanding of the structural basis for the molecular regulation of the transcription machinery is essential for the development of new approaches to the prevention and treatment of disease.
- Detailed information about peptide processing and biotransformation as well as their concentration may provide the basis for clinical assays and illuminate the pathophysiology related to psychiatric as well as neurologic disorders using the lymphocyte as a model reflecting brain activities related to various psychosomatic disturbances.

ACKNOWLEDGMENT

The contributions of co-workers and fellow scientists are acknowledged. The authors acknowledge financial support from the Swedish Foundation for Strategic Research, the Fredrik and Ingrid Thuring Foundation, the Wilhelm and Martina Lundgren Foundation, the Magnus Bergvall Foundation, the Swedish Alzheimer Foundation, the Syskonen Svensson Foundation, the Gamla Tjänatinnor Foundation, the Knut and Alice Wallenberg Foundation, the Swedish Lundbeck Foundation, the Swedish Society for Medical Research, the Swedish Natural Science Research Council (Grant K-AA/KU 12003-300), and the Swedish Medical Research Council (Grant 13123, 07517).

REFERENCES

1. D. A. Weigent, D. J. J. Carr, and E. J. Blalock, *Ann. NY Acad. Sci.* **579**, 17–27 (1990).
2. J. Gobom, E. Nordhoff, R. Ekman, and P. Roepstorff, *Int. J. Mass Spectrom. Ion Process.* **169/170**, 153–163 (1997).
3. J. Gobom, E. Nordhoff, E. Mirgorodskaya, R. Ekman, and P. Roepstorff, *J. Mass Spectrom.* **34**, 105–116 (1999).
4. A. L. Burlingame, R. K. Boyd, and S. J. Gaskell, *Anal. Chem.* **70**, 647R–716R (1998).
5. T. Johnson, J. Bergquist, R. Ekman, E. Nordhoff, M. Schurenberg, K.-D. Klöppel, M. Muller, H. Lehrach, and J. Gobom, *Anal. Chem.* **73**, 1670–1675 (2001).
6. J. Bergquist, J. Gobom, A. Blomberg, P. Roepstorff, and R. Ekman, *J. Neurosci. Meth.* **1109**, 3–11 (2001).
7. T. Johnson, M. Palmblad, R. Ekman, and J. Bergquist, *Desorption 2000, Saint Malo*, O-41.
8. T. Hökfelt, *Neuron.* **7**, 867–879 (1991).
9. R. Ekman, B. Servenius, M. G. Castro, P. J. Lowry, A.-S. Cederlund, O. Bergman, and H. O. Sjögren, *J. Neuroimmunol.* **44**, 7–14 (1993).
10. G. Morel, *Biochem Pharmacol* **47**, 63–76 (1994).
11. C. L. Nilsson, C. M. Murphy, and R. Ekman, *Rapid Commun. Mass Spectrom.* **11**, 610–612 (1997).
12. C. L. Nilsson, E. Brodin, and R. Ekman, *J. Chromatogr. A* **800**, 21–27 (1998).
13. M. Blomqvist, J. Bergquist, A. Westman, K. Håkansson, P. Håkansson, P. Fredman, and R. Ekman, *Eur. J. Biochem.* **263**, 312–318 (1999).
14. R. Ekman, J. Gobom, R. Persson, P. Mecocci, and C. L. Nilsson, *Peptides* **22**, 67–72 (2001).
15. D. Görlich and U. Kutay, *Annu. Rev. Cell Devel. Biol.* **15**, 607–660 (1999).
16. P. Turpin, B. Ossareh-Nazari, and C. Dargemont, *FEBS Lett.* **452**, 82–86 (1999).
17. S. R. Wente, *Science* **288**, 1374–1377 (2000).
18. P. E. Lobie, *Thesis*, Karolinska Institute, Stockholm, Sweden, 1994, pp. 7–75.
19. K. J. Ryan and S. R. Wente, *Curr. Opin. Cell Biol.* **12**, 361–371 (2000).
20. T. Pederson, *Nature Cell Biol.* **2**, E73–E74 (2000).
21. M. Carmo-Fonseca, L. Mendes-Soares, and I. Campos, *Nature Cell Biol.* **2**, E107–E112 (2000).

22. G. L. Schieven, J. M. Kirihara, D. E. Myers, J. A. Ledbetter, and F. M. Uckun, *Blood* **82**, 1212–1220 (1993).
23. R. E. Ballieux, *Eur. J. Clin. Invest.* (Suppl. 1), 6–9 (1992).
24. J. E. Blalock, *Immunol. Today* **15**, 504–511 (1994).
25. J. Bergquist, A. Tarkowski, A. Ewing, and R. Ekman, *Immunol. Today* **19**, 562–567 (1998).
26. R. Barouch, E. Appel, G. Kazimirsky, A. Braun, H. Renz, and C. Brodie, *J. Neuroimmunol.* **103**, 112–121 (2000).
27. B. C. Wu, W. H. Spohn, and H. Bush, *Cancer Res.* **41**, 336–342 (1981).
28. D. I. Stott, *Immunol. Commun.* **12**, 605–623 (1983).
29. N. Feurenstein and J. J. Mond, *J. Biol. Chem.* **262**, 11389–11397 (1987).
30. S. M. Hanash and D. Teichroew, *Electrophoresis* **19**, 2004–2009 (1998).
31. J. Gobom, K.-O. Kraeuter, R. Persson, P. Roepstorff, and R. Ekman, *Anal. Chem.* **72**, 3320–3326 (2000).
32. F. L. Strand, Bradford Books MIT Press, Cambridge, UK, 1999.
33. J. J. Legros, M. Ansseau, and M. Timsit-Berthier, *Regul. Pept.* **45**, 133–138 (1993).
34. L. V. Scott and T. G. Dinan, *Life. Sci.* **62**, 1985–1998 (1998).
35. A. W. Zhou, W. X. Li, and Y. C. Du, *Peptides* **18**, 1179–1187 (1997).
36. A. Prochiantz, *Curr. Opin. Cell Biol.* **12**, 400–406 (2000).
37. M. P. Rout, J. D. Aitchison, A. Suprapto, K. Hjertaas, Y. Zhao, and B. T. Chait, *J. Cell Biol.* **148**, 635–651 (2000).

Appendix

AGNIESZKA SCIUBISZ

Faculty of Chemistry and Regional Laboratory, Jagiellonian University, Krakow, Poland

Mass Spectrometry and Hyphenated Techniques in Neuropeptide Research, Edited by Jerzy Silberring and Rolf Ekman.
ISBN 0-471-35493-7

Exact Masses and Isotopic Abundances of Elements

Name	Symbol	Mass	Abundance	Symbol	Mass	Abundance	Symbol	Mass	Abundance
Aluminum	Al(27)	26.981541	100.00	—	—	—	—	—	—
Antimony	Sb(121)	120.903824	57.30	Sb(123)	122.904222	42.70	—	—	—
Argon	Ar(36)	35.967546	0.34	Ar(38)	37.962732	0.063	Ar(40)	39.962383	99.60
Arsenic	As(75)	74.921596	100.00	—	—	—	—	—	—
Barium	Ba(130)	129.906277	0.11	Ba(132)	131.905042	0.10	Ba(134)	133.904490	2.42
	Ba(135)	134.905668	6.59	Ba(136)	135.904556	7.85	Ba(137)	136.905816	11.23
	Ba(138)	137.905236	71.70	—	—	—	—	—	—
Beryllium	Be(9)	9.012183	100.00	—	—	—	—	—	—
Bismuth	Bi(209)	208.980388	100.00	—	—	—	—	—	—
Boron	B(10)	10.012938	19.80	B(11)	11.009305	80.20	—	—	—
Bromine	Br(81)	78.918336	50.69	Br(81)	80.916290	49.31	—	—	—
Cadmium	Cd(106)	105.906461	1.25	Cd(110)	109.903007	12.49	Cd(111)	110.904182	12.80
	Cd(112)	111.902761	24.13	Cd(113)	112.904401	12.22	Cd(114)	113.903361	28.73
	Cd(116)	115.904758	7.49	—	—	—	—	—	—
Calcium	Ca(40)	39.962591	96.95	Ca(42)	41.958622	0.65	Ca(43)	42.958770	0.14
	Ca(44)	43.955485	20.86	Ca(46)	45.953689	0.004	Ca(48)	47.952532	0.19
Carbon	C(12)	12.000000	98.90	C(13)	13.003355	1.10	—	—	—
Cerium	Ce(136)	35.907140	10.19	Ce(138)	137.905996	0.25	Ce(140)	139.905442	88.48
	Ce(142)	141.909249	11.08	—	—	—	—	—	—
Cesium	Cs(133)	132.90543	100.00	—	—	—	—	—	—
Chlorine	Cl(35)	34.968853	75.77	Cl(37)	36.965903	24.23	—	—	—
Chromium	Cr(50)	49.946046	4.35	Cr(52)	51.940510	83.79	Cr(53)	52.940651	9.50
	Cr(54)	53.938882	2.36	—	—	—	—	—	—
Cobalt	Co(59)	58.933198	100.00	—	—	—	—	—	—
Copper	Cu(63)	62.929599	69.17	Cu(65)	64.927792	30.83	—	—	—

Dysprosium	Dy(156)	155.924287	0.060	Dy(158)	157.924412	0.10	Dy(160)	159.925203	2.34
	Dy(161)	160.926999	18.90	Dy(162)	161.926805	25.50	Dy(163)	162.928737	24.90
	Dy(164)	163.929183	28.20	—	—	—	—	—	—
Erbium	Er(162)	161.928787	0.14	Er(164)	163.929211	1.61	Er(166)	165.930305	33.60
	Er(167)	166.932061	22.95	Er(168)	167.932383	26.80	Er(170)	169.935476	14.90
Europium	Eu(151)	150.919860	47.80	Eu(153)	152.921243	52.20	—	—	—
Fluorine	F(19)	18.998403	100.00	—	—	—	—	—	—
Galium	Ga(69)	68.925581	60.10	Ga(71)	70.924701	39.90	—	—	—
Galodinium	Gd(152)	151.919803	0.200	Gd(154)	153.920876	2.18	Gd(155)	154.822629	14.80
	Gd(156)	155.922130	20.47	Gd(157)	156.923967	15.65	Gd(158)	157.924111	24.84
	Gd(160)	159.927061	21.86	—	—	—	—	—	—
Germanium	Ge(70)	69.924250	20.50	Ge(72)	71.922080	27.40	Ge(73)	72.923464	7.80
	Ge(74)	73.921179	36.50	Ge(75)	75.921403	7.80	—	—	—
Gold	Au(197)	196.966560	100.00	—	—	—	—	—	—
Hafnium	Hf(174)	173.940065	0.16	Hf(176)	175.941420	5.20	Hf(177)	176.943233	18.60
	Hf(178)	177.943710	27.10	Hf(179)	178.945827	13.74	Hf(180)	179.946561	35.20
Helium	He(3)	3.016029	0.0001	He(4)	4.002603	100.00	—	—	—
Holmium	Ho(165)	164.930332	100.00	—	—	—	—	—	—
Hydrogen	H(1)	1.007825	99.99	H(2)	2.014102	0.015	—	—	—
Indium	In(113)	112.904056	4.30	In(115)	114.903875	95.70	—	—	—
Iodine	I(127)	126.904477	100.00	—	—	—	—	—	—
Iridium	Ir(191)	190.960603	37.30	Ir(193)	192.962942	62.70	—	—	—
Iron	Fe(54)	53.939612	5.80	Fe(56)	55.934939	91.72	Fe(57)	56.935396	2.20
	Fe(58)	57.933278	0.28	—	—	—	—	—	—
Krypton	Kr(78)	77.920397	0.35	Kr(80)	79.916375	2.25	Kr(82)	81.913483	11.60
	Kr(83)	82.914134	11.50	Kr(84)	83.911506	57.00	Kr(86)	85.910614	17.30
Lanthanum	La(138)	137.907114	0.09	La(139)	138.906355	99.91	—	—	—

(*Continued*)

Exact Masses and Isotopic Abundances of Elements (*Continued*)

Name	Symbol	Mass	Abundance	Symbol	Mass	Abundance	Symbol	Mass	Abundance
Lead	Pb(204)	203.973037	1.40	Pb(206)	205.974455	24.10	Pb(207)	206.975885	22.10
	Pb(208)	207.976641	52.40	—	—	—	—	—	—
Lithium	Li(6)	6.015123	7.42	Li(7)	7.016005	92.58	—	—	—
Lutetium	Lu(175)	174.940785	97.40	Lu(176)	175.942694	2.60	—	—	—
Magnesium	Mg(24)	23.985045	78.90	Mg(25)	24.985839	10.00	Mg(26)	25.982595	11.10
Manganese	Mn(55)	54.938046	100.00	—	—	—	—	—	—
Mercury	Hg(196)	195.965812	0.15	Hg(198)	197.966760	10.10	Hg(199)	198.968269	17.00
	Hg(200)	199.968316	23.10	Hg(201)	200.970293	13.20	Hg(202)	201.970632	29.65
	Hg(204)	203.973481	6.80	—	—	—	—	—	—
Molybdenum	Mo(92)	91.906809	14.84	Mo(94)	93.905086	9.25	Mo(95)	94.905838	15.92
	Mo(96)	95.904676	16.68	Mo(97)	96.906018	9.55	Mo(98)	97.905405	24.13
	Mo(100)	99.907473	9.63	—	—	—	—	—	—
Neodymium	Nd(142)	141.907731	27.13	Nd(143)	142.909823	12.18	Nd(144)	143.910096	23.80
	Nd(145)	144.912582	8.30	Nd(146)	145.913126	17.19	Nd(148)	147.916901	5.76
	Nd(150)	149.920900	5.64	—	—	—	—	—	—
Neon	Ne(20)	19.992439	90.60	Ne(21)	20.993845	0.26	Ne(22)	21.991384	9.20
Nickel	Ni(58)	57.935347	68.27	Ni(60)	59.930789	26.10	Ni(61)	60.931059	1.13
	Ni(62)	61.928346	3.59	Ni(64)	63.927968	0.91	—	—	—
Niobium	Nb(93)	92.906378	100.00	—	—	—	—	—	—
Nitrogen	N(14)	14.003074	99.63	N(15)	15.000109	0.37			
Osmium	Os(184)	183.952514	0.02	Os(186)	185.953852	1.58	Os(187)	186.955762	1.60
	Os(188)	187.955850	13.30	Os(189)	188.958156	16.10	Os(190)	189.958455	26.40
	Os(192)	191.961487	41.00	—	—	—	—	—	—
Oxygen	O(16)	15.994915	99.76	O(17)	16.999131	0.038	O(18)	17.999159	0.20
Palladium	Pd(102)	101.905609	1.02	Pd(104)	103.904026	11.14	Pd(105)	104.905075	22.33
	Pd(106)	105.903475	27.33	Pd(108)	107.903894	26.46	Pd(110)	109.905169	11.72
Phosphorus	P(31)	30.973763	100.00	—	—	—	—	—	—

Platinum	Pt(190)	189.959937	0.010	Pt(192)	191.961049	0.79	Pt(194)	193.962679	32.90
	Pt(195)	194.964785	33.80	Pt(196)	195.964947	25.30	Pt(193)	197.967879	7.20
Potassium	K(39)	38.963708	93.20	K(40)	39.963999	0.012	K(41)	40.961825	6.73
Praseodyium	Pr(141)	140.907657	100.00	—	—	—	—	—	—
Rhenium	Re(185)	184.952977	37.40	Re(187)	186.955765	62.60	—	—	—
Rhodium	Rh(103)	102.905503	100.00	—	—	—	—	—	—
Rubidium	Rb(85)	84.911800	72.17	Rb(87)	86.909184	27.84	—	—	—
Ruthenium	Ru(96)	95.907596	5.52	Ru(98)	97.905287	1.88	Ru(99)	98.905937	12.70
	Ru(100)	99.904218	12.60	Rb(101)	100.905581	17.00	Ru(102)	101.904348	31.60
	Ru(104)	103.905422	18.70	—	—	—	—	—	—
Samarium	Sm(144)	143.912009	3.10	Sm(147)	146.914907	15.00	Sm(148)	147.914832	11.30
	Sm(149)	148.917193	13.80	Sm(150)	149.917285	7.40	Sm(152)	151.919741	25.70
	Sm(154)	153.922218	22.70	—	—	—	—	—	—
Scandium	Sc(45)	44.955914	100.00	—	—	—	—	—	—
Selenium	Se(74)	73.922477	0.90	Se(76)	75.919207	9.00	Se(77)	76.919908	7.60
	Se(78)	77.917304	23.50	Se(80)	79.916521	49.60	Se(82)	81.916709	9.40
Silicon	Si(28)	27.976928	92.23	Si(29)	28.976496	4.67	Si(30)	29.973772	3.10
Silver	Ag(17)	106.905095	51.84	Ag(109)	108.904754	48.16	—	—	—
Sodium	Na(23)	22.989770	100.00	—	—	—	—	—	—
Strontium	Sr(84)	83.913428	0.56	Sr(86)	85.909273	9.86	Sr(87)	86.908902	7.00
	Sr(88)	87.905625	82.58	—	—	—	—	—	—
Sulfur	S(32)	31.972072	95.02	S(33)	32.971459	0.75	S(34)	33.967868	4.21
	S(36)	35.967079	0.020	—	—	—	—	—	—
Tantalum	Ta(180)	179.947489	0.012	Ta(181)	180.948014	99.99	—	—	—
Tellurium	Te(122)	121.903055	2.60	Te(123)	122.904278	0.91	Te(124)	123.902825	4.82
	Te(125)	124.904435	7.14	Te(126)	125.903310	18.95	Te(128)	127.904464	31.69
	Te(130)	129.90229	33.80	—	—	—	—	—	—
Terbium	Tb(159)	158.925350	100.00	—	—	—	—	—	—
Thallium	Tl(203)	202.972336	29.52	Ti(205)	204.974410	70.48	—	—	—
Thorium	Th(233)	232.038054	100.00	—	—	—	—	—	—

(*Continued*)

Exact Masses and Isotopic Abundances of Elements (*Continued*)

Name	Symbol	Mass	Abundance	Symbol	Mass	Abundance	Symbol	Mass	Abundance
Thulium	Tm(169)	168.934225	100.00	—	—	—	—	—	—
Tin	Sn(116)	115.901744	14.70	Sb(117)	116.902954	7.70	Sn(118)	117.901607	24.30
	Sn(119)	118.903310	8.60	Sn(120)	119.902199	32.40	Sn(122)	121.903440	4.60
	Sn(124)	123.905271	5.60	—	—	—	—	—	—
Titanium	Ti(46)	45.952633	8.00	Ti(47)	46.951765	7.30	Ti(48)	47.947947	73.80
	Ti(49)	48.947871	5.50	Ti(50)	49.944786	5.40	—	—	—
Tungsten	W(180)	179.946727	0.13	W(182)	181.948225	26.30	W(183)	182.950245	14.30
	W(184)	183.950953	30.67	W(186)	185.954377	23.60	—	—	—
Uranium	U(234)	234.040947	0.006	U(235)	235.043925	0.72	U(238)	238.050786	99.27
Vanadium	V(50)	49.947161	0.25	V(51)	50.943963	99.75	—	—	—
Xenon	Xe(128)	127.903531	1.91	Xe(129)	128.904780	26.40	Xe(130)	129.903510	4.10
	Xe(131)	130.905076	21.20	Xe(132)	131.904148	26.90	Xe(134)	133.905395	10.40
	Xe(136)	135.907219	8.90	—	—	—	—	—	—
Yiterbium	Yb(168)	167.933908	0.13	Yb(170)	169.934774	3.05	Yb(171)	170.936338	14.30
	Yb(172)	171.936393	21.90	Yb(173)	172.938222	16.12	Yb(174)	173.938873	31.80
	Yb(176)	175.942576	12.70	—	—	—	—	—	—
Yitrium	Y(89)	88.905856	100.00	—	—	—	—	—	—
Zinc	Zn(64)	63.929145	48.60	Zn(66)	65.926035	27.90	Zn(67)	66.927129	4.10
	Zn(68)	67.924846	18.80	Zn(70)	69.925325	0.60	—	—	—
Zirconium	Zr(90)	89.904708	51.45	Zr(91)	90.905644	11.27	Zr(92)	91.905039	17.17
	Zr(94)	93.906319	17.33	Zr(96)	95.908272	2.78	—	—	—

Autolysis of Bovine and Porcine Trypsin

Fragment	Mass		Sequence
From–To	Monoisotopic	Average	
Porcine Trypsin			
52–53	261.14	261.28	SR
54–57	514.32	514.63	IQVR
108–115	841.50	842.01	VATVSLPR
209–216	905.50	906.05	NKPGVYTK
148–157	1005.48	1006.15	APVLSDSSCK
98–107	1044.56	1045.16	LSSPATLNSR
134–147	1468.72	1469.68	SSGSSYPSLLQCLK
217–231	1735.84	1736.97	VCNYVNWIQQTIAAN
116–133	1767.79	1768.99	SCAAAGTECLISGWGNTK
158–178	2157.02	2158.48	SSYPGQITGNMICVGFLEGGK
58–77	2210.10	2211.42	LGEHNIDVLEGNEQFINAAK
78–97	2282.17	2283.63	IITHPNFNGNTLDNDIMLIK
179–208	3012.32	3014.33	DSCQGDSGG . . . SWGYGCAQK
9–51	4488.11	4491.07	IVGGYTCAA . . . VVSAAHCYK
Bovine Trypsin			
110–111	259.19	259.35	LK
157–159	362.20	362.49	CLK
238–243	632.31	632.67	QTIASN
64–69	658.38	658.76	SGIQVR
112–119	804.41	804.86	SAASLNSR
221–228	905.50	906.05	NKPGVYTK
160–169	1019.50	1020.17	APILSDSSCK
229–237	1110.55	1111.33	VCNYVSWIK
146–156	1152.57	1153.25	SSGTSYPDVLK
207–220	1432.71	1433.65	LQGIVSWGSGCAQAK
191–206	1494.61	1495.61	DSCQGDSGGPVVCSGK
70–89	2162.05	2163.33	LGEDNINVVEGNEQFISASK
170–190	2192.99	2194.47	SAYPGQITSNMFCAGYLEGGK
90–109	2272.15	2273.60	SIVHPSYNSNTLNNDIMLIK
120–145	2551.24	2552.91	VASISLPTS . . . LISGWGNTK
21–63	4550.12	4553.14	IVGGYTCGA . . . VVSAAHCYK

Posttranslational Modifications

Modification	Reagent	Site	Mass Difference	
			Monoisotopic	Average
Acetylation	—	*N*-term, K	42.01057	42.037
Biotinylation	—	*N*-term, K	226.07760	226.293
Carbamidomethyl	Iodoacetamide	*C*	57.03404	57.072
Carbamyl	Cyanate from alkaline decomposition of urea	*N*-term	43.00581	43.025
Carboxymethyl	Iodoacetic acid	*C*	58.00548	58.037
Formylation	—	*N*-term	27.99492	28.010
Methyl ester	—	*C*-term, D, E	14.01565	14.027
NIPCAM	*n*-Isopropyl iodoacetamide	*C*	99.06842	99.132
Oxidation	—	H, M, W	15.99492	15.999
Phosphorylation	—	S, T, Y	79.96633	79.980
Propionamide	Acrylamide	*C*	71.03712	71.079
Pyro-glu	—	Q at *N*-term	–17.02655	–17.030
Pyro-glu	—	E at *N*-term	–18.01057	–18.015
S-Pyridylethyl	4-Vinylpyridine	*C*	105.05785	105.139
SMA	*N*-Succinimidyl (3-morpholine) acetate	*N*-term, K	127.06333	127.143
Sodiation	—	*C*-term, D, E	21.98194	21.982
Suphone	—	M	31.98983	31.999

The Masses and Compositions of Commonly Occuring Amino Acid Residues

Symbol	Residue	CGCY Residue Structure	Mass	
			Monoisotopic	Average
AlaA	Alanine C_3H_5NO	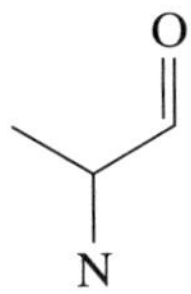	71.03711	71.0788
ArgR	Arginine $C_6H_{12}N_4O$	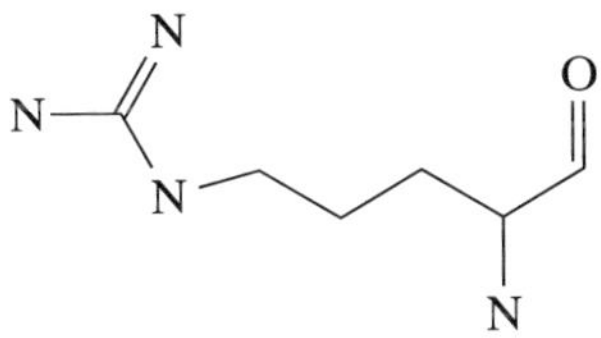	156.10111	156.1875
AsnN	Asparagine $C_4H_6N_2O_2$	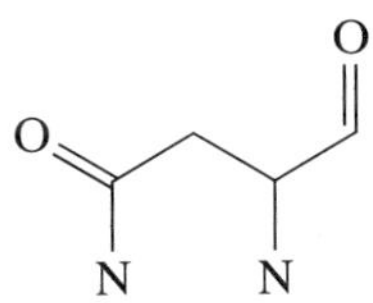	114.04293	114.1038
AspD	Aspartic acid $C_4H_5NO_3$	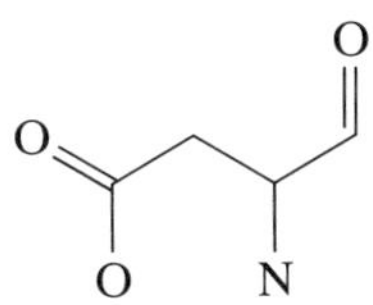	115.02694	115.0886
CysC	Cysteine C_3H_5NOS	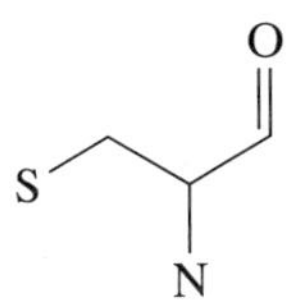	103.00919	103.1388
GluE	Glutamic acid $C_5H_7NO_3$	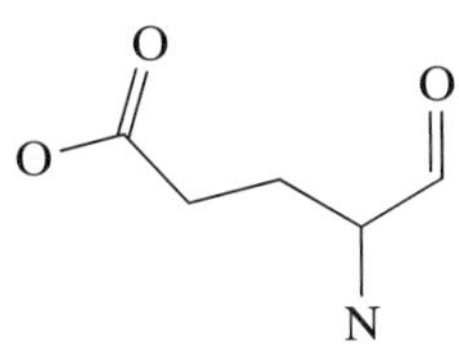	129.04259	129.1155
GlnQ	Glutamine $C_5H_8N_2O_2$	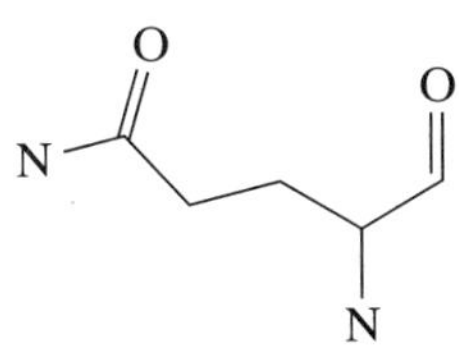	128.05858	128.1307

(*Continued*)

The Masses and Compositions of Commonly Occuring Amino Acid Residues (*Continued*)

Symbol	Residue	CGCY Residue Structure	Mass	
			Monoisotopic	Average
GlyG	Glycine C_2H_3NO	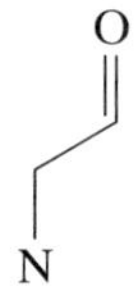	57.02146	57.0519
HisH	Histidine $C_6H_7N_3O$	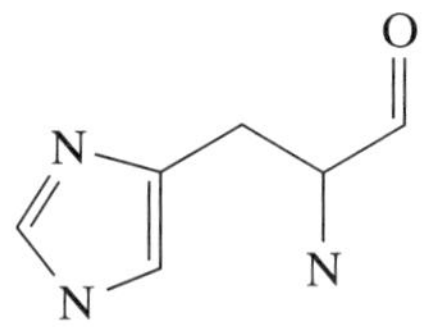	137.05891	137.1411
IleI	Isoleucine $C_6H_{11}NO$	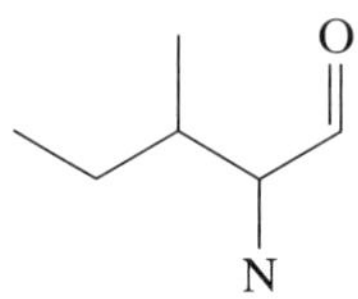	113.08406	113.1594
LeuL	Leucine $C_6H_{11}NO$	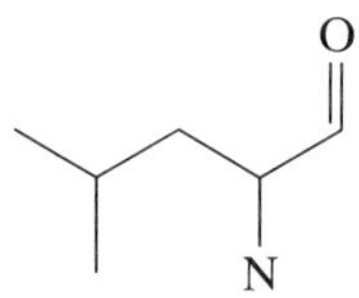	113.08406	113.1594
LysK	Lysine $C_6H_{12}N_2O$	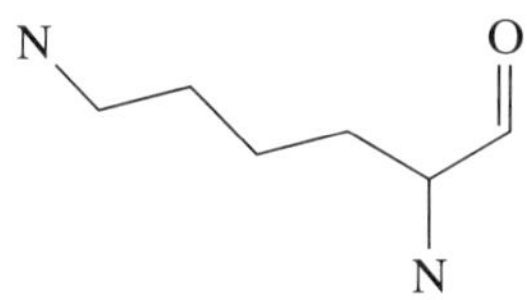	128.09496	128.1741
MetM	Methionine C_5H_9NOS	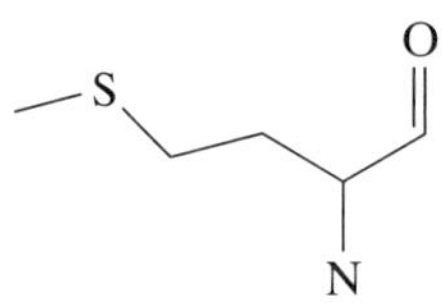	131.04049	131.1926
PheF	Phenylalanine C_9H_9NO	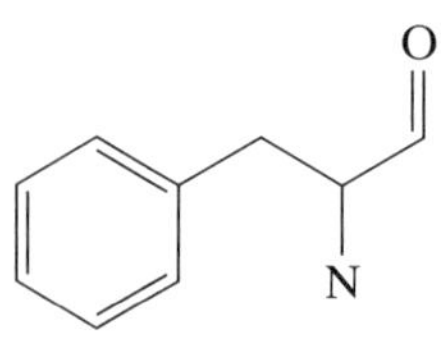	147.06841	147.1766

The Masses and Compositions of Commonly Occuring Amino Acid Residues (*Continued*)

Symbol	Residue	CGCY Residue Structure	Mass	
			Monoisotopic	Average
ProP	Proline C_5H_7NO	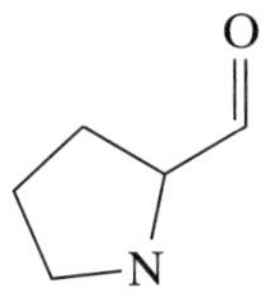	97.05276	97.1167
SerS	Serine $C_3H_5NO_2$	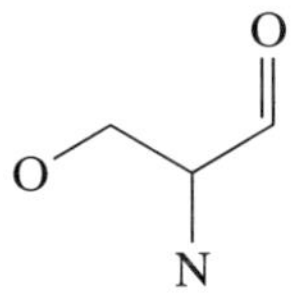	87.03203	87.0782
ThrT	Threonine $C_4H_7NO_2$	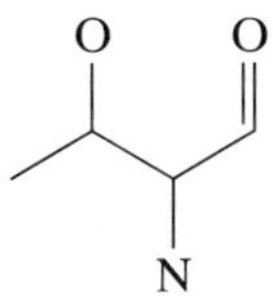	101.04768	101.1051
TrpW	Tryptophan $C_{11}H_{10}N_2O$	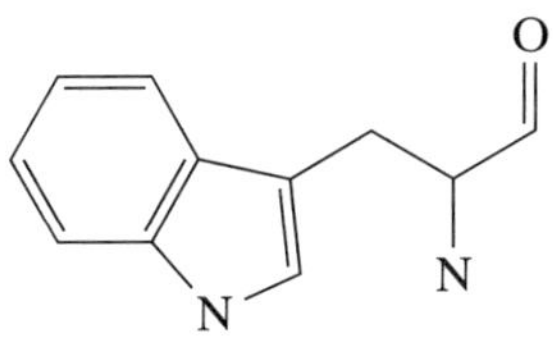	186.07931	186.2132
TyrY	Tyrosine $C_9H_9NO_2$	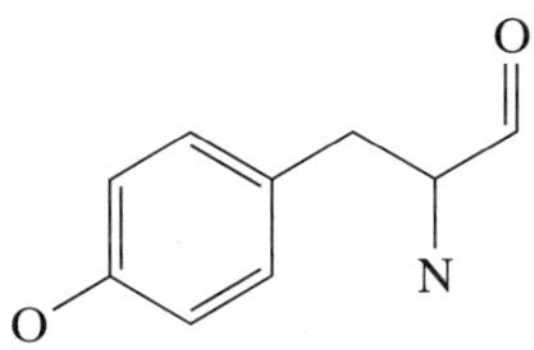	163.06333	163.1760
ValV	Valine C_5H_9NO	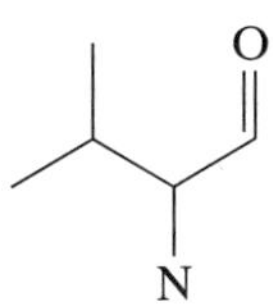	99.06841	99.1326

An AA Mutation Mass Shift

		Gly	Ala	Ser	Pro	Val	Thr	Cys	Leu/ Ile	Asn	Asp	Gln/ Lys	Glu	Met	His	Phe	Arg	Tyr	Trp
		57	71	87	97	99	101	103	113	114	115	128	129	131	137	147	156	163	186
Gly	57		14	30	40	42	44	46	56	57	58	71	72	74	80	90	99	106	129
Ala	71	−14		16	26	28	30	32	42	43	44	57	58	60	66	76	85	92	115
Ser	87	−30	−16		10	12	14	16	26	27	28	41	42	44	50	60	69	76	99
Pro	97	−40	−26	−10		2	4	6	16	17	18	31	32	34	40	50	59	66	89
Val	99	−42	−28	−12	−2		2	4	14	15	16	29	30	32	38	48	57	64	87
Thr	101	−44	−30	−14	−4	−2		2	12	13	14	27	28	30	36	46	55	62	85
Cys	103	−46	−32	−16	−6	−4	−2		10	11	12	25	26	28	34	44	53	60	83
Leu/ Ile	113	−56	−42	−26	−16	−14	−12	−10		1	2	15	16	18	24	34	43	50	73
Asn	114	−57	−43	−27	−17	−15	−13	−11	−1		1	14	15	17	23	33	42	49	72
Asp	115	−58	−44	−28	−18	−16	−14	−12	−2	−1		13	14	16	22	32	41	48	71
Gln/ Lys	128	−71	−57	−41	−31	−29	−27	−25	−15	−14	−13		1	3	9	19	28	35	58
Glu	129	−72	−58	−42	−32	−30	−28	−26	−16	−15	−14	−1		2	8	18	27	34	57
Met	131	−74	−60	−44	−34	−32	−30	−28	−18	−17	−16	−3	−2		6	16	25	32	55
His	137	−80	−66	−50	−40	−38	−36	−34	−24	−23	−22	−9	−8	−6		10	19	26	49
Phe	147	−90	−76	−60	−50	−48	−46	−44	−34	−33	−32	−19	−18	−16	−10		9	16	39
Arg	156	−99	−85	−69	−59	−57	−55	−53	−43	−42	−41	−28	−27	−25	−19	−9		7	30
Tyr	163	−106	−92	−76	−66	−64	−62	−60	−50	−49	−48	−35	−34	−32	−26	−16	−7		23
Trp	186	−129	−115	−99	−89	−87	−85	−83	−73	−72	−71	−58	−57	−55	−49	−39	−30	−23	

Dipeptide Masses

		Gly	Ala	Ser	Pro	Val	Thr	Cys	Leu/Ile	Asn	Asp	Gln/Lys	Glu	Met	His	Met-ox	Phe	Arg	Cys-cm	Tyr	Cys-am	Trp
		57	71	87	97	99	101	103	113	114	115	128	129	131	137	147	147	156	161	163	174	186
Gly	57	115																				
Ala	71	129	143																			
Ser	87	145	159	175																		
Pro	97	155	169	185	195																	
Val	99	157	171	187	197	199																
Thr	101	159	173	189	199	201	203															
Cys	103	161	175	191	201	203	205	207														
Leu/Ile	113	171	185	201	211	213	215	217	227													
Asn	114	172	186	202	212	214	216	218	228	229												
Asp	115	173	187	203	213	215	217	219	229	230	231											
Gln/Lys	128	186	200	216	226	228	230	232	242	243	244	257										
Glu	129	187	201	217	227	229	231	233	243	244	245	258	259									
Met	131	189	203	219	229	231	233	235	245	246	247	260	261	263								
His	137	195	209	225	235	237	239	241	251	252	253	266	267	269	275							
Met-ox	147	205	219	235	245	247	249	251	261	262	263	276	277	279	285	295						
Phe	147	205	219	235	245	247	249	251	261	262	263	276	277	279	285	295	295					
Arg	156	214	228	244	254	256	258	260	270	271	272	285	286	288	294	304	304	313				
Cys-cm	161	219	233	249	259	261	263	265	275	276	277	290	291	293	299	309	309	318	323			
Tyr	163	221	235	251	261	263	265	267	277	278	279	292	293	295	301	311	311	320	325	327		
Cys-am	174	232	246	262	272	274	276	278	288	289	290	303	304	306	312	322	322	331	336	338	349	
Trp	186	244	258	274	284	286	288	290	300	301	302	315	316	318	324	334	334	343	348	350	361	373

K22e_human: Keratin, Type Cytoskeletal 2 Epidermal (Cytokeratin 2e) (k2e) (ck 2e)—*Homosapiens* (Human): Cleavage by Trypsin (Most Important Signals)

Fragments From–To	Observed Mass	Mass	Sequence
46–61	1320.60	1319.58	HGGGGGGFGGGGFGSR
71–92	1838.90	1837.91	SISISVAGGGGGFGAAGGFGGR
93–128	2831.20	2830.19	GGGFGGGSGFGGGSGFGGGSGFSG GGFGGGGFGGGR
245–253	1037.50	1036.52	YLDGLTAER
381–390	1193.60	1192.61	YEELQVTVGR

PPG Calibration Table (Selected Peaks)

Mass	Int	Mass	Int
56.02621	170	1085.75391	230
251.18585	100	1143.79578	220
309.22771	140	1201.83764	210
331.20967	100	1259.87951	200
367.26958	180	1317.92137	190
389.25153	200	1375.96324	180
425.31140	300	1434.00510	170
447.29340	400	1492.04697	160
505.33526	1000	1550.08883	150
563.37713	800	1608.13070	140
621.41899	600	1666.17256	130
679.46086	400	1724.21443	120
737.50272	300	1782.25629	110
795.54459	290	1840.29816	100
853.58645	280	1898.34002	99
911.62832	270	1956.38189	85
969.67018	250	2014.42375	80
1027.71205	240		

Important Internet Addresses (Selected)

Internet Address	Description
http://*base-peak.wiley.com/links/*	A useful link containing many options
http://*falcon.ludwig.ucl.ac.uk/mshome3.2.htm*	Protein Prospector
http://*masspec.scripps.edu/massspec.html*	Good resource including a history of mass spectrometry
http://*www.asms.org/*	American Society for Mass Spectrometry (discussions groups, positions, links etc.)
http://*www.bmb.leeds.ac.uk/esms/*	European Society for Mass Spectrometry
http://*www.latrobe.edu.au/www/anzsms/Journals.html*	Mass spectrometry journals
http://www.expasy.ch/	ExPASy Molecular Biology Server (databases and tools for proteins/peptide identification; many important links)
http://*www.matrixscience.com/cgi/index.pl?page =../home.html*	Mascot—a search engine from Matrix Science for identification of proteins and peptides; this software is commonly applied by manufacturers of mass spectrometers
http://*www.ncbi.nlm.nih.gov/BLAST//blast_over view.html*	BLAST—a search similarity program designed to explore all the available sequence databases, regardless of whether the query is protein of DNA; also contains links to free MEDLINE
http://*www.chromatographyonline.com/*	LC/GC (an important free journal for those solving various analytic problems)
http://*www.public.iastate.edu/~pedro/research_to ols.html*	Pedro's Biomolecular Research Tools (a collection of links to information and services useful to molecular biologists, biochemists, protein chemists, and mass spectrometrists)
http://*www.chem.umn.edu/orgs/ampepsoc/apslin ks.html*	The American Peptide Society

(*Continued*)

Important Internet Addresses (Selected) (*Continued*)

Internet Address	Description
http://*eps.chem.elte.hu/*	The European Peptide Society
http://*www.abrf.org/*	ABRF—an email-based discussion group covering a broad range of topics of interest to its members; topics include protein/peptide and nucleic acid chemistry, sequencing, synthesis, mass spectrometry, amino acid analysis, biocomputing, and laboratory management
http://*prowl.rockefeller.edu/*	PROWL—a resource for protein chemistry and mass spectrometry; another very useful tool for protein identification

Index